THE EARLY TRANSITION METALS

THE EARLY TRANSITION METALS

D. L. KEPERT
University of Western Australia,
Perth, Australia

1972

ACADEMIC PRESS
LONDON AND NEW YORK

ACADEMIC PRESS INC. (LONDON) LTD.
24/28 Oval Road,
London NW1

United States Edition published by
ACADEMIC PRESS INC.
111 Fifth Avenue
New York, New York 10003

Library of Congress Catalog Card Number: 75–170765
ISBN: 0–12–404350–X

PRINTED IN GREAT BRITAIN BY
UNWIN BROTHERS LIMITED
WOKING AND LONDON

Preface

The chemistry of those elements towards the left hand side of the transition metal block in the periodic table extends over a wide range of oxidation states and is fairly complex, even if somewhat neglected. It is the purpose of this book to present a broad survey of the general chemistry of these elements.

It is clear that these elements are sufficiently well related to each other, and yet sufficiently different from elements in other areas of the periodic table that it is appropriate to treat them in the one volume. The characteristic properties of these large electropositive metals containing few *d* electrons are discussed in some detail in Chapter 1. It may be inevitable that this treatment reflects the author's research interests. Parts of this chapter have appeared in other forms in the Proceedings of the Royal Australian Chemical Institute, and in a forthcoming publication written by the author and Professor Dr. K. Vrieze.

The remaining chapters cover the metals of the titanium, vanadium and chromium columns. For a broad survey of this type, where the gaps in our knowledge are best highlighted by a systematic approach, it is hoped that a single-author work is more appropriate than a more piece-meal multi-author work, even if greater expertise is brought to certain areas by the latter approach.

Mention should also be made of other books which may be profitably used in conjunction with this one. Those which can be highly recommended include "The Organic Chemistry of Titanium," by R. Feld and P. L. Cowe, "The Chemistry of Titanium and Vanadium," by R. J. H. Clark, "The Chemistry of Niobium and Tantalum," by F. Fairbrother, and "Halides of the Transition Elements" (two volumes) by R. Colton and J. H. Canterford.

This work was completed while on leave in the University Chemical Laboratory, Cambridge, and thanks are due to Professor J. Lewis and his colleagues for their hospitality.

D. L. KEPERT

Cambridge
February, 1972

CONTENTS

CHAPTER 1

General Properties of the Early Transition Metals

CHAPTER 2

Group IV—Titanium, Zirconium, Hafnium

CHAPTER 3

Group V—Vanadium, Niobium, Tantalum

CHAPTER 4

Group VI—Chromium, Molybdenum, Tungsten

CHAPTER 1

General Properties of the Early Transition Metals

1. INTRODUCTION

The early transition metals are sufficiently well related to each other, and yet sufficiently different to elements in other areas of the periodic table, that it appears justifiable to treat the chemistry in some detail in the one volume.

The early transition metals have relatively few electrons outside the rare gas electron configurations, leading to relatively well screened metal atoms with low effective nuclear charges. This leads to large sizes with the *d* orbitals in particular being well expanded, and low ionization potentials with the consequent possibility of high valence states containing few or no *d* electrons.

Even a brief look at the chemistry of the early transition metals will immediately reveal a number of important general differences to the rest of the periodic table. To this writer it appears that there are three distinctive features:

(i) The stereochemistry is notable for the proliferation of coordination numbers greater than six.
(ii) The chemistry of the low valence states is very interesting due to the prevalence of metal–metal bonding.
(iii) The aqueous chemistry of the higher valence states is dominated by the formation of isopoly and heteropolyanions.

Each of these properties is dealt with in turn in this Chapter, before dealing systematically with Groups IV, V and VI of the transition elements. It is realized, of course, that these distinctive properties are not solely restricted to these Groups of the transition metals. For example high coordination numbers are also common to elements to the left of the Group IV elements in the periodic table, including the lanthanides and actinides, and the occurrence of metal–metal bonding spills over to elements to the right of the Group VI elements, particularly to technetium and rhenium. Thus it can be argued that in some respects the chemistries of, say, lanthanum or rhenium, are more "early transition metal like" than the chemistry of chromium. It is for this reason that these distinctive properties are dealt with separately so as they can be discussed in a wider context, and also why the chemistry of chromium is dealt with in less detail than that of the other elements.

Other authors might legitimately choose other features to define as characteristic of an early transition metal. One such characteristic could be the remarkable range of nonstoichiometric oxides found, but at this stage it is not easy to decide if this finding will be restricted only to these elements, or if so, to what atomic properties it owes its origin. A second such characteristic could be that these electropositive metals in high oxidation states, with no or few electrons capable of backbonding to ligands, are typical "hard" or "*a* class" acceptors; that is they form stronger complexes with the more electronegative fluoride than with chloride, bromide or iodide, with oxygen donors than with sulphur donors, and with nitrogen donors rather than with phosphorus donors. Similarly their chlorides are readily hydrolysed in water, and for example, only oxochloro complexes of the type $M^{V}OCl_5^{2-}$ (where M is V, Nb, Ta, Mo, W or Re) are obtained from concentrated hydrochloric acid.

2. HIGH COORDINATION NUMBERS

One of the most characteristic properties of the early transition metals, and of the left-hand side of the periodic table in general, is the common occurrence of coordination numbers greater than six. High coordinate compounds are most prolific with the lanthanides and actinides, where they are more common than those with coordination number six.

It is to be expected that a metal atom will continue to accept additional ligands as long as the increase in bond energy due to the increasing number of metal–ligand bonds is greater than the sum of the unfavourable energy terms due to increasing repulsions between the ligands and/or the bonding pairs of electrons, that is increasing steric strain, and to unfavourable electronic factors concerned mainly with the metal atom.

The requirements of the metal atom and of the ligand in order to attain high coordination numbers will be considered separately. Additional factors, such as solvation and crystal packing forces are also important, but it is difficult to generalize about these. For example, the eight coordinate complexes $Na_3[TaF_8]$ and $[TiCl_4(diars)_2]$ (where diars is *o*-phenylene-bisdimethylarsine) dissociate in solution to the six coordinate $[TaF_6]^-$ (1345) and $[TiCl_4(diars)]$ (505, 506) respectively.

A. Requirements of Metal Atom

The two main requirements of a metal ion in order to attain a high coordination number are its size and its charge.

The larger metal ions can obviously fit more ligands around them than the smaller metal ions. A *qualitative* measure of this tendency is the radius ratio (r/R) of the metal ion radius (r) to the donor atom radius (R). Assuming hard spherical atoms, values for the radius ratio can be calculated which correspond to the largest metal ion which can be accommodated in the hole formed by the appropriate number of donor atoms touching each other (Table 1).

TABLE 1

Radius ratios of common high coordinate stereochemistries

Coordination number	Stereochemistry	Radius ratio (r/R)
3	Triangular	0·157
4	Tetrahedral	0·225
6	Octahedral	0·414
7	Capped octahedron	0·591
	Capped trigonal prism	0·619
	Pentagonal bipyramid	0·701
8	Square antiprism	0·645
	Dodecahedron	0·668
	Cube	0·732
	Planar hexagonal bipyramid	1·000
9	Tricapped trigonal prism	0·732
	Capped square antiprism	0·742
10	Bicapped square antiprism	0·848
12	Icosahedron	0·902
	Cubooctahedron	1·000

These figures must not be interpreted too literally, but they do qualitatively show the effect of the increasing number of ligands upon the size of the central cavity holding the metal atom, and the effect of changing the stereochemistry for a given coordination number (which is sometimes greater than increasing the coordination number). The metal and ligand atoms should not, of course, be considered as rigid spheres, and it is found that the effective radii of metal atoms are dependent upon their coordination number. For example the titanium-chlorine distances in the four, six and eight coordinate $TiCl_4$, $[TiCl_6]^{2-}$ and $[TiCl_4(diars)_2]$ are 2·18, 2·34 and 2·46 Å respectively, while the zirconium-fluorine distances in the six, seven, and eight coordinate $[ZrF_6]^{2-}$, $[Zr_2F_{13}]^{5-}$ and $(ZrF_6^{2-})_n$ are 2·04, 2·06 and 2·18 Å respectively. Neither can most of these high coordinate polyhedra be considered as rigid, and, for example, it is found that the real polyhedra observed for the three seven coordinate stereochemistries have closely similar radius ratios.

Certainly it is true that higher coordination numbers are common for the larger metal ions to the left-hand side of the transition metals (Table 2). For example the oxides of zirconium and hafnium are eight coordinate, those of titanium, niobium, tantalum and tungsten are six coordinate, those of vanadium and molybdenum are five or six coordinate, and those of chromium are four coordinate. Similarly the coordination number of the third row transition metal chlorides progressively falls from nine for lanthanum, through six and four for tungsten and platinum respectively, to two for mercury.

TABLE 2

Eight coordinate transition metals

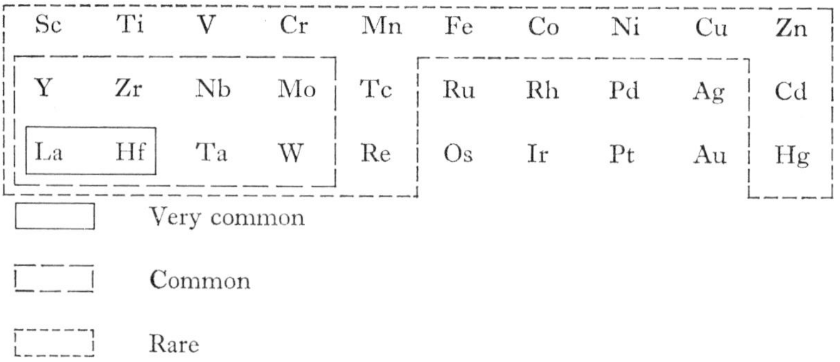

Sc	Ti	V	Cr	Mn	Fe	Co	Ni	Cu	Zn
Y	Zr	Nb	Mo	Tc	Ru	Rh	Pd	Ag	Cd
La	Hf	Ta	W	Re	Os	Ir	Pt	Au	Hg

Very common

Common

Rare

The charge on the metal atom is also important. It is normally high, +3, +4 or +5, in order to prevent too much negative charge accumulating on the metal atom, stopping the accepting of additional lone pairs of electrons from additional ligands. Most complexes of high coordination numbers are accordingly neutral or anionic. The build up of negative charge on the metal can be alleviated to a certain extent by using ligands capable of metal-to-ligand π bonding, as discussed in the next section, and in these cases cationic complexes can be obtained, for example $[WBr(CO)_2(diars)_2]Br$. The high oxidation state requires low ionization potentials as found in the early part of each transition series, but in very high oxidation states, for example +6 and +7, the small size of the resulting ion prevents high coordination numbers due to steric reasons.

For compounds containing spin paired ions of the second and third row transition metals, and/or ligands of high ligand field strength, there is a large ligand field splitting (or large energy gap between bonding molecular orbitals and antibonding molecular orbitals), and the valence bond approximation can be used. In these cases the maximum coordination number is determined by how many of the nine orbitals in the valence shell are vacant to accept electron pairs from the ligands. Thus d^0 ions have a maximum coordination number of nine (d^5sp^3 hybrids) as in $[Nd(H_2O)_9]^{3+}$ or $[ReH_9]^{2-}$, d^1 and d^2 ions have a maximum coordination number of eight (d^4sp^3 hybrids) as in $[Mo(CN)_8]^{4-}$, d^3 and d^4 ions have a maximum coordination number of seven (d^3sp^3 hybrids) as in $[WBr(CO)_2(diars)_2]^+$, and so on.

The high coordinate complexes obtained with first row transition metals and bidentate ligands such as nitrate, have the metal atom in the high spin configuration, and in these cases different factors are clearly important as discussed in the next section.

B. Requirements of Ligand

Two ligand properties which can lead to an increase in the coordination number of the metal in a complex are small size, in order to decrease the steric repulsions between the donor atoms, and metal-to-ligand π bonding capacity, to prevent the accumulation of too much negative charge on the metal atom. These two ligand requirements often appear to be mutually exclusive, for example donor atoms such as sulphur and arsenic are able to stabilize the eight coordinate $[Zr(S_2CNEt_2)_4]$ (286) and $[TiCl_4(diars)_2]$ (605), whereas corresponding ligands with oxygen and nitrogen donors do not form complexes of this type. The monodentate ligand which appears most able to combine both favourable features is the cyanide ion, which forms eight coordinate molybdenum and tungsten complexes which have not been obtained with any other ligand. In addition to $[Re^{V}(CN)_8]^{3-}$ and $[Re^{VI}(CN)_8]^{2-}$ (1569), rhenium also forms high coordinate complexes with hydride, $[Re^{VII}H_9]^{2-}$ (6, 1431), and fluoride, $[Re^{VI}F_8]^{2-}$ (1465).

The most effective way to decrease the ligand–ligand repulsions is to incorporate two or more donor atoms in a more or less rigid chelating ligand where the donor atoms are held close together. Examples include the bidentates peroxide (O–O $\sim$ 1·4 Å), nitrate (O–O $\sim$ 2·1 Å), and carboxylate (O–O $\sim$ 2·2 Å), which are discussed below. These ligands are outstanding in being able to form seven and eight coordinate compounds with the first row transition elements from titanium to zinc, and ten, eleven and twelve coordinate compounds with the lanthanides and actinides. Another severely constrained ligand is the tridentate BH_4^- (H–H $\sim$ 1·6 Å) which forms the twelve coordinate $Zr(BH_4)_4$ and $Hf(BH_4)_4$ (232).

Peroxide. The bidentate peroxo group, O_2^{2-}, in which both oxygen atoms are equidistant from the metal atom, is extremely small for a bidentate and forms a number of high coordinate complexes. Examples from the early transition metals include the eight coordinate, dodecahedral $[M^{V}(O_2)_4]^{3-}$ (where M is V, Nb, Ta or Cr), and the seven coordinate, pentagonal bipyramidal $[Cr^{IV}(O_2)_2(Ligand)_3]^{x-}$ and $[Cr^{VI}(O_2)_2O(bidentate)]$. These, and a number of other examples, are dealt with later in the appropriate chapters.

The O–O stretching frequencies and bond lengths in these complexes indicate a strong O–O bond, which is incompatible with the donation of π electrons from the oxygen to the metal. The existence of a bidentate peroxo group with bent σ bonds has therefore been suggested (1061), as

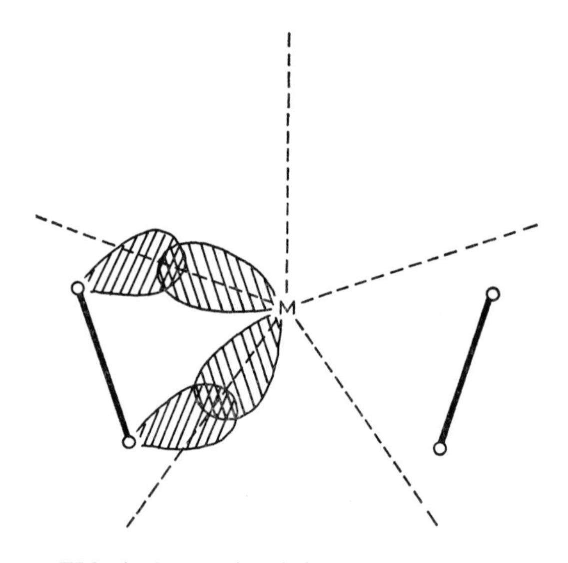

FIG. 1 Bent σ-bonds in peroxo complexes.

indicated in Fig. 1, where two peroxo groups lie in the pentagonal plane of a pentagonal bipyramid.

Nitrate. The nitrate ion forms tetranitrato complexes of titanium(IV), manganese(II), iron(II), cobalt(II), nickel(II), copper(II), gold(III), zinc(II), and tin(IV) (16, 17). In $Ti(NO_3)_4$ the nitrate ion behaves as a true bidentate ligand, both bound oxygen atoms being equidistant from the metal atom (Table 3). In the manganese compound however one bound oxygen atom of each nitrate group is significantly further away than the other, while in the cobalt and gold compounds this difference is accentuated considerably more.

The visible spectra and magnetic susceptibility of $[Mn(NO_3)_4]^{2-}$ and $[Co(NO_3)_4]^{2-}$ however are consistent with tetrahedrally coordinated metal atoms, and these ions can be considered to have tetrahedral structures if the "centre of gravity" of the electrostatic potential of each nitrate ion is considered. It might also be noted that the point symmetry of the dodecahedral $[Co(NO_3)_4]^{2-}$ (D_{2d}) is a subgroup of the point symmetry of the tetrahedron. Alternatively the bonding could be envisaged as a type of three centre bonding, involving the two oxygen atoms but only one orbital from the metal atom. In the extreme case of $[Au(NO_3)_4]^-$, there is little doubt that the gold atom is best considered as being surrounded by a square planar arrangement of four monodentate nitrate ions.

TABLE 3

Metal–oxygen bond distances in tetranitrato complexes (Å)

	$Ti(NO_3)_4$	$(Ph_4As)_2[Mn(NO_3)_4]$	$(Ph_4As)_2[Co(NO_3)_4]$	$(Ph_4As)[Au(NO_3)_4]$
$M–O_A$ (average)	2·07	2·27	2·06	2·00
$M–O_B$ (average)	2·06	2·34	2·45	2·87
$\frac{M–O_A}{M–O_B}$	1·01	0·97	0·83	0·70
Reference	977	695	215	978

Even higher coordination numbers are obtained if the nitrate is bonded to large metal atoms. Examples include the ten coordinate $(Ph_3EtP)_2[Ce(NO_3)_5]$ (31), $[La(NO_3)_3(dipy)_2]$ (32), and $[Tb(NO_3)_3(dipy)_2]$ (1724), the eleven coordinate $[Th(NO_3)_4(H_2O)_3],2H_2O$ (2266, 2311), and the twelve coordinate $Mg[Th(NO_3)_6],8H_2O$ (2003), $(NH_4)_2[Ce(NO_3)_6]$ (200), and $Mg_3[Ce(NO_3)_6]_2,24H_2O$ (2434). In these last examples the bond distances to both ends of the nitrate group are approximately equal, for example 2·49 and 2·51 Å in $(NH_4)_2[Ce(NO_3)_6]$. The stereochemistry of the twelve oxygen atoms about the metal atom is icosahedral, but if the nitrate ion is considered to be functioning as a monodentate, the stereochemistry is octahedral. Similarly the other complexes can be considered as a normal stereochemistry of lower coordination number.

Other bidentates similar to nitrate behave in the same manner. For example nitrite forms $K_3[Hg(NO_2)_4](NO_3)$ where each nitrite group is bidentate (O–O $\sim$ 2·1 Å, Hg–O = 2·34–2·58 Å) (1123).

Carboxylate. Carboxylate groups are somewhat less efficient in forming high coordinate complexes. The trifluoracetato cobalt(II) complex $(Ph_4As)_2[Co(OOC.CF_3)_4]$ is much closer to a tetrahedral structure (rather than an eight coordinate dodecahedral structure) than is the above $(Ph_4As)_2[Co(NO_3)_4]$. The cobalt–oxygen distances to each trifluoroacetato group are 2·00 and 3·11 Å. Similarly the tetracetato complexes

$Ca[Cu(OAc)_4],6H_2O$ and $Ca[Cd(OAc)_4],6H_2O$ can be considered as four or eight coordinate, as one end of the bidentate group is closer than the other (Cu–O = 1·97 and 2·79 Å, and Cd–O = 2·29 and 2·69 Å) (1492).

Thorium again forms the twelve coordinate, icosahedral $[C(NH_2)_3]_2[Th(OAc)_6]$ (1517).

C. Seven Coordination

There are a variety of ways in which seven ligands can be envisaged as being arranged around a central metal atom, but unlike stereochemistries of lower coordination number, the various arrangements are rather non-uniform, there being significant gaps between the ligands in some places.

Minimization of the repulsion energy (U) between the seven metal-ligand bonds (or alternatively between the ligands themselves), assuming that it is some inverse power (n) of the distance (r) between the ligands, $U \propto r^{-n}$, leads to no less than four minima on the potential energy surface, each corresponding to a different stereochemistry (339, 512, 999, 1352, 2278). These four stereochemistries are the capped octahedron (C_{3v} symmetry), the capped trigonal prism (C_{2v} symmetry), the pentagonal bipyramid (D_{5h} symmetry), and a more irregular polyhedron of C_2 symmetry (Fig. 2).

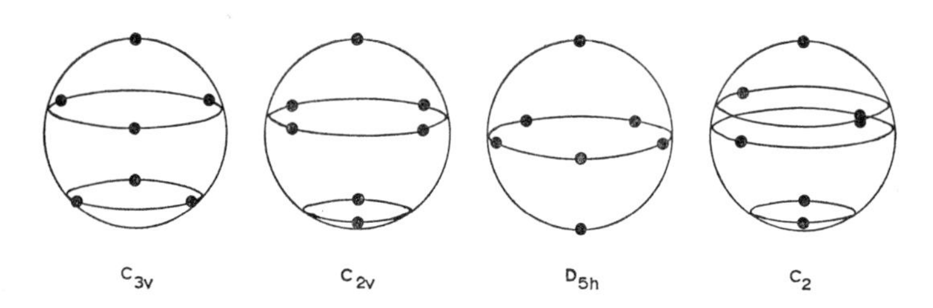

FIG. 2 The four known seven coordinate stereochemistries. Capped octahedron (C_{3v}). Capped trigonal prism (C_{2v}). Pentagonal bipyramid (D_{5h}). Irregular (C_2).

The exact stereochemistry corresponding to the minimum energy for each structure, the "most favourable polyhedron", is dependent upon the exact form of the assumed repulsive law, that is, upon the value of n. For the capped octahedron, a decrease in n leads to a decrease in the angle each metal–ligand bond makes with the three-fold axis (for example: $n = 12$, $\theta = 75{\cdot}8$ and $131{\cdot}4°$; $n = 6$, $\theta = 74{\cdot}6$ and $130{\cdot}3°$; $n = 1$, $\theta = 73{\cdot}3$ and $128{\cdot}8°$). For the capped trigonal prism, a decrease in n leads to an increase in the angle each metal-ligand bond makes with the two-fold axis (for example: $n = 12$, $\theta = 78{\cdot}3$ and $142{\cdot}6°$; $n = 6$, $\theta = 79{\cdot}4$ and $143{\cdot}3°$; $n = 1$, $\theta = 80{\cdot}8$ and $144{\cdot}2°$), with a simultaneous change so that the four coplanar points more closely approach a square.

The most favourable polyhedron for the stereochemistry of C_{2v} symmetry is essentially the same as the pentagonal bipyramid, with a slight puckering of the pentagonal plane. There is no potential energy barrier between these two stereochemistries. Puckering of the pentagonal plane by several degrees changes the repulsive energy by only about one part in 10^6.

The stereochemical difference between all these structures is very small. The relation between the C_{2v} and the D_{5h} and C_2 structures is fairly obvious from Fig. 2, while the C_{2v} structure can be considered to be a distorted C_{3v} structure if the three-fold axis is taken through one of the pair of lower points of the C_{2v} structure.

In agreement with this stereochemical similarity, the differences in the repulsive energies between these four stereochemistries is also very small, amounting, in the most unfavourable case, to no more than $0{\cdot}1\%$ of the total repulsive energy term. It is therefore expected that in an isolated seven coordinate molecule, unrestricted by solvation or crystal packing forces, the stereochemistries would be interconvertible by the normal small bending vibrations. To put this in perspective, the energy required to distort a six coordinate octahedron by less than 1° is more than the energy difference between these seven coordinate polyhedra.

Since no seven coordinate compound with equivalent monodentate ligands has had its structure determined with a high degree of precision, considerable caution is required in interpreting published structural results.

Neutron diffraction (367) and X-ray data (1198) show a slightly distorted C_{2v} structure for NbF_7^{2-}, with large vibrational distortions. Other heptafluoroanions have been determined with less precision, and different stereochemistries proposed.

Single crystal X-ray diffraction work on IF_7 is not sufficiently precise to show its exact stereochemistry (383, 678). Electron diffraction, infra red, and Raman spectroscopy on gaseous IF_7 and ReF_7 are consistent with a distorted D_{5h} structure but with a non planar belt of five fluorine atoms (481, 1511). The ^{19}F-nmr spectra of these molecules show that there is a rapid intramolecular rearrangement (1733).

The situation is considerably clearer for seven coordinate compounds with nonequivalent ligands, and with multidentate ligands. For example in compounds with one metal–ligand bond different to the other six, it would be expected that this ligand would occupy the unique position in the C_{2v} or C_{3v} structures, as is found in $K_3[NbOF_6]$ (2395). Similarly for compounds with two equivalent ligands different to the other five, the D_{5h} structure would be expected, and is found for example in $K_3[UO_2F_5]$ (2429). The ^{19}F-nmr spectra of $WF_6.Me_3P$ and $WF_6.Me_3N$ in solution show rapid intramolecular rearrangement (2267). Another relevant molecule is XeF_6, in which the seventh coordination position can be considered as being occupied by a nonbonding pair of electrons. The most precise electron diffraction work on the gas shows that the instantaneous molecular configuration is in the broad vicinity of C_{3v} with the lone pair of electrons occupying the unique axial site, but exhibiting large amplitudes of bending vibrations (155, 381, 986).

Considerably greater constraints upon choice of stereochemistry are obtained by using multidentate ligands. The heptadentate amine $N(CH_2.CH_2.N = CH.C_5H_4N)_3$ forms $[Ni(heptadentate)](PF_6)_2$, in which the unique nitrogen atom occupies the unique position of the capped octahedron (C_{3v}) structure, which also allows three equivalent imino nitrogen atoms and three equivalent heterocyclic nitrogen atoms, so C_{3v} symmetry for the whole complex is retained (2400). Similarly C_{3v} stereochemistry is observed in $[Y(BzAc)_3(H_2O)]$ (568) and $[Ho(DBM)_3(H_2O)]$ (2437), where BzAc and DBM are the anions of benzoylacetone and dibenzoylmethane respectively. In the corresponding acetylacetonate complexes $[Yb(acac)_3(H_2O)]$ (616) and $[Yb(acac)_3(H_2O)],\frac{1}{2}C_6H_6$ (2346) however, the stereochemistry appears closest to a capped trigonal prism (C_{2v}), with the water molecule occupying one of the four equivalent prismatic corners rather than the unique position on the two-fold rotation axis.

Pentagonal bipyramidal stereochemistry is forced in $[Fe(pentaamine)(NCS)_2](ClO_4)$ and $[(pentaamine)(H_2O)Fe–O–Fe(H_2O)(pentaamine)](ClO_4)_2$ by the use of a planar, cyclic, pentaamine (884).

The two nitrogen atoms of ethylenediaminetetraacetic acid (H_4Y) and of trans-1,2-diaminocyclohexane-N,N′-tetraacetic acid (H_4Z) occupy the two adjacent equivalent positions in the C_2 structure in $Li[Fe(Y)(H_2O)],2H_2O$ (1539), the $[Mn(Y)(H_2O)]$ unit in $(H_3O)_2Mn_3(Y)_2,8H_2O$ (1921), and $Ca[Fe(Z)(H_2O)]_2,9H_2O$ (520), with the water molecule occupying the unique axial position in each case.

Other seven coordinate structures described later in this book include $K_2[Mo(O_2)F_4(H_2O)]$, $K_3[Cr(O_2)_2(CN)_3]$, and $[Cr(O_2)_2(NH_3)_3]$.

The orbital hybridization consistent with these seven coordinate stereochemistries has been calculated (1391, 1403, 2401). Apart from the pentagonal bipyramid obtained by $sp^3d_{xz}d_{x^2-y^2}d_{z^2}$ and equivalent hybridization schemes, each of the other seven coordinate polyhedra can be obtained from more than one sp^3d^3 hybridization scheme, using different combinations of d orbitals. For example the capped trigonal prism can be obtained from $sp^3d_{xy}d_{xz}d_{z^2}$ or $sp^3d_{xy}d_{xz}d_{x^2-y^2}$ hybrid orbitals, and the capped octahedron from $sp^3d_{xy}d_{x^2-y^2}d_{z^2}$ or $sp^3d_{yz}d_{xz}d_{z^2}$ hybrid orbitals.

The d orbital splittings under the influence of the seven coordinate C_{2v} stereochemistry have been calculated, and found to be very sensitive to the detailed geometry (1909).

D. Eight Coordination

There have been six different stereochemistries noted for monomeric eight coordinate complexes (Fig. 3), and at least two others postulated.

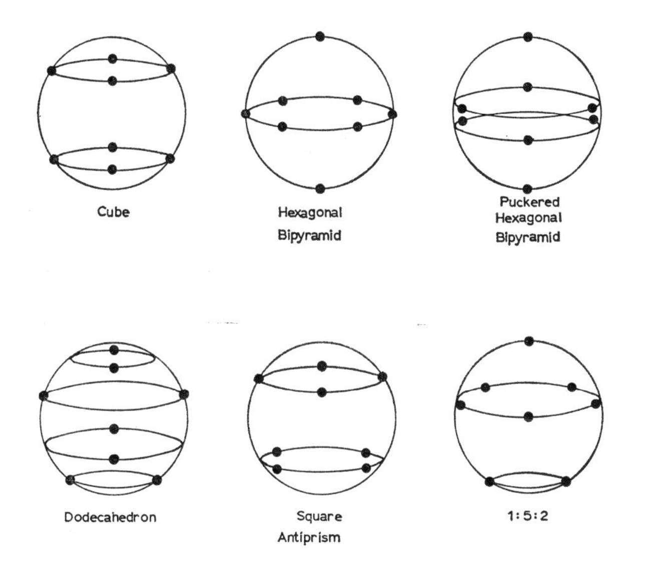

FIG. 3 The six known eight coordinate stereochemistries.

They can be conveniently subdivided into two groups:

Group I

The cube, hexagonal bipyramid, and puckered hexagonal bipyramid form a group of three related structures which differ from the other three in several ways:

(a) They each can be considered to consists of four linear ligand–metal–ligand units, whereas there are no such units present in the dodecahedral, square antiprismatic, or 1 : 5 : 2 stereochemistries.
(b) Hybrid orbitals cannot be constructed towards the corners of these polyhedra using only combinations of s, p and d atomic orbitals (1391).

(c) They are energetically unfavourable from the point of view of minimizing the repulsion between bonding pairs of electrons (1353).

Cubic stereochemistry is found only in ionic structures such as caesium chloride and calcium fluoride. The structure of Na_3PaF_8 is closely related to the fluorite structure, with sodium and protactinium occupying the cubic sites. Taking the position of Pa^{5+} as (0,0,0), then F^- is at (0·236, 0·236, 0·113) rather than (0·25, 0·25, 0·125) of the fluorite cell, so that (Pa–F = 2·21 Å) < (Na–F = 2·39 and 2·46 Å) (354).

The hexagonal bipyramidal and puckered hexagonal bipyramidal stereochemistries are observed only in uranyl (and post uranium dioxo cations) complexes, where six other ligands are arranged around the linear UO_2^{2+} group. In monomeric complexes, the hexagonal plane is normally puckered by a few degrees (Fig. 4) (Table 4). It is not until the

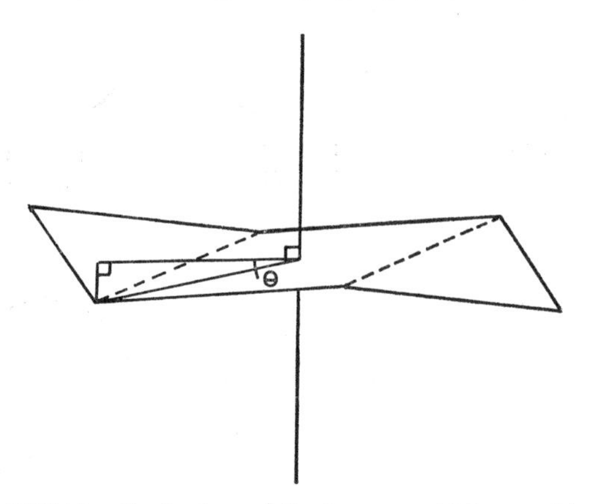

FIG. 4 Puckering of the hexagonal bipyramid.

puckering angle θ approaches 19·53° that the structure approaches a cube, which can also be considered to be derived from an octahedron by the addition of ligands outside two opposite triangular faces.

TABLE 4

Puckering angle in puckered hexagonal bipyramidal molecules

		Reference
$Na[UO_2(OAc)_3]$	$\theta = 1°$	2431
$Rb[UO_2(NO_3)_3]$	$\theta = 2°$	143
$[UO_2(NO_3)_2(H_2O)_2],4H_2O$	$\theta = 3°$	1124, 2265
$[UO_2(NO_3)_2\{(EtO)_3PO\}_2]$	$\theta = 3°$	1580
$(Me_4N)[UO_2(S_2CNEt_2)_3]$	$\theta = 5°$	277

Group II

The second group, comprising the dodecahedron, square antiprism and 1 : 5 : 2 stereochemistry, is the more important and more interesting, and each of these will be considered in turn. However, an idealized stereochemistry cannot always be unambiguously assigned, even when the structure is precisely known, as intermediate stereochemistries are also possible (1562).

1: 5: 2 Stereochemistry. The 1 : 5 : 2 stereochemistry has been observed only in the dithiocarbamate complex $(Et_4N)[Np(S_2CNEt_2)_4]$, which is the only known example where eight sulphur atoms surround a metal atom (355, 357). Figure 5 also shows the relationship between this structure and that of a dodecahedron, where the five sites closest to forming a planar regular pentagon are emphasized, which in the dodecahedron are puckered from the least square plane by an average of about 12°.

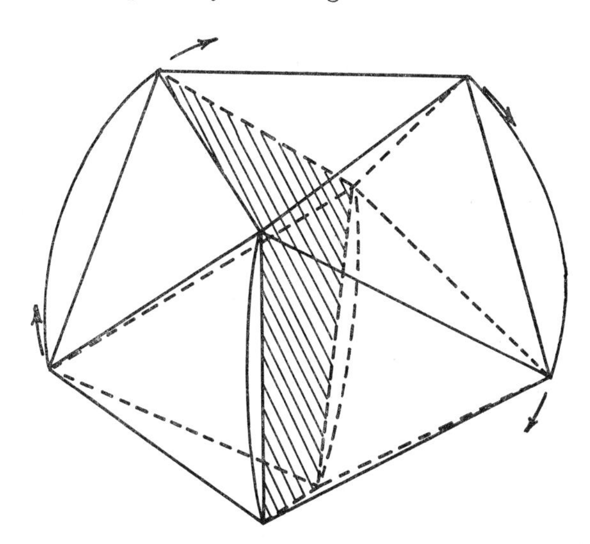

FIG. 5 The 1 : 5 : 2 structure of $[Np(S_2CNEt_2)_4]^-$, and its relation to the dodecahedron.

Dodecahedron. The dodecahedron (D_{2d} symmetry) contains two different ligand sites. There are four "A" sites forming a tetrahedron elongated along the four-fold inversion axis, and four "B" sites between the pairs of A sites, which form a tetrahedron flattened along the same axis (Fig. 6). It is conventional to specify the shape of the dodecahedron by three parameters; θ_A and θ_B, the angles the M–A and M–B bonds respectively

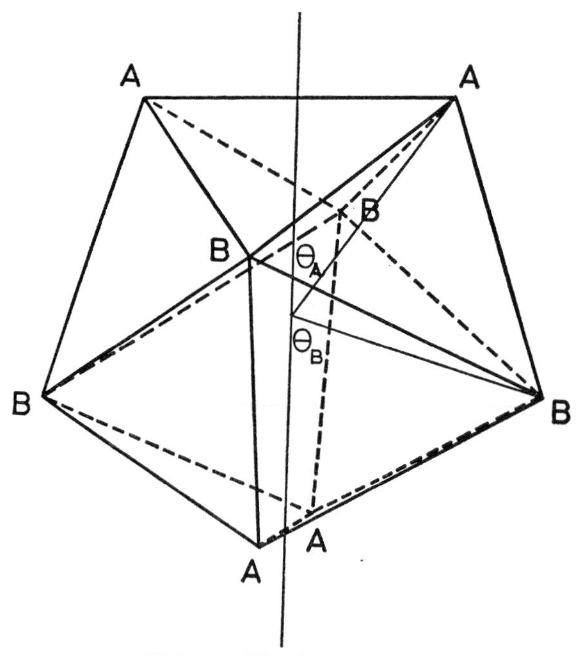

FIG. 6 The dodecahedron.

make with the four-fold inversion axis, and M–A/M–B, the ratio of the respective bond lengths.

The values of θ_A and θ_B corresponding to the theoretically most stable dodecahedron depend upon the method of calculation. For a model based on all donor atoms being represented by hard spheres (HSM) so that each A sphere is in contact with one A and three B spheres, and all A–A and A–B edges are equal to twice the radius of the sphere, then $\theta_A = 36 \cdot 8_5°$ and $\theta_B = 69 \cdot 5°$ (and the B–B edge is equal to 1·250 A–A). For a calculation in which the covalent bonding is maximized using d^4sp^3 hybrids (d_{xy} orbital omitted), an increase in θ_B (that is a further flattening of the B tetrahedron) and a decrease in θ_A is expected (that is a further elongation of the A tetrahedron) with respect to the hard sphere model (1904). Finally for the model which minimizes ligand-ligand repulsion energies, in which it is assumed that the repulsive energy is of the form $U \propto r^{-n}$, where n can have values between 1 (for a purely Coulombic interaction) and about 12 (for repulsion between fairly rigid donor atoms), there is again predicted a slight increase in θ_B, but now a slight increase in θ_A (1207, 1353, 1840). These results are summarized in Table 5, together with those parameters for known molecules. This model also predicts that increasing the M–A/M–B ratio will decrease θ_A and increase θ_B.

TABLE 5

Structural data for dodecahedral molecules

	θ_A	θ	$\frac{M–A}{M–B}$	Reference
Hard sphere model	36·8$_5$°	69·5°	1·00	
d^4sp^3 hybridization	34·6	72·8	1·00	
Ligand–ligand repulsion: $n = 1$	38·5	71·7	1·00	
$n = 6$	37·3	71·4	1·00	
$n = 12$	37·1	70·8	1·00	
$K_4[Mo(CN)_8],2H_2O$	36·0	72·9	1·00	1202
$Li_6[BeF_4][ZrF_8]$	43·0	65·5	1·05	2087
$(NH_4)[Pr(TTA)_4],H_2O$	41·4	65·7	1·00	1488
$Cs[Y(HFA)_4]$	38·6	68·5	1·01	206
$[Th(S_2CNEt_2)_4]$	44	66	1·00	356
$Na_4[Zr(C_2O_4)_4],3H_2O$	35·2	73·5	1·03	1013
$[Y(acac)_3(H_2O)_2]$	37·1	73·5	1·00	615, 1562
$[Cr(O_2)_4]^{3-}$	43·3	89·5	0·95	2225, 2234
$[Ti(NO_3)_4]$	37·2	80·8	1·01	977
$[Sn(NO_3)_4]$	38·9	81·7	1·01	976
$[TiCl_4(diars)_2]$	36·4	72·5	—	505, 506, 1846

However, probably the most important result from calculations of covalent bonding and ligand–ligand repulsions is that it requires very little energy to distort significantly the idealized stereochemistry while retaining the D_{2d} symmetry.

The structure of $K_4[Mo(CN)_8],2H_2O$ is the only example of a dodecahedron with eight monodentate ligands which has been determined with high precision. The deviations by a few degrees from the calculated polyhedra emphasize the ease of distortion of this polyhedron, although it may be noted that the structure adopted is considerably more stable than that of the hard sphere model.

The structure of the $[ZrF_8]^{4-}$ ion in $Li_6[BeF_4][ZrF_8]$ has abnormally high values of θ_A and low values of θ_B, which may be attributed to lithium ions lying outside all twelve A–B edges, so that the two A–A and four B–B edges are abnormally long.

In $(NH_4)[Pr(TTA)_4],H_2O$ (where TTA is the thenoyltrifluoroacetate ion, $C_4H_3S.CO.CH.CO.CF_3^-$), $Cs[Y(HFA)_4]$ (where HFA is the hexafluoroacetylacetonate ion, $CF_3.CO.CH.CO.CF_3^-$), and the diethyldithiocarbamate complex $[Th(S_2CNEt_2)_4]$, the bidentate ligands span the "g" edges of the dodecahedron (Fig. 7), which would be expected to decrease the ligand–ligand repulsions along these edges leading to the observed increase in θ_A and decrease in θ_B.

In the other examples with four bidentate ligands listed in Table 5, the ligands span the "m" edges of the dodecahedron (Fig. 7), which for

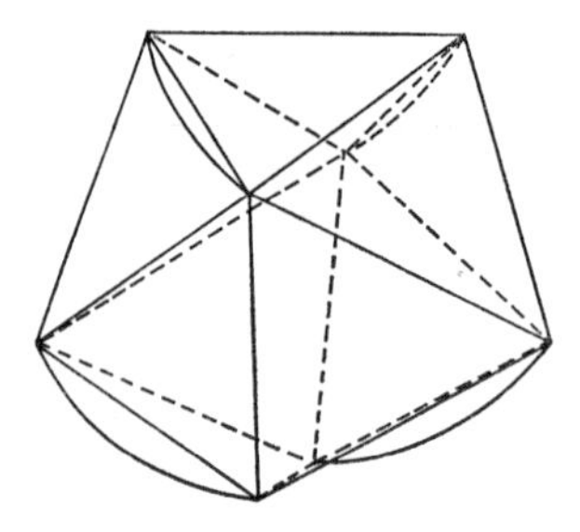

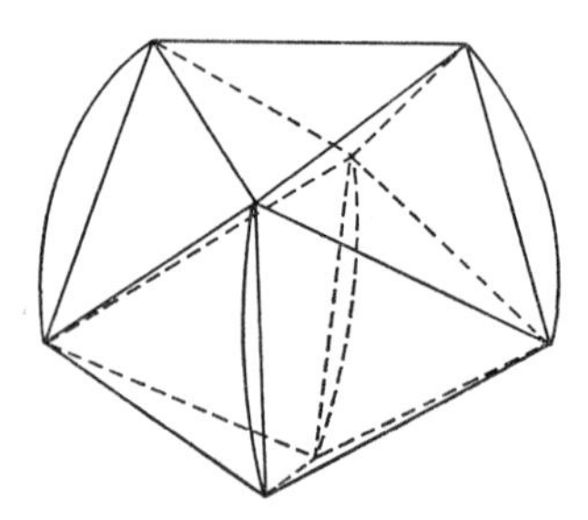

FIG. 7 The dodecahedron formed by four bidentate ligands. (a) Spanning of "g" edges. (b) Spanning of "m" edges.

sterically constrained ligands would be expected to increase both θ_A and θ_B.

The acetylacetonate complex $[Y(acac)_3(H_2O)_2]$ has a distorted dodecahedral structure, and is best considered as intermediate between a dodecahedron and a square antiprism. $[Eu(acac)_3(H_2O)_2],H_2O$ (1251) and $[La(acac)_3(H_2O)_2]$ (1867) appear similar.

It appears that for a complex containing four sterically small bidentate ligands, the dodecahedron is favoured relative to the square antiprism. This is because the square antiprism has two sets of eight equal edges which are unable to accommodate four bidentates having a short span, whereas the dodecahedron has a greater variety of edges available.

The dodecahedron is also favoured if four donor atoms are different to the other four, as in $[TiCl_4(diars)_2]$, as additional stability may be obtained by sorting the different ligands into the appropriate A and B sites.

For tridentate and tetradentate ligands in which the distances between the donor atoms are similar, the ligand will span those edges of the dodecahedron which have similar lengths, that is the A–A and A–B edges rather than the B–B edges. Thus in the bis(nitrilotriacetato) complex $[Zr\{N(CH_2COO)_3\}_2]^{4+}$ the three N- -O "bites" of each chelate span one A–A and two A–B edges of the dodecahedron ($\theta_A = 34\cdot7_5$, $\theta_B = 73\cdot8°$, M–A/M–B = 1·06) (1206, 1208).

The splittings of the five *d* orbitals under the influence of a dodecahedral crystal field shows that the d_{xy} orbital is of considerably lower energy than the other four *d* orbitals. The z-axis is taken as the major four-fold inversion axis and the xz and yz planes as incorporating the B sites. The d_{xy} orbital then has lobes directed towards the middle of the four long B–B edges, and is clearly the best orientated to avoid the ligand positions.

The calculated relative energies of the other d orbitals is sensitive to the exact shape parameters of the dodecahedron, and also, less importantly, to the assumed radial integrals. For example an increase in θ_A lowers the energy of the d_{z^2} orbital relative to the other d orbitals, while an increase in θ_B increases the relative energies of the d_{xy} and $d_{x^2-y^2}$ orbitals (655, 1071, 1840, 1908). The order shown in Fig. 8 has been experimentally

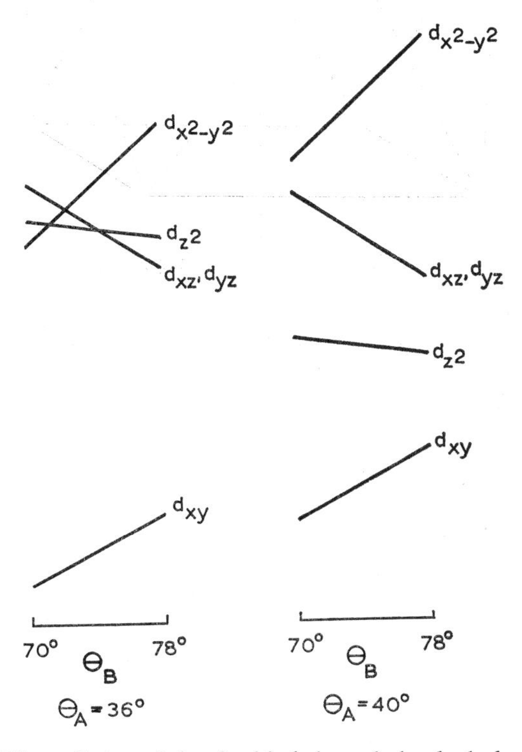

FIG. 8 The splitting of the d orbitals by a dodecahedral crystal field.

confirmed by studying a series of d^1 compounds of the type NbX_4(bidentate)$_2$ (656). Variation of the field along the z axis by varying the bidentate ligand which spans the two A–A edges of a dodecahedron alters the lowest energy band of the three d–d bands observed in the visible spectrum, while changing the field in the xy plane by varying the halogen atom X situated at the B sites of a dodecahedron alters the highest energy band.

Square antiprism. The square antiprism is shown in Fig. 9, and the detailed shape is specified by θ, the angle the eight metal–ligand bonds make with the eight-fold inversion axis.

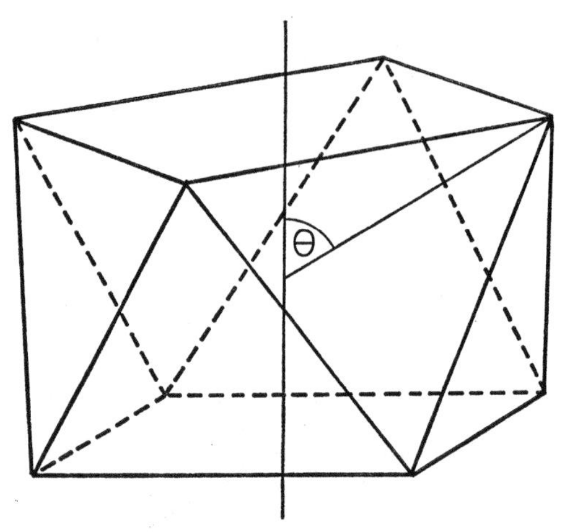

FIG. 9 The square antiprism.

It is again found that the calculated shape of the theoretically ideal square antiprism depends upon the assumptions made in the calculation. For a model based on the ligands being represented by hard spheres, so that all edges of the square antiprism are equal, $\theta = 59 \cdot 3°$. For a model based on d^4sp^3 hybrid orbitals in which the d_{z^2} orbital is omitted, $\theta = 57 \cdot 6°$, and for a less favourable one based on p^3d^5 hybrid orbitals, $\theta = 60 \cdot 9°$, corresponding to an elongation and a compression respectively of the square antiprism along the major axis (704). For a model which minimizes ligand–ligand repulsion, or repulsions between the bonding

TABLE 6

Structural data for square antiprismatic molecules

	θ	Reference
Hard sphere model	59·3°	
d^4sp^3 hybridization	57·6°	
Ligand–ligand repulsion: $n = 1$	55·9°	
$n = 6$	57·1°	
$n = 12$	57·9°	
$Li_4[UF_8]$	57·1°	377
$Na_3[TaF_8]$	59·0°	1205
$K_2[ReF_8]$	58·0°	1465
[Zr(acac)$_4$]	57·3°	2145
[Th(acac)$_4$]	58·2°	1048
[Ce(acac)$_4$]	58·5°	1665, 2283
[Y(acac)$_3$(H_2O)$_2$],H_2O	57·1°	615
[La(acac)$_3$(H_2O)$_2$]	57·1°	1867

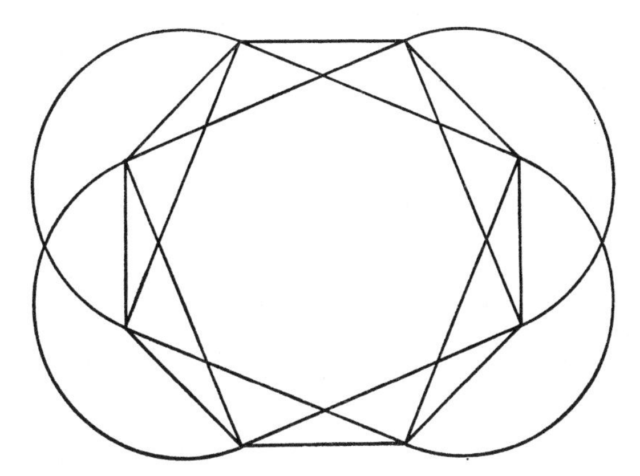

FIG. 10 The structure of $[M(acac)_4]$.

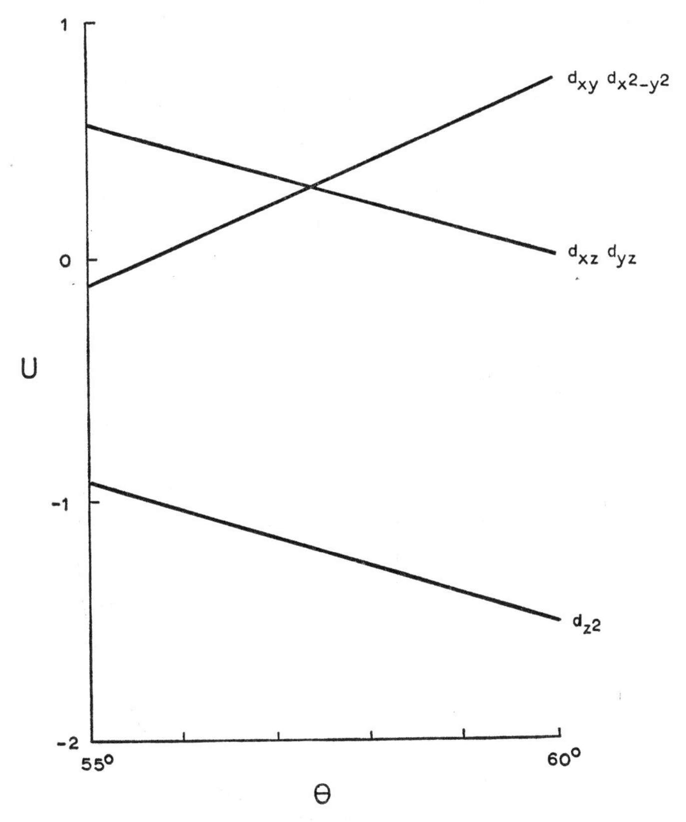

FIG. 11 The splitting of the *d* orbitals by a square antiprismatic crystal field.

pairs of electrons, an elongation of the square antiprism along the eight-fold inversion axis relative to the hard sphere model is predicted, $\theta = 55 \cdot 9$, $57 \cdot 1$ and $57 \cdot 9°$ for n = 1, 6 and 12 respectively.

Again it is found that the potential minima are very shallow, very little energy being required for a considerable distortion.

Values of θ for known square antiprismatic molecules are shown in Table 6.

A class of compounds which have been particularly extensively studied are the β-diketonates. The tetrakisacetylacetonate complexes are square antiprismatic with the bidentates spanning the edges of the square faces (Fig. 10). The trisacetylacetonate diaquo complexes are intermediate between square antiprismatic and dodecahedral stereochemistries as noted above. The tetrakis hexafluoracetylacetonate, thenoyltrifluoroacetonate and dibenzoylmethanate complexes however are dodecahedral.

Under the influence of a square antiprismatic crystal field, the d_{z^2} orbital, which projects along the major axis towards the centres of the square faces, is of lowest energy (1014, 1355, 1840, 1807). Which of the two remaining degenerate pairs, d_{xz}, d_{yz} or d_{xy}, $d_{x^2-y^2}$ is of lower energy depends upon the shape of the square antiprism (Fig. 11).

One important point which emerges from the above discussion is that both the square antiprism and the dodecahedron are easily distorted, and

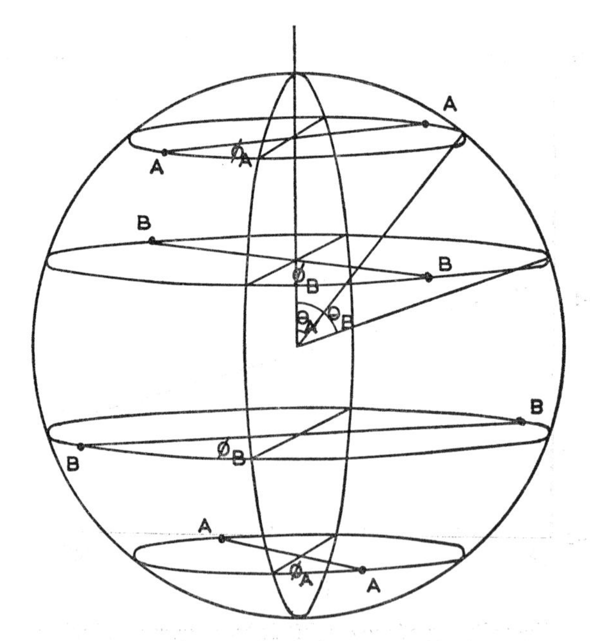

FIG. 12 Generalized dodecahedral-square antiprismatic stereochemistry.

are closely similar in both energy and stereochemistry. It even appears that a continuous interconversion between these two stereochemistries is possible.

The dodecahedron and square antiprism are both particular examples of the more general stereochemistry shown in Fig. 12. In addition to θ_A and θ_B which have the same meaning as before, the positions of the A and B sites are defined by their "longitudes" ϕ_A and ϕ_B. Evaluation of the ligand–ligand repulsion energies as before, shows that there is no potential energy barrier to prevent an isolated dodecahedron from being converted into an isolated square antiprism (247). Fig. 13 shows the projection of a potential energy surface as the A–A edges are twisted relative to each other.

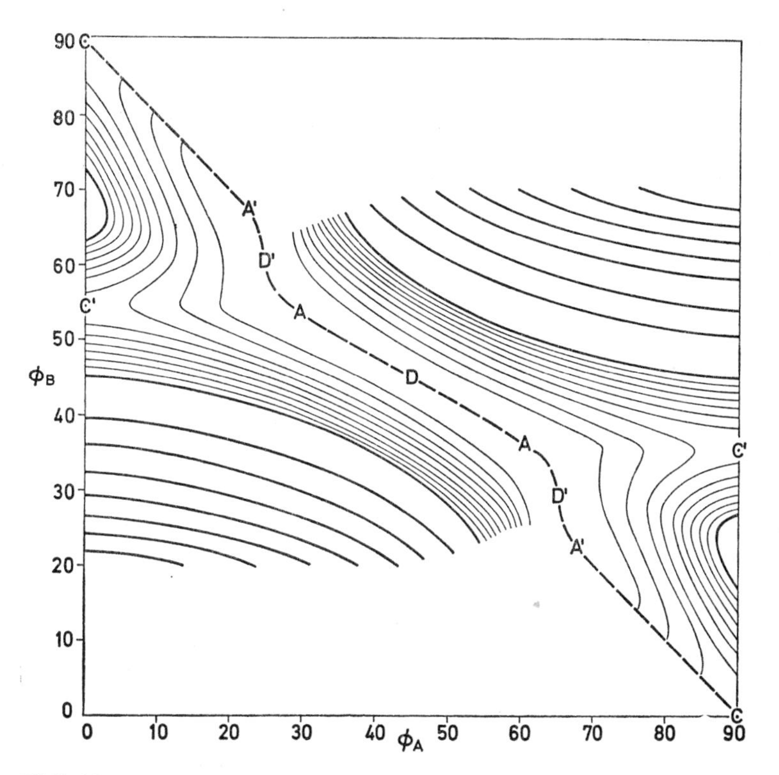

FIG. 13 Projection of the potential energy surface for the ligand–ligand repulsion for eight monodentate ligands onto the ϕ_A–ϕ_B plane. The difference in energy between the faint contour lines is 10% of that between the dark contour lines. The positions of the regular stereochemistries are shown by D (for dodecahedron), A (for square antiprism), and C (for cube). Reproduced from reference 247, Blight and Kepert (1968).

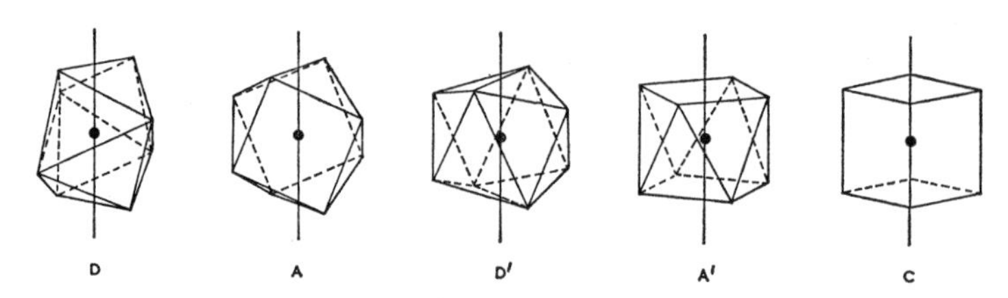

FIG. 14 Polyhedra corresponding to positions in Fig. 13.

Fig. 14 shows the polyhedra formed along the "reaction coordinate". The coordinate axes can be changed each time a square antiprismatic or dodecahedral stereochemistry is traversed, and so these changes could be envisaged as being continuous in contrast to the normal reciprocatory bending vibrations.

E. Coordination Numbers above Eight

Coordination numbers greater than eight have not been found for monomeric complexes of the Groups IV, V and VI transition metals, but there appears little doubt that they will be prepared. These higher coordination numbers are found for larger ions, particularly the lanthanides and actinides, while technetium and rhenium form the nine coordinate $[TcH_9]^{2-}$ and $[ReH_9]^{2-}$ with the small hydride ion (6, 1002, 1431). The metal atoms in clusters of the type $[(Nb_6Cl_{12})Cl_6]^{4-}$ and $[(Mo_6Cl_8)Cl_6]^{2-}$ are also formally nine coordinate.

Two stereochemistries have been observed for nine coordination. The first is the tricapped trigonal prism in which additional ligands are outside each of the square faces of a trigonal prism. This polyhedron is defined by θ, the angle the six "prismatic" metal–ligand (A) bonds make to the three-fold axis, and M–B/M–A, the ratio of the "equatorial" bond lengths to the "prismatic" bond lengths. Evaluation of the ligand–ligand repulsion energies shows that a more stable structure is obtained by increasing θ by about 3° relative to the hard sphere model (1353). In addition to $[TcH_9]^{2-}$ and $[ReH_9]^{2-}$, this stereochemistry is found in a number of rare earth nonahydrates, $[Nd(H_2O)_9](BrO_3)_3$ (1170) and $[M(H_2O)_9](EtSO_4)_3$ (where M is Y, Pr or Er) (883), and also in $[Eu(terpyridyl)_3](ClO_4)_3$ (948).

The second stereochemistry, the capped square antiprism, is observed in $[Th(oxine)_4(Me_2SO)],Me_2SO$, where the dimethylsulphoxide ligand occupies the unique coordination position (2158). However, the stoicheiometrically similar $(NH_4)[Y(C_2O_4)_4(H_2O)]$ has a stereochemistry closer to a tricapped trigonal prism rather than a capped square antiprism. The stereochemistry of the ethylenediaminetetraacetic acid (H_4A) derivative

$K[La(A)(H_2O)_3],5H_2O$ is also rather irregular, and the stereochemistry is best regarded as being determined by the constraints of the polydentate ligand (1203).

Apparently only the tricapped trigonal prism can be obtained from sp^3d^5 hybridization (705, 1403).

Crystal field splittings for the tricapped trigonal prism as a function of θ are shown in Fig. 15. A similar order is obtained for the capped square

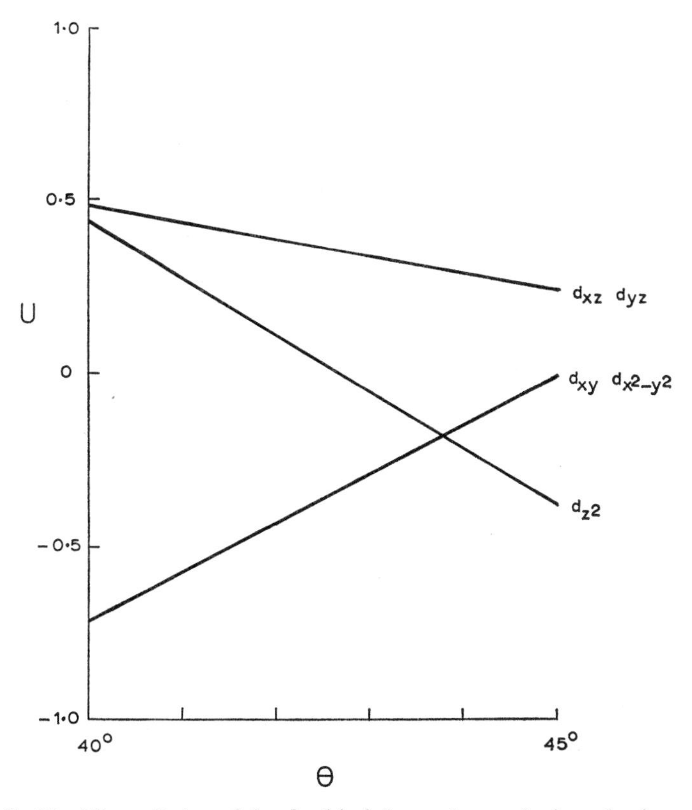

FIG. 15 The splitting of the d orbitals by a tricapped trigonal prismatic crystal field. Reference 1355, Kepert (unpublished).

antiprism, the order for the hard sphere model polyhedron being $d_{z^2} < d_{xy}, d_{x^2-y^2} < d_{xz}, d_{yz}$.

Apart from the nitrato complexes mentioned above, the only monomeric ten coordinate complexes appear to be the monohydrogenethylene-diaminetetraacetate complex $[La(HA)(H_2O)_4],3H_2O$ (1540) and the novel gold cluster $[Au\{Au(PPh_3)\}_7\{Au(SCN)\}_3]$ in which the central metal atom is surrounded by ten other metal atoms (1611).

B

3. METAL–METAL BONDING

A. Introduction

The common occurrence of compounds which contain metal–metal bonds of considerable strength (that is, an implied strength of at least 20 kcal mol^{-1}) is mainly restricted to coordination compounds, including oxides and halides, of the early transition metals, and to organometallic compounds in general. For metal–metal bonding between transition elements to occur, the major prerequisites are the presence of relatively expanded *d* orbitals on the metal atoms to allow overlap with each other, the availability of electrons which can be used for metal–metal bonding, and the absence of unfavourable steric effects.

Expanded *d* orbitals are normally present with elements towards the left-hand side of the transition element block of the periodic table, particularly in the second and third rows, but in the case of the remaining transition elements only by the use of ligands which are capable of forming metal-to-ligand $d\pi$–$p\pi$ bonds.

The absence of electrons available for metal–metal bonding means that no such bonding can occur for example, in the d^0 TiO_2 or $[Cr(C_5H_5)(CO)_3]^-$ which have the argon and krypton closed shell electron configurations respectively. However addition of an electron in the former case forms the formally d^1 Ti_2O_3 or VO_2, both of which contain pairs of metal atoms joined by a simple electron pair bond (pages 122 and 190 respectively), whereas if one electron is removed from $[Cr(C_5H_5)(CO)_3]^-$, the dimeric chromium–chromium bonded $Cr_2(C_5H_5)_2(CO)_6$ is formed (page 363).

The crowding of too many groups about the metal atom hinder the close approach and hence bonding of metal atoms. For example the tetrahalides $NbCl_4$ and WCl_4 contain bonded pairs of metal atoms, but metal-metal bonding is absent in complexes such as the eight coordinate $[NbCl_4(diars)_2]$ or the six coordinate $[WCl_4(py)_2]$ (246, 656).

A metal–metal bond can be most reliably recognized either from a study of properties such as magnetism and electrical conductivity, or from an X-ray determination of the structure.

By far the greatest majority of metal–metal bonded compounds are diamagnetic, but there are a few which contain an odd number of electrons, and are of course paramagnetic. These are $(Ta_6Cl_{12})^{3+}$, $Ru_2(OOC.Pr)_4Cl$ (2217), Nb_3Cl_8, $(Nb_6I_8)^{3+}$, $Tc_2Cl_8{}^{3-}$(556), $Co_3(CO)_6(Ph_3P)_3$ (1895), $(Nb_6Cl_{12})^{3+}$, and $(C_5H_5)_3Ni_3(CO)_2$ (870). Electron spin resonance results of the last two compounds indicate that the unpaired electron is delocalized over all six and all three metal atoms respectively.

Although a metal–metal distance is readily obtained from an X-ray structural determination, a close approach of metal atoms, for example to a distance similar to that found in the metal itself, does not necessarily lead to metal–metal bonding. Conversely there are many compounds where such bonding does occur which have metal–metal bond lengths considerably greater than in the metal itself. Nevertheless provided additional criteria are used, an X-ray structural determination remains one of the most useful techniques of recognizing a metal–metal bond.

A metal–metal bond can be recognized either by the close approach of two metal atoms in the absence of bridging groups, as in $[Cl_4Mo–MoCl_4]^{4-}$ (Fig. 192, page 357), or $(CO)_3(C_5H_5)Mo–Mo(C_5H_5)(CO)_3$ (page 336), or if bridging groups are present, by the closer approach of metal atoms than appears necessary from the bridging geometry, as indicated by molecules containing similar groups, but in which metal–metal bonding is absent. This last criterion will be illustrated by considering a number of examples:

(i) The structures of $Cr_2Cl_9^{3-}$ and $W_2Cl_9^{3-}$ are based on pairs of octahedra sharing a common face. In the $Cr_2Cl_9^{3-}$ anion, the chromium atoms are displaced from the centres of the octahedra away from one another (3·12 compared with 2·83 Å), whereas in $W_2Cl_9^{3-}$, in spite of the greater size of the metal atom, the tungsten atoms are displaced towards one another (2·42 compared with 2·56 Å) (Fig. 16). In the caesium salts, the

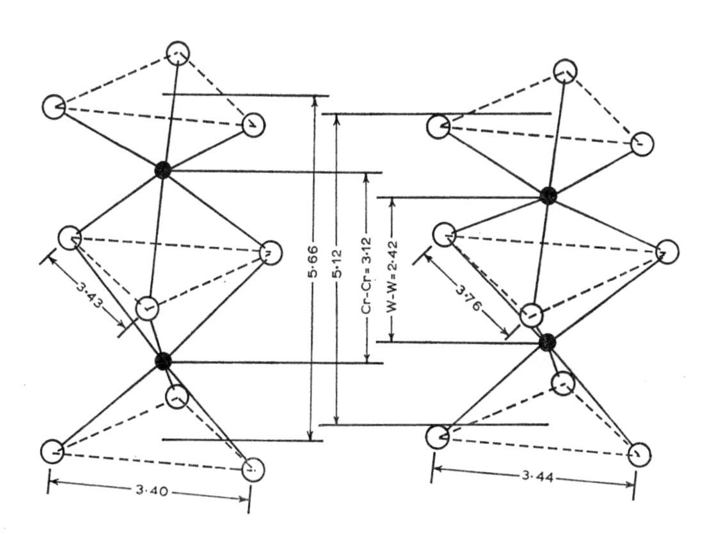

FIG. 16 A comparison of the structures of $Cr_2Cl_9^{3-}$ (left) and $W_2Cl_9^{3-}$ (right).

caesium and chloride ions constitute a close packed lattice, and the tungsten–tungsten bond results in a contraction in the *c*-axis of the hexagonal unit cell, and an increase in the *a*-axis because of the bulging out of the three bridging chlorine atoms. Thus the formation of metal–metal bonds is observed from the cell dimensions, without a complete structural determination.

	$Cs_3Cr_2Cl_9$	$Cs_3W_2Cl_9$
c	17·93 Å	17·08 Å
a	7·22 Å	7·41 Å
c/a	2·48	2·32

The metal–metal bonding is also confirmed by the paramagnetism of the chromium salt but the diamagnetism of the tungsten salt.

(ii) The structures of many halides and oxides of the early transition metals can be considered to be based on a close packed anion lattice with metal atoms occupying the octahedral interstices. Metal–metal bonds can be formed either between adjacent atoms in the same layer (that is between octahedra sharing edges), or between atoms in adjacent layers (that is between octahedra sharing faces).

The structures of the tetrachlorides of niobium, tantalum, molybdenum (α-isomer) and tungsten have one half the octahedral holes filled in every

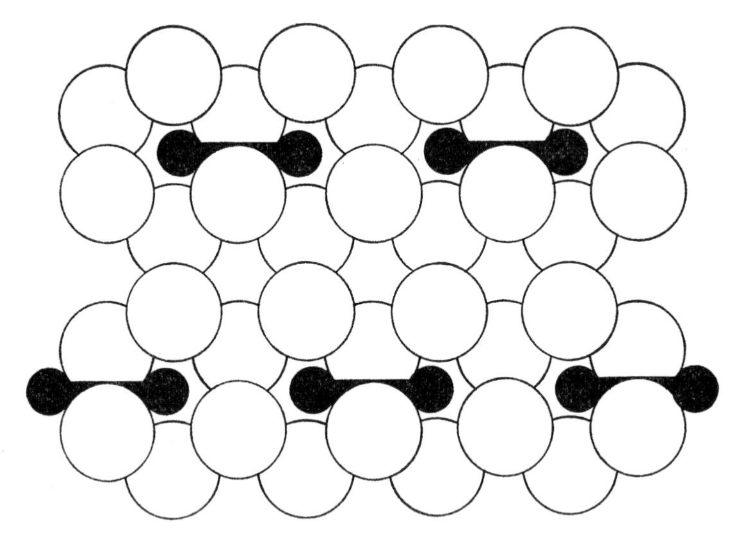

FIG. 17 The structure of $NbCl_4$, $TaCl_4$, α-$MoCl_4$ and WCl_4, showing the formation of metal–metal bonds between pairs of metal atoms in octahedral sites between close packed layers of chloride ions.

second layer so that the structure can alternatively be considered to be composed of infinite linear strings of octahedra sharing opposite edges (Fig. 17). The formation of metal–metal bonds is shown by the displacement of the metal atoms so that the metal–metal distances are alternately short (3.06 Å) and long (3.76 Å), clearly indicating the existence of discrete M_2 pairs.

In molybdenum trichloride, two thirds of the sites in every alternate layer of octahedral holes are occupied by metal atoms, but the lattice is again distorted due to the formation of bonds between pairs of molybdenum atoms (Fig. 18). Each molybdenum atom has one molybdenum

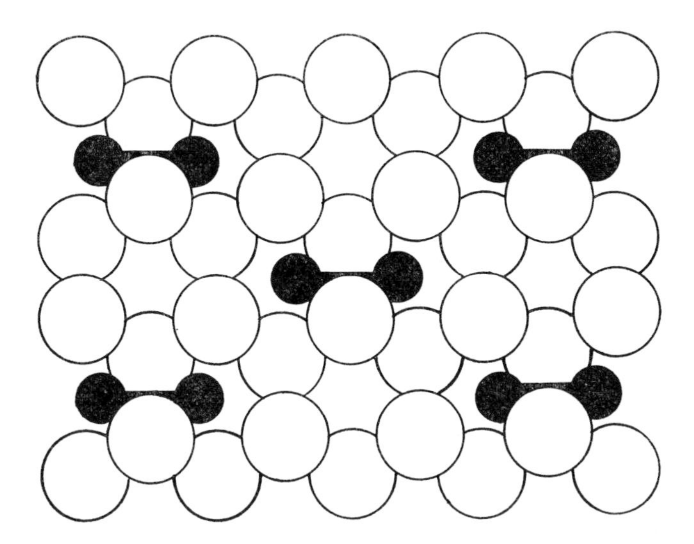

FIG. 18 The structure of $MoCl_3$ showing the formation of metal–metal bonds between pairs of molybdenum atoms in octahedral sites between close packed layers of chloride ions.

neighbour at 2·76 Å and two at 3·71 Å, compared with three equidistant neighbours in the undistorted structure found for $CrCl_3$, $AlCl_3$, etc.

The dioxides which have the rutile structure also have linear strings of metal atoms occupying half the octahedral holes in any one layer, but now the change in stoicheiometry means that every layer of octahedral interstices is half filled, so that the strings formed by edge sharing of octahedra also share corners with each other. The structure of VO_2 shows that the metal–metal distances along the strings are alternately short (2·65 Å) and long (3·12 Å), whereas in TiO_2 the structure shows that all titanium–titanium distances are equal (2·95 Å). Similar bonding occurs in the dioxides of niobium, molybdenum, tungsten, technetium and rhenium (page 191). On heating VO_2 to about 70° C, the structure changes to that of TiO_2, and the freeing of electrons formally used for vanadium–vanadium bonding is reflected in a dramatic increase in magnetic susceptibility and electrical conductivity (page 192).

In Nb_3Cl_8, $Zn_2Mo_3O_8$ and FeS, a somewhat different distortion from the ideal close packed structure is observed. In these cases the metal–metal bonding is between groups of three metal atoms, which form triangular clusters. Only $\frac{3}{4}$ of the sites in any one layer are occupied in Nb_3Cl_8 and $Zn_2Mo_3O_8$ (Fig. 19). The structure of FeS is related to the nickel arsenide

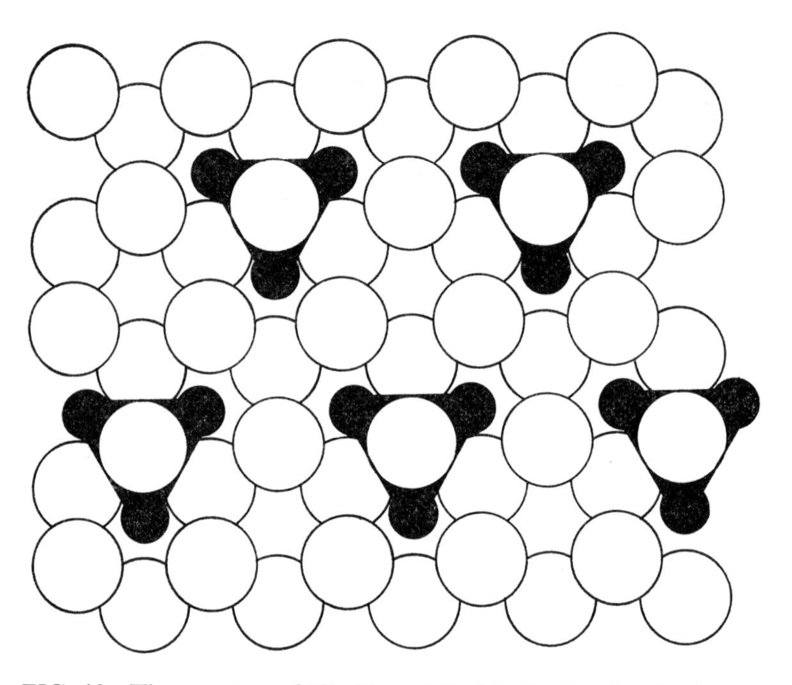

FIG. 19 The structure of Nb_3Cl_8 and $Zn_2Mo_3O_8$ showing the formation of triangular clusters of metal atoms in octahedral sites between close packed anion layers.

structure, and is based on double hexagonal close packing (ABCB) of sulphide ions with ferrous ions in all the octahedral interstices. That is, FeS_6 octahedra share opposite faces forming strings along the *c*-axis (in the direction normal to the plane of the close packed layers), and three edges in the plane of the close packed layers. The structure is then distorted by the formation of more closely bonded Fe_3 groups (Fig. 20) (72, 220, 1111):

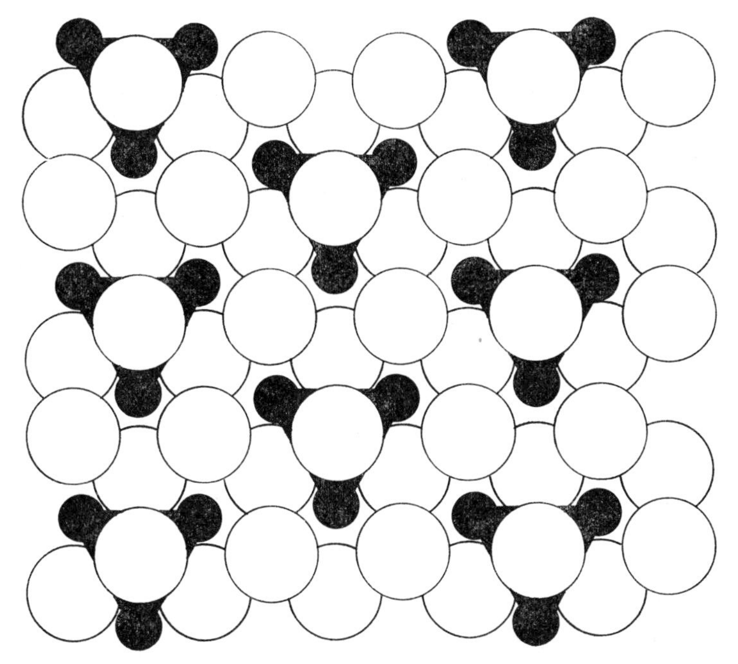

FIG. 20 The structure of FeS showing the formation of triangular clusters of iron atoms between layers of sulphur atoms.

	Reference 72	Reference 220
Fe–Fe (within triangular clusters)	3·00 Å	3·08 Å
Fe–Fe (between triangular clusters)	3·73 Å	3·64 Å
Fe–Fe (along *c*-axis)	2·94 Å	2·94 Å

On heating to 140° C the metal–metal bonding is broken, and the relative freeing of electrons is shown by a sharp increase in magnetic susceptibility and electrical conductivity (1193, 1195, 1330, 1743, 2192). On heating through this transition temperature, the disappearance of the superstructure associated with metal–metal bonding is accompanied by an expansion of the *a*-axis by 0·5% and a contraction of the *c*-axis by 0·7% (2192). Measurements of the heat capacity (1196) through this transition temperature show $\Delta H = 550$ cal mol^{-1} of FeS.

(iii) A comparison of the structures of $(C_5H_5)_2Ni_2(PPh_2)_2$, in which each metal atom formally has the inert gas configuration, and $(C_5H_5)_2Co_2(PPh_2)_2$, which has one less electron, is shown in Fig. 21. The structure is folded in the latter case as a consequence of cobalt–cobalt bonding, the metal–metal distance being reduced from 3·36 Å in the nickel compound to 2·56 Å in

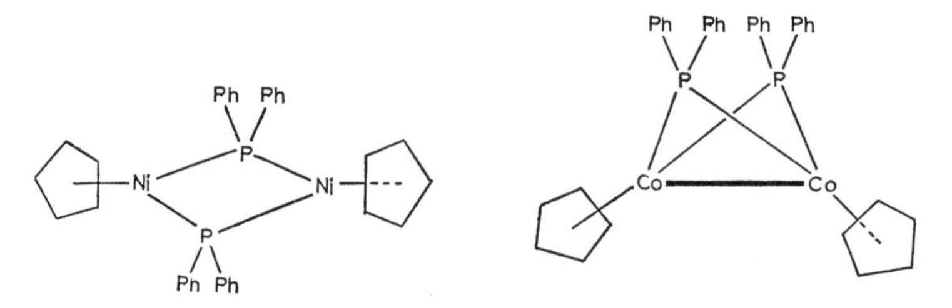

FIG. 21 A comparison of the structures of $(C_5H_5)_2Ni_2(PPh_2)_2$ and $(C_5H_5)_2Co_2(PPh_2)_2$.

the cobalt compound. This folding is accompanied by considerable steric strain in the bridging groups, the metal–phosphorus–metal angle being reduced from 102·4° to only 72·5° (522).

(iv) A comparison of the structures of $(Nb_6Cl_{12})^{2+}$ and (Pt_6Cl_{12}) (346) is shown in Fig. 22, viewed as a projection along one of the four-fold axes

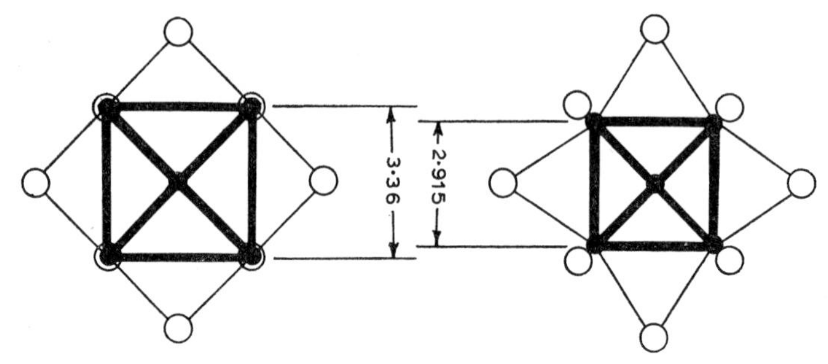

FIG. 22 A comparison of the structures of $(Nb_6Cl_{12})^{2+}$ and (Pt_6Cl_{12}). The octahedron of metal atoms is shown by the bold lines, and the chlorine atoms by the open circles.

of the molecule. (A perspective view is shown on page 237.) It is clear that the niobium atoms are drawn inwards from the square faces of the cubo-octahedron formed by the twelve bridging chlorine atoms.

(v) A number of divalent transition metal acetates form $M_2(OAc)_4$ where the two metal atoms are bridged by four acetate groups:

	M–M distance	Reference
$Mo_2(OAc)_4$	2·11 Å	1512
$Ru_2(OOC.Pr)_4Cl$	2·28 Å	204
$Rh_2(OAc)_4(H_2O)_2$	2·45 Å	1881
$Cr_2(OAc)_4(H_2O)_2$	2·64 Å	1786
$Cu_2(OAc)_4(H_2O)_2$	2·64 Å	1785

In the hydrated copper acetate, magnetic studies show that the copper–copper interactions are very weak (848, 1028, 1320), and the copper atoms lie outside the plane of the four bonded oxygen atoms (Fig. 23). At the other limit, the diamagnetic molybdenum acetate has an extremely short molybdenum–molybdenum distance, and the molybdenum atoms are inside the plane of the four bonded oxygen atoms. The molybdenum–molybdenum bond length of 2·11 Å is even shorter than in complexes such as $Re_2Cl_8^{2-}$ (Re–Re = 2·24 Å) which has a similar structure and in which the two rhenium atoms are joined by a quadruple bond ($\sigma + 2\pi + \delta$) (555, 564).

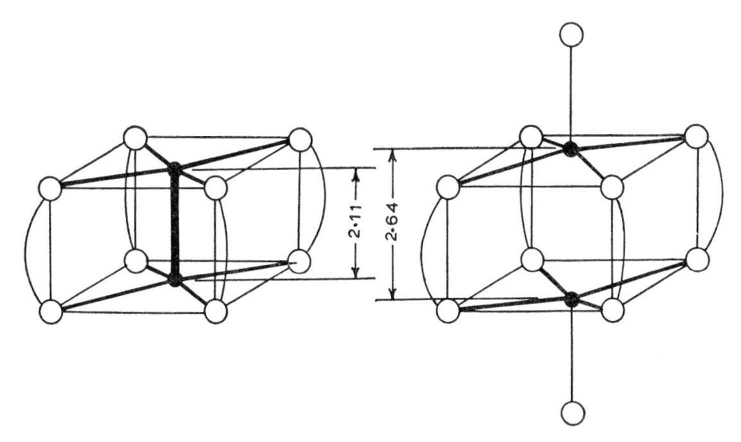

FIG. 23 A comparison of the structures of $Mo_2(OAc)_4$ (left) and $Cu_2(OAc)_4(H_2O)_2$ (right). The metal atoms are shown by the filled circles and the oxygen atoms on each end of the bidentate acetate groups as open circles.

The observation that the stronger metal–metal bonded molybdenum compound does not add additional ligands (H_2O or Cl^-) to complete an octahedral stereochemistry about each metal atom is also significant, as it is a general observation that an increase in metal–metal bonding weakens the bonding of the ligand *trans*- to the metal–metal bond (845).

The additional three electrons as in the paramagnetic $Ru_2(OOC.Pr)_4Cl$ are accommodated in nonbonding and slightly antibonding orbitals, which together with weaker bonding due to the contraction of the *d* orbitals due to an increase in nuclear charge on moving from molybdenum to ruthenium, results in weaker bonding and a longer metal–metal bond (2217). A further three electrons fills these orbitals, as in $Rh_2(OAc)_4(H_2O)_2$, with an equal increment in bond length (1311, 1414, 1766).

The δ bond in $Re_2Cl_8^{2-}$ keeps the two $ReCl_4$ units in a sterically unfavourable eclipsed configuration, whereas in the diacetato complexes the eclipsed configuration is forced in all the complexes due to the constraints of the bridging ligands. A similar eclipsed configuration is found in $Ni^{II}_2(Ph.N_3.Ph)_4$, in which the nickel–nickel bond length is 2·38 Å (551), but in $Ni^{II}_2(PhCOS)_4.EtOH$ the larger span of the bidentate ligand is partially overcome by screwing the NiO_4 plane relative to the NiS_4 plane so that the eclipsed configuration is destroyed (Ni–Ni = 2·49 Å) (265).

It can be seen that in certain borderline cases such as Pt_6Cl_{12} and $Cr_2(OAc)_4(H_2O)_2$ above, where the stereochemistry of the molecule places the metal atoms so close together that the internuclear distance is comparable to that found in the metal itself, some chemical judgement must be exercised to determine whether there is a significant metal–metal bonding interaction between partially filled orbitals on the metal atoms. The origin of these partly filled orbitals is often far from obvious. In cases such as $Tl^{I}_4(OR)_4$, $Pb^{II}_6O(OH)_6{}^{4+}$, and $Bi^{III}_6(OH)_{12}{}^{6+}$ where metal–metal bonding appears important, one of the electrons in the formally nonbonding 6*s* subshell is presumably promoted to a higher orbital. Similarly in InCl, the sodium chloride type structure observed is severely distorted, so that of the twelve indium atoms surrounding a given indium atom, three are much closer to it than the other nine (3·65 and 4·70 Å respectively) (213).

Tetrahedral groups of metal atoms, with a bridging ligand over each triangular face so that the eight atoms form a cubane structure, are found in the structures of $Li_4(CH_3)_4$ (Li–Li ~2·5 Å) (663, 1607, 2263) and $K_4(OBu)_4$ (K–K ~ 3·8 Å) (2361). Lithium–lithium bonding in $Li_4(CH_3)_4$ via the delocalization of the electron pair in each carbon–lithium bond over a four centre CLi_3 bond has been proposed (592); this increases the total overlap population per lithium atom, increasing the covalent bonding in the molecule.

Related bonding between metal atoms possessing closed shell electron configurations apparently occurs in $Li^{I}_3Nb^{V}O_4$, which has a structure related to that of sodium chloride. However instead of the pentavalent metal atoms being distributed in the cation sites so as they would be as far away as possible from each other, as in Li_3SbO_4 and Li_3TaO_4, the niobium atoms occupy adjacent sites in such a way that each niobium atom has other niobium atoms as nearest neighbours (Nb–Nb = 2·98 Å). It was suggested that the *p* orbitals of the oxide ions partially feed their electrons into the empty t_{2g} orbitals of the niobium atom, which bond with a similar niobium atom forming an electron deficient multicentred bond (242, 243).

The factors influencing the strength of metal–metal bonding will now be considered in a little more detail for the following two general classes of compounds:

(a) Inorganic compounds in which the metal atom is attached to essentially σ donor ligands, in which the heat of atomization is the most important factor in determining the extent of metal–metal bonding.

(b) Organometallic compounds in which the unfavourable nonbonding electron repulsions are considerably decreased by delocalization of these electrons onto the ligands by metal-to-ligand $d\pi$–$p\pi$ bonding, and the formation of metal–metal bonding is primarily governed by the number of electrons available for bonding.

B. Metal–Metal Bonded Inorganic Compounds

With ligands which can be classified as being mainly bonded by the σ donation of electron pairs to the metal atom, such as oxide or halide, metal–metal bonding is most pronounced in the second and third row transition elements, being centred on rhenium, molybdenum, tungsten, niobium and tantalum. It is just these elements which have the strongest metal–metal bonding in the metals themselves, as indicated by the high melting points, boiling points, hardness, and atomization energies of this group of elements.

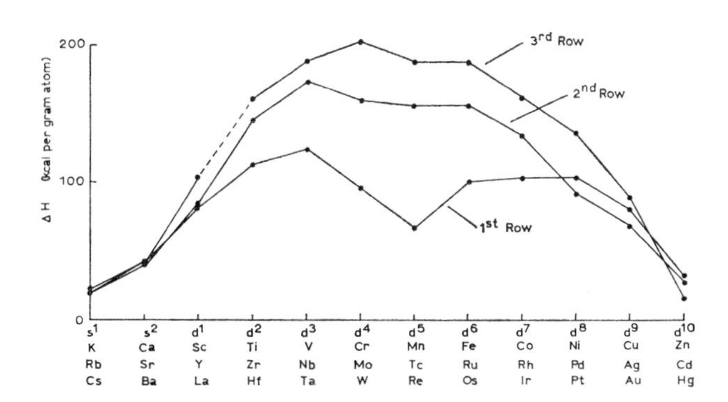

FIG. 24 Atomization energies of the transition elements.

The atomization energies are shown in Fig. 24 and Table 7. These comprise a metal–metal bond energy per metal atom together with a significant correction involving a promotion energy and a spin coupling energy to allow for the change from the sd^{n+1} electron configuration in the metallic state to the s^2d^n electron configuration of the gaseous state (1060). This effect of changing electron configuration is not relevant to metals with the s^1 and $d^{10}s^1$ electron configuration, and in these cases there is a

TABLE 7

Heats of atomization of the elements in kcal per gram atom

H														
52·1														
Li	Be											B	C	N
38·6	78·3											(135)	170·9	113·0
Na	Mg											Al	Si	P
25·9	35·3											78·0	107·4	79·8
K	Ca	Sc	Ti	V	Cr	Mn	Fe	Co	Ni	Cu	Zn	Ga	Ge	As
21·4	42·4	82·3	112·5	123·2	94·9	66·8	99·6	101·5	102·7	80·5	31·3	69·0	90·0	>69·0
Rb	Sr	Y	Zr	Nb	Mo	Tc	Ru	Rh	Pd	Ag	Cd	In	Sn	Sb
19·6	39·1	85·7	145·4	172·5	158·7	(155)	154·9	133·1	91·0	68·1	26·8	58·0	72·1	62·7
Cs	Ba	La	Hf	Ta	W	Re	Os	Ir	Pt	Au	Hg	Tl	Pb	Bi
18·7	41·7	104·0	160·0	186·9	201·8	186·1	187·4	159·9	135·2	88·0	14·7	42·7	46·6	49·5

clear relation between the atomization energies and the bond strength of the gaseous diatomic molecules (1361) (Fig. 25).

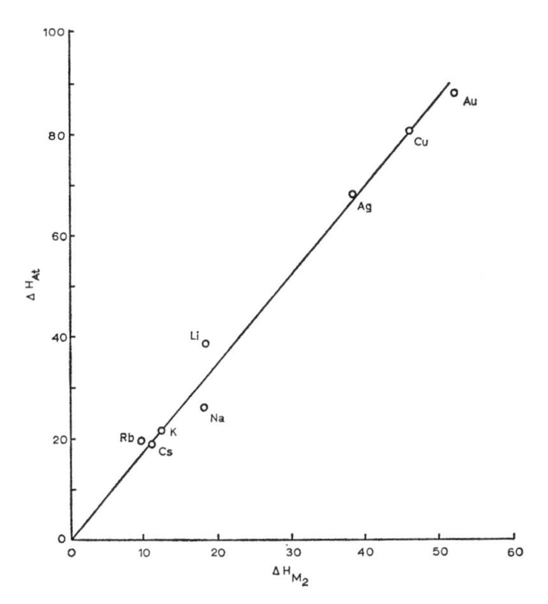

FIG. 25 The relation between atomization energies and bond energies of the gaseous diatomic molecules (both in kcal mol^{-1}).

As we move from left to right across the periodic table, the atomization energies increase from potassium, rubidium and caesium as electrons are progressively added to the metal, go through a maximum, and fall again to a minimum at zinc, cadmium and mercury. The maximum occurs near the Group VI elements chromium, molybdenum and tungsten, corresponding to a filling of the bonding delocalized metallic orbitals formed from the six *s* and *d* atomic orbitals. The lower atomization energies of the first row transition metals may suggest partial electron localization on the metal atom, and can be correlated with the lower third and subsequent ionization potentials and the observed increase in oxidation states as the *d* block portion of the periodic table is descended, for example the highest chlorides are VCl_4, $NbCl_5$, $TaCl_5$, and $CrCl_3$, $MoCl_5$, WCl_6. It has been shown (1060, 1361) that the low values for chromium and manganese in the first row elements reflects differences in that part of the atomization energy term involving the change in electron configuration between the metallic and gaseous states, rather than a change in metallic bonding energy itself.

The enhanced formation of metal–metal bonded compounds towards the left-hand side of the transition elements in the periodic table, and to the second and third row transition metals can also be ascribed to the enhanced overlap of the *d* orbitals which is possible in this area of the periodic table. Figs. 26 and 27 show two attempts at calculating the radii corresponding to maximum orbital electron density of the n*d* and n*p* electrons as an increasing number of *d* electrons are added around the nucleus. Fig. 26, based on a relativistic self consistent field calculation using Slater orbitals (2326), shows that the *d* orbitals contract more than the filled *p* orbitals (and the filled *s* orbitals which are more contracted than the filled *p* orbitals), and hence become less accessible for bonding.

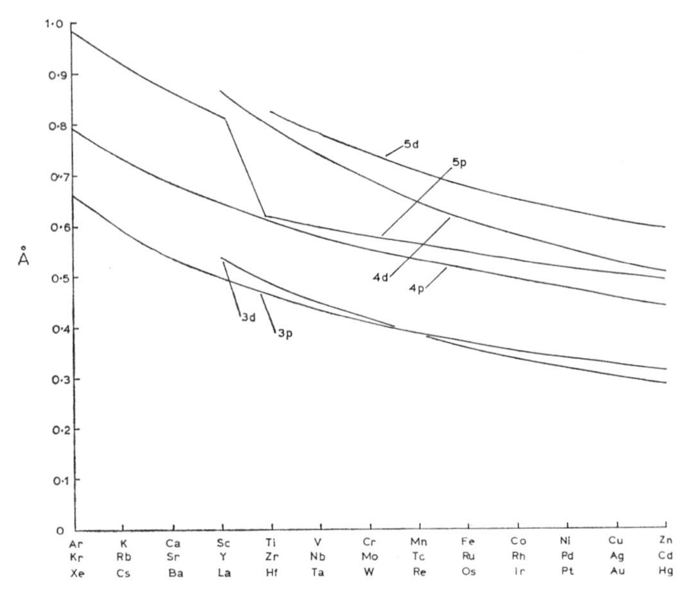

FIG. 26 Radii corresponding to maximum orbital electron density for the n*p* and n*d* electrons for the three transition series. From Reference 2326, Waber and Cromer (1965).

This is consistent with the observation of paramagnetic compounds towards the right-hand side of the transition metals, but metal–metal bonding towards the left-hand side of the transition metals. It is also apparent that the 4*d* and 5*d* orbitals are relatively more available for bonding than the 3*d* orbitals, which is again consistent with the increase in metal–metal bonding for the second and third row transition elements.

Alternative calculations of self consistent field functions using Slater orbitals, and radii corresponding to maximum orbital electron density (515) or overlap integrals for reasonable internuclear distances (1838) again show that the $4d$ and $5d$ bonding orbitals extend more beyond the nonbonding $4\,s^2p^6$ and $5\,s^2p^6$ cores, than do the $3d$ orbitals extend beyond the $3\,s^2p^6$ core. Unexpectedly, however, a maximum is shown in the d orbital radius (Fig. 27), and a corresponding maximum in the overlap integrals, at niobium, molybdenum, tantalum and tungsten. Whether this is real or is an artifact of the calculations it is difficult to decide, as the nonexistence of metal atom clusters of divalent zirconium and hafnium could equally well be due to there being insufficient electrons available for metal–metal bonding.

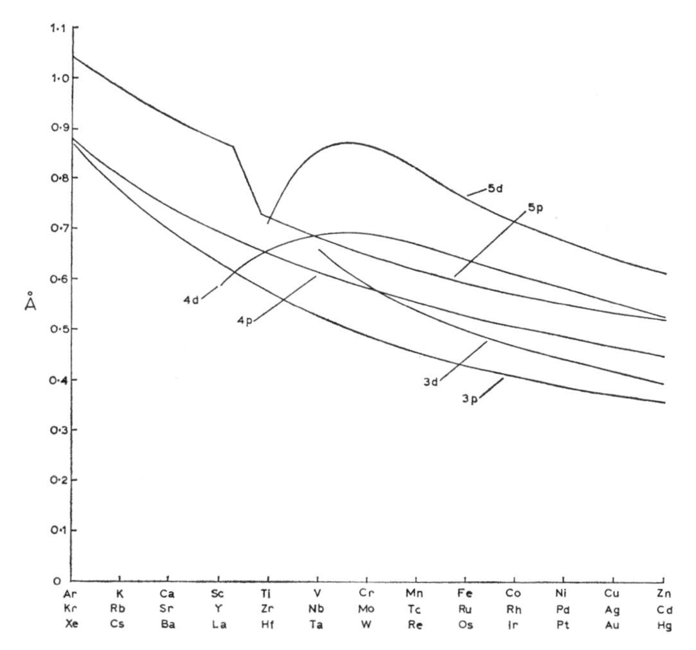

FIG. 27 Radii corresponding to maximum orbital electron density for the np and nd electrons for the three transition series. From Reference 515, Clementi *et al.* (1967).

There have been a number of attempts (1027, 2128) to correlate experimentally the relative size of orbitals as calculated by methods similar to those used above, and the extent of metal–metal bonding, but such schemes have little predictive ability because of the large effect that the ligands present have upon those orbitals.

The general distribution of metal–metal bonds over the transition metal dichlorides, trichlorides, tetrachlorides, and other halides and oxides will now be considered in more detail.

Dichlorides

The importance of atomization energies in the stability of metal–metal bonded compounds can be further demonstrated through Born-Haber cycle calculations of the thermodynamic stability of the hypothetical "ionic dichlorides" of the third row transition elements, which show that such "ionic dichlorides" are unstable with respect to the elements and/or with respect to disproportionation to a higher chloride and the metal (1362, 2325). The observed existence of dichlorides of some of these elements is because they are more stable than the hypothetical "ionic dichlorides", due to extra stability gained in metal–metal bond formation and/or formation of covalent metal–chlorine bonds stronger than purely ionic bonds. The structures of $NbCl_{2\cdot33}$ (Nb_6Cl_{14}) (page 241) and WCl_2 (W_6Cl_{12}) (page 353) show octahedral clusters of six metal atoms surrounded by halogen atoms, and clearly involve strong metal–metal bonding. At the other end of the third row, $PtCl_2$ also exists as the cluster Pt_6Cl_{12} (346), but here the predominant stabilizing influence must be considered to lie in the platinum–chlorine bonds. There can be little platinum–platinum bonding either from the point of view of the filled and contracted *d* subshell in square planar platinum(II), or from a molecular orbital approach as both the bonding and antibonding molecular orbitals formed from overlap of the orbitals on the six metal atoms are filled (1370).

Similarly for the second row elements, niobium and molybdenum dichlorides (Nb_6Cl_{14} and Mo_6Cl_{12}) are stabilized by metal–metal bonding, while there are two forms of palladium dichloride, one of which (2040) is analogous to Pt_6Cl_{12} while the other (2366) is clearly a linear covalent polymer with very long metal–metal distances of 3·34 Å.

The situation for the intermediate Group VII and Group VIII dihalides is not yet clear, and although it might be thought that there could be a continuous series of octahedral clusters from tantalum and tungsten right across to platinum, this is not necessarily true as there may be insufficient stability attainable from either metal–metal bonding or from metal–halogen covalent bonding. Thus the dichlorides of technetium, ruthenium, rhodium, rhenium, osmium and iridium do not appear to be known (417).

Trichlorides

For the trichlorides metal–metal bonded compounds may be divided into two groups; those which have structures essentially based on close packing of chloride ions with some distortion caused by metal–metal bonding between metal atoms in adjacent octahedral holes, and those which have considerably stronger bonding in which the structures are based on clusters of metal atoms (Table 8).

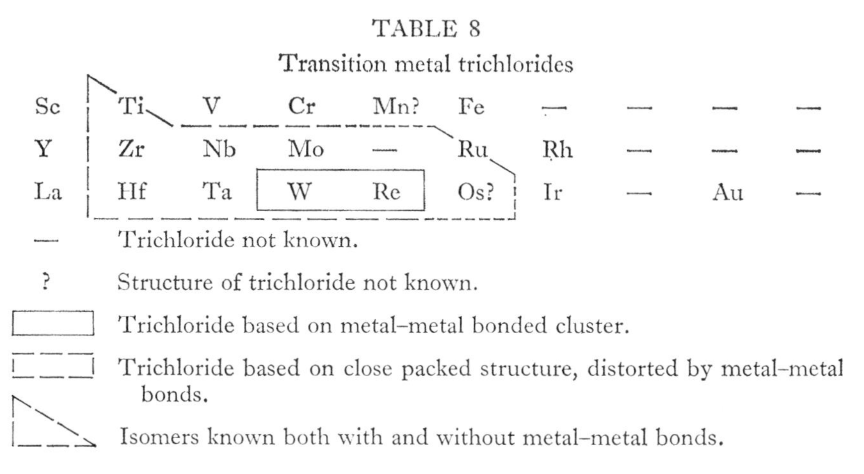

TABLE 8

Transition metal trichlorides

Sc	Ti	V	Cr	Mn?	Fe	—	—	—	—
Y	Zr	Nb	Mo	—	Ru	Rh	—	—	—
La	Hf	Ta	W	Re	Os?	Ir	—	Au	—

— Trichloride not known.

? Structure of trichloride not known.

[solid box] Trichloride based on metal–metal bonded cluster.

[dashed box] Trichloride based on close packed structure, distorted by metal–metal bonds.

[diagonal arrow] Isomers known both with and without metal–metal bonds.

Tungsten trichloride, or $(W_6Cl_{12})Cl_6$ (page 337) has a structure based on an octahedral cluster of six tungsten atoms. The tungsten–tungsten distance is 2·92 Å which is the same as the niobium–niobium distance in $[(Nb_6Cl_{12})Cl_6]^{4-}$, which has 16 electrons available for metal–metal bonding compared with 18 for $(W_6Cl_{12})Cl_6$. There are again chlorine atoms above each of the twelve octahedral edges which bridge two tungsten atoms, and also an additional chlorine atom bonded to each of the six tungsten atoms in a "centrifugal" position.

The structure of rhenium trichloride is based on a triangular cluster of three rhenium atoms (Re–Re = 2·49 Å) with three coplanar bridging chlorine atoms across each edge of the triangle (569). Each rhenium atom is also bonded to three additional chlorine atoms, one above the Re_3Cl_3 plane, one below the Re_3Cl_3 plane, and one in the plane in a "centrifugal" position to form $[(Re_3Cl_3)Cl_9]^{3-}$ (Fig. 28). In $ReCl_3$ itself (Fig. 29) these clusters are linked to form infinite sheets by sharing of the three halogen atoms on one side of the Re_3Cl_3 plane, and the three "centrifugal" halogen atoms; the chlorine atoms on the other side of the plane are not shared, and so C_{3v} symmetry for the group is retained.

The structure of $MoCl_3$ has been briefly referred to above (page 27) and is dealt with in more detail later (page 336), and contains pairs of molybdenum atoms between distorted close packed sheets of chloride ions. The structures of $ScCl_3$, α-$TiCl_3$, VCl_3, $CrCl_3$, $FeCl_3$, α-$RuCl_3$, $RhCl_3$ and $IrCl_3$ are similar, but now the close packed sheets are undistorted by metal–metal bonding. The absence of metal–metal bonding in $RhCl_3$ (147)

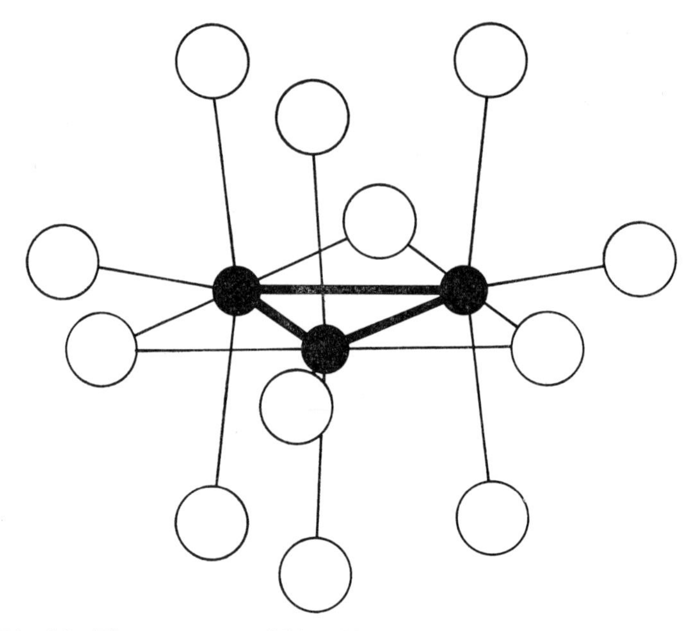

FIG. 28 The structure of $[(Re_3Cl_3)Cl_9]^{3-}$. The filled circles denote rhenium atoms and the open circles chlorine atoms.

and $IrCl_3$ (105, 345) can be attributed to the filled d_ε^6 subshell which is therefore unavailable for metal–metal bonding, as well as to the normal contraction in *d* orbitals on moving across the periodic table. The "trichlorides" of niobium and tantalum are also referred to above (page 28) and in more detail later (page 223). They are in reality the nonstoicheiometric compounds $Nb_{3-x}Cl_8$ (x = 0–0·45) and $Ta_{3-x}Cl_8$ (x = 0–0·3), and

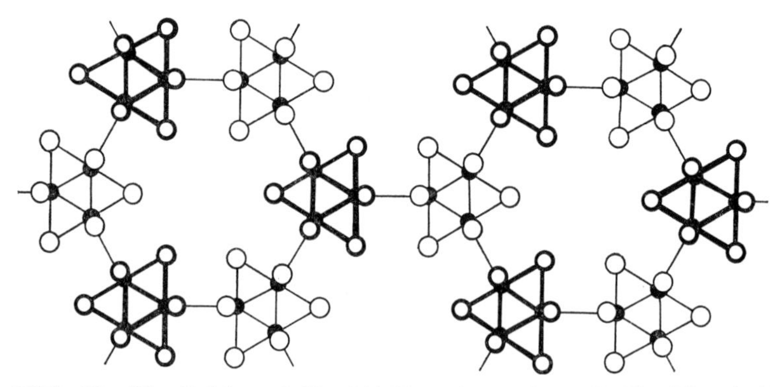

FIG. 29 The linking of $(Re_3Cl_3)Cl_9$ units to form $ReCl_3$. The dark clusters are above the plane of the light clusters.

have triangular clusters of metal atoms between close packed layers of chloride ions. In contrast to these compounds in which only every second layer of close packed holes is occupied, the structures of β-$TiCl_3$, $ZrCl_3$ and $HfCl_3$ (page 126) have metal atoms between every close packed layer so that infinite polymers are formed based on strings of octahedra sharing opposite faces. Metal–metal bonding is confirmed by the diamagnetism of these compounds. In β-$RuCl_3$ a similar structure is found, but now the ruthenium–ruthenium distances are alternately short and long with the formation of dimeric units (343).

The structural change from a non-close packed structure to a structure based on the close packing of chloride ions distorted by metal–metal bonds, as observed on oxidation of molybdenum dichloride, Mo_6Cl_{12}, to $MoCl_3$, is also paralleled by the chloroanions. Molybdenum(II) forms the $Mo_2Cl_8^{2-}$ cluster in which the strong molybdenum–molybdenum bonding results in an unfavourable arrangement of the chloride ions (page 357), but on oxidation to $Mo^{II}Mo^{III}Cl_8^{3-}$ the structure changes to a sterically more favourable one based on octahedral coordination of the metal atom (page 340). Similarly $ReCl_3$ (569) and $Re_2^{III}Cl_8^{2-}$ (555, 564) are based on sterically unfavourable clusters, but the oxidized $ReCl_4$ (205) and $Re_2^{IV}Cl_9^{-}$ (266) are based on the sterically favourable octahedral coordination, the rhenium–rhenium bonding being through pairs of octahedra sharing a common face with short rhenium–rhenium bonds (2·73 and 2·71 Å respectively).

Tetrachlorides

The known transition metal tetrachlorides are shown in Table 9. Those containing metal–metal bonds are again those of niobium, tantalum, molybdenum, tungsten and rhenium.

TABLE 9

Transition metal tetrachlorides

—	Ti	V	Cr*	—	—	—	—	—	—
—	Zr	Nb	Mo	Tc	Ru*	—	—	—	—
—	Hf	Ta	W	Re	Os?	—	Pt?	—	—

— Tetrachloride not known.

* Stable in gas phase only.

? Structure of tetrachloride not known.

(box) Tetrachloride containing metal–metal bonds.

(diagonal) Isomers known both with and without metal–metal bonds.

The tetrachlorides $NbCl_4$, $TaCl_4$, α-$MoCl_4$ and WCl_4 are isomorphous, and consist of close packed layers of chloride ions with infinite linear strings of metal atoms occupying one half the octahedral holes in every alternate layer; the metal–metal distances are alternately long and short with the formation of discrete pairs of metal atoms, with consequent distortion of the lattice (pages 26 and 198).

In β-$ReCl_4$ (205) the structure is again based on close packing of chloride ions but now the rhenium–rhenium bonding is between pairs of face-sharing octahedra. The second form of $ReCl_4$ may be related to the other four tetrachlorides, but the strength of the metal–metal interaction is not known (54). Although edge sharing of octahedra also occurs in the paramagnetic $TcCl_4$ ($\mu_{eff} = 3{\cdot}48$ B.M.), the metal atoms are repelled away from each other with the formation of zig-zag chains with Tc–Tc = 3·62 Å.

No metal atom clusters occur for the tetravalent state, and the extent of electron delocalization progressively decreases as the metal is oxidized. For example complete delocalization occurs in Nb and NbO (57, 276, 318), while the electron is delocalized over 6, 3 and 2 metal atoms in Nb_6Cl_{14}, Nb_3Cl_8 and $NbCl_4$ respectively. The decrease in metal–metal bonding as the oxidation state is increased is due to a combination of several factors, including the contraction of the *d* orbitals used for metal–metal bonding, the decrease in the number of electrons available for metal–metal bonding, and the unfavourable steric effects as the Cl : M ratio increases.

Other halides and oxides

It is expected that as the halide anion becomes smaller and the lattice energy increases, it will become more difficult to distort the lattice through metal–metal bonding. Thus the only metal–metal bonded fluoride known is Nb_6F_{15} whose structure is shown on page 239. It is generally found that the distance a metal–metal bond can reach increases as the polarizability of the anion sublattice increases, due to increasing screening of the *d* electrons from the nucleus allowing the *d* orbitals to expand. However, the metal atoms are simultaneously forced further apart by the larger anions. This can be seen by the metal–metal distances in some typical halides (Table 10). Likewise the iron–iron bond length of 3·00 Å in FeS

TABLE 10

Metal–metal bond distances in some halides

	M–M(Å)		M–M(Å)		M–M(Å)
Nb_3Cl_8	2·81	β-$TiCl_3$	2·96	$NbCl_4$	3·06
Nb_3Br_8	2·88	$TiBr_3$	3·05	$NbBr_4$	—
Nb_3I_8	3·00	TiI_3	3·23	NbI_4	3·31

(page 29) and the vanadium–vanadium bond length of 2·83 Å in $V^{IV}(S_2)_2$ (1423) are much longer than metal–metal bond lengths in oxides.

In comparing oxides with halides, although the polarizability of the anion sublattice is unfavourable for metal–metal bonding, the occurrence of metal–metal bonding actually appears slightly more widespread for the same oxidation state. This is due to the combined effect of the change in stoicheiometry which increases the probability of metal atoms occupying adjacent octahedral holes, and the smaller size of the oxide ions which means that these octahedral holes are closer together. Thus metal–metal bonding occurs in VO_2 and V_2O_3, but not in VCl_4 or VCl_3.

The only metal–metal bonds between metals having formal oxidation states greater than four appear to be with oxygen or sulphur bridging atoms, as in $La_4Re_6O_{19}$ (1572), $Mo_2^VO_4(C_2O_4)_2(H_2O)_2{}^{2-}$ (page 319), and $Mo_2^VO_2S_2(C_5H_5)_2$ (page 321).

C. Metal–Metal Bonded Organometallic Compounds

In contrast to the oxides and halides, metal carbonyls and their derivatives readily form metal–metal bonds for almost all transition metals, in which the metal atoms have much lower formal oxidation states. This field has been reviewed in some detail, and this should be consulted for many of the original literature references (1362).

TABLE 11

Transition metal carbonyls

$V(CO)_6$	$Cr(CO)_6$	$Mn_2(CO)_{10}$	$Fe(CO)_5$ $Fe_2(CO)_9$ $Fe_3(CO)_{12}$	$Co_2(CO)_8$ $Co_4(CO)_{12}$ $Co_6(CO)_{16}$	$Ni(CO)_4$
	$Mo(CO)_6$	$Tc_2(CO)_{10}$	$Ru(CO)_5$ $Ru_3(CO)_{12}$	$Rh_2(CO)_8$ $Rh_4(CO)_{12}$ $Rh_6(CO)_{16}$	
	$W(CO)_6$	$Re_2(CO)_{10}$	$Os(CO)_5$ $Os_3(CO)_{12}$	$Ir_2(CO)_8$ $Ir_4(CO)_{12}$ $Ir_6(CO)_{16}$	$Pt_n(CO)_n$

In transition metal carbonyls (Table 11), metal–metal bonds commonly occur when the metal atom formally has an odd number of electrons. One of these is then used to form a metal–metal single bond with a similar electron on another metal atom. The eighteen-electron inert gas configuration is attained by donation of pairs of electrons from the requisite number of carbonyl groups. Thus in moving left across the first row of the transition

TABLE 12

Transition metal carbonyl anions

$[V(CO)_6]^-$	$[Cr_2(CO)_{10}]^{2-}$	$[Mn(CO)_5]^-$	$[Fe_2(CO)_8]^{2-}$	$[Co(CO)_4]^-$	$[Ni_2(CO)_6]^{2-}$
$[Nb(CO)_6]^-$	$[Mo_2(CO)_{10}]^{2-}$	$[Tc(CO)_5]^-$			
$[Ta(CO)_6]^-$	$[W_2(CO)_{10}]^{2-}$	$[Re(CO)_5]^-$			

TABLE 13

Transition metal cyclopentadienyl carbonyls

$(C_5H_5)V(CO)_4$	$(C_5H_5)_2Cr_2(CO)_6$	$(C_5H_5)Mn(CO)_3$	$(C_5H_5)_2Fe_2(CO)_4$	$(C_5H_5)Co(CO)_2$	$(C_5H_5)_2Ni_2(CO)_2$
	$(C_5H_5)_2Mo_2(CO)_6$		$(C_5H_5)_2Ru_2(CO)_4$	$(C_5H_5)Rh(CO)_2$	
$(C_5H_5)Ta(CO)_4$	$(C_5H_5)_2W_2(CO)_6$	$(C_5H_5)Re(CO)_3$	$(C_5H_5)_2Os_2(CO)_4$	$(C_5H_5)Ir(CO)_2$	$(C_5H_5)_2Pt_2(CO)_2$

metals, an alternate formation of metal–metal bonds is observed, the simplest carbonyls being $Ni(CO)_4$, $Co_2(CO)_8$, $Fe(CO)_5$, $Mn_2(CO)_{10}$ and $Cr(CO)_6$.

One carbonyl group can be replaced by two electrons, for example in $Mn_2(CO)_9{}^{2-}$, with retention of the metal–metal bond. However, protonation of these metal–metal bonded anions to form the corresponding hydride is often accompanied by the replacement of the metal–metal bond with a three centre metal–hydrogen–metal bond.

The exception to this eighteen-electron rule is $V(CO)_6$, which is an air-sensitive paramagnetic monomer, with only seventeen electrons in the valence shell. It may be noted however that a vanadium–vanadium bond does occur in the closely related $[V(CO)_4(diars)]_2$ (where diars is *o*-phenylenebisdimethylarsine).

The effect of changing the ligand is not at all well understood, as in addition to the stabilization of the low metal oxidation states and the reduction of unfavourable nonbonding electron repulsions by delocalization onto the organic ligands, there are also effects upon the size and electronegativity of the orbitals used in metal–metal bonding. It may be noted that in contrast to $V(CO)_6$ and $[V(CO)_4(diars)]_2$, the reverse effect is noted for $[FeI(CO)_4]_2$ and $[FeI(CO)_2(diars)]$, where the iron–iron bond is broken by substitution of two carbonyl groups by the diarsine.

Examples of carbonyls in oxidation states $-$I and $+$I are given in Tables 12 (the carbonyl anions) and 13 (the cyclopentadienyl carbonyls). For any element there is an alternate formation of metal–metal bonded compounds as the formal oxidation state is increased, for example $[Cr_2^{-I}(CO)_{10}]^{2-}$, $[Cr^{0}(CO)_6]$, $[(C_5H_5)_2Cr_2^{I}(CO)_6]$, and $[Fe^{-II}(CO)_4]^{2-}$, $[Fe_2^{-I}(CO)_8]^{2-}$, $[Fe^{0}(CO)_5]$, $[Fe_2^{I}I_2(CO)_8]$ and $[Fe^{II}Br_2(CO)_4]$.

Metal carbonyls containing an even number of electrons can form metal–metal bonds with retention of the 18-electron configuration, if the gain attained by sharing electrons in the formation of each metal–metal bond is compensated by the loss of one donor carbonyl group. It may be that this process is favoured by the simultaneous formation of more favourable higher coordination numbers. Thus in addition to $Fe(CO)_5$, iron forms $Fe_2(CO)_9$ with one iron–iron bond (which is isoelectronic with $Fe_2^{-I}(CO)_8{}^{2-}$ above), and also $Fe_3(CO)_{12}$ with three iron–iron bonds, the three metal atoms being at the corners of a triangle.

In the same way the Group VI carbonyls could be imagined to form, for example, $Mo_2(CO)_{11}$ (one metal–metal bond), $Mo_3(CO)_{15}$ (three metal–metal bonds), $Mo_4(CO)_{18}$ (six metal–metal bonds), and so on, and although these neutral molecules are not known, the isoelectronic dinegative anions $Mo_2(CO)_{10}{}^{2-}$ (which is also isoelectronic with $Mn_2(CO)_{10}$), $Mo_3(CO)_{14}{}^{2-}$ and $Mo_4(CO)_{17}{}^{2-}$ have been prepared.

In the case of metal atoms with an odd number of electrons, larger polymers can be formed provided there is an even number of metal atoms retained to lead to the eighteen-electron configuration. Thus two molecules of $Co_2(CO)_8$, each containing one cobalt–cobalt bond, can be formally envisaged as forming $Co_4(CO)_{12}$ with the cobalt atoms arranged at the corners of a tetrahedron, with six cobalt–cobalt bonds. The additional four metal–metal bonds are compensated by the reduction in the number of carbonyl groups by four. An alternative way of arranging four metal atoms, which involves only five metal–metal bonds is found in $[Re_4(CO)_{16}]^{2-}$, in which one of the tetrahedral edges is broken and the structure opened out.

However, when we consider $Co_6(CO)_{16}$, the metal atoms are arranged in the form of an octahedron, and if it is assumed that each of the twelve edges of the octahedron corresponds to an electron pair bond, the molecule has $12(2) + 16(2) + 6(9) = 110$ electrons available for bonding, compared with the expected $6(18) = 108$ electrons. Other octahedral clusters with two electrons in excess of the rare gas configuration include $Co_6(CO)_{15}^{2-}$, $Co_6(CO)_{14}^{4-}$, $Rh_6(CO)_{16}$, $Rh_6(CO)_{14}^{4-}$, $Ir_6(CO)_{16}$, $Ru_6C(CO)_{17}$, $Ru_6C(CO)_{14}$ (arene) and $Rh_{12}(CO)_{30}^{2-}$ (in which two octahedral Rh_6 clusters are joined by a single rhodium–rhodium bond). Similarly the seven atom cluster $Rh_7(CO)_{16}^{3-}$ can be considered to be based on the hypothetical $Rh_6(CO)_{15}^{2-}$ (compare with $Co_6(CO)_{15}^{2-}$ above), where an additional eighteen electrons are contributed by the twelve-electron $Rh(CO)^-$ moiety to a triangular face of the Rh_6 octahedron, with the formation of three additional rhodium–rhodium bonds.

Finally multiple bonding between metal atoms also appears possible. For example although cyclopentadienyl forms $(CO)_3(C_5H_5)Mo–Mo(C_5H_5)(CO)_3$, an additional carbonyl group is lost from each molybdenum atom with pentamethylcyclopentadienyl, which must therefore be formulated as having a triple molybdenum–molybdenum bond in order to achieve the eighteen-electron configuration, that is $(CO)_2(C_5Me_5)Mo \equiv Mo(C_5Me_5)(CO)_2$ (page 364). An example of a grossly electron deficient compound is the π-allyl complex $Cr_2(C_3H_5)_4$, which contains the shortest metal–metal bond distance known (Cr–Cr = 1·97 Å) (page 360).

4. ISOPOLY AND HETEROPOLYANIONS

In aqueous solution the Group V and Group VI ions in their highest oxidation states form a group of polymers which are termed the isopolyanions. Specific examples include $V_{10}O_{28}^{6-}$, $Nb_6O_{19}^{8-}$, $Ta_6O_{19}^{8-}$, $Cr_2O_7^{2-}$, $Mo_7O_{24}^{6-}$, $Mo_8O_{26}^{4-}$, $HW_6O_{21}^{5-}$, and $H_2W_{12}O_{40}^{6-}$. Mixtures of these ions with different ions form the so-called heteropolyanions, for example vanadium(V) forms $PV_{12}O_{36}^{7-}$ and $H_2NiV_{13}O_{39}^{7-}$, while molybdenum(VI) forms $PMo_{12}O_{40}^{3-}$, $H_4Co_2Mo_{10}O_{38}^{6-}$, $NiMo_9O_{32}^{6-}$, $AlMo_6O_{24}^{9-}$ and $CeMo_{12}O_{48}^{8-}$.

These individual ions will be dealt with in detail in Chapters 3 and 4. In this section attention will be confined to viewing the formation of these polyanions in relation to other processes occurring in aqueous solution, and to point out some structural relationships between these complexes, and to suggest reasons for them.

For the general case of a water molecule bound to an ion of charge Z+, the formation of the M←O bond draws electrons away from the O–H bonds, and weakens them. These coordinated water molecules then behave as stronger acids than the solvent water molecules:

$$M^{Z+}\leftarrow O\begin{smallmatrix}H\\H\end{smallmatrix} \rightleftarrows M^{Z+}\text{—}OH + H^+ \rightleftarrows M^{Z+}{=}O + 2H^+$$

The acidity therefore increases as the strength of the M–O bond increases.

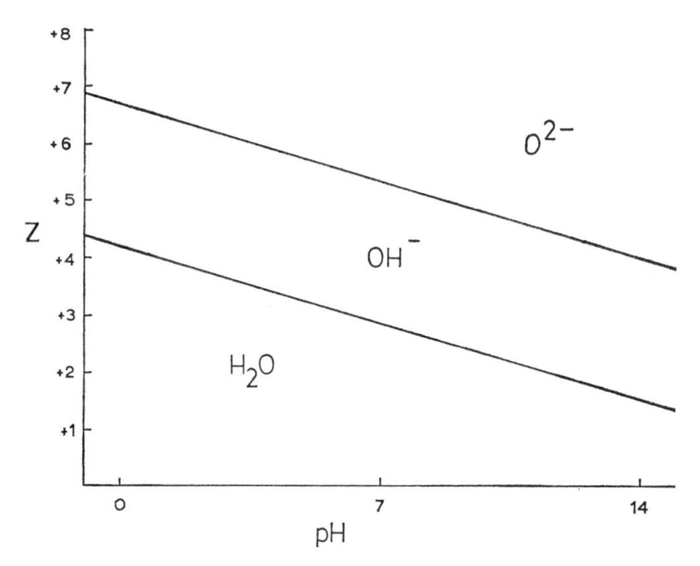

FIG. 30 The pH intervals in which water, hydroxide, and oxide are common ligands to a central atom of average size and of oxidation number Z.

A convenient rule-of-thumb (1318) is shown in Fig. 30. This is a diagram indicating the importance of Z and pH in determining whether the ligand is present as H_2O, OH^- or O^{2-} when bound to a central atom of average size. Such a diagram rationalizes the observations that Cr^{VI} and

Mn^{VII} form CrO_4^{2-} and MnO_4^- respectively whereas Cr^{III} and Mn^{II} form $[Cr(H_2O)_6]^{3+}$ and $[Mn(H_2O)_6]^{2+}$ respectively, and why the acid strength decreases along the series $Cl^{VII}O_3(OH)$, $S^{VI}O_2(OH)_2$, $P^VO(OH)_3$, $Si^{IV}(OH)_4$. However, there are clearly other factors, such as size, which determine the strength of the M–O bond. To take the single example of $Z = 4$, Th^{IV}, Hf^{IV}, Zr^{IV} and Ti^{IV} ($r = 0{\cdot}95$–$0{\cdot}68$ Å) are normally present as aquo-hydroxo complexes, Si^{IV} ($r = 0{\cdot}41$ Å) is present as hydroxo species at low pH and hydroxo-oxo species at high pH, while the very small C^{IV} ($r = 0{\cdot}15$ Å) can also have coordinated hydroxo groups, as in alcohols, or coordinated oxo groups, as in carbonyl compounds.

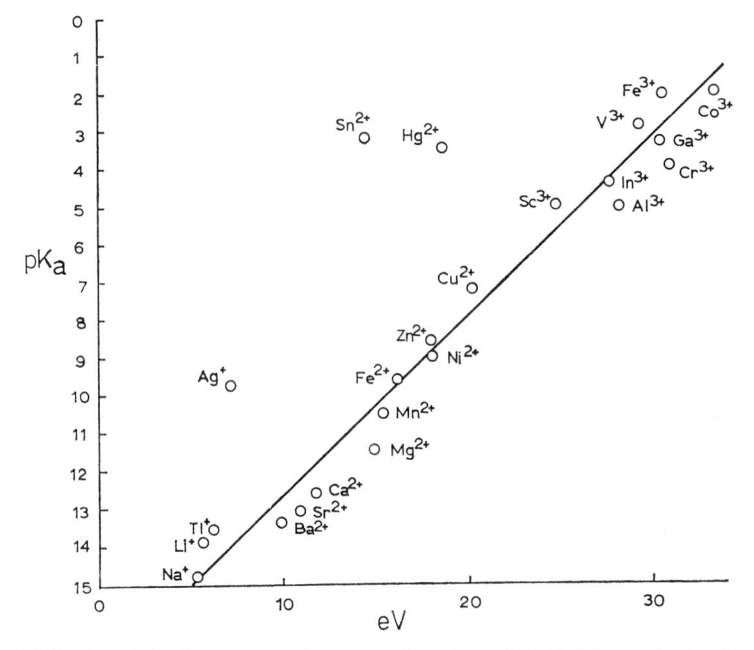

FIG. 31 pK_a of aquo complexes as a function of the "ultimate ionization potentials" of these ions.

A more quantitative correlation is shown in Fig. 31, which is a plot of the acid strength (pK_a) against the power of the gaseous central atom to attract electrons to itself. This is measured by the ultimate ionization potentials of these ions, for example the first ionization potential of sodium, the third ionization potential of aluminium, and so on. Note the range of the abscissa, 0–30 eV, which is the ionization potential corresponding to the conversion of some of the aquo groups to hydroxo groups by the loss of a proton.

It is almost universally observed that the resultant monomeric hydroxo complexes are not stable in solutions at normal concentrations. For example ions such as $[Al(OH)(H_2O)_5]^{2+}$ are not observed as they immediately dimerize through bridging hydroxo groups (1302):

$$[Al(OH)(H_2O)_5]^{2+} \rightarrow [(H_2O)_4Al\begin{matrix} OH \\ \diagup \diagdown \\ \diagdown \diagup \\ OH \end{matrix}Al(H_2O)_4]^{4+} + 2H_2O$$

This dimeric ion has a greater overall positive charge than the monomeric ion, and the dimeric ion will therefore preferentially lose protons forming further hydroxo groups. The polymerization process will therefore continue until reasonably large polymers are formed. In this particular case the ion which is ultimately formed is $[Al_{13}O_4(OH)_{24}(H_2O)_{12}]^{7+}$ (Fig. 32)

FIG. 32 The $[Al_{13}O_4(OH)_{24}(H_2O)_{12}]^{7+}$ ion. Twelve AlO_6 octahedra are joined by common edges. The unshaded tetrahedron in the centre of the structure contains the thirteenth aluminium atom.

(1303). The structure is composed of twelve aluminium atoms each surrounded by an octahedral arrangement of oxygen atoms, and which are grouped around the thirteenth aluminium atom which is only tetrahedrally coordinated. The four oxo groups are each bound to four aluminium atoms, the twenty-four hydroxo groups are each bridging two aluminium atoms, and the twelve water molecules are each bound to a single aluminium atom.

This polymerization process appears to be a general phenomenon, which is illustrated by examples shown in Figure 33 (water molecules

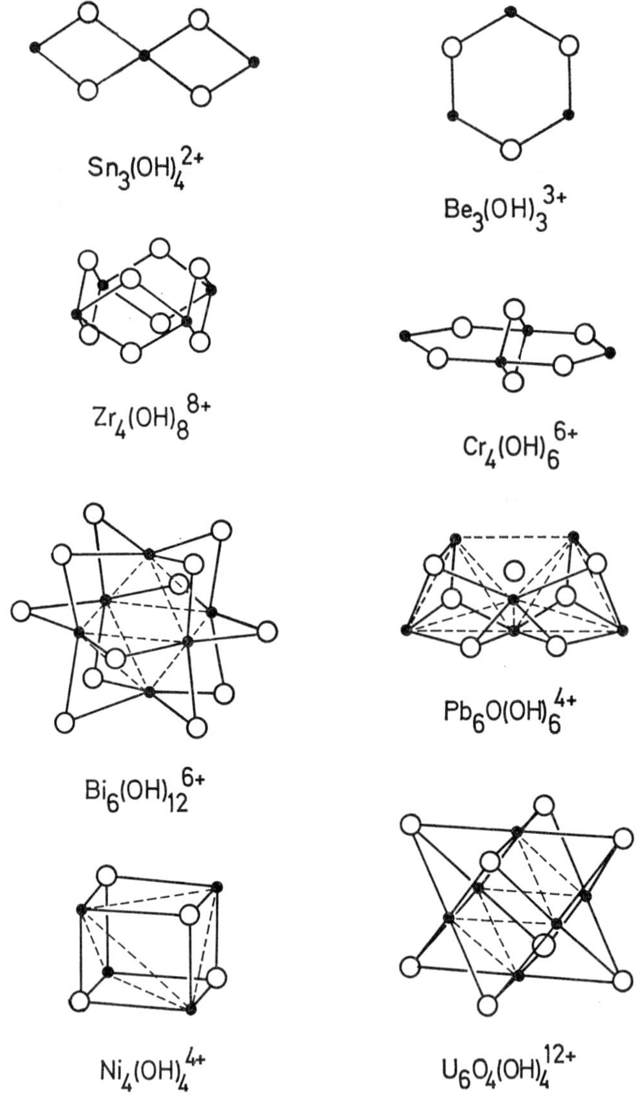

FIG. 33 Some polymeric hydroxo complexes.

are omitted for clarity) (139, 387, 888, 890, 1191, 1522, 1576, 1640). These last examples are relatively unusual in having hydroxo groups bonded to three lead, nickel or uranium atoms. Other complexes containing this feature include $Ni_6(CF_3.CO.CH.CO.CH_3)_{10}(OH)_2(H_2O)_2$ (577), $[Cu_3(C_5H_4N.CH{=}NO)_3(OH)]^{2+}$ (182), $[Mo_4(OH)_4(CO)_8(NO)_4]_4,Ph_3PO$ (24), and $Me_{12}Pt_4(OH)_4$ (2199).

In the literature such complex ions are often given confusingly simple names, for example $Na[Al_{13}O_4(OH)_{24}(H_2O)_{12}](SO_4)_4,xH_2O$ is described as basic aluminium sulphate, $[Zr_4(OH)_8(H_2O)_{16}]Cl_8,12H_2O$ is described as hydrated zirconyl chloride, $ZrOCl_2,8H_2O$, and $[Bi_6(OH)_{12}]^{6+}$ as the bismuthyl ion BiO^+.

If the charge on the metal atom is now increased, for example if the Bi^{III} is replaced by Th^{IV}, the resulting $Th_6(OH)_{12}{}^{12+}$ is of such high charge that it adds further hydroxo groups forming $[(Th_6(OH)_{12})(OH)_2]^{10+}$ and $[(Th_6(OH)_{12})(OH)_3]^{9+}$ (1192, 1304).

These polymeric species are rather difficult to study, and the work can be simplified if some of the water molecules attached to the metal atom are replaced by other groups. As long ago as 1907 Werner (2374) understood this problem, and prepared and studied such complexes as

$$[(NH_3)_4Co\begin{smallmatrix}\diagup OH \diagdown \\ \diagdown OH \diagup\end{smallmatrix}Co(NH_3)_4]^{4+},\quad [(NH_3)_3Co\begin{smallmatrix}\diagup OH \diagdown \\ -OH- \\ \diagdown OH \diagup\end{smallmatrix}Co(NH_3)_3]^{3+},\ \text{and}$$

$$[Co\left\{\begin{smallmatrix}\diagup OH \diagdown \\ \diagdown OH \diagup\end{smallmatrix}Co(NH_3)_4\right\}_3]^{6+}.$$

An additional complication must now be considered. Many of these hydroxo bridged complexes lose water to form oxo bridged complexes:

$$M\begin{smallmatrix}\diagup OH \diagdown \\ \diagdown OH \diagup\end{smallmatrix}M \xrightarrow{-H_2O} M\overset{O}{\diagup\ \diagdown}M \quad \text{or} \quad M-O-M$$

Such processes in solution are often difficult to observe experimentally.

It is not obvious why this reaction occurs in some cases but not in others. For example Cr^{III} forms the hydroxo bridged $Cr\begin{smallmatrix}\diagup OH \diagdown \\ \diagdown OH \diagup\end{smallmatrix}Cr$ (1436), or $Cr_4(OH)_6{}^{6+}$ (139, 889), whereas Fe^{III} forms the oxo bridged Fe–O–Fe (or $Fe\overset{O}{\diagup\ \diagdown}Fe$ depending upon the other groups present) (884, 989, 1563).

Two reasons can be suggested which might explain this difference in behaviour, although they are undoubtedly not the only factors. The first is that hydroxo bridging allows a direct bonding interaction between the metal atoms, $\mathrm{M}\genfrac{}{}{0pt}{}{\diagup\mathrm{OH}\diagdown}{\diagdown\mathrm{OH}\diagup}\!\!\!\!\!\!\!\!\!\!\!\text{———}\mathrm{M}$. It appears that bonding of this type may be significant in $Tl_4(OR)_4$, $Pb_6O(OH)_6^{4+}$ and $Bi_6(OH)_{12}^{6+}$ (1650, 2198, 2200), but such bonding is probably not important in $\mathrm{Cr}\genfrac{}{}{0pt}{}{\diagup\mathrm{OH}\diagdown}{\diagdown\mathrm{OH}\diagup}\mathrm{Cr}$. The second reason is that the oxo bridged structure allows, in addition to the normal M–O–M σ bonding, very favourable π bonding between the filled p_π orbitals on the oxygen atom and the empty d_ε orbitals of an octahedrally coordinated metal atom. This bonding is particularly favourable if the M–O–M system is close to linear, and also if the ligands on one metal atom are in an eclipsed configuration relative to those on the other metal atom (238, 709). Again there is very strong evidence from visible spectra, infra red spectra and magnetic measurements for this type of bonding in Fe–O–Fe complexes (1378, 1523, 1563). Similarly Table 27 (page 289) shows that the structure of $Cr_2O_7^{2-}$ is significantly different in different salts, and as the Cr–O–Cr unit straightens, the chromium-bridging oxygen bond becomes shorter due to π bonding of this type.

Oxygen-to-metal p_π–d_π bonding is of considerable importance in many transition metal oxo anions. In complexes such as $MoOCl_5^{2-}$, both the electron pairs in the filled p_π orbitals on the oxygen atom are donated into empty molybdenum d_ε orbitals, so that the molybdenum–oxygen bond approaches a triple bond. In dioxo and trioxo complexes of transition metals containing no *d* electrons, the *cis*-configuration is found as this is the one which makes maximum utilization of these empty *d* orbitals, for example *cis*-$[MoO_2Cl_4]^{2-}$ and *cis*-$[MoO_3F_3]^{3-}$. However, for metals containing *d* electrons, or for metals having vacant *f* orbitals available for bonding, the *cis*-structure is not preferred, for example in *trans*-$[RuO_2Cl_4]^{2-}$ and *trans*-$[UO_2F_5]^{3-}$ respectively (1069).

In an analogous manner to the polymerization by the sharing of hydroxo groups between two or more metal atoms referred to above, polymerized oxo compounds can be formed by the sharing of oxo groups between more than two metal atoms. Examples of three metal atoms surrounding a central oxygen atom in a triangular arrangement include $OCr_3(CH_3COO)_6^+$, $OFe_3(CH_3COO)_6^+$ (83, 849) and $[O(HgCl)_3]^+$ (2001, 2359). Examples of four metal atoms attached to a single oxygen atom in a tetrahedral arrangement include $OM_4(RCOO)_6$ (where M = Be, Zn or Co) (237, 1464, 2309), $OBe_4(NO_3)_6$ (19), $OCu_4Cl_6L_4$ (where L = Cl^-, Ph_3PO,

C_5H_5N, etc.) (224, 253, 1388), and $OMg_4Br_6(Et_2O)_4$ (2243). Some of the polyanions which will be described later have five and even six metal atoms attached to a single oxygen atom.

Summarizing, for metal ions with charges of +4 or less, or ultimate ionization potentials less than 50 eV, some of the bound water molecules can lose protons to form hydroxo bridged or oxo bridged polymers. At the other end of the scale we can consider metal ions with charges of +7 and +8 and final ionization potentials in the approximate region 80–120 eV, and these polarize the bound water molecules sufficiently to form the tetraoxo species $Mn^{VII}O_4^-$, $Tc^{VII}O_4^-$, $Re^{VII}O_4^-$, $Ru^{VIII}O_4$ and $Os^{VIII}O_4$ respectively; these are the anions of very strong acids and do not add protons and polymerize under normal conditions. Ions in the intermediate range, that is with charges of +5 or +6 and with final ionization potentials in the approximate region 50–80 eV, form tetraoxo anions under alkaline conditions, but under acid conditions can form mixed oxo/hydroxo species which immediately polymerize to form the Group V and Group VI isopolyanions, the actual species formed being dependent upon experimental conditions.

$$Cr^{VI}O_4^{2-} \xrightarrow{H^+} CrO_3(OH)^-, Cr_2O_7^{2-} \text{ or } Cr_3O_{10}^{2-}$$

$$Mo^{VI}O_4^{2-} \xrightarrow{H^+} Mo_7O_{24}^{6-} \text{ or } Mo_8O_{26}^{4-}$$

$$W^{VI}O_4^{2-} \xrightarrow{H^+} H_2W_{12}O_{42}^{10-} \text{ or } H_2W_{12}O_{40}^{6-}$$

$$V^{V}O_4^{3-} \xrightarrow{H^+} V_{10}O_{28}^{6-}$$

$$(Nb^{V}O_4^{3-} \text{ and } Ta^{V}O_4^{3-}) \xrightarrow{H^+} Nb_6O_{19}^{8-} \text{ and } Ta_6O_{19}^{8-}$$

With the exception of the polychromates, all these large anions are composed of octahedrally coordinated MO_6 units which share edges with each other and are discussed in more detail in Chapters 3 and 4 later.

Acidification of mixtures containing these ions results in similar reactions. To confine attention here to the single example of MoO_4^{2-}, as long ago as 1826 Berzelius recognized that acidification of mixtures of MoO_4^{2-} and PO_4^{3-} formed a new species which is now known to be $PMo_{12}O_{40}^{3-}$. The molybdenum-oxygen framework of this analytically important ion is the same as the tungsten-oxygen framework in $H_2W_{12}O_{40}^{6-}$, but now the tetrahedral holes in the centre of the skeleton is occupied by the phosphorus atom. The anion $P_2Mo_{18}O_{62}^{6-}$ is structurally related. For larger ions, the molybdenum-oxygen framework surrounds an octahedral hole containing the metal atom, as in $M^{IV}Mo_9O_{32}^{6-}$ (M^{IV} =

Mn, Ni), and $Te^{VI}Mo_6O_{24}{}^{6-}$ and $M^{III}Mo_6O_{24}{}^{9-}$ (M^{III} = Al, Ga, Cr, Fe, Co, Rh). The very large Zr^{IV}, Ce^{IV} and Th^{IV} form $M^{IV}Mo_{12}O_{42}{}^{8-}$ in which the metal atom in the centre of the molybdenum-oxygen skeleton is surrounded by twelve oxygen atoms in the form of an icosahedron. In some cases the "hetero atom" is not situated in the centre of the large anion as in the above examples, but on the periphery, as in $[(en)(H_2O)Co^{III}(Nb_6O_{19})]^{5-}$. Even further complications appear with ternary mixtures, for example the formation of $PMMo_{11}O_{40}{}^{x-}$ appears to be a general reaction on acidification of mixtures of phosphate, molybdate and a large number of metal ions M.

It is particularly instructive to compare the structures of $[Al_{13}O_4(OH)_{24}(H_2O)_{12}]^{7+}$ and $[PMo_{12}O_{40}]^{3-}$, both of which contain a tetrahedrally coordinated central atom surrounded by twelve octahedrally coordinated metal atoms. The $[PMo_{12}O_{40}]^{3-}$ structure can be converted to the $[Al_{13}O_4(OH)_{24}(H_2O)_{12}]^{7+}$ structure by rotating each of the four M_3O_{13} groups through 60° around its three-fold axis (Fig. 34, the M_3O_{13}

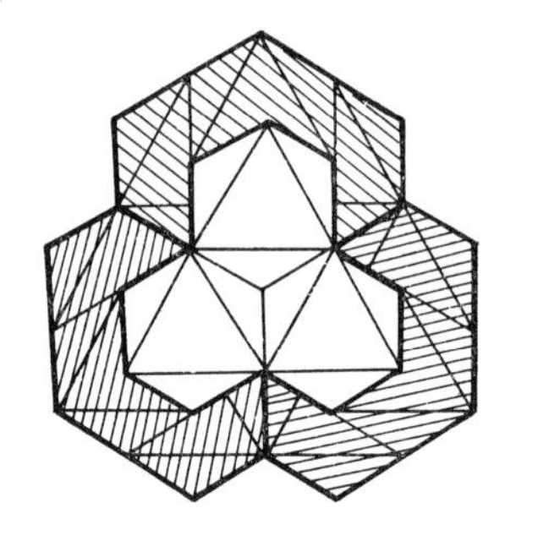
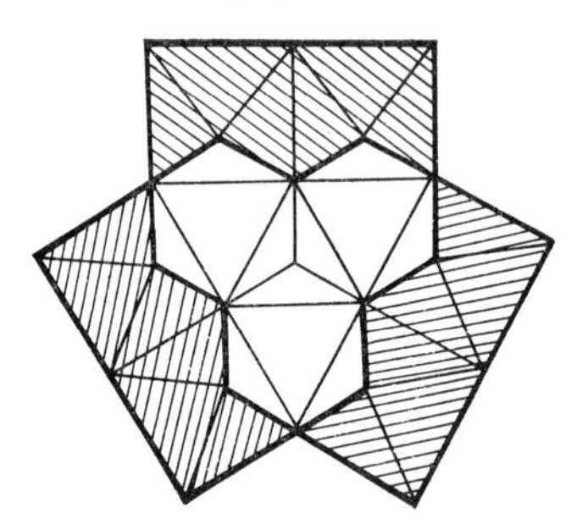

FIG. 34 The conversion of the $[Al_{13}O_4(OH)_{24}(H_2O)_{12}]^{7+}$ structure (left) to the $[PMo_{12}O_{40}]^{3-}$ structure (right) by rotation of each of the four M_3O_{13} groups through 60° around its three-fold axis. In both structures the M_3O_{13} group in the plane of the paper is left unshaded.

group in the plane of the paper is left unshaded). In this conversion the metal : oxygen stoicheiometry and the size of the central tetrahedral hole remain unchanged. However, the number of edges shared by the octahedra decrease thereby lowering the unfavourable Coulombic repulsion between the metal atoms, and this change is presumably more favourable for the larger and more highly charged Mo^{VI} than for Al^{III}. This structural change simultaneously increases some of the M–O–M angles from 90° to 141° (both angles for idealized octahedra) allowing enhanced metal–oxygen–metal π bonding, which is again more important for molybdenum, a transition metal, compared with aluminium, a non-transition metal.

The main structural features of these polyanions are summarized:

(i) These polymeric anions are basically constructed by the edge-sharing of MO_6 octahedra.
(ii) These formidable-looking ions are not chemical oddities (as might be sometimes assumed from the literature), but are merely the culmination of the inevitable processes which occur in aqueous solution.
(iii) The shapes of these polymeric ions are approximately spherical.
(iv) Under any one set of conditions, and in so far as most experimental techniques available are not capable of detecting minor species in solution, only a single polymerized species is formed. This is in sharp contrast to, for example, organic polymerization reactions where a broad spectrum of different sized units is formed. It is clear that this polymerization process encounters some energy barrier and is abruptly stopped.
(v) The size of the anion found is different for different metal atoms, indicating that this barrier depends upon the properties of the metal atom.

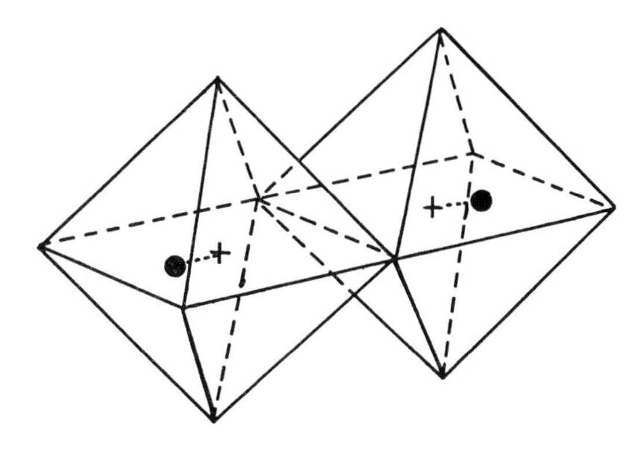

FIG. 35 Mutual repulsion of two metal atoms resulting from two MO_6 octahedra sharing a common edge.

An attempt to rationalize the size and structure of these ions has been made by considering only the electrostatic interactions between the metal atoms (1354). The formation of polymers by the edge sharing of octahedra introduces unfavourable Coulombic repulsions between the metal atoms which will depend upon the charge on the metal atoms. This repulsion will be decreased if the metal atoms move away from the centres of their octahedral cages of metal atoms (Fig. 35). This repulsion will be least for the ions of lower change, and the distortion will be easier for the smaller metal ions, that is in the order V^{5+}, Nb^{5+}, Ta^{5+}, and Mo^{6+}, W^{6+}. As the size of the polymer increases it will become increasingly difficult to overcome such Coulombic repulsions by such distortions, and polymerization *by edge sharing* will cease, as borne out by the decreasing sizes in the order $V_{10}O_{28}{}^{6-}$, $Nb_6O_{19}{}^{8-}$, $Ta_6O_{19}{}^{8-}$ and $Mo_8O_{26}{}^{4-}$, $Mo_7O_{24}{}^{6-}$, W_3O_{13} (the *edge shared* unit in $H_2W_{12}O_{42}{}^{10-}$ and $H_2W_{12}O_{40}{}^{6-}$). (Similarly the structure of the partially hydrolysed titanium ethoxide $Ti_7O_4(OEt)_{20}$ is the same as that of $Mo_7O_{24}{}^{6-}$ with seven octahedra sharing edges, but the larger and more highly charged niobium(V) forms $Nb_8O_{10}(OEt)_{20}$ which contains only groups of three octahedra sharing edges.)

The shapes of the polyanion can also be predicted. If a third octahedron is added to a pair of edge shared octahedra, the M–M–M angle can be 60°, 90°, 120° or 180° depending upon the particular edges shared by the central octahedron (Fig. 36). The 180° interaction is clearly the most

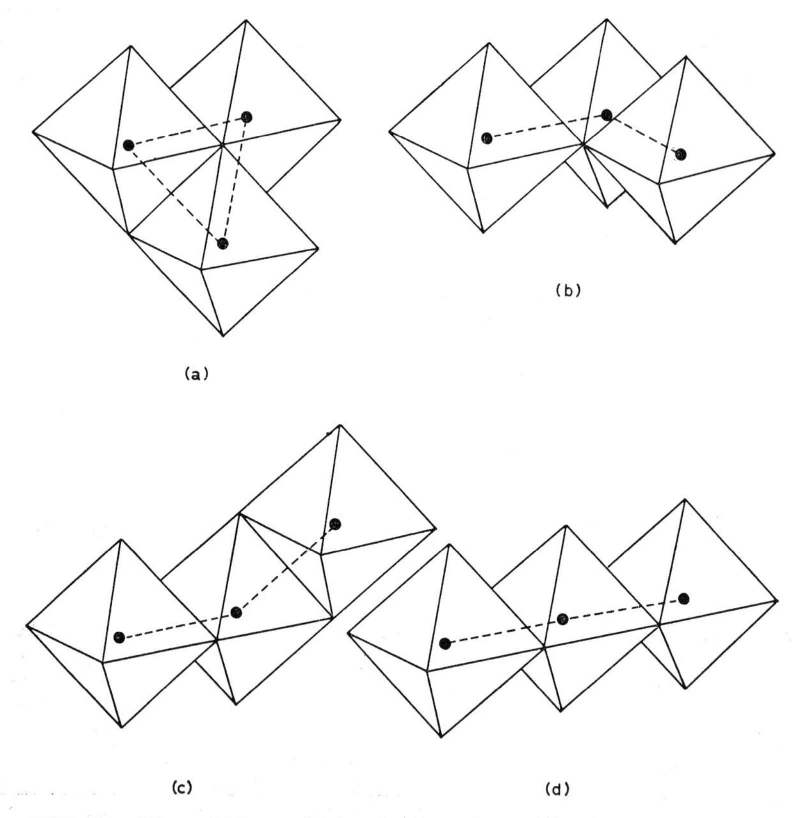

FIG. 36 Three MO_6 octahedra sharing edges with M–M–M angles of (a) 60°, (b) 90°, (c) 120°, and (d) 180°.

unfavourable as the central metal atom is subjected to opposing Coulombic forces which cannot be readily relieved by distortion. The 60° interaction is clearly the most favourable, and this is the edge shared unit found in, for example, $H_2W_{12}O_{40}^{6-}$ and $PMo_{12}O_{40}^{3-}$. There are about twenty different ways in which four octahedra can share edges, but again only one stands out as the most able to accommodate metal–metal repulsion, and this is the structure of $W_4O_{16}^{8-}$ in which the metal atoms are placed approximately at the corners of a tetrahedron (Fig. 37). Similarly the structure which has

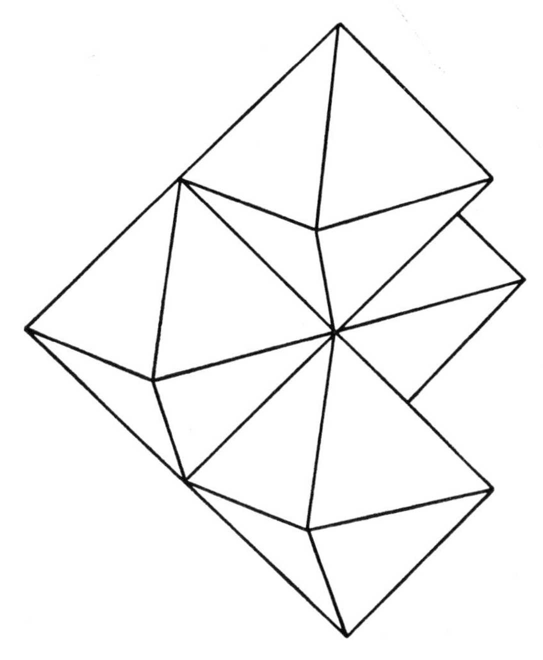

FIG. 37 Four MO_6 octahedra sharing edges with the metal atoms situated at the corners of a distorted tetrahedron.

octahedral arrangement of metal atoms is uniquely stable among the very large number of structures which can be envisaged for six octahedra sharing edges, and this is the structure found for $Nb_6O_{19}^{8-}$ and $Ta_6O_{19}^{8-}$.

Fig. 38 shows that the structures of $Mo_8O_{26}^{4-}$, $Mo_7O_{24}^{6-}$, $Nb_6O_{19}^{8-}$ and $Ta_6O_{19}^{8-}$ are derived from the $V_{10}O_{28}^{6-}$ structure by expulsion of octahedra to remove the excess strain resulting from the more highly charged and/or larger Mo^{VI}, Nb^{V} and Ta^{V} compared with V^{V}. This strain also results in the central V–V–V angles in $V_{10}O_{28}^{6-}$ being reduced from 180° to 175° due to the most Coulombically strained central metal atoms moving away from

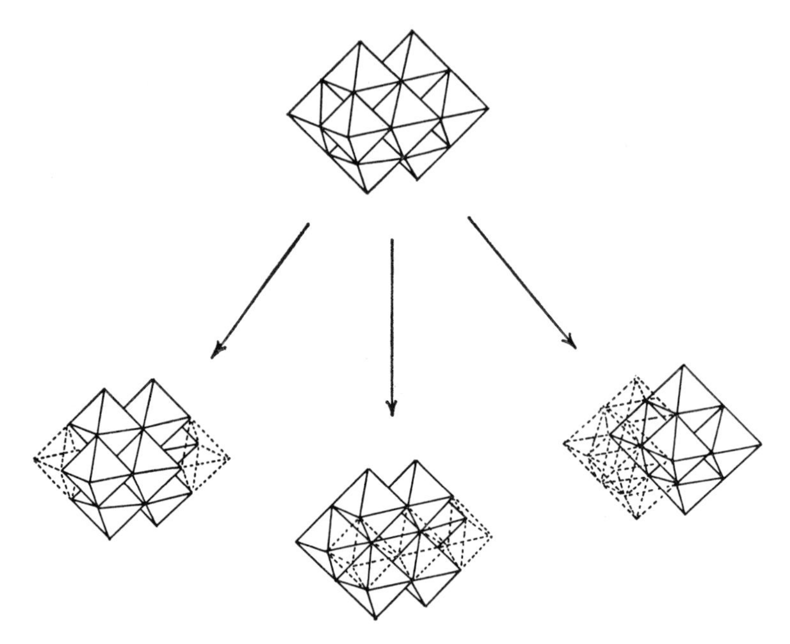

FIG. 38 The structural relation between $V_{10}O_{28}{}^{6-}$ (top) and $Mo_8O_{26}{}^{4-}$ (bottom left), $Mo_7O_{24}{}^{6-}$ (bottom centre), and $Nb_6O_{19}{}^{8-}$ and $Ta_6O_{19}{}^{8-}$ (bottom right).

the centre of the molecule. Similarly the central Mo–Mo–Mo angle in $Mo_7O_{24}{}^{6-}$ is reduced from 180° to 160–170°.

In the same way Fig. 39 shows that $[M^{III}Mo_6O_{24}]^{9-}$ and $[M^{x+}Mo_9O_{32}]^{(10-x)-}$ (in "exploded" form) are derived from the hypothetical $[M^{x+}Mo_{12}O_{38}]^{(4-x)-}$ (in both real and "exploded" forms) by removal of six octahedra from the corners of an approximate octahedron and three octahedra from the corners of a triangle respectively. Not only would the highly symmetrical $[M^{x+}Mo_{12}O_{38}]^{(4-x)-}$ (formed by the central octahedron sharing all twelve edges with neighbouring octahedra) be highly electrostatically strained, it would also have a uniquely low negative charge for a polyanion, or even a highly unfavourable positive charge.

Larger heteropolymolybdates which are built up from edge sharing MoO_6 octahedra but avoiding the presence of linear units containing three MoO_6 octahedra, can only be achieved if there are two heteroatoms incorporated in the centre of the structure, as in $H_4Co_2^{III}Mo_{10}O_{38}{}^{6-}$ (Fig. 40) (798).

As noted above, vanadium can form larger polyanions than molybdenum.

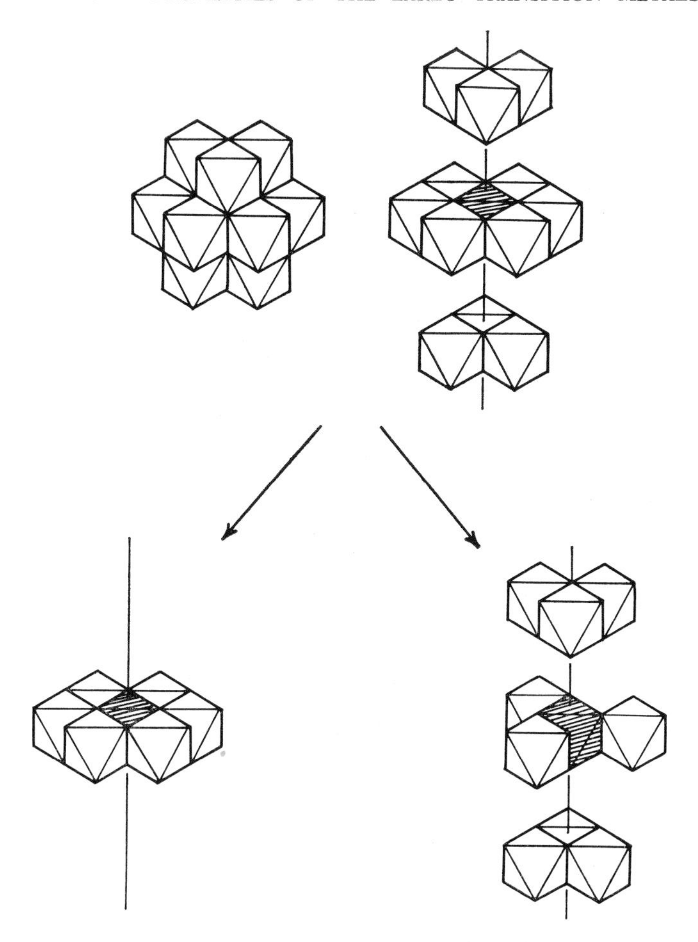

FIG. 39 The structural relation between $[M^{x+}Mo_{12}O_{38}]^{(4-x)-}$ (top, in both real and "exploded" forms), and $[M^{III}Mo_6O_{24}]^{9-}$ (bottom left) and $[M^{x+}Mo_9O_{32}]^{(10-x)-}$ (bottom right, in "exploded" form).

Manganese(IV) and nickel(IV) form the $H_2M^{IV}V_{13}O_{39}{}^{7-}$ ion, in which the suggested structure is that shown in Fig. 39 (top), with an additional VO_6 octahedron inserted into one of the six sites along the four-fold axes (Fig. 41) (892).

Finally the face sharing of MoO_6 octahedra found in $CeMo_{12}O_{42}{}^{8-}$ would be even less favourable for WO_6 octahedra, and this is presumably the reason that tungsten does not form complexes of this type.

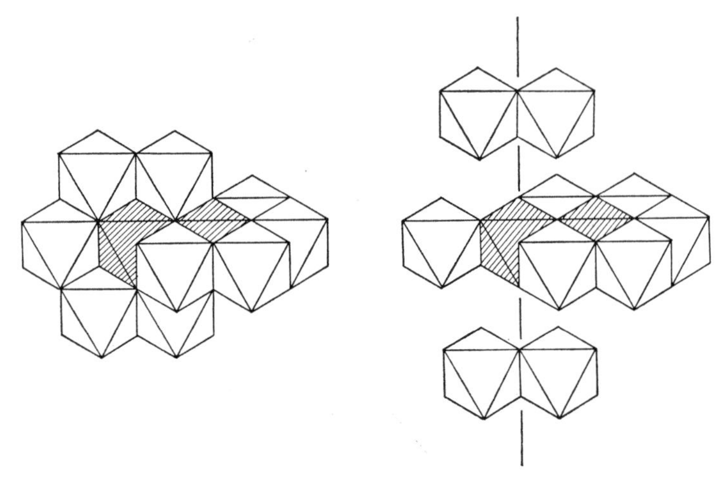

FIG. 40 The structure of $H_4Co_2Mo_{10}O_{38}^{6-}$ showing linking of CoO_6 octahedra (shaded) and MoO_6 octahedra (unshaded). In real form at left and "exploded" form at right.

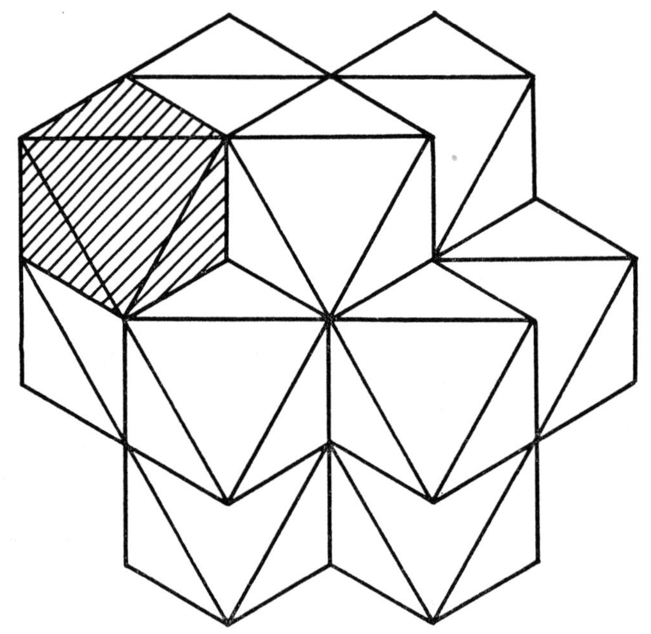

FIG. 41 The structure of $H_2MnV_{13}O_{39}^{7-}$ and $H_2NiV_{13}O_{39}^{7-}$ showing linking of MO_6 octahedra. The manganese and nickel heteroatoms are at the centre of structure and cannot be seen. Removal of the shaded VO_6 octahedron would leave the $[M^{x+}Mo_{12}O_{38}]^{(4-x)-}$ structure shown in Fig. 39.

CHAPTER 2

Group IV — Titanium, Zirconium, Hafnium

1. OXIDATION STATE (IV)

A. Oxides

Titanium dioxide can exist in four different crystallographic modifications, all of which contain an octahedrally coordinated titanium atom. The larger zirconium and hafnium atoms form dioxides in which the metal atom is seven or eight coordinate. These are described first, followed by the somewhat reduced titanium oxides whose structures are related to that of TiO_2. Finally some titanates are briefly described.

1. *Dioxides*

Zirconium dioxide (or baddeleyite) and hafnium oxide have distorted fluorite structures, but instead of there being eight equidistant oxygen atoms around each metal atom in cubic coordination, seven are at 2·04–2·26 Å distance, the eighth being much further away at 3·77 Å (13, 1596). A high temperature form of zirconium dioxide has an eight coordinate metal atom with four zirconium–oxygen bond lengths of 2·46 Å and four of 2·07 Å, in a stereochemistry which could be described as a grossly distorted dodecahedron (2272).

The most common form of titanium dioxide is rutile, the structure of which is based on a distorted hexagonal close packing of oxygen atoms. Each cation layer is half filled, the metal atoms within each layer being arranged in parallel straight rows, separated by similar rows of octahedral vacancies (Fig. 42a). The structure therefore consists of strings of octahedra sharing opposite edges running parallel to the *c*-axis, which are linked to adjacent strings by corner sharing of the octahedra to form a three-dimensional network (Fig. 42b and c). Two of the titanium–oxygen bonds are slightly longer (1·98 Å) than the other four (1·95 Å) (606).

The other three forms of TiO_2 are again based on close packing of oxide ions with half the octahedral holes between every layer of oxide ions occupied, but now the metal atoms are arranged in zig-zag rows (Fig. 43) rather than the linear rows as in rutile (Fig. 42a). The octahedra in these zig-zag rows then share corners and/or edges with octahedra in the adjacent layers, depending upon how these layers are stacked together.

The structure of the anatase form of TiO_2 is based on cubic close packing of oxide ions. Fig. 44 shows the relationship to the sodium chloride structure; the structure is considerably elongated however, the c/a axial ratio of the tetragonal unit cell being 2·53 rather than 2·00 (606). Two of the titanium–oxygen bonds are again longer than the other four (1·96 and 1·94 Å respectively).

Rutile or anatase can be obtained by ignition of the hydroxide from aqueous solutions using the appropriate crystalline form as a seed. The anatase form is also favoured if the hydrolyzed solutions contain sulphate or phosphate.

The third form, brookite, is formed under hydrothermal conditions in the presence of sodium hydroxide (1341). The structure is again based on cubic close packing of oxide ions with zig-zag strings of metal atoms in each layer (Fig. 45) (165). The octahedral coordination is considerably more distorted than in rutile, the titanium–oxygen distances varying from 1·87 to 2·04 Å.

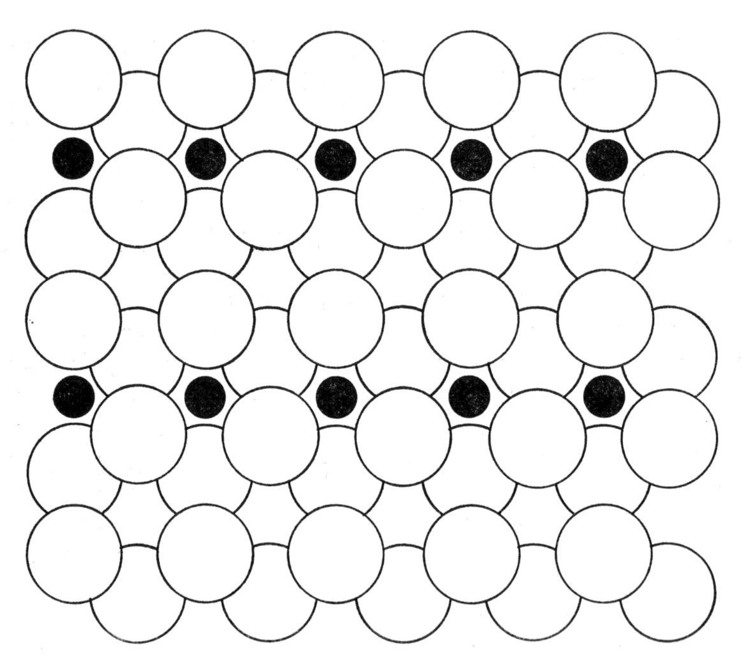

FIG. 42(a) The structure of rutile. (a) Rows of titanium atoms between close packed layers of oxide ions. (b) Linking of TiO_6 octahedra. (c) View down the c-axis.

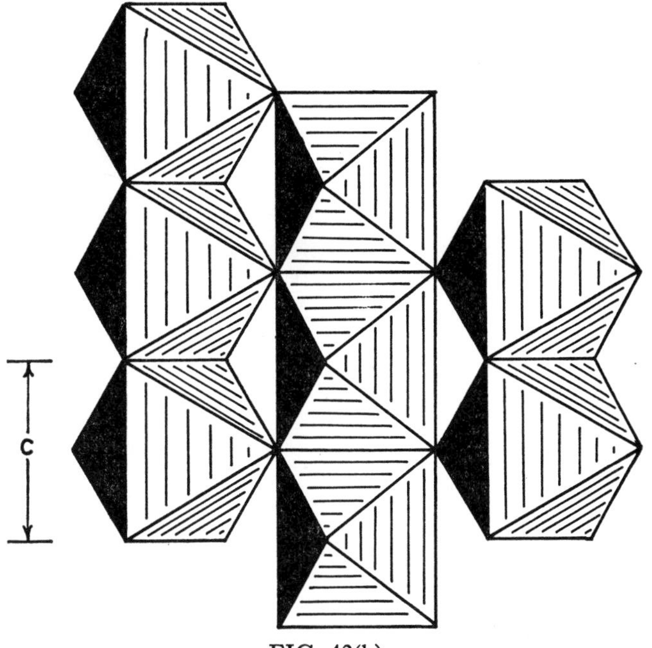

FIG. 42(b)

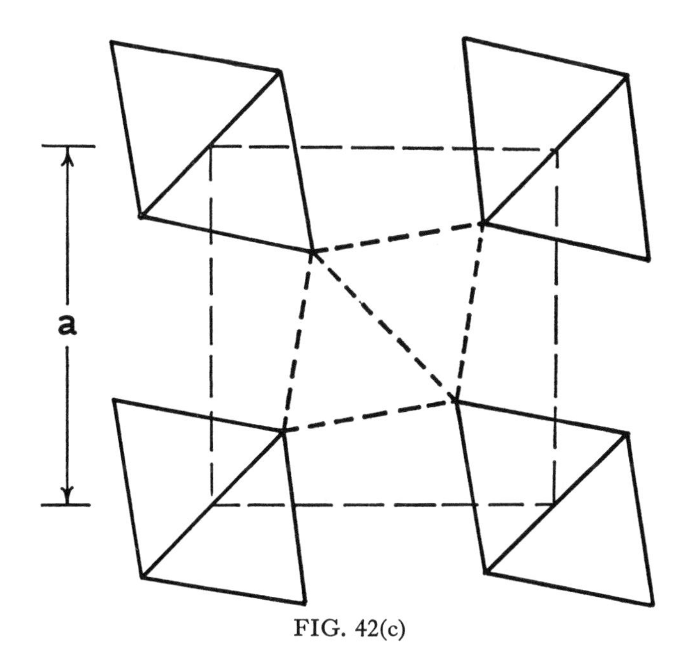

FIG. 42(c)

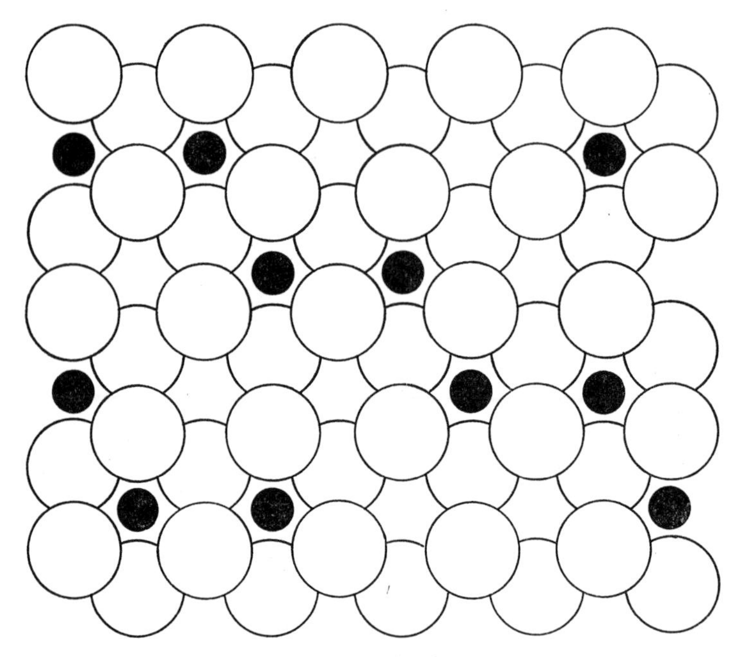

FIG. 43 Zig-zag rows of titanium atoms between close packed layers of oxide ions as found in anatase, brookite, and the α-PbO_2 form of titanium dioxide.

Rutile is denser, harder, and has a higher refractive index than either anatase or brookite:

	Rutile	Anatase	Brookite
Density	4·27	3·90	4·13
Hardness (Mohr's scale)	7·0–7·5	5·5–6·0	5·5–6·0

At high pressures (40 kbar and 450° C) titanium dioxide adopts the α-PbO_2 structure, which is based on hexagonal close packing of oxide ions, with again zig-zag strings of metal atoms. Two titanium–oxygen bonds are again longer than the other four (2·05 and 1·91 Å respectively) (2156).

This α-PbO_2 structure is also observed in the mixed oxides $ZrTiO_4$, $ZrSnO_4$ and $HfTiO_4$. Replacement of the M^{4+} ions with M^{3+} plus M^{5+} or M^{2+} plus M^{5+} forms similar oxides such as $GaTaO_4$ and $FeNb_2O_6$ (the columbite structure) respectively (244).

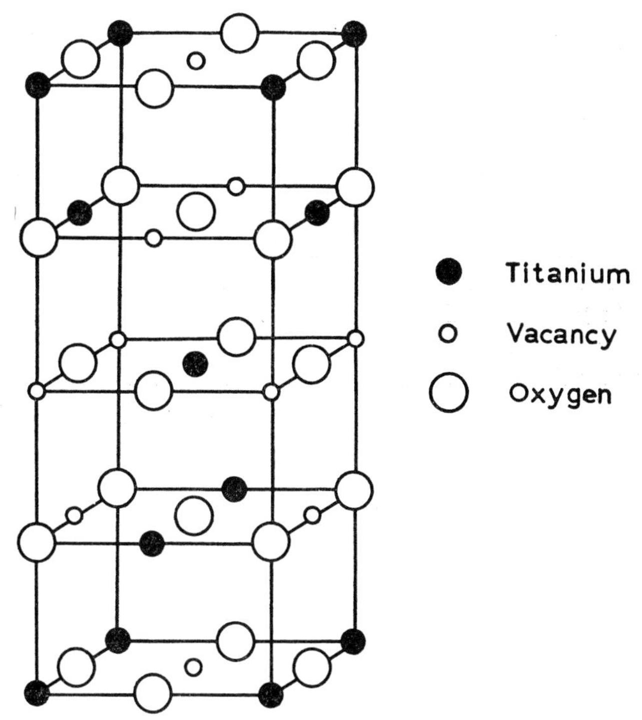

FIG. 44 The structure of anatase showing the relationship to the sodium chloride structure.

2. *Reduced titanium oxides*

The reduction of rutile leads to a large number of compounds which can be represented by the general formula Ti_nO_{2n-1}. These compounds are all based on the rutile structure, but the perfect rutile structure is disrupted by periodic shear planes running through the crystal formed by the removal of planes of oxygen atoms, so that the TiO_6 octahedra share faces with one another as well as edges and corners. The type of shear plane depends upon the exact stoicheiometry of the oxide.

At high values of n, from about 36 to about 16, that is from $TiO_{1 \cdot 97}$ to $TiO_{1 \cdot 93}$, the shear planes are parallel to the crystallographic $\{132\}$ planes of rutile (389). Fig. 46 shows the structure of $Ti_{16}O_{31}$. The value of n equals the number of rutile cells which extend along the crystallographic a-axis between the parallel shear planes. Only even values of n have been found for these $\{132\}$ shear planes. They are also normally twinned, for example so as alternating $(\bar{1}32)$ and (132) planes are observed. For higher values of n, that is for compositions approaching $TiO_{2 \cdot 00}$, the distance between the shear planes becomes somewhat disordered, with a tendency for several to group together.

For values of n between 9 and 4, that is $TiO_{1\cdot89}$ to $TiO_{1\cdot75}$, the shear planes are parallel to {121} (58, 63, 65). The value of n is again the number of rutile unit cells between the shear planes (Fig. 47).

In any one reduced rutile crystal there can be a mixture of {132} and {121} shear planes, together with {011} antiphase boundaries (Fig. 48) which do not change the composition of the crystal.

The next lower oxide, Ti_3O_5, is not obviously related to the above Ti_nO_{2n-1} ($n > 4$), and exists in two forms. One form contains metal–metal bonds whereas in the other this bonding has been destroyed either by heating or by the addition of impurity atoms (90). The former structure can be considered to be based on zig-zag layers of edge sharing octahedra

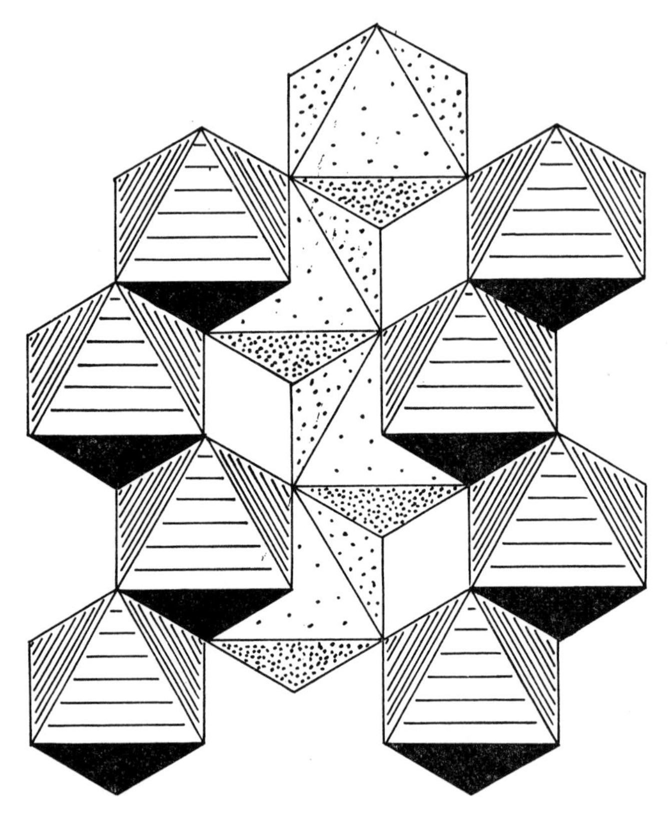

FIG. 45 The structure of brookite, showing the linking of two zig-zag strings of TiO_6 octahedra by a third string behind the other two.

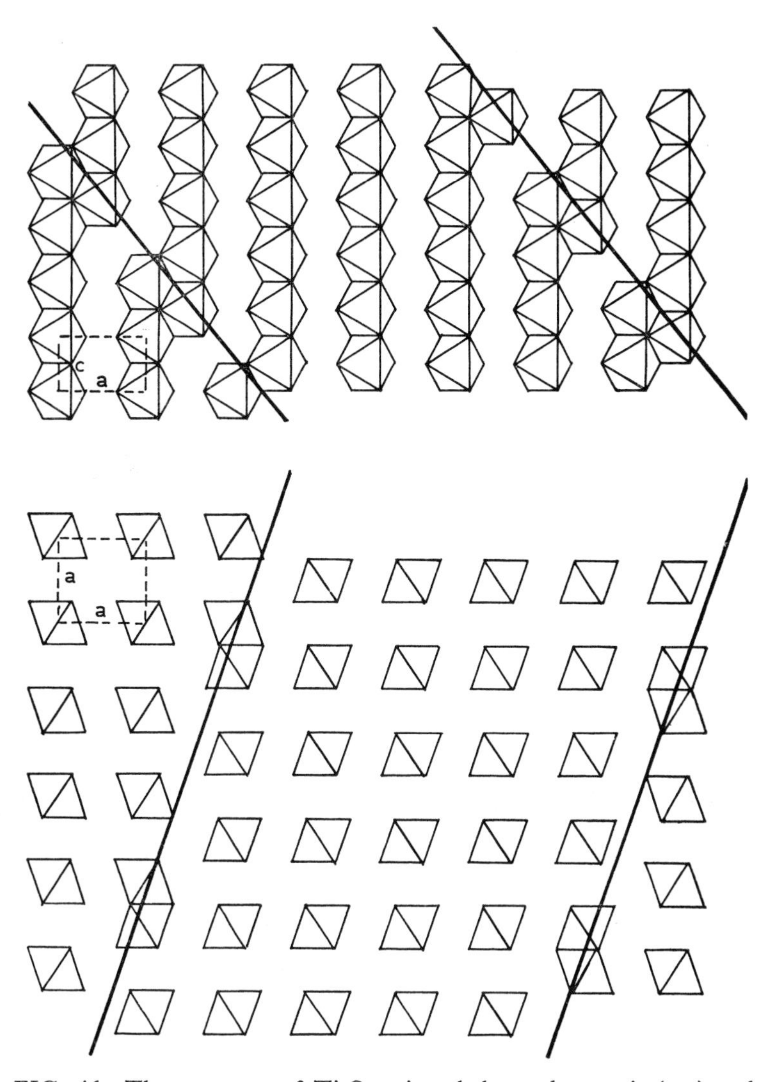

FIG. 46 The structure of Ti_6O_{31} viewed down the *a*-axis (top) and along the *c*-axis (bottom) of the rutile structure showing the linking of the TiO_6 octahedra. The shear planes (heavy lines) formed by octahedra sharing additional edges are shown parallel to {132} of the rutile unit cell (broken lines).

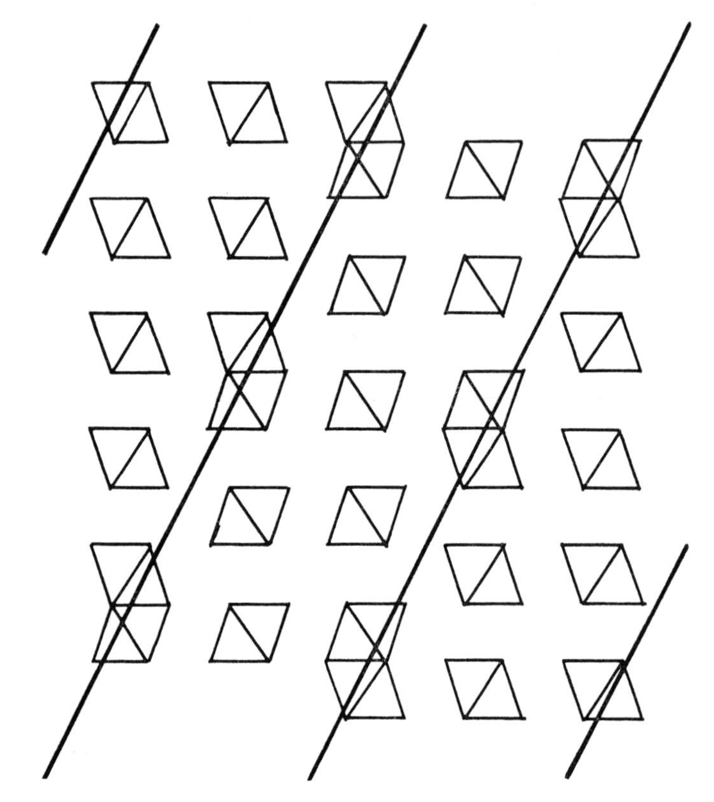

FIG. 47 The structure of Ti_5O_9 viewed along the *c*-axis of the rutile unit cell. The shear planes parallel to {121} of the rutile structure are shown.

(Fig. 49a). The titanium–titanium distances which are shorter than 2·9 Å (compare with 2·96 Å in TiO_2) are shown (Fig. 49b).

$$Ti_1\text{–}Ti_1 = 2\cdot 61 \text{ Å}$$
$$Ti_2\text{–}Ti_3 = 2\cdot 82 \text{ Å}$$
$$Ti_2\text{–}Ti_2 = 2\cdot 77 \text{ Å}$$

It appears that Ti_1–Ti_1 is a normal electron pair bond, while the other available electron pair is distributed over the bent four-centre bond Ti_3–Ti_2–Ti_2–Ti_3. Each layer shares edges (but with fairly long Ti–Ti distances) with similar layers (Fig. 49c), and in addition each layer is joined to a similar layer in the *y* direction (vertical to page) by corner sharing of the octahedral vertices.

On heating to 100–200° C, Ti_3O_5 changes from diamagnetic to paramagnetic (1373), the a and c axes expand, and the unit cell changes from monoclinic to orthorhombic:

Low temperature phase: $a = 9 \cdot 75$, $b = 3 \cdot 80$, $c = 9 \cdot 44$ Å, $\beta = 91 \cdot 6°$.
High temperature phase: $a = 9 \cdot 82$, $b = 3 \cdot 75$, $c = 9 \cdot 73$ Å, $\beta = 90 \cdot 0°$.

The structure is now composed of wavy layers as shown in Fig. 50, which are linked to each other as before. The relation between the phases is also indicated.

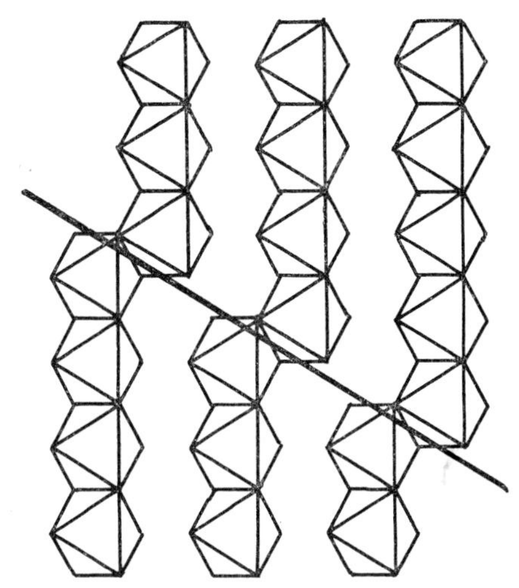

FIG. 48 A {011} antiphase boundary in rutile.

Compounds which contain no electrons potentially available for metal–metal bonding, for example $Mg^{II}Ti_2^{IV}O_5$, $Al_2^{III}Ti^{IV}O_5$ and $Fe_2^{III}Ti^{IV}O_5$, have the high temperature structure. However, this form is also stabilized at room temperature by the addition of only small amounts of impurities as in $(Fe_{0 \cdot 03}\,Ti_{0 \cdot 97})_3O_5$.

3. *Titanates*

Reaction between many salts and titanium dioxide at elevated temperatures forms complexes of the type A_2^IO, $n$$TiO_2$ (where A is an alkali metal and n is $\frac{1}{2}$, 1, 2, 3, 4 or 6), or $A^{II}O$, $n$$TiO_2$ (where A is a divalent metal and n is $\frac{1}{3}$, $\frac{1}{2}$, $\frac{2}{3}$, 1, 2 or 4). Many of these compounds have standard structures, for example $BaTiO_3$ has the perovskite structure (but distorted with ferroelectric properties), $Sr_3Ti_2O_7$ has a structure composed of double perovskite layers with interleaved layers of composition SrO, $A^{II}TiO_3$ (where A^{II} is Mg, Mn, Fe, Co, Ni, or Cd) have the ilmenite structure, and $A_2^{II}TiO_4$ (where A^{II} is Mg, Mn, Co or Zn) have the spinel structure.

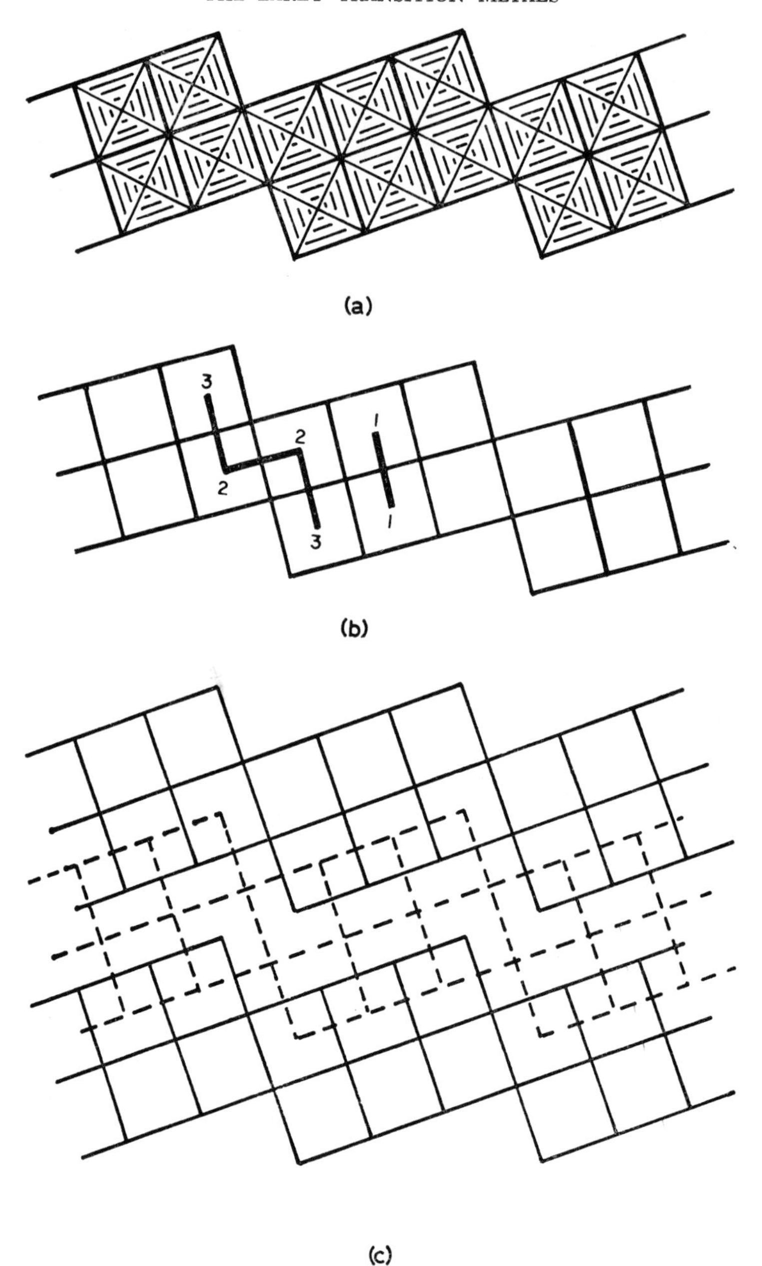

FIG. 49 The structure of the low temperature form of Ti_3O_5. (a) TiO_6 octahedra sharing edges. (b) Titanium–titanium bonding between octahedra. (c) Linking of units parallel to the *y*-axis.

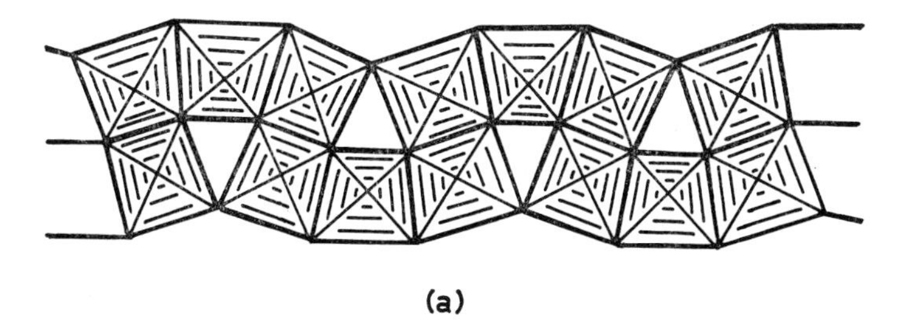

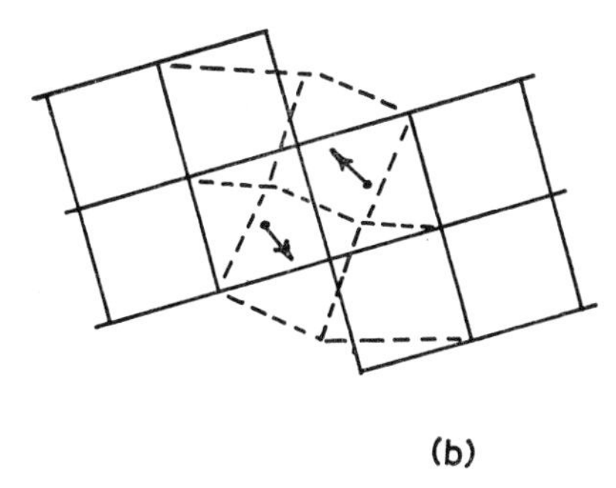

FIG. 50 The structure of the high temperature form of Ti_3O_5. (a) TiO_6 octahedra sharing edges. (b) Relation between low temperature form (full lines, compare with Fig. 49 (a)) and high temperature form (broken lines).

Of particular interest are the structures of some of the alkali metal titanates in which the alkali metal is situated between sheets or in channels formed by a rigid titanium–oxygen framework in an analogous manner to some of the vanadium and tungsten bronzes.

In $K_2Ti_2O_5$, strings of edge sharing trigonal bipyramids (Ti–O = 1·57–2·00 Å) share corners forming sheets which are separated by the potassium ions (Fig. 51) (67). In $Na_2Ti_3O_7$ some of the titanium atoms can be considered to be five coordinate (five titanium–oxygen bond lengths in the range 1·71–1·94 Å) or six coordinate (sixth titanium–oxygen distance of 2·34 Å). Fig. 52 shows the structure considered as being built up from zig-zag ribbons three octahedra wide which share corners to form layers which are separated by the cations (68). The relationship to $Li_{1+x}V_3O_8$ (page 146) should be noted. The sodium atoms can be readily exchanged with other cations, even by long chain alkyl ammonium ions (2360). As the length of the alkyl chain increases, the sheets are forced further apart, and a linear relationship exists between the separation and the number of carbon atoms. At C_{26} the sheets are about 40 Å apart, but

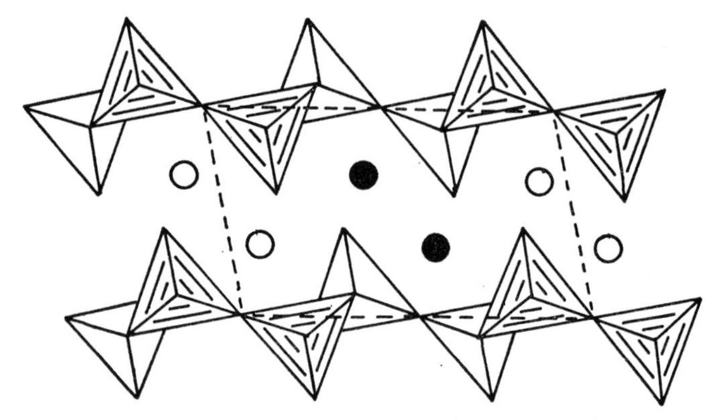

FIG. 51 The structure of $K_2Ti_2O_5$, showing linking of TiO_5 trigonal bipyramids.

the crystallographic *b*-axis remains unchanged. In $Na_2Ti_6O_{13}$ and $K_2Ti_6O_{13}$ there is no tendency for the titanium atom to be five coordinate (69, 480). The structures are built up from three edge-sharing octahedra (Fig. 53) which share further edges (above and below the plane of the paper) to form zig-zag ribbons which share corners with identical ribbons. There is an obvious relation between $Na_2Ti_3O_7$ and $Na_2Ti_6O_{13}$, and it is not surprising that when $Na_2Ti_3O_7$ is heated to 950° C in air it loses Na_2O to form $Na_2Ti_6O_{13}$.

The reduction of these titanates forms compounds which are called "titanium bronzes" by analogy with the tungsten bronzes (page 264). For example reduction of $Na_2Ti_3O_7$ forms the blue-black $Na_{\sim 0 \cdot 8}Ti_4O_8$ which

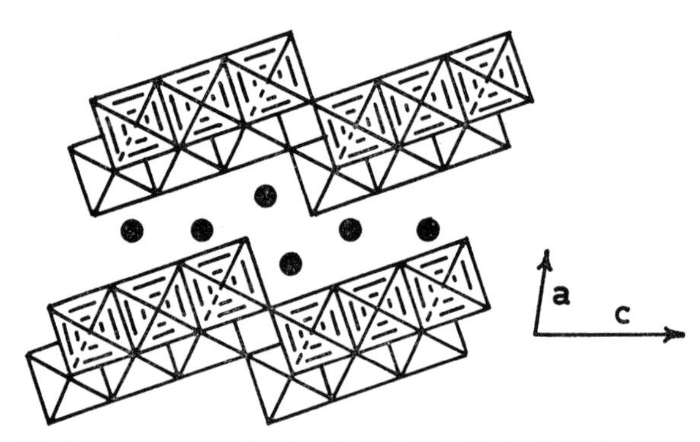

FIG. 52 The structure of $Na_2Ti_3O_7$ showing infinite ribbons normal to the page and three octahedra wide formed by the edge sharing of TiO_6 octahedra. The longer titanium–oxygen distances are indicated by the broken lines.

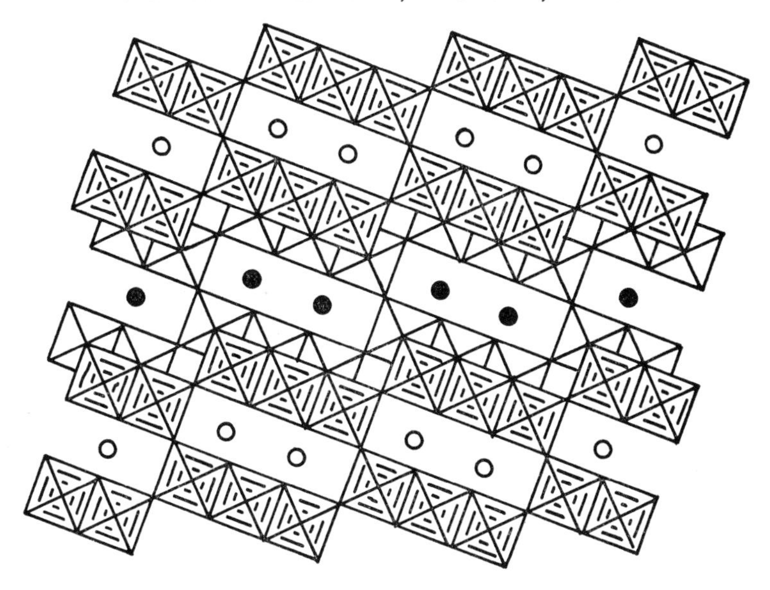

FIG. 53 The structure of $Na_2Ti_6O_{13}$.

has a metallic appearance, is electrically conducting, and chemically very inert (no reaction with boiling concentrated sulphuric or hydrofluoric acids). The structure (Fig. 54) shows double zig-zag rows of octahedra, compared with the triple rows in $Na_2Ti_3O_7$ and $Na_2Ti_6O_{13}$ (70, 1914). The mineral freudenbergite can be formulated $Na_{\sim 0 \cdot 8}(Ti_{4-x}Fe_x)O_8$ where 10–20% of the titanium atoms have been replaced by iron atoms (167, 2330).

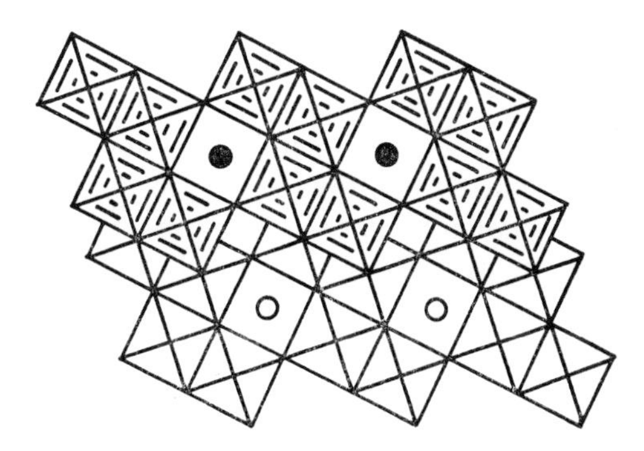

FIG. 54 The structure of $Na_{\sim 0 \cdot 8}Ti_4O_8$. Each octahedron shares edges with similar octahedra above and below the plane of the page forming an infinite structure.

Another type of chemically inert, electrically conducting bronze is $La_{\frac{2}{3}+x}TiO_3$, which has the perovskite structure analogous to $Sr_{\frac{1}{2}+x}NbO_3$ and Na_xWO_3 discussed elsewhere (1367, 1699).

B. Tetrahalides

1. *Tetrafluorides*

(a) *Titanium.* Titanium tetrafluoride may be prepared by the direct fluorination of the metal using fluorine at 200° C, chlorine trifluoride at 350° C (1747), or hydrogen fluoride at red heat. The fluorination of titanium dioxide similarly produces the tetrafluoride, for example by using fluorine at 350° C (1109). A very convenient synthesis is by the halogen exchange reaction between the tetrachloride and anhydrous hydrogen fluoride:

$$TiCl_4 + 4\,HF \rightleftarrows TiF_4 + 4\,HCl$$

The reaction is reversible and proceeds smoothly at room temperature. The product may be purified by sublimation in a nickel apparatus.

Titanium tetrafluoride is a white solid with the relatively high melting point of 284° C indicating a polymeric structure which has apparently not yet been determined. It appears to be monomeric in the vapour. The compound is soluble in hydrofluoric acid forming complex fluorides (page 87), and also in many donor solvents from which adducts such as TiF_4, $2H_2O$, TiF_4, EtOH and TiF_4, py may be obtained (page 78).

The mixed halofluorides TiF_3Cl (1809, 2324), TiF_3Br (1490) and TiF_2Cl_2 (643) have also been characterized.

(b) *Zirconium and hafnium.* The direct fluorination of zirconium or hafnium dioxides yields the white ZrF_4 or HfF_4 (1109). Fluorination of the metals begin at much lower temperatures (200° C), but the protective coating of tetrafluorides prevent complete conversion. Very much higher temperatures are required if anhydrous hydrogen fluoride is used as the fluorinating agent. In a similar manner to titanium tetrafluoride, zirconium tetrafluoride can be more conveniently prepared by the halogen exchange reaction between hydrogen fluoride and zirconium tetrachloride.

Zirconium tetrafluoride is monomeric in the vapour. The vapour pressure has been measured as a function of temperature; it is 1 mm of mercury at 650° C (2108).

At room temperature it forms white crystals with a complex polymeric structure (Fig. 55) (382). Each zirconium atom is eight coordinate, with square antiprismatic stereochemistry. These square antiprisms then share

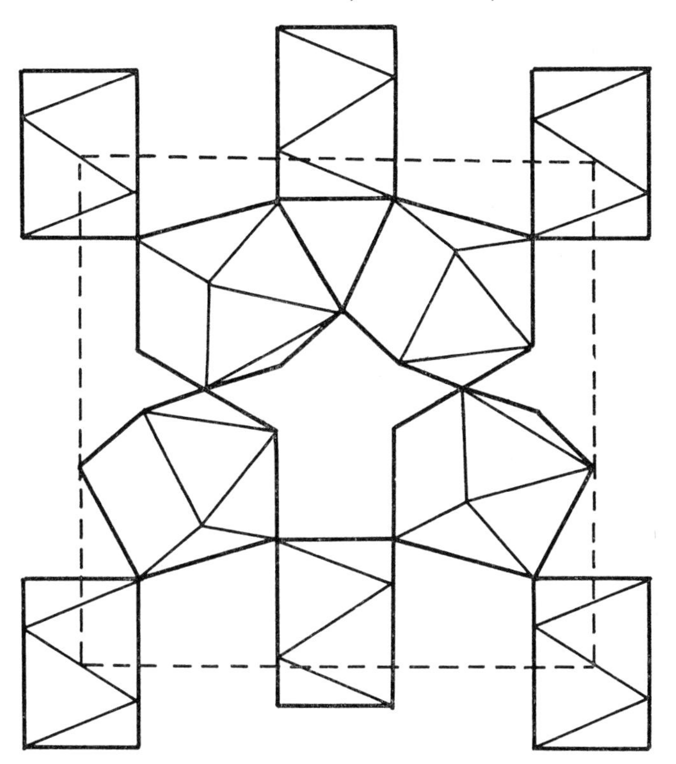

FIG. 55 The structure of ZrF_4 showing the corner sharing of ZrF_8 square antiprisms.

all corners, but no edges, with surrounding units forming a three-dimensional structure. The average zirconium–fluorine distance is 2·10 Å.

Zirconium also forms the mixed halide ZrF_2Cl_2 (1439).

2. *Tetrachlorides, bromides and iodides*

(a) *Titanium.* Titanium tetrachloride is of considerable importance and may be prepared from the oxide by heating with charcoal and chlorine:

$$TiO_2 + 2\,Cl_2 + 2C \rightarrow TiCl_4 + 2\,CO$$

A large number of other chlorinating agents have been used on the laboratory scale, although many are less convenient than the above. Chlorinated hydrocarbons may be sometimes convenient, for example reaction of the dioxide with chloroform at 440° C in sealed tubes proceeds according to the equation:

$$TiO_2 + 2\,CHCl_3 \rightarrow TiCl_4 + 2\,CO + 2\,HCl$$

The sealed tube and high pressures can be dispensed with simply by refluxing with a high boiling point chlorinated hydrocarbon at atmospheric pressure. Octachlorocyclopentene, which has a boiling point of 285° C, and is self drying due to the replacement of the chlorine by oxygen, has been found to be particularly suitable (144). Chlorine reacts with titanium itself at about 350° C to form the tetrachloride. Anhydrous gaseous hydrogen chloride and the heated metal also form the tetrachloride, which is in contrast to the action of concentrated aqueous hydrochloric acid on the metal, where there is the slow production of the violet titanium(III) and hydrogen.

Titanium tetrachloride is a colourless, non-conducting and non-viscous liquid at room temperature. The melting point has been quoted in the range —23° to —30° C depending upon the worker. It has a tetrahedral structure in the solid state (311, 312). The vapour pressure is about 1 cm of mercury at 20° C (1851), and it boils at 136° C to give a vapour of normal density. Electron diffraction of the vapour shows a titanium–chlorine bond length of 2·18 Å (1392, 1715). The Raman spectrum of the liquid shows that it is partially associated, presumably through $\mathrm{Ti}\langle{}^{\mathrm{Cl}}_{\mathrm{Cl}}\rangle\mathrm{Ti}$ bridges, but is monomeric when diluted with carbon tetrachloride (1070).

Titanium tetrachloride is hydrolysed in moist air and its vapour produces dense white fumes of the oxide plus hydrogen chloride. Nevertheless it can be quite conveniently handled with reasonably simple apparatus, and a very large number of complexes have been prepared with various ligands which are discussed later.

The preparation of titanium tetrabromide has been achieved using similar methods to the tetrachloride preparation, for example by heating a mixture of titanium dioxide, carbon and bromine to about 600° C, or by the direct reaction of titanium and bromine at 360° C. Alternatively it may be prepared by halogen displacement from the tetrachloride, for example by reaction with a large excess of liquid hydrogen bromide or boron tribromide (694).

Titanium tetraiodide has been prepared by the direct union of the elements, by the reaction of titanium tetrachloride with boron triiodide (694) or liquid hydrogen iodide at its boiling point (—35° C), and by the iodination of titanium dioxide with aluminium triiodide at 230° C (451):

$$3\,TiO_2 + 4\,AlI_3 \rightarrow 3\,TiI_4 + 2\,Al_2O_3$$

The tetrabromide is an orange to yellow crystalline compound, whereas the tetraiodide is a dark reddish brown solid. The melting points and boiling points increase from the chloride to the bromide to the iodide, but are lower than the polymeric tetrafluoride as expected:

	M.P. (°C)	B.P. (°C)
TiF_4	284	>400
$TiCl_4$	−24	136
$TiBr_4$	39	229
TiI_4	144	360

Even the iodide may be distilled unchanged, and appears monomeric in the vapour.

Titanium tetrabromide, like titanium tetrachloride, is soluble in donor solvents such as diethyl ether, from which solid adducts may be obtained (page 78). The tetraiodide is much less soluble in such solvents, and few complexes have been characterized.

(b) *Zirconium and hafnium.* Zirconium and hafnium tetrachlorides are prepared in a similar manner to the titanium compound, for example by passing chlorine over the heated metal and collecting the tetrachloride as a white sublimate. Similarly the dioxides may be mixed with carbon and chlorinated as before, or the chlorination may be carried out with S_2Cl_2, PCl_5, CCl_4 or $COCl_2$.

Zirconium tetrachloride is much less volatile than the titanium compound, subliming at about 330° C and melting at 437° C. The structure is shown in Fig. 56 (1470), and can be considered to be based on a cubic

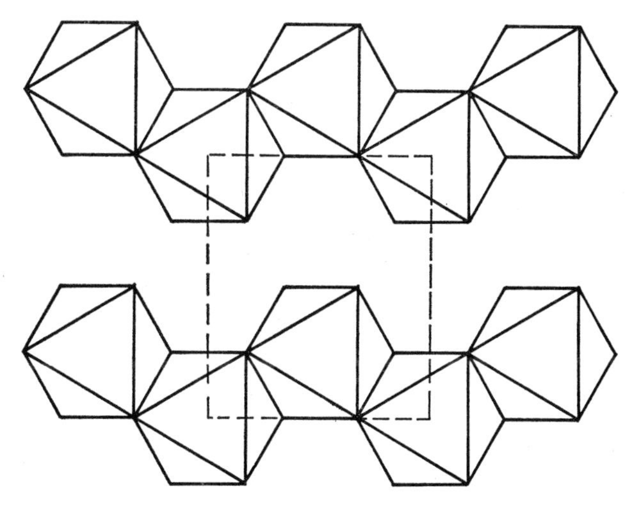

FIG. 56 The linking of $ZrCl_6$ octahedra in $ZrCl_4$.

close packing of chloride ions with one quarter of the octahedral holes occupied by zirconium atoms. Each octahedrally coordinated metal atom shares two octahedral edges with adjacent octahedra to form an infinite one dimensional polymer (Zr-Cl(bridge) = 2·58 (av.), Zr-Cl(terminal) = 2·31 Å). In contrast to the closely related structure of the metal–metal bonded niobium tetrachloride, the terminal halogen atoms are *cis* to each other, so that the zirconium atoms in the polymer are arranged in zig-zag strings.

Thorium tetrachloride on the other hand does not begin to sublime *in vacuo* until 750° C, and the polymeric structure contains the metal atoms in eight coordinate, dodecahedral, environments (1703).

Zirconium tetrachloride is insoluble in nonpolar solvents but soluble in solvents such as anhydrous diethyl ether. It is hydrolysed even in concentrated hydrochloric acid to form the so-called "$ZrOCl_2,8H_2O$" (page 108).

Zirconium tetrabromide is similarly prepared directly from the elements, or from the dioxide, carbon, and bromine. A particularly convenient method is the halogen exchange between $ZrCl_4$ and BBr_3, which gives essentially quantitative yields since the BCl_3 (B.P. $\sim$ 12° at 760 mm Hg) is readily eliminated (694). It is again a volatile solid, probably isostructural with the tetrachloride.

Zirconium tetraiodide has been prepared from the metal and hydrogen iodide, or from the dioxide and carbon plus iodine, or aluminium triiodide (451).

C. Adducts of the Tetrahalides

The Group IV tetrahalides, particularly titanium tetrachloride, have been reacted with an enormous number of ligands, and it is not proposed to mention all the products here. Further examples from the literature may be found in other books (496, 882, 1553). The products obtained may be five, six, seven, or eight coordinate. Adducts containing only one monodentate ligand may be five coordinate or may be chlorine-bridged dimers, as in $[TiCl_4(EtOAc)]_2$. That is, titanium–chlorine–titanium bridges occur in these 1 : 1 adducts, but not in the octahedral 1 : 2 adducts such as $[TiCl_4(EtOAc)_2]$ or in (solid) $TiCl_4$ itself.

The tetrachlorides of zirconium and hafnium form similar complexes to those of titanium. In contrast tin tetrachloride apparently forms octahedral monomeric 1 : 2 adducts. One difference between titanium and the heavier metals is the easier reduction to titanium(III) with some ligands, particularly with the bromide.

It is generally found that titanium tetraiodide forms much weaker complexes, if any at all, than the tetrachloride or tetrabromide. For example with many ethers the tetrafluoride, tetrachloride and tetrabromide form TiX_4(ether) and $TiX_4(\text{ether})_2$, but no reaction is observed with the tetraiodide. From the decrease in the carbonyl stretching frequency in the titanium tetrahalide-ethylacetate addition compounds, it has also been deduced that the relative acceptor strengths lie in the order $TiCl_4 \sim TiBr_4 > TiI_4$ (1493). Titanium tetrafluoride also often does not form complexes readily, for example it may be recrystallized unchanged from trifluoroacetic acid (1729).

The nature of the product obtained is very sensitive to the ligand. For example titanium tetrafluoride forms $TiF_4(Me_2CO)_2$ with acetone, but $TiF_4(Et_2CO)$ with diethylketone. The relation between base strength of a substituted pyridine and the stoicheiometry of its adduct with TiF_4 is shown on page 83. Similarly the features of the diarsine ligands which control whether six coordinate compounds, for example $TiCl_4(o\text{-}C_6H_4(AsEt_2)_2)$, or eight coordinate compounds, for example $TiCl_4(o\text{-}C_6H_4(AsMe_2)_2)_2$, are obtained are discussed on page 84. Ligands such as triphenylarsine on the other hand form the five coordinate $TiCl_4, Ph_3As$.

These complexes will now be treated in more detail as a function of ligand type.

1. *Oxygen donors*

Titanium tetrafluoride reacts with ethers to form the presumably six coordinate $TiF_4((CH_2)_4O)_2$, $TiF_4(C_4H_8O_2)$ and $TiF_4(CH_3O.CH_2.CH_2.OCH_3)$. These compounds are monomeric in tetrahydrofuran, although the dioxane adduct is probably polymeric in the solid state (1729).

Titanium tetrachloride forms both 1 : 1 and 1 : 2 red adducts with ethers. For example $TiCl_4$(ether) is obtained with anisole, tetrahydropyran, tetrahydrofuran and phenetole (517, 614, 1126, 1825). On heating the 1 : 1 tetrahydropyran adduct, the 1 : 2 compound $TiCl_4(C_5H_{10}O)_2$ and $TiCl_4$ are formed (1126). Similarly for the anisole and phenetole derivatives there is spectral evidence for $TiCl_4(PhOMe)_2$ and $TiCl_4(PhOEt)_2$ in solutions in these solvents (517, 1825). However, more complex reactions may also occur; diisopropylether is particularly interesting in forming $TiCl_3(OPr)$, while trioxane ($C_3H_6O_3$) is decomposed to formaldehyde(H.CHO) (1126). Zirconium and hafnium tetrachlorides form $MCl_4(Me_2O)_2$, $MCl_4(Et_2O)_2$ and $MCl_4(C_4H_8O_2)$ (1045).

Similarly the red ether adducts $TiBr_4(C_4H_8O_2)$, $TiBr_4(C_4H_8O)_2$ and $TiBr_4(C_5H_{10}O)_2$ are formed with titanium tetrabromide; the infrared evidence indicates that both oxygen atoms are coordinated in the 1 : 1

dioxane adduct, although it is monomeric as expected in this solvent (1940, 1941).

In contrast, ethers do not appear to react with titanium tetraiodide (1940).

The titanium tetrahalides have been reacted with a very large number of carbonyl compounds, and only some general points will be mentioned here. In many cases the adducts obtained are liquids or intractable oils, but nevertheless others are crystalline compounds, and the structure of $TiCl_4(EtOAc)$ has been determined. Titanium tetrachloride forms yellow or red 1 : 1 adducts with dialkyl ketones such as acetone, mixed alkyl-aryl ketones, and benzophenone (614, 2082). Similar adducts are obtained with zirconium and hafnium tetrachlorides (1045, 1844).

Reactions with aldehydes have also been studied. Titanium tetrafluoride forms a 1 : 1 adduct with acetaldehyde (776). Liquid products which could not be characterized were found for titanium tetrachloride with acetaldehyde and higher straight chain aliphatic aldehydes (2081). Whether or not this behaviour is due to enolization and subsequent elimination of hydrogen chloride is not clear. With benzaldehyde however the yellow

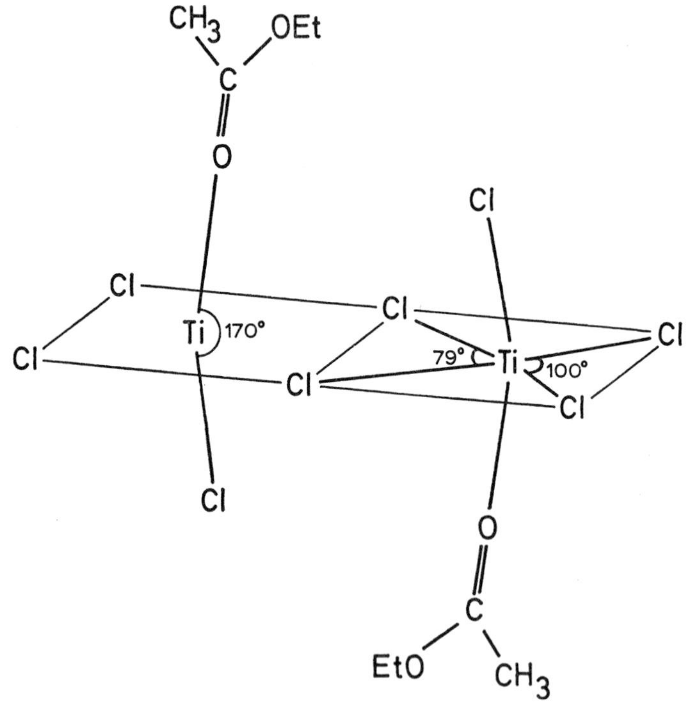

FIG. 57 The structure of $TiCl_4(EtOAc)$.

$TiCl_4(PhCHO)_2$ is obtained (668, 2081). Similar behaviour was observed with other aromatic and heterocyclic aldehydes (2081). Similarly titanium, zirconium and hafnium tetrachlorides form 1 : 2 adducts with dimethylformamide (1045).

Esters similarly form 1 : 1 adducts, or less commonly 1 : 2 adducts, with titanium tetrachlorides (1713). For example ethylacetate forms the monomeric $TiCl_4(EtOAc)_2$ and also $TiCl_4(EtOAc)$ (288), which retains octahedral coordination by dimerizing through two bridging chlorine atoms (Fig. 57) (375). Zirconium and hafnium tetrachlorides form similar complexes (1045, 1245). Thermochemical measurements show that the reaction of ethyl benzoate with zirconium tetrachloride is more exothermic than with hafnium tetrachloride; methyl benzoate is similar to ethyl benzoate, whereas the reactions with phenylbenzoate are much less exothermic (1240). These zirconium and hafnium compounds, $MCl_4(PhCOOMe)_2$ and $MCl_4(PhCOOEt)_2$ decompose when heated between 100° and 200° C to form the metal benzoate and the corresponding alkyl chloride (461). Reaction of diesters of aliphatic α, ω-dicarboxylic acids, and diesters of phthalic acid, with titanium, zirconium or hafnium tetra-

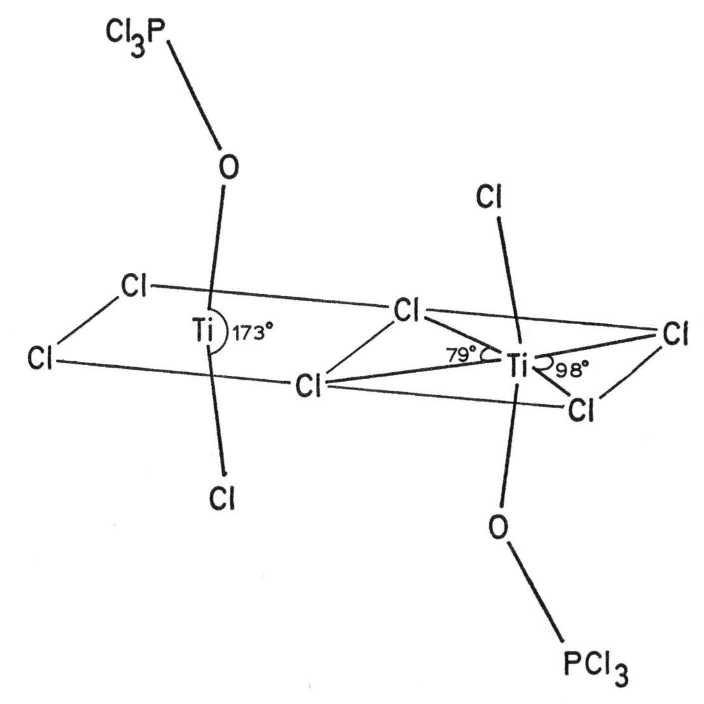

FIG. 58 The structure of $TiCl_4(OPCl_3)$.

chlorides lead to the formation of the analogous compounds MCl_4, $\frac{1}{2}$ Diester and MCl_4,Diester (99, 1707, 1929).

Other carbonyl compounds which have been reacted with titanium tetrachloride include acyl chlorides (223, 614), amides (85), and some diimides (1281).

Phosphorus oxytrichloride forms both $TiCl_4(OPCl_3)$ and $TiCl_4(OPCl_3)_2$ (313, 1847). The former has a dimeric structure with octahedral coordination about the metal atom (Fig. 58) in a completely analogous manner to $TiCl_4(EtOAc)$ (Fig. 57). The titanium-terminal chlorine and titanium-bridging chlorine distances are 2·23 and 2·49 Å respectively. A number of other phosphine oxides, for example Ph_3PO, behave similarly (181, 1498, 2129).

Titanium tetrafluoride forms the air-stable 1 : 2 adducts $TiF_4(Ligand)_2$ with a number of oxygen donor ligands, including $Me_2N.COMe$, $Me_2N.CHO$, Me_2SO, $Me_2N.CONMe_2$ and $Me_2C = NOH$ (1729). The first three are monomeric in solution, whereas the other two dissociate in acetonitrile and tetrahydrofuran respectively. It has been established from ^{19}F-nmr spectroscopy that $TiF_4(Me_2N.COMe)_2$ (1729) and $TiF_4(Me_2N.COMe)(C_5H_4NO)$ (713) have the *cis*-configuration. The *trans*-configuration can be stabilized in compounds of this type by either:

(a) Reducing the p_π–d_π bonding between the fluorine and metal. For example $TiF_4(EtOH)_2$ exists only in the *cis*-configuration, but *cis*- and *trans*-isomers are in equilibrium in $SnF_4(EtOH)_2$ (1905).

(b) Increasing the steric interaction between the oxygen-donor ligands. To take the pyridine-1-oxides as examples, $TiF_4(C_5H_5NO)_2$ has the *cis*-structure, an equilibrium exists between the *cis* and the *trans* structures of $TiF_4(2\text{-}Me.C_5H_4NO)_2$, and only *trans*-$TiF_4(2,6\text{-}Me_2.C_5H_3NO)_2$ can be obtained (712).

Zirconium tetrafluoride forms $ZrF_4(Me_2SO)_2$ analogous to $TiF_4(Me_2SO)_2$, but only the 1 : 1 compounds $ZrF_4(Me_2N.CHO)$ and $ZrF_4(Me_2N.CONMe_2)$; the latter is monomeric in acetonitrile compared with the 1 : 2 titanium analogue which dissociates in this solvent (see above). It is not known whether these compounds are five coordinate in the solid state or are polymeric with bridging fluorine atoms.

2. *Sulphur and selenium donors*

Titanium tetrachloride reacts with dialkyl sulphides R_2S (where R is Me, Et, Pr^n, Bz) or cyclic sulphides $(CH_2)_3S$, $(CH_2)_4S$ and $(CH_2)_5S$ to form the yellow to red $TiCl_4(Ligand)$ or $TiCl_4(Ligand)_2$ depending upon which reagent is in excess (22, 125, 126, 1377, 2379). Similar compounds of the type $TiCl_4(Bidentate)$ are obtained with $RS.CH_2.CH_2.SR$ (where R is

Me, Et, Bu, Ph), 1,4-dithian, diethylphosphinyl P,P′-disulphide and *cis*-dimethylthiomaleodinitrile MeS.C(CN) = C(CN).SMe, although in the last case no definite conclusion could be drawn concerning whether the ligand was bonded through the sulphur or nitrogen atoms (125, 126, 497, 1377). All compounds are monomeric in benzene.

Some interesting conclusions were drawn from the monothio- and monoseleno-dioxane compounds $TiCl_4(C_4H_8OS)_2$, $TiCl_4(C_4H_8OSe)_2$ and $TiCl_4(C_4H_8OSe)$. From infrared and nmr evidence it was deduced that these ligands are sulphur or selenium-bonded to the metal, rather than oxygen-bonded (127, 938). This result emphasizes that the division of the transition metals into "Class *a*" and "Class *b*" electron acceptors is primarily restricted to studies in aqueous solution. In contrast to $TiCl_4(Et_2S)_2$ above, diethylselenide forms only the dark red $TiCl_4(Et_2Se)$ (2379).

The behaviour of titanium tetrabromide is significantly different to the tetrachloride. For example dimethylsulphide forms $TiBr_4(Me_2S)_2$ which is monomeric in benzene, diethylsulphide forms $TiBr_4(Et_2S)_2$ which has only half the formula weight in benzene due to dissociation, whereas dipropylsulphide forms only $TiBr_4(Pr_2S)$ (125).

3. *Nitrogen donors*

Titanium tetrafluoride forms only 1 : 1 complexes with a variety of nitrogen-donor ligands, including pyridine, trimethylamine and N,N-dimethylaniline (1729). The last complex appeared to be significantly associated in acetonitrile. A particularly interesting series of complexes was obtained with substituted pyridines in dimethoxyethane; the weakest bases formed *cis*-TiF_4(Ligand)$_2$ whereas the strongest formed the apparently polymeric TiF_4(Ligand) (Table 14) (683). As has been noted above,

TABLE 14

Stoicheiometry of complex obtained between titanium tetrafluoride and substituted pyridines

Ligand	Complex (TiF_4: Ligand)	$pK_a(BH^+)$
2-Cl.C_5H_4N	1:2	1·3
2-Br.C_5H_4N	1:2	1·6
3-Br.C_5H_4N	1:1	2·2
3-Cl.C_5H_4N	1:1	2·2
C_5H_5N	1:1	5·1
2-CH_3.C_5H_4N	1:1	6·1
4-CH_3.C_5H_4N	1:1	6·2
2,6-$(CH_3)_2$.C_5H_3N	1:1	6·7
2,4-$(CH_3)_2$.C_5H_3N	1:1	6·7

oxygen donor ligands, which are weaker bases, form *cis*-1:2 complexes. The bidentates 2,2′-dipyridyl and *o*-phenanthroline form TiF_4(Bidentate), whereas *o*-phenylenebisdimethylamine forms TiF_4, $\frac{1}{2}$Diamine (498).

Zirconium tetrafluoride can form complexes containing more ligand. For example $4\text{-}CH_3.C_5H_4N$ forms a 1:2 adduct (1729), whereas 2,2′-dipyridyl forms the apparently eight coordinate $ZrF_4(dipy)_2$ (498).

Pure trimethylamine and triethylamine reduced titanium tetrachloride to, for example, the blue green $TiCl_3(Me_3N)_2$. Dilution of the ligand with benzene or petroleum ether allowed the isolation of the diamagnetic yellow $TiCl_4(Me_3N)$ and black $TiCl_4(Et_3N)$ (82, 908). The former was monomeric in benzene indicating five coordination, which is very unusual for this type of complex. The heterocycles pyridine, γ-picoline, pyrazine and 2,6-dimethylpyrazine form yellow $TiCl_4(Ligand)_2$, but again the use of excess ligand causes some reduction to titanium(III) (175, 776, 908, 938, 1729). The bidentates 2,2′-dipyridyl and *o*-phenanthroline similarly form the yellow $TiCl_4$(Bidentate) whereas *o*-phenylenebisdimethylamine forms $TiCl_4$, $\frac{1}{2}$Diamine as was observed above for TiF_4 (463, 498, 936). Mainly on the basis of infrared evidence, molecules containing more than one type of potentially donating atom appear to prefer to bond through the nitrogen atom only. Examples include $CS(NH_2)_2$ (compared with the oxygen bonded $CO(NH_2)_2$) (1928), $o\text{-}HO.C_6H_4.CH{=}N.Ph$ (and a number of similar substituted Schiff bases) (1435), and $(Me_3PN)_3$ (1494).

The behaviour of titanium tetrabromide appears analogous to titanium tetrachloride, except for the expected greater ease of reduction (776, 908, 936, 938).

Under the appropriate conditions zirconium and hafnium tetrachlorides and tetrabromides accommodate more ligands than the corresponding titanium tetrahalides. For example $ZrCl_4$,1$\frac{1}{2}$Bidentate and $HfCl_4$,1$\frac{1}{2}$Bidentate are obtained with *excess* 2,2′-dipyridyl and *o*-phenylenebisdimethylamine, and pyridine similarly forms $ZrCl_4(py)_3$ and $HfCl_4(py)_3$, although one molecule of pyridine is easily lost at 70° C under vacuum (498, 1045, 1911).

The alkyl nitriles and Group (IV) tetrahalides very easily form pure crystalline adducts of the type $MX_4(RCN)_2$, and this has proven to be a very popular field of study (776, 937, 1045, 1187, 1364, 1729). Bidentate nitriles behave likewise (938, 1280, 1477).

8-Hydroxquinoline (oxH) in tetrahydrofuran at room temperature forms $ZrF_4(oxH)$, $ZrCl_4(oxH)$, and $HfCl_4(oxH)_2$, while at 65° C in the same solvent hydrogen chloride is lost forming $Zr(ox)_4$ and $Hf(ox)_4$ (892).

4. *Phosphorus and arsenic donors*

Phosphine, trialkylphosphines, triphenylphosphine, arsine, and triphenylarsine form $TiCl_4(Ligand)$ or $TiCl_4(Ligand)_2$ depending on ligand

and upon experimental conditions (412, 463, 690, 938, 2379). Reduction to titanium(III) can occur under more vigorous conditions (936, 938).

The behaviour of bidentate phosphines and arsines is particularly interesting. Ligands such as $Et_2P.CH_2.CH_2.PEt_2$, $Me_2P.CH_2.CH_2.PMe_2$, *cis*-$Me_2As.CH = CH.AsMe_2$, *o*-$C_6H_4(PEt_2)_2$ and *o*-$C_6H_4(AsEt_2)_2$ form $TiCl_4$(Diphosphine) (463) and $TiCl_4$(Diarsine) (495, 509). However, *o*-$C_6H_4(AsMe_2)_2$ and *o*-$C_6H_4(PMe_2)_2$ form the eight coordinate $TiCl_4$(Diarsine)$_2$ (505, 506) and $TiCl_4$(Diphosphine)$_2$ (510) respectively. The structure of the former shows dodecahedral stereochemistry, with the two bidentate ligands spanning the two A-A edges of dodecahedron (pages 13 and 15) and the four chlorine atoms occupying the four B vertices. It is clear that eight coordinate adducts are formed only with phosphorus or arsenic donor atoms, as replacement of only one arsenic atom by a nitrogen atom as in *o*-$C_6H_4(AsMe_2)(NMe_2)$ leads to only the 1:1 adduct $TiCl_4$(Ligand) (605). Excess of *o*-$C_6H_4(NMe_2)_2$ yields a black oily solid of indefinite composition (498). These arsine and phosphine ligands may differ from the amine ligands in being weaker σ-donors and also in possessing empty d_π orbitals which can interact with the d orbitals of the titanium atom. Although these d orbitals are formally empty for titanium(IV), they will be partly filled by chlorine-to-titanium p_π-d_π bonding. The appropriate d orbitals on the titanium atom which can be used for this chlorine–titanium–arsenic p_π–d_π–d_π bonding are the d_{xy}, d_{xz} and d_{yz} which are directed midway between the chlorine and arsenic atoms, and which also possess appropriate symmetry properties. The aryl substituted 1,2-bis-dimethylarsine-4-methyl benzene also forms $TiCl_4$(Bidentate)$_2$ showing that the eight coordination is not due to a uniquely favourable lattice energy (605). From a study of the deuterated ligand *o*-$C_6H_4(As(CD_3)_2)_2$ it was concluded that the main factor in preventing eight coordination was the steric properties of the alkyl groups attached to the donor atoms (605).

Similar work has been carried out on titanium tetrabromide, and on zirconium and hafnium tetrachlorides and tetrabromides (495, 499, 505, 506, 1911). Titanium tetraiodide forms TiI_4(*o*-$C_6H_4(AsMe_2)_2)_2$ which is not isomorphous with the eight coordinate chloro and bromo complexes, whereas titanium tetrafluoride forms only TiF_4, $\frac{1}{2}$(*o*-$C_6H_4(AsMe_2)_2$) (499).

This work has also been extended to tridentate ligands. The $MeAs(o\text{-}C_6H_4.AsMe_2)_2$ and $MeC(CH_2.AsMe_2)_3$ form the monomeric nonelectrolytes $TiCl_4$(Triarsine), which are formulated as seven coordinate complexes since nmr evidence indicates that all arsenic atoms are coordinated (501). However $MeAs(CH_2.CH_2.CH_2.AsMe_2)_2$ forms $TiCl_4$, $\frac{1}{2}$(Triarsine) (142). Other titanium and zirconium tetrahalides have similarly been reacted with these ligands (142, 501).

D. Oxohalides and Adducts

1. *Oxohalides*

The only known oxohalides of the Group IV metals which have been well characterized are $TiOCl_2$, $TiOBr_2$, $TiOI_2$, $ZrOCl_2$, and oxofluorides. In addition complexes corresponding to the parent hypothetical halides Ti_2OCl_6 and Ti_2OBr_6 have been prepared. The corresponding $TiSCl_2$ is also known (759).

In general partial hydrolysis of the tetrahalides has not proven to be a particularly satisfactory method for the preparation of the oxohalides, and other sources of the oxygen atom have proven to be more convenient. For example the partial hydrolysis of $ZrCl_4$ has not yielded $ZrOCl_2$, although the so-called octahydrate "$ZrOCl_2$,$8H_2O$" may be obtained from aqueous solution, but this in reality is the tetramer $[Zr_4(OH)_8(H_2O)_{16}]Cl_8,12H_2O$ (page 108). Similarly when the hydrolysis product of titanium(IV) in aqueous hydrofluoric acid is dried, the so-called "$TiOF_2$" is obtained (2324). However, it is not possible to obtain products free of hydroxyl groups by this method, and the product has been formulated as $TiO(OH)F$ (644). The compound has the cubic ReO_3 structure (2323). Partial hydrolysis however has been successfully used for the preparation of some complexes.

The oxofluoride Ti_2OF_6 has been obtained by heating the reaction product of TiF_3Br and $ClNO_3$ at 180° C ; an infrared band at 964 cm^{-1} suggests that the oxygen atom does not bridge different metal atoms (1490). The thermal decomposition of ZrF_4, H_2O and its partially hydrolysed products forms Zr_4OF_{14}, $Zr_3O_2F_8$ and $ZrOF_2$ respectively, all of which presumably contain Zr-F-Zr bridging groups (1441).

Pure $TiOCl_2$ is readily obtained by dissolving As_2O_3, Sb_2O_3 or Bi_2O_3 in $TiCl_4$ (750), or by the reaction of $TiCl_4$ with Cl_2O (639):

$$TiCl_4 + Cl_2O \rightarrow TiOCl_2 + 2\,Cl_2$$

Although of lower reactivity, ozone reacts with $TiCl_4$ at its boiling point to yield $TiOCl_2$ of high purity:

$$TiCl_4 + O_3 \rightarrow TiOCl_2 + O_2 + Cl_2$$

The action of chlorine monoxide on TiF_2Cl_2 yields $TiOF_2$ rather than $TiOCl_2$, which is not unexpected (641). The dark yellow $TiOBr_2$ is similarly prepared by passing chlorine monoxide, or preferably ozone, with oxygen carrier gas, into molten $TiBr_4$ (M.P. = 39° C) (645). The melting point of titanium tetraiodide is too high to consider the analogous reactions by this method, as at the required temperature $TiOI_2$ is found to disproportionate into $Ti^{III}OI$ and I_2. However, the reaction proceeds satisfactorily in cyclohexane solution, yielding the dark brown $TiOI_2$ (647).

This decomposition of $TiOI_2$ into TiOI and I_2 is in contrast to the action of heat on $TiOF_2$, $TiOCl_2$ and $TiOBr_2$, which form TiO_2 and TiX_4.

The addition of stoicheiometric amounts of chlorine monoxide to a slurry of zirconium tetrachloride in cold carbon tetrachloride yields $ZrOCl_2$ (648). It is not isomorphous with $TiOCl_2$.

2. *Oxohalide adducts*

The addition of neutral ligands to $TiOCl_2$ in general forms $TiOCl_2L_2$ (where L is MeCN, Me_3N, $(CH_2)_4O$, $(CH_2)_5O$, 4-$CH_3.C_5H_4N$, $\frac{1}{2}MeO.CH_2.CH_2.OMe$) (919). The trimethylamine compound is isomorphous with $VOCl_2(Me_3N)_2$ (page 211) confirming the existence of five coordinate titanium. Pyridine and dioxane however form $TiOCl_2,2\frac{1}{2}py$ and $TiOCl_2,1\frac{1}{2}Diox$ respectively. Apart from the trimethylamine (Ti = 0 at 976 cm^{-1}) and γ-picoline adducts (Ti = 0 at 978 cm^{-1}), the titanium-oxygen stretching frequencies are below 900 cm^{-1} suggesting bridging oxygen atoms.

The addition of phosphorus oxychloride or pyridine to $TiOBr_2$ yields $TiOBr_2,2POCl_3$ and $TiOBr_2,2\frac{1}{2}py$ respectively (645). Similarly $TiOI_2$ forms $TiOI_2,3py$ (647). The addition of $POCl_3$ to $ZrOCl_2$ forms $ZrOCl_2$, $2POCl_3$ (648).

The partial hydrolysis of $TiCl_4,2MeCN$ in acetonitrile has yielded $Ti_2OCl_6,4MeCN$, which on the basis of infrared and molecular weight evidence can be described as $(MeCN)_2Cl_3Ti\text{-}O\text{-}TiCl_3(MeCN)_2$. Similar compounds are formed with benzonitrile or dioxane as ligands (824, 825). The analogous partial hydrolysis of $ZrCl_4$, 2MeCN in acetonitrile forms a compound formulated as $Zr_3Cl_9O(OH)$, 7MeCN which on heating gives $ZrCl_4$, MeCN and amorphous $ZrOCl_2$ (827).

E. Anionic Halo and Oxohalo Complexes

1. *Fluoro and oxofluoro complexes*

Many complex fluorides of titanium, zirconium and hafnium have been obtained from aqueous solution of hydrofluoric acid, and most of the well characterized compounds contain only metal–fluorine bonds. Little is known about the partially hydrolysed complex oxofluorides which are found at lower fluoride ion concentration. The hydrates $ZrF_4,3H_2O$ and $HfF_4,3H_2O$, which are also obtained from aqeuous hydrofluoric acid, are also discussed in this section.

There is little that can be said about those complex fluorides whose structures are not known, since the stoicheiometries by themselves provide little information about the nature of the compounds. Before starting a more detailed treatment, this point is illustrated briefly by the following examples. In the complex fluorides of titanium, the metal atom appears to be invariably six coordinate with a regular octahedral structure, even those with such unlikely stoicheiometries as Na_3HTiF_8. The reverse is true for the salts of the zirconium and hafnium complex fluorides. A few salts such as Rb_2ZrF_6 contain a six coordinate metal atom, whereas the zirconium atom is seven coordinate in $K_2[Co(H_2O)_6][ZrF_6]_2$ and eight coordinate in K_2ZrF_6, which contain dimeric and infinite polymeric units respectively, with edge-sharing of the coordination polyhedra. The salt $Na_5Zr_2F_{13}$ contains a dimeric anion with a single bridging fluorine atom in which the metal atoms are seven coordinate, whereas monomeric seven and eight coordinate structures occur in $K_3[ZrF_7]$ and $Cu_2[ZrF_8],12H_2O$.

(a) *Titanium.* Aqueous hydrofluoric acid solution of TiF_6^{2-} can be readily obtained by dissolving TiO_2 or TiF_4, or, in the presence of an oxidizing agent such as nitric acid, titanium itself. The stable alkali metal salts can be readily obtained by the addition of the appropriate alkali metal fluoride. The hexafluorotitanate anion hydrolyses in less acid solution to $[TiF_5(H_2O)]^-$ and $[Ti_2(OH)_4F_4(H_2O)_2]$. The *cis*-$TiF_4(H_2O)_2$ is obtained from the tetrafluoride in dilute hydrofluoric acid (394).

The potassium, rubidium, caesium and ammonium compounds $M^I_2TiF_6$ obtained from aqueous solution have the rhombahedral K_2GeF_6 structure, which may be considered to be based on hexagonal close packed layers of potassium and fluoride ions, with the metal atom in octahedral holes formed by the six fluoride ions (2111, 2134). The titanium–fluorine distance in K_2TiF_6 is 1·92 Å (1248, 2134). On heating, K_2TiF_6 changes to the cubic K_2SiF_6 structure and the hexagonal K_2MnF_6 structure (2111), in which the (KF_3) layers are cubic close packed and double hexagonal close packed respectively. The easily dehydrated hydrate K_2TiF_6, H_2O is also known.

The sodium salt Na_2TiF_6 has a hexagonal structure again containing octahedrally coordinated titanium atoms, as in Na_2GeF_6 (593). Sodium also forms the interesting Na_3HTiF_8, which contains octahedral TiF_6^{2-} and linear HF_2^- as the only fluoroanions, and which is correctly written as $Na_3(HF_2)(TiF_6)$ (2365). Similarly the titanium atoms remain six coordinate in $NH_4CuTiF_7, 4H_2O$ which is better written $(NH_4)[Cu(H_2O)_4][TiF_6]F$ (with Ti-F = 1·86 Å) (636); similar salts are obtained with potassium, rubidium and caesium. Other divalent cations form the straight-forward $Ba[TiF_6]$, $[Mg(H_2O)_6][TiF_6]$, $[Zn(H_2O)_6][TiF_6]$ and $[Cu(H_2O)_4][TiF_6]$ (879, 2111).

(b) *Zirconium and hafnium.* The fluorozirconates, and presumably also the fluorohafnates, show a much greater complexity, and the anions ZrF_6^{2-}, $Zr_2F_{12}^{4-}$, $Zr_2F_{13}^{5-}$, ZrF_7^{3-}, $Zr_2F_{14}^{6-}$ and ZrF_8^{4-} have been structurally characterized. Most compounds have been formed by the addition of stoicheiometric quantities of the alkali metal fluoride to hydrofluoric acid solutions of zirconium.

The hexafluoro salts Rb_2ZrF_6 and Cs_2ZrF_6 are isomorphous with the titanium analogues (rhombahedral K_2GeF_6 structure) and simply contain octahedral ZrF_6^{2-} anions.

The analogous salts K_2ZrF_6, Tl_2ZrF_6 and $(NH_4)_2ZrF_6$ (and the hafnium analogues) have a different and more interesting structure (255, 256). The zirconium atom is eight coordinate with dodecahedral stereochemistry. The ZrF_8 polyhedra share opposite A-B edges (page 14) to form infinite strings of composition $(ZrF_6^{2-})_\infty$ (Fig. 59). The average zirconium–fluorine bond

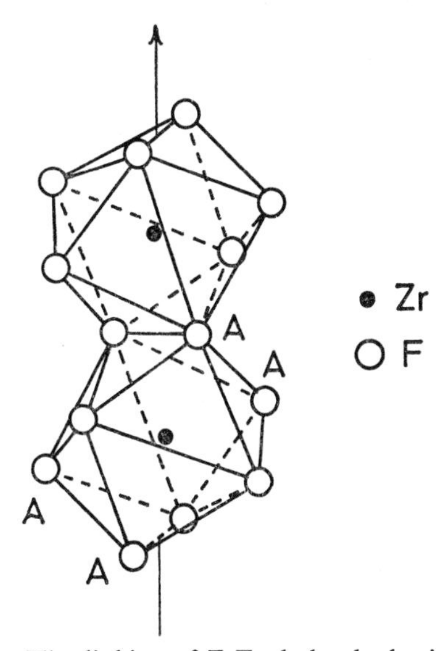

FIG. 59 The linking of ZrF_8 dodecahedra in K_2ZrF_6.

length of 2·18 Å for the eight coordinate structure is greater than for the six coordinate structure, 2·04 Å, which is as expected. The fractional crystallization of $(NH_4)_2ZrF_6$ and $(NH_4)_2HfF_6$ is one of the methods used for the separation of these elements.

Intermediate between the discrete six coordinate ZrF_6^{2-} and the infinite eight coordinate $(ZrF_6^{2-})_\infty$, is the dimeric $Zr_2F_{12}^{4-}$ anion obtained as $K_2[Cu(H_2O)_6][Zr_2F_{12}]$. The structure shows each zirconium atom to be seven coordinate, being bonded to two bridging fluorine atoms

(Zr-F = 2·16 Å) and five non-bridging fluorine atoms (Zr-F = 2·03 Å) (881).

The heptafluoro complexes K_3ZrF_7 and $(NH_4)_3ZrF_7$ contain seven coordinate ZrF_7^{3-} anions, but the exact stereochemistry is uncertain (1127, 2430).

The metal atom is also seven coordinate in the dimeric $Na_5Zr_2F_{13}$ (1175). The single bridging fluorine atom occupies the unique position in the capped trigonal prismatic stereochemistry (page 22) about each metal atom (Fig. 60). The average zirconium–fluorine distance of 2·06 Å is

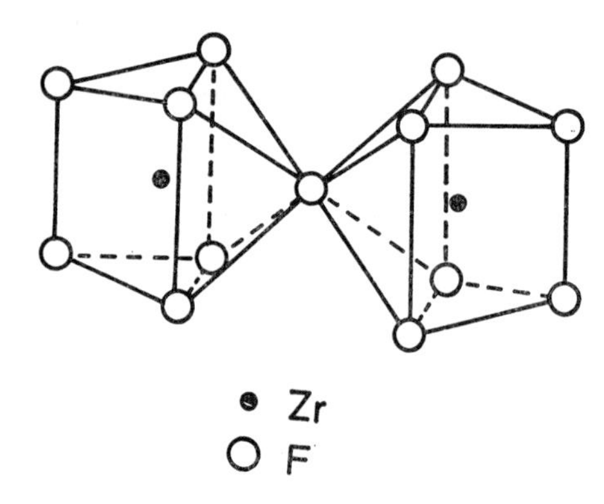

FIG. 60 The structure of $(Zr_2F_{13})^{5-}$ in $Na_5Zr_2F_{13}$.

between that for the six and eight coordinate structures.

A dimeric eight coordinate structure is found in $Cu_3[Zr_2F_{14}],16H_2O$ (880). Each zirconium atom is bonded to two bridging fluorine atoms (Zr-F = 2·18 Å) and six terminal fluorine atoms (Zr-F = 2·07 Å) in the form of a square antiprism (Fig. 61).

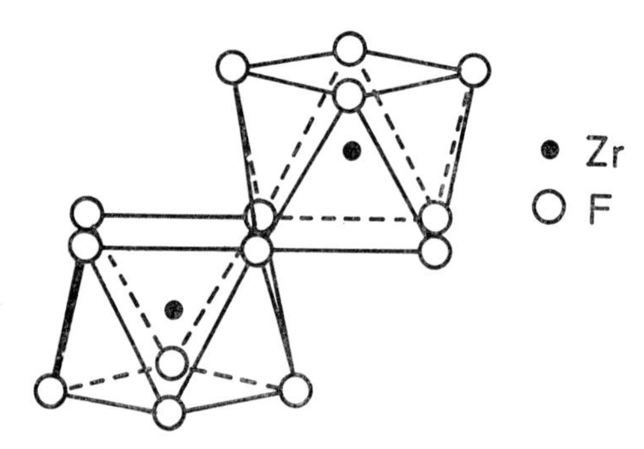

FIG. 61 The structure of $(Zr_2F_{14})^{6-}$ in $Cu_3[Zr_2F_{14}],16H_2O$.

In addition to these dimeric copper salts $K_2[Cu(H_2O)_6][Zr_2F_{12}]$ and $Cu_3[Zr_2F_{14}],16H_2O$, the eight coordinate monomeric $Cu_2[ZrF_8],12H_2O$ can also be obtained, with zirconium–fluorine bond lengths of 2·08 Å (878). A number of salts of composition $M_2^{II}ZrF_8$ are obtained with bivalent metal cations, while lithium fluoride similarly forms Li_4ZrF_8.

An investigation of the ternary system LiF-BeF_2-ZrF_4 produced the compound $Li_6[BeF_4][ZrF_8]$. The structure (2087) shows that the ZrF_8^{4-} ion is eight coordinate with dodecahedral stereochemistry (page 13), with Zr-$F_A = 2{\cdot}16$ Å and Zr-$F_B = 2{\cdot}05$ Å. However, the dodecahedron is severely distorted; $\theta_A = 43{\cdot}0°$ and $\theta_B = 65{\cdot}5°$, compared with normal values of about 37° and 72° respectively (page 15). This distortion is apparently due to the lithium cations which lie outside all twelve A-B edges in the ionic lattice, so that the A-A and B-B edges are unusually long, leading to a large value for θ_A and a small value for θ_B.

In the binary system CsF-ZrF_4, the compounds $CsZrF_5$, Cs_2ZrF_6 and Cs_3ZrF_7 have been observed (1930). The first two can appear in different polymorphic modifications, while the last shows variable composition.

The crystalline trihydrates $ZrF_4,3H_2O$ and $HfF_4,3H_2O$ are obtained by evaporation from aqeuous 5–20 M hydrofluoric acid. Most unexpectedly, the structure of the zirconium compound is different to that of the hafnium compound. The zirconium compound is dimeric, the two zirzirconium atoms being linked by two fluorine atoms so that the structure can be best represented by the formula

$$(H_2O)_3F_3Zr\langle{}^{F}_{F}\rangle ZrF_3(H_2O)_3$$

(2345). The two bridging fluorine atoms occupy one of the "A-B" edges

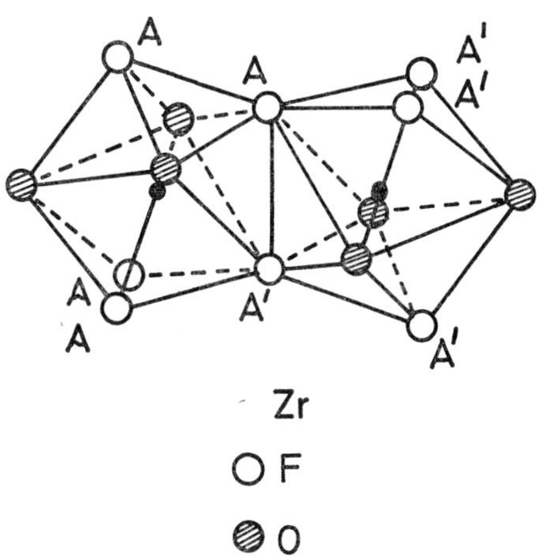

FIG. 62 The structure of $ZrF_4,3H_2O$.

of a dodecahedron (page 14). The bridging atom which is in the "A" position of the first dodecahedron is in the "B" position of the second dodecahedron, and vice versa (Fig. 62), with Zr-F_A(bridging) = 2·37 Å and Zr-F_B(bridging) = 2·20 Å. The three remaining fluorine atoms occupy the three remaining "A" sites of the dodecahedron (Zr-F_A(non-bridging) = 1·97 Å), while the water molecules are grouped in the remaining three "B" sites (Zr-O = 2·22 Å). The monohydrate ZrF_4,H_2O is readily formed at 100° C. The metal atom in $HfF_4,3H_2O$ is again eight coordinate, but in this case the square antiprisms are linked together by two edges into infinite polymeric strings of composition $HfF_4,2H_2O$ (Fig. 63) (1125). It is then simply coincidence that the presence of an extra molecule of water of crystallization leads to the same composition as the zirconium analogue.

The phase diagrams of the ZrF_4-AF-H_2O and HfF_4-AF-H_2O systems (where A is Na, K, Rb, or Cs), show in addition to $A_3M^{IV}F_7$, $A_5M_2^{IV}F_{13}$ and $A_2M^{IV}F_6$ described above, the aquo complex AMF_5, H_2O (2261). The controlled alkaline hydrolysis of $K[ZrF_5(H_2O)]$ produces

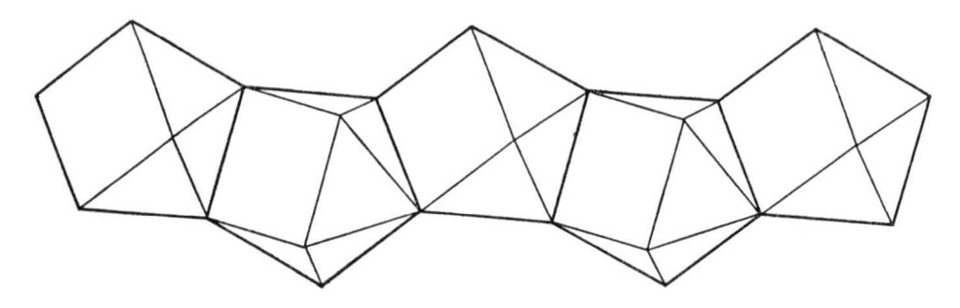

FIG. 63 The structure of $HfF_4,3H_2O$ showing edge sharing of square antiprisms to form infinite strings of composition $HfF_4(H_2O)_2$.

$K[Zr(OH)_2F_3(H_2O)]$ (1440) or $K[ZrOF_3(H_2O)_2]$ (2118). Heating this compound initially forms $K[Zr(OH)_2F_3]$ or $K[ZrOF_3(H_2O)]$, and finally $K[ZrOF_3]$. The acid hydrolysis of $K[ZrF_5(H_2O)]$ was found to form $K_{1\frac{1}{2}}H_{1\frac{1}{2}}[Zr_2OF_8]$. The ammonium salt behaved in a related manner.

2. *Chloro, bromo, iodo and oxochloro complexes*

In contrast to the variety of complex fluorides which have been observed, the chemistry of the other haloanions is much more simple. The best characterized compounds are those containing the octahedral anions $TiCl_6^{2-}$, $TiBr_6^{2-}$ and $ZrCl_6^{2-}$. The haloanions become progressively more difficult to prepare in the series fluoride, chloride, bromide and iodide. Although TiI_6^{2-} is known, it is unstable with respect to decomposition into titanium(III) and iodine (1973).

(a) *Titanium.* Compounds containing the yellow $TiCl_6^{2-}$ are formed by reacting titanium tetrachloride with aqueous or ethanolic hydrogen chloride and isolating as $A^I_2TiCl_6$ (where A^I is ammonium, substituted ammonium, pyridinium or quinolinium) (924). The compounds are stable to air while moist with hydrochloric acid, which however cannot be removed under vacuum without decomposition of the complexes since constant boiling point hydrochloric acid is insufficiently acidic to prevent hydrolysis. The final drying is best carried out by washing with a 5% solution of thionyl chloride in diethyl ether, followed by diethyl ether itself.

Alternatively the reaction between alkali metal chlorides and titanium tetrachloride to form $A^I_2TiCl_6$ has been carried out in fused antimony trichloride (1080).

The deep red salt containing the analogous $TiBr_6^{2-}$ are likewise prepared from aqueous hydrobromic acid (924) (for the ammonium or pyridinium salts) or fused antimony tribromide (1081) (for the alkali metal salts). These bromo complexes are more easily hydrolysed than the chloro or fluoro complexes. The alkali metal salts are isomorphous with K_2PtCl_6, showing the presence of discrete octahedral anions.

The mixed halogen anions *cis*-$[TiBr_2Cl_4]^{2-}$, *cis*-$[TiBr_4Cl_2]^{2-}$ and *cis*-$[TiCl_4I_2]^{2-}$ have also been obtained by reaction of the appropriate tetrahalide with a different halide anion (397, 508).

It has been noted above that these ions are only stable with respect to hydrolysis to hydroxo and/or oxo complex halides in the presence of concentrated hydrohalic acids (12, 629). The reaction between $TiOCl_2$ or $Ti_2OCl_6,4MeCN$ and univalent chlorides in acetonitrile or dichloromethane yields $A^I_2TiOCl_4$ or $A^I_3TiOCl_5$ depending upon the cation (826, 829, 919). The titanium-oxygen stretching frequency for the tetrachloro complexes depends upon the cation; the tetraethylammonium and tetraphenylarsonium salts absorb at 960–970 cm^{-1} indicating a terminal Ti=0 structure, while rubidium and the alkyl substituted ammonium salts absorb below 900 cm^{-1} indicating a polymeric Ti-O-Ti structure. $(Me_4N)_3TiOCl_5$ absorbs at 960–970 cm^{-1} indicating nonbridging oxygen atoms as expected.

(b) *Zirconium and hafnium.* Salts containing the colourless $ZrCl_6^{2-}$ anion are prepared from aqueous or ethanolic solutions saturated with hydrogen chloride in a similar manner to the titanium complexes (688, 2288). The binary systems NaCl–$ZrCl_4$, KCl–$ZrCl_4$ and CsCl–$ZrCl_4$, but not LiCl–$ZrCl_4$, show only the formation of the 1 : 2 compounds $A^I_2ZrCl_6$ (1567, 2168). The $ZrCl_6^{2-}$ anion in Cs_2ZrCl_6 is octahedral with zirconium-chlorine bond lengths of 2·45 Å. The extraction of zirconium(IV) and hafnium(IV) from hydrochloric acid with amines dissolved in organic solvents has been suggested as a feasible procedure for the separation of zirconium from hafnium. For example with tribenzylamine in chloroform, the zirconium is extracted as $(Bz_3NH)_2(ZrCl_6)$ more readily than the hafnium analogue (1005). The visible spectra (334) and vibrational spectra (335) of these hexahalocomplexes has been assigned.

The zirconium and hafnium complex chlorides are more stable to hydrolysis than the titanium complex chlorides. In a similar manner to the titanium complex above, the reaction between Zr_2OCl_6,4MeCN and Et_4NCl in acetonitrile forms $(Et_4N)_2[Zr_2OCl_{10}]$ (832).

In addition to hydrolysis by water, these compounds react in the same manner with liquid ammonia. Ammonolysis of one Zr–Cl, two Ti–Cl and two Ti–Br bonds occurs at —33° C (688, 689, 923).

F. Alkoxides

1. *Introduction*

Titanium tetrachloride reacts vigorously with anhydrous alcohols, ROH, forming compounds of the general type $TiCl_{4-x}(OR)_x$ with the evolution of hydrogen chloride. Completion of the reaction is ensured by the addition of a suitable base. The tetraalkoxides $Ti(OR)_4$ are often referred to as titanium esters, that is, as derivatives of the hypothetical titanic acid $Ti(OH)_4$.

These compounds are polymeric with bridging alkoxy groups Ti(μ-OR)₂Ti (diagram: two Ti atoms bridged by two O–R groups, each O donating to the second Ti),

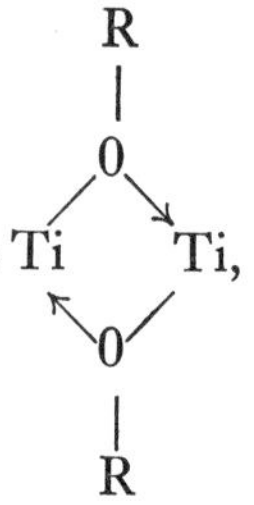

although in spite of a great deal of experimental work, there is still doubt about the exact degree of molecular complexity in many cases. The titanium is octahedrally coordinated by oxygen atoms in $Ti(OEt)_4$ and $Ti(OMe)(OEt)_3$, which are tetrameric molecules in the solid state, compared with the monomeric, tetrahedral structure of $TiCl_4$. The $TiCl_2(OPh)_2$ is intermediate, the molecule being dimeric and the titanium atom five coordinate.

These alkoxides have found uses in heat resistant paint manufacture, for example paints containing $Ti(OBu)_4$ can withstand temperatures up to 650° C. The resistance to water at these temperatures is improved if the alkoxide is partially hydrolysed before use, for example by passing moist air through the hot alkoxide forming linear and crosslinked polymers of the general type:

$$2\,\text{-}\overset{|}{\underset{|}{\text{Ti}}}\text{-OR} + H_2O \rightarrow \text{-}\overset{|}{\underset{|}{\text{Ti}}}\text{-O-}\overset{|}{\underset{|}{\text{Ti}}}\text{-} + 2ROH$$

The alkoxide ligand is more electronegative than chloride, and the properties of the alkoxides are considerably different to those of the chlorides. For example they are considerably more stable to hydrolysis, and the formation of adducts $Ti(OR)_4(Ligand)_x$ and complex anions $[Ti(OR)_{4+x}]^{x-}$ is less common than for the halides.

The stability towards hydrolysis is further enhanced if polyhydroxy compounds such as glycols are used; many such compounds appear stable even in aqueous solution. The titanium alkoxides are capable of cross linking polyhydroxy compounds such as vegetable oils producing gels and resins for which numerous uses have been suggested.

The literature about the preparation, properties and uses of these titanium alkoxides is vast, and has been reviewed (280, 1001). A more recent book (822) is particularly useful as the enormous patent literature in this field is critically examined, and also includes a full tabulation of the physical properties of these compounds.

Only the more essential points will be referred to in the following brief survey, which expands the topics mentioned above, keeping them in approximately the same order.

The reactions of the other titanium tetrahalides, and of the zirconium and hafnium tetrahalides appear analogous, although they have been studied in less detail.

2. *Preparation*

The strongly exothermic reaction between aliphatic alcohols and titanium tetrachloride yields products of the type $TiCl_3(OR)$ (if excess halide is used), $TiCl_2(OR)_2$ and $TiCl(OR)_3$. In many cases additional molecules of the alchohol used are present, as in $TiCl_2(OEt)_2(EtOH)$. Nearly all compounds prepared have an integral number of halogen atoms (and alkoxy groups) per metal atom, although since these compounds are polymeric there appears to be no obvious reason for this observation.

The yield decreases with the higher alcohols because the hydrogen chloride formed during the reaction reacts with the excess alcohol forming the alkyl chloride and water, or alternatively the corresponding olefine and water. The water then decomposes the chloro-alkoxide. These undesirable side reactions can be diminished by the addition of base to remove the hydrogen chloride as it is generated.

With many alcohols, for example ethanol, only two of the chlorides are replaced by the ethoxy groups, and further replacement does not occur even after prolonged boiling in ethanol. This is in contrast to the reaction with silicon tetrachloride which readily forms $Si(OEt)_4$. The titanium tetra-alkoxides can be readily prepared however by the removal of the hydrogen chloride by blowing through an inert gas, or by addition of bases such as ammonia, amines, amides, nitriles, or the corresponding sodium alkoxide. Not all bases are equally applicable, for example the reaction of sodium ethoxide with zirconium tetrachloride in ethanol forms $NaZr_2(OEt)_9$ (see later).

The reaction between the metal halide and the alcohol is at least partially reversible. Thus titanium tetraethoxide reacts with hydrogen chloride, chlorine, or titanium tetrachloride forming $TiCl_2(OEt)_2(EtOH)$, $TiCl_2(OEt)_2$ or $TiCl_3(OEt)$ depending on conditions. Similarly bromine or titanium tetrabromide form $TiBr_2(OEt)_2$ or $TiBr_3(OEt)$. If the chlorination is carried out with acetyl chloride, esters are formed which may form adducts with the product, for example $TiCl_3(OEt)(AcOEt)$ and $TiCl_4(AcOEt)$. In the same way $Zr(OPr)_4$ reacts with hydrogen bromide in benzene to form $ZrBr_3(OPr)$, and with acetyl bromide in the cold to form $ZrBr_4(AcOPr)_2$.

The titanium, zirconium and hafnium alkoxides may be conveniently prepared by alcohol exchange:

$$Ti(OEt)_4 + 4\,ROH \rightarrow Ti(OR)_4 + 4\,EtOH$$
$$ZrBr_2(OPr)_2 + 2\,BuOH \rightarrow ZrBr_2(OBu)_2 + 2\,PrOH$$

This method is particularly suitable for the large aliphatic alkoxy groups. In many cases mixed alkoxides are formed by this method, for example zirconium tetramethoxide and butanol form $Zr(OMe)(OBu)_3$, but this disproportionates on heating to $Zr(OBu)_4$ plus $Zr(OMe)_3(OBu)$. Other mixed alkoxides, for example $Zr(OEt)_2(OBu)_2$ can be distilled unchanged. Exchange will also occur with the enol form of aliphatic aldehydes. For example:

$$Ti(OPr)_4 + 4\,Ph.CH_2.CHO \rightarrow Ti(Ph.CH = CH.O)_4 + 4\,PrOH$$
$$Ti(OPr)_4 + 2\,CH_3.CHO \rightarrow Ti(OPr)_2(CH_2 = CH.O)_2$$

This is in contrast to the aluminium alkoxides which catalyse the reaction:

$$R.CHO \rightarrow R.COO.CH_2.R$$

Mixed alkoxides have also been prepared by the reaction of the tetra-halide with a mixture of alcohols, for example titanium tetrachloride with propanol-butanol mixtures forms $Ti(OPr)_3(OBu)$ or $Ti(OPr)(OBu)_3$ depending on conditions, while $Ti(OEt)(OAm)_3$ can be formed from $TiCl(OAm)_3$ and ethanol.

Exchange also readily occurs with aliphatic esters. Thus $Ti(OPr)_2(OBu)_2$ and $Ti(OBu)_4$ are formed from $Ti(OPr)_4$ and butylacetate (1674).

3. *Molecular weight and structure*

It has always been recognized that these alkoxides are polymeric with bridging alkoxy groups, but there has been a good deal of conflict in the literature concerning the exact degree of molecular complexity of the species.

There now appears little doubt that in benzene solution, $Ti(OEt)_4$, $Ti(OPr^n)_4$ and $Ti(OBu^n)_4$ are trimeric, for which three structures have been suggested. The first (440) involves octahedrally coordinated metal atoms joined by face sharing of the octahedra. The second (1653) involves the titanium atoms in trigonal prismatic coordination, the trigonal prisms sharing edges to form a cyclic structure. The third (1653) is again a cyclic structure, this time formed by the corner sharing of three trigonal bipyramids. A fourth alternative is a linear polymer containing both five and six coordinated titanium atoms.

A determination of the structure of solid $Ti(OEt)_4$ showed it to be *tetrameric* (Fig. 64) (1246), with each titanium atom octahedrally surrounded by oxygen atoms. The titanium atoms are displaced from the centres of their octahedra away from each other, and the ethyl groups, which are not shown in Fig. 64, spread out from the oxygen atoms leading to a molecule that is approximately cylindrical in shape.

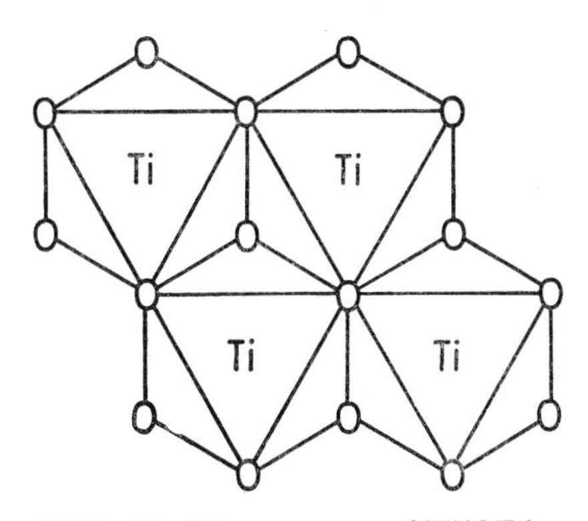

FIG. 64 The structure of $Ti(OEt)_4$.

Solid $Ti(OMe)(OEt)_3$ (2406) is also tetrameric with a similar structure to $Ti(OEt)_4$.

Titanium tetramethoxide exists in two modifications (1329). Form B is obtained as large crystals from toluene, and the structure is the same as the tetraethoxide above. The other isomer, Form A, is insoluble in most organic solvents and has a different infrared spectrum to form B. Form A has bands at 1085 and 1040 cm^{-1} which are assigned to methoxy groups attached to one and two titanium atoms respectively, while form B has an additional band at 995 cm^{-1} which is assigned to methoxy groups attached to three titanium atoms. This suggests that form A has an infinite linear or cyclic structure, compared with the compact structure known for form B.

The structural determination of the solids has not completely resolved the molecular weights in solution, which have been further pursued. A reinvestigation (1655) again indicated that the degree of polymerization in benzene was dependent on the concentration at low concentration, and the data appeared to indicate a trimeric ceiling at higher concentrations. Finally (290) it was shown that the decrease in complexity at low concentrations, and also the decrease in molecular weight with time observed by a number of workers, was due to hydrolysis reactions from imperfectly dried benzene. However, even using improved techniques, it was found that the molecular complexity over the range 2–100 $\times$ 10^{-3} M was constant at 2·82.

Bearing in mind the experimental difficulties and the resultant uncertainty about the absolute meaning of the results, it is still worthwhile to look at the molecular weights of related compounds. The following trends are readily discernible and there is little doubt about their authenticity:

1. The molecular complexity increases as the tetrachloride is progressively converted to the tetraalkoxide. For example (1654):

$TiCl_4$	monomer
$TiCl_3(OBu)$	monomer
$TiCl_2(OBu)_2$	dimer
$TiCl(OBu)_3$	trimer
$Ti(OBu)_4$	trimer (approximately)

However, the situation may be reversed with the fluoroalkoxides, where the molecular complexity appears to decrease as we go from the tetrafluoride to the tetraethoxide (1507):

$TiF_3(OEt)$	hexamer
$TiF_2(OEt)_2$	tetramer
$TiF(OEt)_3$	tetramer
$Ti(OEt)_4$	trimer

The greater bridging capacity of fluoride is also supported by infrared evidence. A band in the infrared spectrum at 1045 cm^{-1} in $Ti(OEt)_4$,

$TiF(OEt)_3$ and $TiF_2(OEt)_2$, is attributed to bridging ethoxy groups, and is found to be concentration dependent and to disappear in dilute solutions. However, it is not present in $TiF_3(OEt)$, which is therefore probably fluorine bridged only.

2. The degree of polymerization is greater for the zirconium (and hafnium) alkoxides than for the corresponding titanium alkoxides, as indicated in Tables 15 and 16.

TABLE 15

Molecular complexity (determined ebullioscopically in benzene) and boiling points (0·01–0·2 mm Hg) of titanium and zirconium amyloxides, $M(OR)_4$

	$Ti(OC_5H_{11})_4$		$Zr(OC_5H_{11})_4$	
R	Molecular complexity	B.P. °C	Molecular complexity	B.P. °C
$CH_3.CH_2.CH_2.CH_2.CH_2$	1·4	175	3·2	256
$(CH_3)_2CH.CH_2.CH_2$	1·2	148	3·3	247
$(C_2H_5)(CH_3)CH.CH_2$	1·1	154	3·7	238
$(CH_3)_3C.CH_2$	1·3	105	2·4	188
$(C_2H_5)_2CH$	1·0	112	2·0	178
$(C_3H_7)(CH_3)CH$	1·0	135	2·0	175
$(CH_3)_2CH.CH(CH_3)$	1·0	131	2·0	156
$(CH_3)_2(C_2H_5)C$	1·0	98	1·0	95

3. The degree of polymerization decreases as the branching of the aliphatic chain increases (Table 15). Similar results are observed for the amyloxides of, for example, aluminium, iron, cerium, thorium, niobium and tantalum.

4. The degree of polymerization decreases as the molecular weight increases (Table 16). This is also related to point 3 above, but gives rise to the unusual observation that, for a given metal, the boiling point of the alkoxide *decreases* as the formula weight increases.

5. The degree of polymerization determined ebullioscopically decreases with increasing boiling point of the solvent, and decreases with increasing donor power of the solvent.

4. *Adducts*

In contrast to the tetrahalides, the alkoxides do not readily form adducts with typical donor molecules, partly because the coordination number of the metal in the alkoxide is already five or six. The alkoxides however, readily add additional alkoxide groups to form complex alkoxide anions as indicated later.

TABLE 16

Molecular complexity (determined ebullioscopically in benzene) and boiling points (0·01–0·2 mm Hg) of titanium, zirconium and thorium alkoxides

	$Ti(OR)_4$		$Zr(OR)_4$		$Th(OR)_4$	
R	Molecular complexity	B.P. °C	Molecular complexity	B.P. °C	Molecular complexity	B.P. °C
CH_3, H, H >C	2·4	102	3·4	180	—	involatile
CH_3, CH_3, H >C	1·4	49	3·0	160	3·8	210
CH_3, CH_3, CH_3 >C	1·0	52	1·0	50	3·4	160

Most of the adducts which have been isolated are with oxygen donor ligands. As noted above adducts containing the parent alcohol are sometimes obtained during the preparation of the alkoxides, examples including $Ti(OBu)_4(BuOH)$, $Zr(OPr)_4(PrOH)$ and $ZrCl_3(OEt)(EtOH)$. Many of these adducts are thermally stable, for example the addition compound $TiCl_3(OEt)(AcOEt)$ obtained by treating titanium tetraethoxide with acetyl chloride may be distilled unchanged. Zirconium tetraethoxide similarly forms $ZrCl(OEt)_3$, $ZrCl_2(OEt)_2(AcOEt)$, $ZrCl_3(OEt)(AcOEt)$ and $ZrCl_4(AcOEt)$ (287).

The formation of adducts is enhanced if electron attracting groups are substituted within the alkoxide group. For example when zirconium isopropoxide was treated with anhydrous chloral, the product $Zr(OCH_2.CCl_3)_4(Me_2CO)_2$ contained two molecules of solvent acetone (302). The tetrapropoxide itself forms compounds with pyridine and propanol but compounds could not be isolated with diethylether, thiourea, triethylamine, ethylenediamine or dipyridyl (302).

Other complexes with nitrogen donors include $Ti(OPr)_4(EtNH_2)$ (544), $Ti(OPr)_4(H_2N.NH_2)$ and $Zr(OPr)_4(H_2N.NH_2)$ (122). Replacement of alkoxide groups to form hydrazinates containing $H_2N.NH$ occurs in solution on standing or on boiling.

The structures of $Ti(OBu)_4(BuOH)$ and $Zr(OPr)_4(PrOH)$ show dimeric molecules, with two octahedrally coordinated metal atoms sharing a common edge (1254).

5. *Condensation and hydrolysis reactions*

The alkoxides of the lower aliphatic alcohols are readily hydrolysed and are generally unstable to air, ultimately yielding hydrated titanium dioxide via a series of intermediates. The resistance to hydrolysis increases as the chain length of the aliphatic group increases.

There are a large number of suggestions in the patent literature about the use of these partially hydrolysed titanium alkoxides (822). Uses have been found in the fields of catalysis, heat resistant paints, polymers, coating materials, and waterproofing and fireproofing agents for fabrics and other materials. The tetraethoxide, tetrapropoxide and tetrabutoxide in particular are prepared commercially on the ton scale for these purposes.

In most cases the exact nature of the hydrolysed polymers is not known with certainty, and the composition and properties of the product are very dependent upon experimental conditions. For example the hydrolysis of one mole of $Ti(OBu)_4$ or $Ti(OPr)_4$ forms liquid polymers with 0·5 mole of water, gelatinous substances with 1·0 mole of water, and solids with 1·5 moles of water. Similarly the reaction of one mole of $Ti(OEt)_4$ with 0·5 mole of water in ethanol under controlled conditions yields a viscous substance of composition $Ti_3O_2(OEt)_8$. Such compounds show many similar reactions to the starting material. For example they can be readily halogenated forming $Ti_3O_2(OEt)_7Cl$ and $Ti_3O_2(OEt)_6Cl_2$, or can exchange with other alcohols forming, for example, $Ti_3O_2(OBu)_8$. Disproportionation of $Ti_3O_2(OEt)_8$ to $Ti(OEt)_4$ and polymeric $TiO(OEt)_2$ occurs at 100°C. The latter further disproportionates at 200° C to form $Ti(OEt)_4$ and solid $Ti_2O_3(OEt)_2$. The same products, $TiO(OEt)_2$ and $Ti_2O_3(OEt)_2$ can be prepared directly from the hydrolysis of $Ti(OEt)_4$ in ethanol using 1·0 and 1·75 moles of water respectively (285).

A preliminary communication (2342) of the crystal structure of one of these partially hydrolysed titanium tetraethoxides shows the presence of seven octahedrally coordinated titanium atoms, the titanium–oxygen framework being the same as the molybdenum–oxygen framework in $Mo_7O_{24}^{6-}$ (page 290). If ethyl groups are attached to the oxygen atoms bonded to one or two metal atoms, but not to those bonded to three or four metal atoms, the formula would be $Ti_7^{IV}O_4(OEt)_{20}$. The formula proposed in the preliminary communication, $Ti_7O_{24}Et_{19}$, is not consistent with titanium(IV).

The titanium alkoxides themselves slowly decompose at about 250° C to polymeric compounds ranging from viscous liquids to solids, with the formation of the corresponding alcohol and alkene. When a mixture of the alkoxide and air is blown onto a heated surface above 600° C, decomposition takes place instantly with the formation of a hard, transparent, continuous film of titanium dioxide.

6. *Anionic alkoxide complexes*

The $NaZr_2(OEt)_9$ is prepared from $Zr(OEt)_4$ and NaOEt (158). A large number of compounds containing other cations and other alkoxide groups have also been prepared. These compounds are particularly interesting in that although they contain an alkali metal they are volatile, and soluble in organic solvents in which they are monomeric (with the exception of $LiZr_2(OEt)_9$ which is dimeric). The addition compound $LiZr_2(OPr)_9(PrOH)$ may also be prepared.

The literature also contains indications of species such as $[Ti(OR)_5]^-$, $[Ti(OR)_6]^{2-}$ and $[Zr(OR)_5]^-$.

In a few cases well defined hydrolysis products may be obtained. For example $NaZr_2(OEt)_9$ forms crystalline $Na_2Zr_3O(OEt)_{12}$ (158).

The reaction of $TiCl_3(OEt)$ with tetraethylammonium chloride in acetonitrile yields $(Et_4N)[TiCl_4(OEt)(MeCN)]$, which loses the acetonitrile molecule on warming (828).

7. *Related compounds*

Many other organic molecules containing acidic hydrogen atoms will react with titanium tetrachloride with the loss of hydrogen chloride. Attention here will be restricted to polyfunctional alcohols, phenolic compounds, siloxanes, and β-diketones. Zirconium and hafnium complexes of the latter, and of chelating carboxylic acids such as oxalic acid, can also be obtained from aqueous solution, and are dealt with in Section 1.G of this chapter.

(a) *Polyfunctional alcohols.* Derivatives of polyfunctional alcohols have usually been prepared by alcohol exchange (2415):

$$Ti(OEt)_4 + R_2C(OH).CH_2.C(OH)R_2 \rightarrow Ti(OEt)_2(R_2C(O).CH_2.C(O)R_2) + 2\,EtOH$$

$$Ti(OEt)_4 + 2\,R_2C(OH).CH_2.C(OH)R_2 \rightarrow Ti(R_2C(O).CH_2.C(O)R_2)_2 + 4\,EtOH.$$

Polyfunctional alcohols such as glycerides, vegetable oils or cellulose produce highly viscous oils, gels or solids of indefinite structure which are generally insoluble in water but often soluble in organic solvents. Similar polymers can often be obtained by direct reaction between hydrated titanium oxide and the alcohol. The large patent literature of this field has been reviewed (822).

(b) *Aryloxides.* The aryloxides $TiCl_3(O\phi)$, $TiCl_2(O\phi)_2$, $TiCl(O\phi)_3$ and $Ti(O\phi)_4$ can be prepared from the tetrachloride and the appropriate phenol in an analogous manner to the alkoxides. The tetraaryloxides are easier to prepare than the tetraalkoxides, and a hydrogen chloride acceptor is not essential if the mixture is heated. Alternatively they may be prepared by alcohol exchange from the tetraethoxide and phenol (1765).

The structure of $TiCl_2(OPh)_2$ (Fig. 65) (2341) shows a dimeric molecule

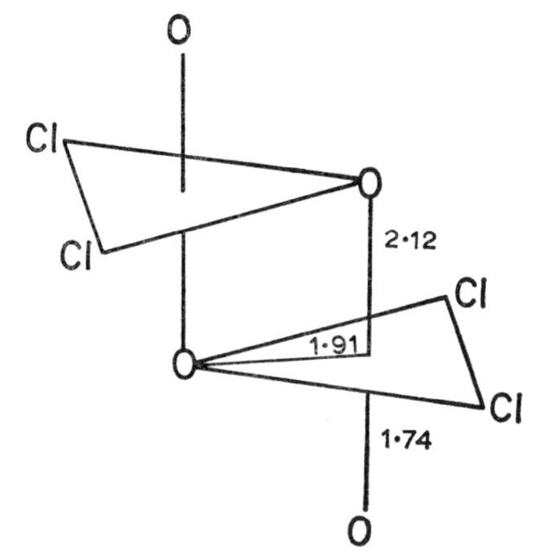

FIG. 65 The structure of $TiCl_2(OPh)_2$.

with two bridging phenoxide groups, but the titanium atom is five coordinate in contrast to the octahedral coordination found in the solid tetrameric tetraalkoxides.

The molecular weight of a number of tetraaryloxides in benzene indicates that they are monomeric (943). However, both $Ti(OPh)_4$ and $TiCl_2(OPh)_2$ readily form adducts with a variety of oxygen and nitrogen donor ligands (943, 955). The aryloxides appear more resistant to hydrolysis than the alkoxides.

The reaction between titanium tetrachloride and 8-hydroxyquinoline in chloroform produces the red $TiCl_4$(oxineH) and $TiCl_4(\text{oxineH})_2$. Heating the latter formed $TiCl_2(\text{oxine})_2$ with the evolution of hydrogen chloride. The compounds $TiCl_3$(oxine) and $TiCl(\text{oxine})_3$ were similarly prepared (944). The structure of $TiCl_2(\text{oxine})_2$ shows an octahedral stereochemistry with a *cis* arrangement of the two chloride ligands, a *cis* arrangement of the two nitrogen atoms, but a *trans* arrangement of the two oxygen atoms (2244).

(c) *Siloxanes.* One particular class of alkoxides which has attracted considerable attention is that of the titanium siloxanes, $Ti(OSiR_3)_4$. The methyl compound is a colourless liquid readily soluble in organic solvents, but in contrast to the corresponding alkoxide, is hydrolysed only slowly in water. The phenyl compound has the high melting point of 501°–

505° C, and is even more stable to water. Due to this stability towards heat and hydrolysis, there have again been numerous suggestions for the commercial use of such compounds as resins, waterproofing materials, coating compositions, etc. Once again it is not proposed to review the enormous patent literature, as excellent reviews are available (822, 2044).

The direct preparation from $TiCl_4$ and R_3SiOH in the presence of base is rather unsatisfactory due to the simultaneous formation of $R_3Si.O.SiR_3$. A more satisfactory method appears to be by alcohol exchange from a tetraalkoxide and the silanol (303, 624, 780, 2442).

The metal-oxygen stretching frequencies are lower than for the corresponding carbon compounds. For example $Ti(OCEt_2Me)_4$ (615 and 576 cm^{-1}) and $Ti(OSiEt_2Me)_4$ (510 cm^{-1}), and also $Zr(OCMe_3)_4$ (557 and 540 cm^{-1}) and $Zr(OSiMe_3)_4$ (521 cm^{-1}) (150).

(d) *β-Diketones.* The direct reaction between titanium tetrachloride and acetylacetone at 0° forms $TiCl_4$(acacH) (44) which on warming loses hydrogen chloride forming $TiCl_3$(acac) and $TiCl_2(acac)_2$. The monomeric nonelectrolytes [$TiX_2(acac)_2$] are known for all four halogens (44, 596, 597, 816, 817, 1835, 1948, 2109), and are assigned the *cis* configuration on the basis of far infrared and low temperature nmr measurements.

The analogous *cis*-[$Ti(OR)_2(acac)_2$] may be prepared from the tetraalkoxides and acetylacetone, or from $TiCl_2(acac)_2$ and the appropriate alcohol (291, 294, 1835). Whereas the dibenzoylmethanate and ethylacetoacetate compounds $Ti(OR)_2$(β-diketonate)$_2$ are monomeric, $Ti(OR)_3$(β-diketonate) are partly associated in solution (1688, 1902).

These acetylacetonate complexes are readily hydrolysed in air. Oxy compounds which have been isolated include [TiOCl(acac)]$_2$, [$TiCl(acac)_2$]$_2$O, and TiO(acac)$_2$ (524). The structure (2343) of the chloroform solvate [$TiCl(acac)_2$]$_2$O,$CHCl_3$ shows titanium–bridging oxygen bond lengths of 1·70 and 1·81 Å and a Ti–O–Ti and angle of 168° suggesting considerable double bond p_π–d_π character. The chlorine atoms are *cis* with respect to the bridging oxygen atom.

G. Complexes with other Anionic Ligands

Complexes containing no hydroxo groups are in general unstable to air, and are obtained only under anhydrous conditions. Examples include $K_2[M(NCS)_6]$ (where M is Ti, Zr or Hf) (2061), the azide $TiCl_3(N_3)$, amido complexes such as [$Ti(NMe_2)_4$] and its derivatives, nitrato complexes such as [$Ti(NO_3)_4$] and [$Zr(NO_3)_6$]$^{2-}$, and the borohydride complex [$Zr(BH_4)_4$]. In [$Ti(NO_3)_4$] the nitrate group behaves as a bidentate and the metal atom is therefore eight coordinate, whereas in [$Zr(BH_4)_4$] each ligand behaves as a tridentate so that the metal atom is formally twelve coordinate.

In aqueous solution the metal atoms are in polymeric hydroxo complexes, with hydroxo groups bridging the metal atoms. The nature of the titanium hydroxo polymers has not yet been elucidated, but for zirconium and hafnium the major species are $[M_4(OH)_8(H_2O)_{16}]^{8+}$ which may be precipitated as the appropriate halide.

However, inorganic oxyanions such as sulphate, nitrate, chromate, periodate and peroxide and organic oxyanions such as carboxylate and β-diketonate displace these hydroxo groups forming complex products.

Each of these major classifications will now be considered in turn.

1. *Amido complexes*

The reaction between titanium tetrachloride (82, 929), titanium tetrabromide (930), titanium tetraiodide (930) and zirconium tetrachloride (922) with anhydrous ammonia results in the formation of amido complexes $MX_{4-x}(NH_2)_x(x = 1, 2 \text{ or } 3)$. Complete displacement of chloride also does not occur with amines such as dimethylamine (82). Some reduction to titanium(III) is also apparent in these reactions. The chlorine-bridged polymers $(R_2N)TiCl_3$ (where R is Me, Et, Pr or Ph) have also been obtained from $TiCl_4$ and $Me_3Si(NR_2)$ (384).

Ammonolysis of titanium tetrahalides at elevated temperatures yields TiNCl, TiNBr and TiNI. These have the FeOCl structure in which layers which may be represented as ClTiNNTiCl are packed so that the chlorine atoms of adjacent layers are close packed (1326). The dark yellow TiNCl is also prepared in the decomposition of the explosive azido compound $TiCl_3N_3$ with solvents such as carbon tetrachloride or titanium tetrachloride (646). Zirconium tetrachloride and tetrabromide similarly form α–ZrNCl and α–ZrNBr with the FeOCl structure, but which can also exist in high temperature β-modifications which possess "random sequences of XZrNNZrX layers" (1325).

Titanium tetrachloride is reduced to titanium(III) by hydrazine in isooctane; this is in contrast to the behaviour in aqueous acid solutions, where titanium(III) reduces the hydrazinium ion to the ammonium ion. Reduction of the tetrachloride does not occur with substituted hydrazines; phenylhydrazine forms $TiCl_3(NH.NHPh)$, whereas dimethylhydrazine forms $TiCl_2(NH.NMe_2)_2$ (1783).

Complete displacement of chloride from titanium tetrachloride can be achieved with lithium dialkylamides to form the yellow to red liquids $Ti(NR_2)_4$ (where R is Me, Et, Pr, Bu) (305). The zirconium and hafnium compounds are obtained in a similar manner. The vibrational spectrum of $Ti(NMe_2)_4$ has been studied, (385) the titanium-nitrogen stretching frequency occurring as a very strong band at 590 cm^{-1}. The corresponding $Ti(NH_2)_4$ has been obtained from $K_2[Ti(SCN)_6]$ and KNH_2 in liquid ammonia (2061). The reactions of these compounds are somewhat unusual as indicated by the following examples:

(i) Reaction with cyclopentadiene forms, for example, the monomeric (π–CP)$Ti(NMe_2)_3$ and (π–CP)$Zr(NMe_2)_3$; similar compounds are obtained with other reactants containing acidic hydrogen, for example alcohols and thiols (305, 457). Of particular interest is the reaction between (π–CP)$Ti(NMe_2)_3$ and $M(CO)_6$ (where M is Cr, Mo or W) which forms (π–CP)$Ti(NMe_2)_3M(CO)_3$ (298). Excess of the acidic Ph_3SnH behaves similarly forming $Ti(SnPh_3)_4$ and $Zr(SnPh_3)_4$ (604). Reaction with only two moles of the hydride followed by hydrolysis or methanolysis forms $[(Ph_3Sn)_2TiO]_n$ and $[(Ph_3Sn)_2Ti(OMe)_2]$ respectively. Similar reactions lead to the preparation of the nine metal atom

$$(Ph_3Sn)_3Zr\text{–}\underset{Ph}{\overset{Ph}{\overset{|}{\underset{|}{Sn}}}}\text{–}Zr(SnPh_3)_3 \text{ and the polymer } \left[-\underset{SnPh_3}{\overset{SnPh_3}{\overset{|}{\underset{|}{Zr}}}}\text{——}\underset{Ph}{\overset{Ph}{\overset{|}{\underset{|}{Sn}}}}- \right]_n.$$

(ii) Insertion reactions occur with substituted acetylenes (456):

$$Ti(NMe_2)_4 + MeOOC.C \equiv C.COOMe \rightarrow (MeO)(Me_2N)_2Ti\text{–}\overset{CONMe_2}{\overset{|}{C}} = \underset{COOMe}{\underset{|}{C}}\text{–}NMe_2$$

$$Zr(NMe_2)_4 + 2MeOOC.C \equiv C.COOMe \rightarrow (MeO)_2Zr \left[-\overset{CONMe_2}{\overset{|}{C}} = \underset{COOMe}{\underset{|}{C}}\text{–}NMe_2 \right]_2$$

The stable carbon-titanium(IV) σ bond is noteworthy, although it is necessary to mix the reagents at $-78°$ C to prevent polymerization of the acetylenedicarboxylic ester. Oxygen abstraction occurs for cinnamaldehyde to form the polymeric $TiO(NMe_2)_2$:

$$Ti(NMe_2)_4 + Ph.CH = CH.CHO \rightarrow Ph.CH = CH(NMe_2)_2 + TiO(NMe_2)_2.$$

(iii) Insertion reactions also occur with carbon disulphide forming the tetrakisdithiocarbamates (286):

$$M(NR_2)_4 + 4\,CS_2 \rightarrow M(S_2CNR_2)_4$$ (where M is Ti or Zr and R is Me, Et or Pr).

2. *Nitrato complexes*

Tetranitrato titanium(IV), $Ti(NO_3)_4$, can be prepared by the reaction of "hydrated titanium nitrate", or more conveniently titanium tetrachloride, with N_2O_5 and warming the resulting addition compounds in vacuo (841, 2055). It has also been prepared from $TiCl_4$ and chlorine nitrate, $ClNO_3$ (2056). The product is a white volatile compound with a melting point of 58° C, and can be sublimed under vacuum at room temperature. It is highly reactive, exploding or inflaming with many organic compounds. Infrared evidence (843) indicates that all nitrate groups are coordinated and bidentate, which has been confirmed by a crystal structure determination (50, 977). The eight coordinate stereochemistry is dodecahedral with each bidentate nitrate group spanning an A–B edge (page 13).

The analogous zirconium compound $Zr(NO_3)_4$ can be similarly obtained from the tetrachloride or the hydrated nitrate with N_2O_5, but hafnium shows an unexpected difference to zirconium in forming the stable adduct $Hf(NO_3)_4,N_2O_5$ (839, 842). Both compounds can be sublimed under vacuum at 100° C. The stronger electron donor N_2O_4 cannot be used for these preparations, as the excess cannot be removed without breakdown of the complex (843).

The compounds $(Me_4N)_2[M(NO_3)_6]$ (where M is Zr or Hf) have been obtained by the action of N_2O_4 on the appropriate $(Me_4N)_2[MCl_6]$ in acetonitrile (115). Corresponding attempts to prepare the titanium compounds were unsuccessful; N_2O_4 in acetonitrile formed $TiO(NO_3)_2$, whereas liquid N_2O_5 formed $Ti(NO_3)_4$ (1102).

3. *Borohydrido complexes*

Although no reaction occurs between $Al(BH_4)_3$ and ZrF_4 and HfF_4, and immediate reduction to titanium(III) occurs with titanium, the tetrakisborohydrido complexes $Zr(BH_4)_4$ and $Hf(BH_4)_4$ can be obtained from $Al(BH_4)_3$ and $NaZrF_5$ and $NaHfF_5$ (1209). These compounds are very volatile (M.P. = 29° C, B.P. = 123° and 118° C respectively,) and the infrared spectra of the gaseous molecules has been interpreted as $Zr\left(\begin{matrix} \diagup H \diagdown \\ \diagdown H \diagup \end{matrix} BH_2\right)_4$ (1284).

However, the structure at —160°C shows $Zr\left(\begin{matrix} \diagup H \diagdown \\ —H—BH \\ \diagdown H \diagup \end{matrix}\right)_4$ with tridentate borohydride groups and a twelve coordinate metal atom (232).

4. *Hydroxo complexes*

Ultracentrifugation measurements on acidic solutions of zirconium(IV) and hafnium(IV) show that the metal atoms exist in a polymerised form, approximately tetrameric (1307, 1308). This has also been confirmed by light scattering (76) and diffusion (2396) measurements, although there has been some controversy about whether the solutions are monodisperse, and about the degree of aggregation of the species.

This has been resolved to a certain extent by a crystal structure of $ZrOCl_2,8H_2O$ which can readily be obtained as well defined crystals from these solutions (514). The corresponding bromo and iodo complexes are obtained under similar conditions. The so-called "zirconyl ion, ZrO^{2+}" is in fact tetrameric with the structure shown in Fig. 66 (and without the

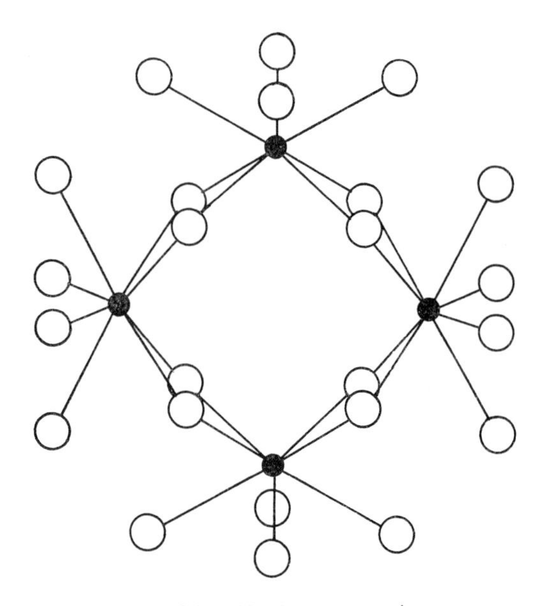

FIG. 66 The structure of $[Zr_4(OH)_8(H_2O)_{16}]^{8+}$. Each zirconium atom (black circles) is surrounded by four aquo and four bridging hydroxo groups (open circles).

water molecules in Fig. 33, page 50). The four zirconium atoms are situated at the corners of a square, and each zirconium atom is linked to each of its two neighbours by two bridging hydroxo groups, one above and one below the plane of the square. Eight coordination about each metal atom is then completed with an additional four water molecules as shown. Several points about this structure may be emphasised:

(i) The halogen atoms are not bound to the metal atoms, and the structure

is best described by the formula $[Zr_4(OH)_8(H_2O)_{16}]Cl_8,12H_2O$. The tetravalent zirconium clearly shows a preference to the more electronegative oxygen atoms of hydroxide and water.
(ii) There is no Zr=O group present. It will be seen in the following sections that this is a general phenomenon, and the so-called zirconyl compounds are invariably polymeric with hydroxo bridges, $Zr\langle^{OH}_{OH}\rangle Zr$.
Two apparent exceptions to this general rule have been claimed. The first is the isothiocyanate complexes $ZrO(NCS)_2$, MNCS, $2H_2O$ where M is K, Rb, Cs or NH_4, which show sharp absorption bands between 913 and 927 cm^{-1} suggesting the presence of Zr=O. However, this conclusion has been questioned (513), and it is possible that these bands are due to the thiocyanate ligands. The second exception is the compound $[ZrO(dipyO_2)_3](ClO_4)_2,2H_2O$ formed by the addition of dipyridyl dioxide to ethanolic solutions of $[Zr_4(OH)_8(H_2O)_{16}]Cl_8,12H_2O$. The existence of a Zr=O bond was suggested on the basis of the presence of a weak band at 964 cm^{-1} (1616).
(iii) In addition to the crystal structure determination, the low angle X-ray scattering curves from concentrated zirconium and hafnium solutions has been interpreted in terms of the $[M_4(OH)_8(H_2O)_{16}]^{8+}$ ions (1735).

Dehydration of $[Zr_4(OH)_8(H_2O)_{16}]Cl_8,12H_2O$ shows in addition to dehydration to $[Zr_4(OH)_8(H_2O)_{16}]Cl_8$, further loss of 2, 6 and 8 water molecules (1035, 1130).

Study of the $CsCl$–$ZrOCl_2$–H_2O system shows the formation of $CsCl, 2ZrOCl_2,18H_2O$ which loses water above 60° C to form $CsCl,2ZrOCl_2,9H_2O$, suggesting loss of some of the bridging hydroxyl groups, or replacement by chloride (201).

Although there is no doubt that at least in aqueous hydrochloric or perchloric acids between 0·5 M and 2·0 M the predominant species is $[Zr_4(OH)_8(H_2O)_{16}]^{8+}$, more complex reactions occur outside these limits. Depolymerization starts to occur above 2 M acid concentration, and above 6 M there are significant amounts of neutral and anionic zirconium complexes, but even in concentrated hydrochloric acid the solution still appears to contain both anionic and cationic species (1034, 1307, 1566, 1748, 1763). The exact nature of these species is not known.

Below 0·5 M acid, the reactions are very slow and the solutions take days, or require heating, to reach equilibrium. Although the nature of the species present is not completely certain, they are clearly of large molecular weight and contain of the order of 30 to 40 metal atoms (1307, 1566). Addition of further base precipitates gelatinous, amorphous precipitates of "zirconium hydroxide" and "hafnium hydroxide", which however retain significant amounts of anion (1034, 1497). The nature of these products has been reviewed (513).

When $[Zr_4(OH)_8(H_2O)_{16}]Cl_8,12H_2O$ is dissolved in methanol, a chloride deficient phase of approximate composition $ZrCl(OH)_3(H_2O)_2$ is obtained, which may consist of $Zr_4(OH)_8{}^{8+}$ units held together by further hydroxo bridges (2424).

5. *Sulphato complexes*

Zirconium sulphato complexes have been studied in some considerable detail. The bridging hydroxo groups between zirconium atoms in aqeuous solution are fairly readily replaced by sulphate, but not so easily with other *inorganic* anions. For example $Zr(SO_4)_2,4H_2O$ is free from hydroxo and oxo groups and is readily obtained from moderately dilute sulphuric acid, in contrast to $[Zr_4(OH)_8(H_2O)_{16}]Cl_8,12H_2O$ obtained from hydrochloric acid. The displacement of hydroxide is also shown by the increase in pH when anions are added to a slurry of "$Zr(OH)_4$". The relative effectiveness is oxalate > monocarboxylic acids > sulphate > nitrate = chloride (2275).

The structure of the above $Zr(SO_4)_2,4H_2O$ shows each zirconium atom linked to four water molecules and four sulphate groups, which bridge to other zirconium atoms forming an infinite chain structure with eight coordinate metal atoms (Fig. 67) (2157). It may be noted that the structure is related to, but not isomorphous with, that of $U(SO_4)_2,4H_2O$, and that both $Ce(SO_4)_2,4H_2O$ and $Pu(SO_4)_2,4H_2O$ can occur with either structure.

Dehydration of $Zr(SO_4)_2,4H_2O$, with fuming sulphuric acid or by heating, forms three different phases of composition $Zr(SO_4)_2$, H_2O, (173) and finally anhydrous $Zr(SO_4)_2$. This anhydrous sulphate also appears to exist in three different anhydrous phases, one of which (α) contains approximately 0·03 mole excess of sulphate, while the other two (β and γ) are approximately 0·03 mole deficient in sulphate (170). The structure of γ–$Zr(SO_4)_2$, H_2O, which is the form initially obtained on dehydration of the tetrahydrate, shows that it contains seven coordinate zirconium. Each zirconium atom is bonded to six oxygen atoms from six different sulphate groups, and also to the water molecule (173). Equilibration of this hydrate with 75% sulphuric acid, or exposing the anhydrous compound to water vapour, forms α–$Zr(SO_4)_2$, H_2O in which the zirconium atom is again seven coordinate. Now each zirconium atom is bonded to six oxygen atoms from only five different sulphate groups, in addition to the water molecule (173).

The rehydration of α–$Zr(SO_4)_2$ forms the *pentahydrate* α–$Zr(SO_4)_2$, $5H_2O$ (171), which is also formed when a supersaturated viscous solution of zirconium sulphate (concentrated above about 60% by weight of $Zr(SO_4)_2$), is rapidly evaporated so that the pentahydrate is obtained before the tetrahydrate can nucleate (169). However when β–$Zr(SO_4)_2$ or γ–$Zr(SO_4)_2$ are exposed to moist air, they rapidly deliquesce dissolving in their absorbed water, but on further standing β–$Zr(SO_4)_2,5H_2O$ crystallizes out (171). Both pentahydrates are only metastable and slowly transform to the tetrahydrate. The structure of α–$Zr(SO_4)_2,5H_2O$ shows that it should be written $[(H_2O)_4(SO_4)Zr(SO_4)_2Zr(SO_4)(H_2O)_4]2H_2O$ with two bidentate sulphato groups bridging the two eight coordinate metal atoms (172). Aqueous solutions at 0° deposit a heptahydrate $Zr(SO_4)_2,7H_2O$, the structure of which shows it to contain the same dimeric unit, and which should therefore be written $[Zr_2(SO_4)_4(H_2O)_8]6H_2O$ (172). Evidence was also presented

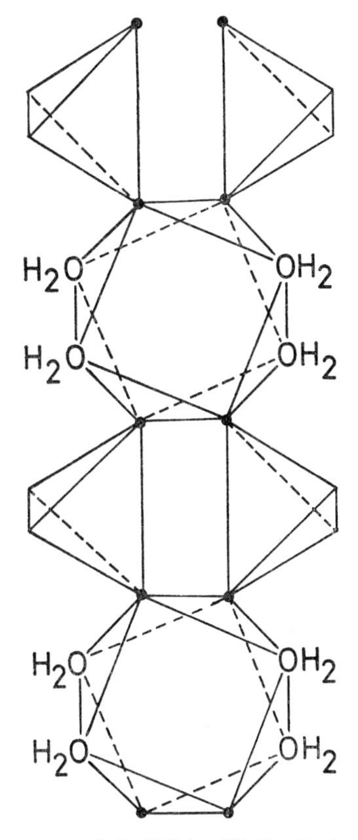

FIG. 67 The structure of $Zr(SO_4)_2,4H_2O$. Each zirconium atom is surrounded by a square antiprismatic arrangement of eight oxygen atoms from the four water molecules and from four different tetrahedral sulphate groups.

to suggest that this species was also present in concentrated solutions of zirconium sulphate (172).

When $Zr(SO_4)_2,4H_2O$ is dissolved in water and heated in a sealed tube to 100°–150° C, $Zr_2(OH)_2(SO_4)_3(H_2O)_4$ is formed. The structure (1615) shows that the zirconium atom is again eight coordinate, each metal atom being bonded to two bridging hydroxo groups, four bridging sulphato groups, and two aquo groups. The stereochemistry appears to be dodecahedral, with $\theta_A = 38{\cdot}8°$, $\theta_B = 72{\cdot}4°$, and (M–A)/(M–B) = 1·01.

Alternatively if the sealed tubes containing aqueous solutions of $Zr(SO_4)_2,4H_2O$ are heated to 175°–225° C, or 250°–300° C, two structural isomers of $Zr(OH)_2(SO_4)$ are obtained (1615). In both cases the metal atom is again eight coordinate.

Both the ZrO_2–SO_3–H_2O and HfO_2–SO_3–H_2O phase diagrams have been studied at 100° C (1726).

6. *Other inorganic anions*

A number of other inorganic oxyanions have been found to form complexes with zirconium, but they have been studied in less detail than sulphate. In general similar structural principles apply. Thus the metal has a high coordination number, usually seven as in $Zr_4(OH)_6(CrO_4)_5(H_2O)_2$ (1577), or eight as in $Zr(OH)_2(NO_3)_2(H_2O)_4$ (1614) and $K_6[(CO_3)_3Zr(OH)_2Zr(CO_3)_3]6H_2O$ (1030). Such compounds are normally obtained simply by precipitation from acid aqueous solutions, and although in many instances they have been formulated as "zirconyl compounds, ZrO^{2+}", there is again no conclusive evidence for these structural units in preference for Zr(OH)₂Zr bridged units:

$$Zr\begin{matrix} \diagup OH \diagdown \\ \diagdown OH \diagup \end{matrix}Zr.$$

In zirconium iodate, $Zr(IO_3)_4$, the zirconium atom has a square antiprismatic stereochemistry with bridging iodate groups (1500). The compound is not isomorphous with $Ce(IO_3)_4$ and $Pu(IO_3)_4$, which however also have the same stereochemistry about the metal atoms.

There are also a large number of ill-defined titanium peroxo compounds in the literature. These have been critically re-examined (1061, 1066), and the following authenticated:

In $K_2[Ti(O_2)_3(H_2O)]$, $3H_2O$ the bidentate nature of the coordinated peroxo group is shown by the single band in the infrared spectrum at 825 cm^{-1}, which has been assigned to the O–O stretching frequency. In addition to this band at 856 cm^{-1}, $TiO_3,3H_2O$ shows weak or broad bands at 1130, 1020 and 620 cm^{-1} which do not change on deuteration and are assigned as being associated with the infinite –Ti–O–Ti–O– group. The compound is therefore formulated as $[TiO(O_2)(H_2O)_3]_n$. The titanium atom is also apparently seven coordinate in $K_2[Ti(O_2)(C_2O_4)_2(H_2O)],2H_2O$ and $K_2[Ti(O_2)(SO_4)_2(H_2O)],4H_2O$. $K_3[Ti(O_2)F_5]$ and $(NH_4)_3[Ti(O_2)F_5]$ are isomorphous with $(NH_4)_3[ZrF_7]$ (page 90). The bonding of the bidentate peroxo group is discussed on page 6.

7. *Organic anions*

A number of bidentate organic ligands have been found to combine with zirconium in a 4 : 1 molar ratio, for example carboxylic acids such as oxalic acid and β-diketones such as acetylacetone. There is no clear division between this class of compounds and those discussed on page 104.

Aqueous solutions of zirconium and hafnium form 4 : 1 complexes with *excess* oxalate, from which salts such as $Na_4[Zr(C_2O_4)_4]$, $3H_2O$ can be obtained. The stereochemistry of the anion is dodecahedral (1013, 1199), and has been discussed in more detail on pages 13–16. In less acid solutions, and/or at lower oxalate concentration, oxalate deficient products with zirconium: oxalate ratios of 1 : 1 (1673) and 1 : 3 (1376) are obtained.

Titanium forms only the 1 : 3 complex $(Et_4N)_2[Ti(C_2O_4)_3]$ (634).

Strong complexes are also formed between zirconium and other polydentate chelating agents containing carboxylic groups. Thus 1 : 1 and 1 : 2 complexes are formed with nitrilotriacetic acid $N(CH_2COOH)_3$,

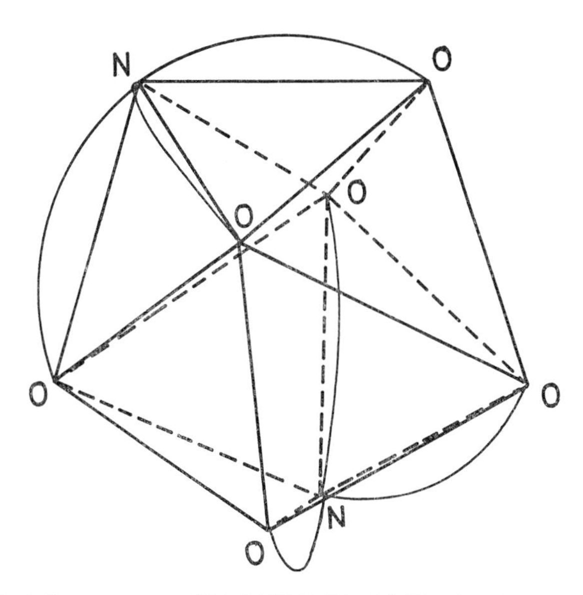

FIG. 68 The structure of $[Zr\{N(CH_2COO)_3\}_2]^{2-}$ showing arrangement of the two tetradentates around the dodecahedrally coordinated metal atom.

ethyliminodiacetic acid $EtN(CH_2COOH)_2$, and ethylenediaminetetra-acetic acid in the pH range 2–8 (1258, 1434, 1712, 1836).

The structure (1206, 1208) of the nitrilotriacetic acid complex $K_2[Zr\{N(CH_2COO)_3\}_2], H_2O$ is shown schematically in Fig. 68. The zirconium atom is eight coordinate with dodecahedral stereochemistry, and the relevant structural parameters are given on page 16. Each nitrogen atom occupies one of the A sites, with the coordinated oxygen atoms of the carboxylic residues on the adjacent A and two of the adjacent B sites. It is interesting, and important, to note that the structure is not that of a puckered hexagonal bipyramid with the nitrogen atoms in the axial positions, even although this ligand appears geometrically ideal for such a structure as full three-fold symmetry for the molecule could be retained. However the dodecahedron is probably about 10 kcal mole^{-1} more stable than the puckered hexagonal bipyramid in ligand–ligand repulsive energy.

Less stable complexes are formed with α-hydroxy carboxylic acids such as glycolic or lactic acid, or with *o*-dihydroxy aromatic molecules such as Tiron (1,2-dihydroxy-benzene 3,5-disulphonate) (513, 1258). The latter will be taken as a typical example. Tiron readily forms polymeric 1 : 1 and 1 : 2 complexes with zirconium with retention of the double hydroxo, bridges:

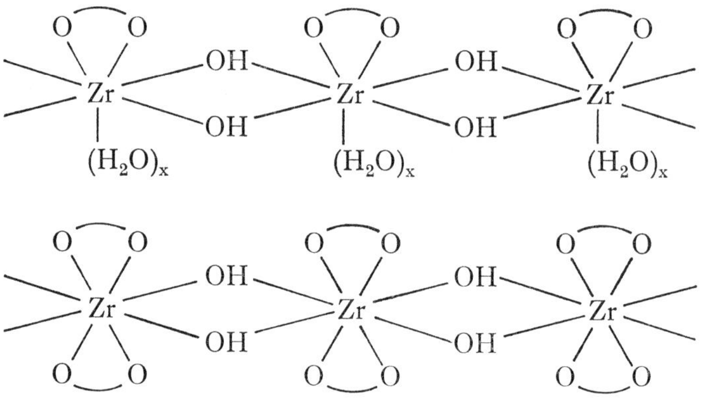

In addition a 1 : 3 complex can also be formed with retention of some hydroxo bridges, but the formation of a hydroxo-free 1 : 4 complex could not be attained.

In aqueous solution the acetylacetonate anion forms the eight co-ordinate $[Zr(acac)_4]$. The structure (2145) shows square antiprismatic stereochemistry, with the bidentate ligand spanning the opposite edges of each of the square faces. The stereochemistry is discussed more fully on page 17. The $[Hf(acac)_4]$ and β–$[Th(acac)_4]$ are isostructural, whereas a different structure, which is also based on square antiprismatic coordination, has been found for α-[Th(acac)$_4$], [Ce(acac)$_4$] and [U(acac)$_4$] (1048). The rapid ligand exchange with other β-diketones has been examined (8, 1868).

The tropolonate ion, which has a smaller "bite" than the β-diketones and is more able to stabilize high coordination numbers, forms tetrakis complexes with titanium as well as zirconium and hafnium (1734), and even a pentakis complex with thorium (1731).

It may be noted that the structure of the tetrakis aldimine complex [Ti{*o*-C_6H_4(O)(CH = NR)}$_4$] shows that only two of the ligands behave as bidentates, the remaining two being bonded through the oxygen atoms only, so that the titanium atom is only six coordinate (295).

H. Organometallic Compounds

Tetravalent Group IV organometallic compounds are generally not very stable, which may be associated with the observation that they do not attain the inert gas configuration. For example as successive methyl groups replace the halogen atoms in $TiCl_4$ the compounds become less thermally stable, until $TiMe_4$ is reached which decomposes above —78° C. The π-cyclopentadienyl compounds are significantly more stable, and have been studied in much more detail. Many compounds of the type (π–C_5H_5)$_2$M(σ-alkyl)$_2$ which contain both types of organometallic group, and which are also stable at room temperature, are discussed with the other cyclopentadienyl compounds, from which they are derived.

Some organometallic compounds derived from the dialkyl amides have been referred to earlier, examples including complexes of the types (π–C_5H_5)M(NMe_2)$_3$, M(OMe)(NMe_2)$_2$(R) and M($SnPh_3$)$_4$.

The important Ziegler–Natta catalysts, formed for example by mixing titanium tetrachloride with trialkylaluminium, are also discussed in this section, although other relevant information may be found under the trivalent Group IV organometallic compounds.

1. σ *Bonded compounds*

The reaction of phenyl lithium with titanium tetrapropoxide yields the colourless PhTi(OPr)$_3$ which is stable at room temperature (1176). The stability of the product decreases rapidly upon further progressive arylation (1177).

More recently there have been a number of successful attempts at the alkylation of titanium tetrahalides. The reaction of titanium tetrachloride with dimethylaluminium chloride or trimethyl aluminium in hexane at room temperature forms $MeTiCl_3$ as a dark violet liquid, and at —80° C forms the dark violet solid Me_2TiCl_2 (185, 186). An infrared study of $MeTiCl_3$ indicates that the carbon–titanium bond is much weaker than the carbon–tin bond in $MeSnCl_3$ (1489). The compound is more stable as its dipyridyl complex, $MeTiCl_3$(dipy) (2273). Similarly tetraethyl lead reacts with excess titanium tetrachloride at —80° C to form the violet solid $EtTiCl_3$, which melts to a red liquid at room temperature and decomposes to ethane, butane and titanium trichloride (166).

With methyl lithium in diethyl ether at —80° C $TiCl_4$ forms the unstable yellow $TiMe_4$, which decomposes above this temperature to metallic titanium (221). Benzyl magnesium chloride and $TiCl_4$ form the red monomeric $TiBz_4$ (992). Titanium tetramethyl forms a number of 1 : 1 and 1 : 2 adducts with nitrogen-donor, phosphorus-donor and oxygen-donor monodentate (1740) and bidenate (2274) ligands.

A titanium–silicon bonded compound $Ti(SiPh_3)_4$ has been obtained from $TiCl_4$ and Ph_3SiK in ether at 0° C, and in contrast to the above titanium–carbon bonded compounds, is stable to air and water (1173).

Zirconium tetramethyl is formed similarly from zirconium tetrachloride. but decomposes with loss of methane above —15° C (222).

2. *π Bonded compounds*

The reaction of titanium tetrachloride with cyclopentadienyl magnesium bromide (2391) or cyclopentadienyl sodium (2389) forms $(C_5H_5)_2TiBr_2$ and $(C_5H_5)_2TiCl_2$ respectively. The compounds $(C_5H_5)_2TiX_2$ (where X is F, Cl, Br, I or OBu) have been prepared by similar reactions from the appropriate titanium tetrahalide (or tetrabutoxide), or by ligand exchange from $(C_5H_5)_2TiCl_2$.

The similar reaction with dicyclopentadienyl magnesium (2165) yields $(C_5H_5)TiCl_3$ in addition to $(C_5H_5)_2TiCl_2$, but this is more conveniently prepared by the reaction of stoicheiometric amounts of $TiCl_4$ and $(C_5H_5)_2TiCl_2$ (1036).

These $(\pi\text{–}C_5H_5)_2TiX_2$ compounds are diamagnetic and of reasonable stability considering that the inert gas configuration has not been attained. Electron diffraction of the gaseous chloride indicates a wedge-shaped molecule, with the cyclopentadienyl rings inclined at an angle of approximately 60° to each other (26). Their chemistry has been studied in some detail. The halide can be readily replaced by NH_2^- in liquid ammonia (50), by pseudo-halide ions (NCS^-, $NCSe^-$ and OCN^-) (585), by phenoxide (71), thiophenoxide (994, 1453), or perfluorocarboxylate ions (1410). It was noted that whereas $(C_5H_5)_2Ti(SMe)_2$ is stable, the corresponding $(C_5H_5)_2Ti(OMe)_2$ could not be prepared, and it was concluded that cyclopentadienyl p_π-titanium d_π-sulphur d_π bonding through the formally empty titanium *d* orbitals, occurred in these complexes (994). Bidentate ligands behave similarly, forming, for example $[(C_5H_5)_2Ti(acac)]^+$ with acetylacetone or other β-diketones (685).

The σ bonded organometallic compounds $(\pi\text{-}C_5H_5)_2Ti(CH_3)_2$ (168, 2392), $(\pi\text{-}C_5H_5)_2Ti(CF_3)_2$ (168), $(\pi\text{-}C_5H_5)_2Ti(C_6H_5)_2$ (168, 2247), and $(\pi\text{-}C_5H_5)_2Ti(C{\equiv}C.C_6H_5)_2$ (2271), as well as the metal–metal bonded $(\pi\text{-}C_5H_5)_2Ti(SnPh_3)_2$ (589) have also been obtained.

The reduction of $(C_5H_5)_2TiCl_2$ to a titanium(III) hydride has provided some interest due to its ability to fix nitrogen. These reactions are discussed in more detail under titanium(III).

The hydrolysis products of $(C_5H_5)_2TiCl_2$ and $(C_5H_5)TiCl_3$ (and likewise the oxidation products of $(C_5H_5)_2Ti^{III}Cl$) have been studied (552, 993, 2160, 2389) and two structures determined. The cyclic tetramer $[-Ti(C_5H_5)_2-O-]_4$ has Ti–O–Ti angles of 160° and 165° and Ti–O distances of 1·79 Å (2160). The dimeric $(C_5H_5)Cl_2Ti\text{–}O\text{–}TiCl_2(C_5H_5)$ contains a linear Ti–O–Ti bond with a Ti–O bond length of 1·78 Å (552). An analogous $(C_5H_5)_2ClTi\text{–}N(Ph)\text{–}TiCl(C_5H_5)_2$ is also known (584).

Zirconium similarly forms $(\pi\text{-}C_5H_5)_2ZrCl_2$ (2389, 2391) and $(\pi\text{-}C_5H_5)ZrCl_3$ (1915, 2165), but these are very sensitive to air in contrast to the relatively stable titanium compounds. Displacement and hydrolysis reactions of $(C_5H_5)_2ZrCl_2$ form similar products to those of the titanium analogue, for example

$(C_5H_5)_2Zr(NCS)_2$ (585), $(C_5H_5)_2Zr(SPh)_2$ (1452),
$(C_5H_5)_2Zr(S_2.C_6H_4)$ (1452), $(C_5H_5)_2Zr(OOC.CF_3)_2$ (1410),
$(C_5H_5)_2Zr(BH_4)_2$ (1752), $(C_5H_5)_2ZrH_2$ (628, 1285),
$(C_5H_5)_2Zr(C_6F_5)_2$ (467), $(C_5H_5)_2ZrCl(CH_3)$ (2249),
$(C_5H_5)_2ClZr\text{–}O\text{–}ZrCl(C_5H_5)_2$, (1913), and
$(C_5H_5)_2(NCO)Zr\text{–}O\text{–}Zr(NCO)(C_5H_5)_2$ (585).

Monocyclopentadienyl derivatives include $(C_5H_5)ZrCl(acac)_2$ (2220) and the hexafluoroacetylacetonate $(C_5H_5)Zr(HFA)_3$ (767). The structure of the latter is shown schematically in Fig. 69 (766). The cyclopentadienyl ring is in a staggered configuration relative to the pentagonal plane containing five of the six oxygen atoms. The zirconium atom is displaced out of the O_5 ring towards the cyclopentadienyl ring by 0·39 Å. The sixth oxygen atom *trans* to the cyclopentadienyl ring is significantly closer to the metal atom (Zr–O = 2·17 Å) than are the other five oxygen atoms (Zr–O = 2·21–2·27 Å, average of 2·23 Å). The ^{19}F–nmr at −30° C shows a 1 : 2 : 2 : 1 pattern consistent with this structure, and it is not until 90° C that a single band is observed when the more strongly bonded out-of-plane bidentate group exchanges with the other two bidentate groups.

The structure of $(\pi\text{-}C_5H_5)M^{IV}Cl(oxine)_2$ (where M is Ti or Zr) has also

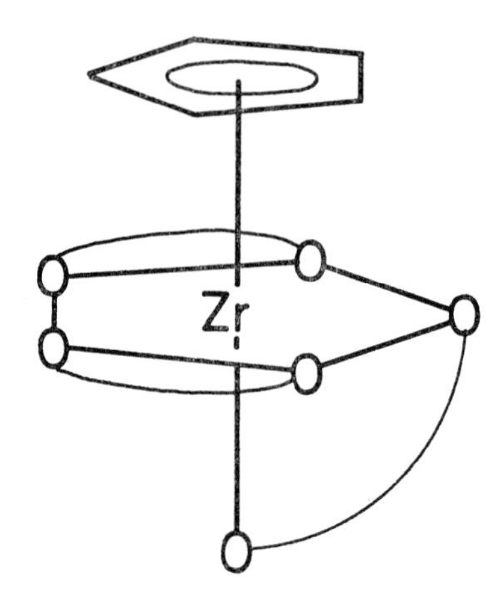

FIG. 69 The structure of $(C_5H_5)Zr(CF_3.CO.CH.CO.CF_3)_3$.

been determined; the two oxygen atoms are *trans* to each other and the nitrogen atoms are *cis* to each other (1668).

Zirconium tetrachloride and allylmagnesium bromide at $-60°$ C form $Zr(C_3H_5)_4$, which is a red solid that sublimes at room temperature (178). The nmr spectrum indicated that all allyl groups are equivalent, and π bonded. However the cyclopentadienyl substituted analogues have been formulated as $(C_5H_5)_2ZrCl(\sigma\text{–}C_3H_5)$ and $(C_5H_5)_2Zr(\sigma\text{–}C_3H_5)(\pi\text{–}C_3H_5)$ (1652).

3. *Ziegler–Natta catalysts*

In 1954 Ziegler reported that a mixture of titanium tetrachloride and triethyl aluminium behaved as a catalyst in the polymerization of olefines. Since this time there have been an enormous number of catalysts reported of this type, most of which have appeared in the patent literature, with little emphasis on the underlying structures and mechanism. An excellent summary of this field is available (822). It now appears that many combinations of Groups I-III alkyls with transition metal (usually titanium) compounds can behave as "Ziegler–Natta" catalysts, the important feature being the formation of unstable metal alkyl compounds. The importance of these Ziegler–Natta catalysts lies in the high degree of stereoregularity with which the olefine is polymerized, and this has a profound effect upon the physical properties of the polymers, which have become commercially important. This stereoregularity presumably arises from the mechanism of the reaction which is still little understood, but may involve an insertion of the olefine into the alkyl-titanium bond via a π bonded alkene-titanium complex.

The structure of $(C_5H_5)_2Ti^{III}\langle\!\begin{smallmatrix}Cl\\ \\ Cl\end{smallmatrix}\!\rangle Al^{III}(C_2H_5)_2$ formed by the reaction of $(C_5H_5)_2TiCl_2$ with aluminium triethyl shows two π bonded cyclopentadienyl groups with titanium-bridging chlorine bond lengths of 2·5 Å (1759, 1762).

An nmr study of $(C_5H_5)_2Ti(CH_3)Cl$ and aluminium trimethyl has been interpreted in terms of the following equilibrium (1912):

$$(CH_3)(C_5H_5)_2Ti^{IV}\langle\!\begin{smallmatrix}CH_3\\ \\ Cl\end{smallmatrix}\!\rangle Al^{III}\langle\!\begin{smallmatrix}CH_3\\ \\ CH_3\end{smallmatrix} \rightleftarrows (CH_3)(C_5H_5)_2Ti\text{–}Cl\text{–}Al\!\begin{smallmatrix}\diagup CH_3\\ —CH_3\\ \diagdown CH_3\end{smallmatrix}$$

The structure of $[(C_5H_5)_2TiAl(C_2H_5)_2]_2$ is particularly complex and interesting (553). This diamagnetic compound contains direct titanium–titanium (Ti–Ti = 3·11 Å) and titanium–aluminium (Ti–Al = 2·79 Å) bonding, the titanium–titanium bond intersecting a two-fold symmetry axis (Fig. 70). The structure shows that one of the carbon atoms of a

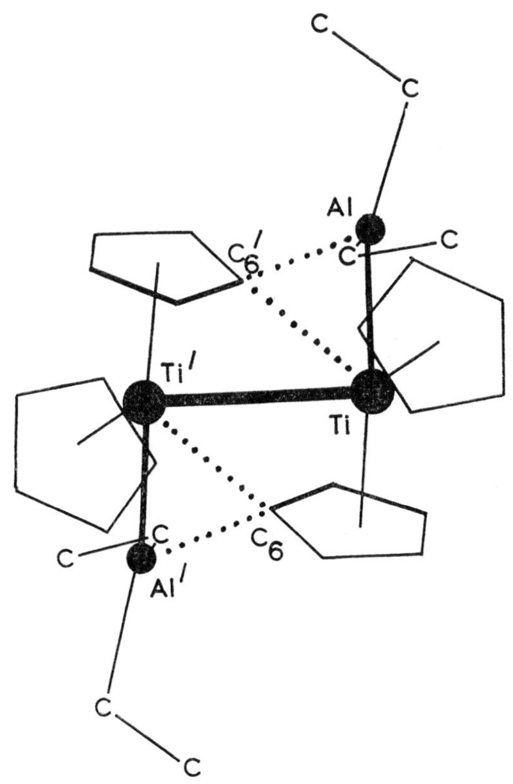

FIG. 70 The structure of $[(C_5H_5)_2TiAl(C_2H_5)_2]_2$.

cyclopentadienyl ring π bonded to one titanium atom, interacts considerably with the other titanium atom and the aluminium atom attached to it (Fig. 70). For example:

(i) $Ti'-C_6 = Ti-C_6' = 2\cdot38$ Å This is not much greater than the average of all the other titanium–carbon distances (2·35 Å).
(ii) $Al'-C_6 = Al-C_6' = 2\cdot11$ Å. This may be compared with the other aluminium–carbon distances of 1·89 and 2·11 Å.
(iii) The cyclopentadienyl plane containing C_6 is tilted 8° away from perpendicular to the titanium–cyclopentadienyl axis, in such a way that C_6 is closer to Al′ and Ti′. The other cyclopentadienyl ring is perpendicular to its titanium–cyclopentadienyl axis. It was therefore suggested that each of the three resonance forms shown in Fig. 71 contribute to the bonding.

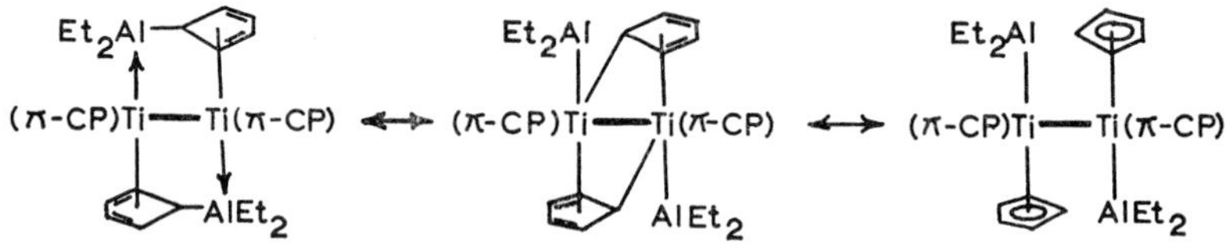

FIG. 71 Bonding in $[(C_5H_5)_2TiAl(C_2H_5)_2]_2$.

2. OXIDATION STATE III

A. Introduction

The chemistry, spectral and magnetic properties of titanium(III) compounds, which have the simple $3d^1$ electron configuration, have been studied in some detail, although the meaning of many of the results is far from clear. Before discussing the detailed chemistry of these compounds, a general note on the spectral and magnetic properties of such compounds appears relevant.

Under the influence of an octahedral field the ground term of the gaseous titanium(III) ion, ^2D, is split into a lower energy $^2T_{2g}$ and a higher energy ^2E_g. One of the historical foundations of the crystal field theory is the assignment (1253) of the weak band at 20 200 cm^{-1} (extinction coefficient of 6) in the visible spectrum of aqueous titanium(III) to the electronic transition between these two levels, that is $^2T_{2g} \rightarrow {}^2E_g$. Similar spectra have been found for titanium(III) in a number of other solvents (1144, 1317).

It is experimentally found that this absorption band contains a shoulder about 2000 cm^{-1} to lower energy, suggesting an environment of lower symmetry for the metal atom, even in those (few) cases where the existence of six equivalent ligands has been established. This effect is commonly attributed to a Jahn–Teller distortion. With a tetragonal distortion the upper ^2E_g state splits into $^2B_{1g}$ and $^2A_{1g}$ components, and the ground state into ^2E_g and $^2B_{2g}$ components (Fig. 72). The two absorptions observed are

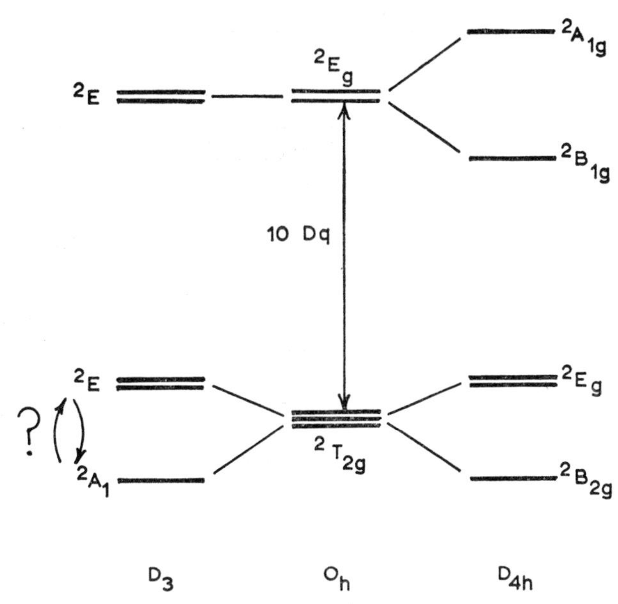

FIG. 72 The splitting of the *d* orbitals for titanium(III) under the influence of dodecahedral (O_h), trigonally distorted (D_3), and tetragonally distorted (D_{4h}) fields.

then attributed to the transitions $^2B_{2g} \rightarrow {}^2B_{1g}$ and $^2B_{2g} \rightarrow {}^2A_{1g}$. It should be noted however that the value of 10Dq is usually taken as tbe energy of the stronger band, rather than as some average of the two bands. Under the influence of a trigonal distortion, the $^2T_{2g}$ state is split into ^2E and $^2A_1$ states, but there is some dispute as to which is of the lowest energy, or whether they are nearly degenerate and both significantly populated (for example, see ref. 670).

The electronic energy levels for TiF_6^{3-} have also been calculated using a molecular orbital scheme (184). Satisfactory agreement with experiment was obtained if metal–ligand π bonding was included in the calculations.

If the aqueous solution of titanium(III) becomes slightly oxidized, the extinction coefficient increases by a factor of 10 and the solution turns dark purple due to the formation of a binuclear complex between titanium(III) and titanium(IV) (1317).

Unfortunately an examination of the visible spectrum does not clearly indicate the coordination number or stereochemistry of the titanium atom. For example the visible spectra of both $TiCl_3(Me_3N)_2$ and $K_5Ti(CN)_8$ were originally interpreted as showing octahedral titanium(III), but the former has since been shown to be five coordinate, and the latter is probably eight coordinate.

The magnetic moment of titanium(III) complexes has been studied in considerable detail by a number of workers (for example, see refs. 245, 504, 847, 1602). The magnetic moment at room temperature of about 1·7 B.M. falls only to about 1·5 B.M. at 80° K, the orbital angular momentum being largely quenched by distortion of the ligand field. These studies indicate that the separation between the orbitals derived from the $^2T_{2g}$ ground state is approximately 400–800 cm^{-1}.

It is unexpectedly found that the magnetic moments of complexes such as $TiBr_3(Dioxane)_3$ ($\mu_{eff} = 1 \cdot 45$ B.M.) and $TiBr_3(MeCN)_3$ ($\mu_{eff} = 1 \cdot 50$ B.M.) appear significantly lower than those for the corresponding chloro complexes (1·69 and 1·58 to 1·68 B.M. respectively).

Even detailed magnetic studies cannot be safely used as a criterion for stereochemistry. For example the susceptibility of the five coordinate $TiBr_3(Me_3N)_2$ can be satisfactorily explained using plausible parameters assuming an *octahedral* complex.

B. Oxide

Titanium sesquioxide has the corundum structure (α–Al_2O_3), which is based on hexagonal close packing of oxygen atoms with $\frac{2}{3}$ of the octahedral holes filled in each layer (Fig. 73). There are pairs of face-sharing octahedra parallel to the *c*-axis, while in the planes normal to the *c*-axis each octa-

TABLE 17

Hexagonal unit cell dimensions and metal–metal distances of some oxides with the corundum structure

	c(Å)	*a*(Å)	*c*/*a*	M–M(Å) (*c*-axis)	M–M(Å) (*a*-axis)
α-Al_2O_3	12·99	4·76	2·73	2·65	2·79
Ti_2O_3	13·64	5·15	2·65	2·59	2·99
V_2O_3	14·00	4·95	2·83	2·70	2·88
Cr_2O_3	13·60	4·96	2·74	2·65	2·89
α-Fe_2O_3	13·75	5·03	2·73	2·89	2·97

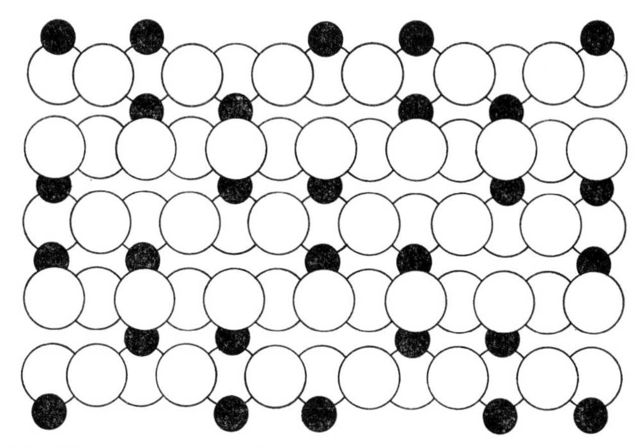

FIG. 73 The structure of Ti_2O_3, showing pairs of metal atoms normal to the close-packed oxide layers.

hedron shares edges with three other octahedra. The face-sharing of the octahedra holds the metal atoms very close together, although the octahedra are severely distorted and the metal layers puckered due to mutual repulsion of the metal atoms.

Relevant hexagonal unit cell dimensions and metal–metal distances of some oxides with this structure are shown in Table 17 (4, 1779, 2237, 2397).

As Ti_2O_3 is heated to 150°–200° C the magnetic susceptibility increases sharply (898, 1852), the material becomes a metallic conductor (4, 897, 1714, 2414), and the ratio c/a of the cell dimensions increases from 2·65 to a more normal value of 2·71 at 300° C (1778, 1852). This change in properties is attributed to the breaking of titanium–titanium bonds between the pairs of titanium atoms (1025). The titanium–titanium bonding is also broken by the addition of a small amount of V_2O_3 (1338).

Single crystal neutron diffraction work at 4° K shows however that although the density maxima of the opposed moments are closer together than the associated nuclei (2·31 compared with 2·59 Å respectively), and are moreover fairly diffuse, this incipient bond formation is not extensive as the electrons remain largely localized on the individual metal atoms (4).

C. Trihalides

1. *Trifluorides*

Titanium trifluoride is conveniently prepared by the action of gaseous hydrogen fluoride on the metal. The formation of the tetrafluoride is suppressed by using hydrogen/hydrogen fluoride mixtures, and by using a moderate temperature (700° C) to prevent disproportionation into the tetrafluoride and difluoride (754). The blue, paramagnetic compound appears stable to air, and is insoluble in water and ethanol.

Titanium trifluoride is isomorphous with vanadium trifluoride, the structure of which is derived from fairly regular octahedra sharing corners (2135).

2. *Trichlorides, bromides and iodides*

The normal method of preparation of titanium trichloride is by the reduction of the tetrachloride with hydrogen, which reversibly forms the violet α-form and hydrogen chloride. A large number of other reductants have also been used. The compound sublimes at 430° C, but at 450° C disproportionates into the tetrachloride and dichloride. However this is much lower than the temperature required to reduce the tetrachloride to the trichloride, which begins at about 600° C. A convenient laboratory method to overcome this problem is to pass hydrogen and titanium tetrachloride vapour through a silica tube at 800° C, and to quench the titanium trichloride formed before disproportionation can take place by collecting it on a cold finger condenser inserted down the centre of the tube furnace. The compound is unstable to air, but forms stable violet solutions in oxygen-free water and ethanol.

The structure of α–$TiCl_3$ shows hexagonal close packing of chloride ions, with one third of the octahedral interstices filled in such a way that the metal atoms are distributed in $\frac{2}{3}$ of the octahedral holes in every alternate layer. Within each layer each $TiCl_6$ octahedron has three neighbouring $TiCl_6$ octahedra with which it shares octahedral edges (Fig. 74) (1422).

Reduction of titanium tetrachloride with organometallic aluminium compounds at 150°–200° C produces a different form of the trichloride (γ–$TiCl_3$) (1757). This has the same colour and chemical behaviour as the α-form, and the arrangement of metal atoms between pairs of close packed chloride sheets is also the same, but in this case the structure is based on cubic close packing of chloride ions. This isomer always contains significant quantities of aluminium, and it may be noted that aluminium trichloride also has this structure.

Prolonged grinding of α–$TiCl_3$ or γ–$TiCl_3$ forms δ–$TiCl_3$, which is again violet (1758). The X-ray pattern becomes less sharp and the layers become disordered, with a statistical distribution of 63% hexagonal close packing and 37% cubic close packing.

The reaction of titanium tetrachloride with trimethylaluminium in cooled hydrocarbon solvents forms a solution of $CH_3.TiCl_3$, which on

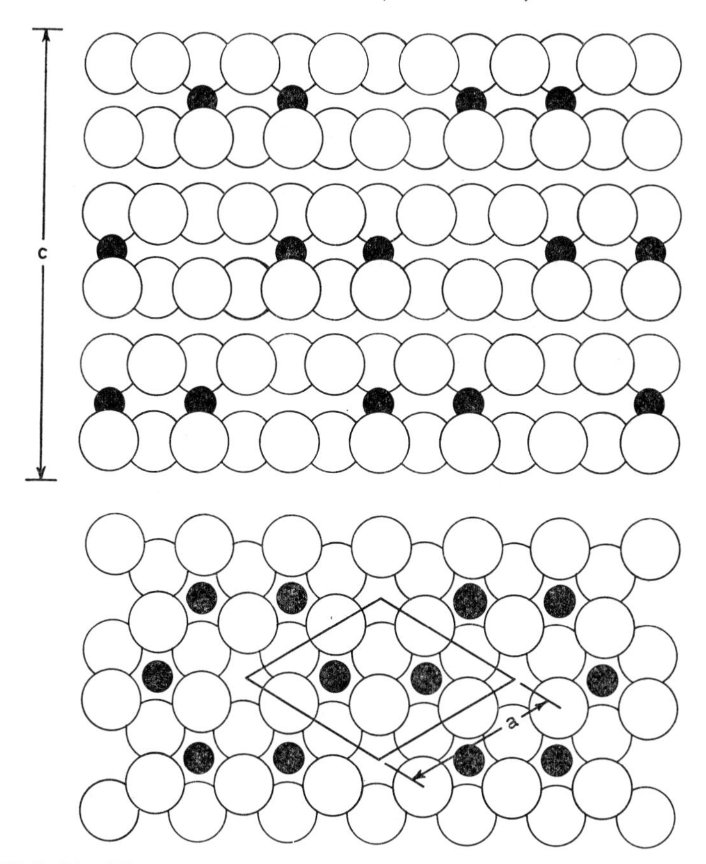

FIG. 74 The structure of α-$TiCl_3$, showing metal atoms occupying $\frac{2}{3}$ of the octahedral holes between alternate layers of close packed chloride ions.

warming to room temperature precipitates the *brown* β–$TiCl_3$ (1760). (On heating to 250°–300° C for 2–3 h the γ-isomer is irreversibly formed). Single crystals can be obtained by slow gaseous diffusion of $TiCl_4$ onto Et_2AlCl in heptane (602). The structure again shows hexagonal close packing of halide ions, but now the titanium atoms are placed between every layer in such a way that the structure can be considered to be a linear polymer formed by $TiCl_6$ octahedra sharing opposite faces (Fig. 75).

These compounds are stereospecific catalysts, for example α–$TiCl_3$ is found to polymerize isoprene as 1,4-*trans* units, whereas β–$TiCl_3$ polymerizes isoprene as 1,4-*cis* units (1760).

Titanium tribromide has similarly been found to exist in the α– and β–forms, whereas the iodide has been found only in the β–form (2063).

Zirconium and hafnium trichlorides, tribromides and triiodides are also known. They are prepared by the reduction of the tetrahalide with the metal or hydrogen at about 500° C (1499, 2052, 2053, 2054, 2078, 2241, 2348), or with hydrogen at lower temperatures in a glow discharge (1777). The zirconium tetrahalides are more easily reduced to the involatile trihalides than are the corresponding hafnium tetrahalides, and this has been suggested as a feasible method for the separation of these elements (1776).

$ZrCl_3$, $ZrBr_3$, ZrI_3 and HfI_3 have been found to have the β–$TiCl_3$ structure (Table 18). Zirconium trichloride has also been prepared by the reduction of the tetrachloride with zirconium metal in a temperature

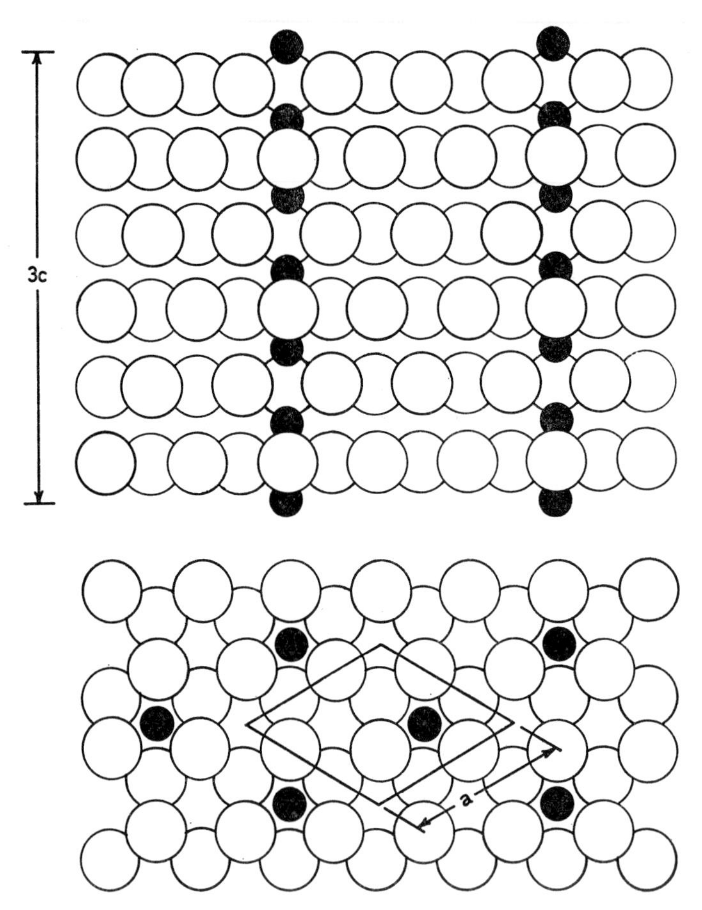

FIG. 75 The structure of β–$TiCl_3$ showing infinite strings of titanium atoms perpendicular to close-packed layers of chloride ions.

gradient of 330°–750° C (2254), for which different hexagonal unit cell dimensions have been quoted (2253).

The effect of metal–metal bonding in these structures is illustrated by a comparison of the *c*/*a* axial ratios of the hexagonal unit cells (Table 18),

TABLE 18

Hexagonal unit cell dimensions and metal–metal distances of halides with structures related to titanium trichloride

	a(Å)	*c*(Å)	*c*/*a*	M–M(Å)	Reference
β-$TiCl_3$ structures:					
β-$TiCl_3$	6·27	5·82 × 3 = 17·46	2·78	2·91	1760
β-$TiBr_3$	6·60	6·10 × 3 = 18·30	2·77	3·05	
TiI_3	7·17	6·47 × 3 = 19·41	2·71	3·24	1942
	7·15	6·50 × 3 = 19·51	2·73	3·25	2063
	7·11	6·49 × 3 = 19·47	2·74	3·25	227
β-$ZrCl_3$	6·36	6·14 × 3 = 18·42	2·89	3·07	618
	6·38	6·14 × 3 = 18·42	2·88	3·07	2351
$ZrBr_3$	6·75	6·32 × 3 = 18·96	2·81	3·19	618
ZrI_3	7·25	6·45 × 3 = 19·92	2·75	3·32	133, 618
	7·25	6·67 × 3 = 20·01	2·76	3·34	227
HfI_3	7·27	6·58 × 3 = 19·74	2·71	3·29	2241
α-$TiCl_3$ structures:					
α-$TiCl_3$	6·16	17·50	2·84		1422, 1811
	6·16	17·62	2·86		2438
γ-$TiCl_3$	6·14	17·40	2·83		1757
δ-$TiCl_3$	6·13	17·40	2·84		1758
α-$TiBr_3$	6·47	18·65	2·88		
$ScCl_3$	6·38	17·78	2·79		1422
VCl_3	6·01	17·34	2·88		1422
$FeCl_3$	6·05	17·38	2·87		
$FeBr_3$	6·42	18·40	2·87		

which are lower in the metal–metal bonded β–$TiCl_3$ structures than in the other structures where there no close metal–metal distances along the *c*-axis (Ti–Ti (*c*-axis) = 5·83 and 5·80 Å in α– and γ–$TiCl_3$ respectively). The *c*-axis in the former structures based on hexagonal close packing has been multiplied by 3 in Table 18 to enable a direct comparison to be made with the other structures (compare Figs. 74 and 75). In this table the data for some isostructural trihalides is also included for convenience. The structures of the Group VI and Group VIII trihalides are also closely related (page 335).

Most of the data in Table 18 is based on X-ray powder patterns only, but single crystal determinations on TiI_3 show additional faint reflections (2063), indicating the presence of a superstructure where the *a*-axes have been doubled. This effect is more pronounced for NbI_3, $MoBr_3$ (106) and $RuBr_3$ (2065).

A more detailed study of HfI_3 shows it to be nonstoicheiometric, the composition being $HfI_{3\cdot0-3\cdot2}$ at 550° C and $HfI_{3\cdot0-3\cdot5}$ at 475° C (Fig. 76).

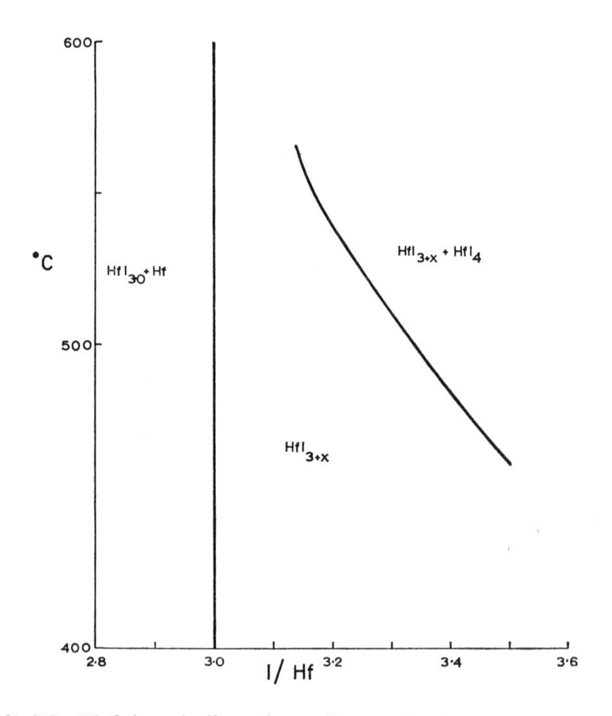

FIG. 76 Hafnium-iodine phase diagram in the region of HfI_3.

It may be noted that this temperature dependence of the homogeneity range is the opposite to that usually observed. The removal of some hafnium atoms breaks the infinite –Hf–Hf–Hf– chains, and the crystal expands along the *c*-axis:

$HfI_{3\cdot00}$	$a = 7\cdot27$	$3c = 19\cdot74$ Å	$3c/a = 2\cdot71$	Hf–Hf = $3\cdot29$ Å
$HfI_{3\cdot35}$	$a = 7\cdot23$	$3c = 20\cdot01$ Å	$3c/a = 2\cdot77$	Hf–Hf = $3\cdot34$ Å

There was some indication that the vacant metal sites were ordered. (Again a number of additional lines indicated a doubling of the *a*-axis and a quadrupling of the *c*-axis).

α–$TiCl_3$ is paramagnetic at room temperature (see later), whereas β–$TiCl_3$ (1524) and TiI_3 (1422) exhibit a small temperature independent paramagnetism, of less than 200×10^{-6} cgsu mole^{-1}, confirming that the $3d$ electrons are paired through bond formation. Nevertheless the metal–metal distances along the c-axis are fairly large (Table 18), for example the titanium–titanium distance in TiI_3 is 3·24 Å compared with 2·94 Å in the metal. This is only one example of the more general observation that metal–metal bonding can extend over greater distances if the anions are more polarizable. For further comments, see Chapter 1. A qualitative molecular orbital scheme incorporating direct metal–metal interactions in these compounds has been proposed (618).

It is of interest to note that this behaviour of titanium in forming paramagnetic, non-metal–metal bonded α–$TiCl_3$ at high temperatures, but diamagnetic metal–metal bonded β–$TiCl_3$ at low temperatures, is paralleled by ruthenium. Ruthenium reacts with chlorine above 350° C to form α–$RuCl_3$, which is paramagnetic with a layer structure containing no ruthenium–ruthenium bonds analogous to α–$TiCl_3$, but below 350° C forms β–$RuCl_3$ having the β–$TiCl_3$ structure and a lower magnetic moment (2065). Trivalent ruthenium has the electron configuration d_ε^5 and

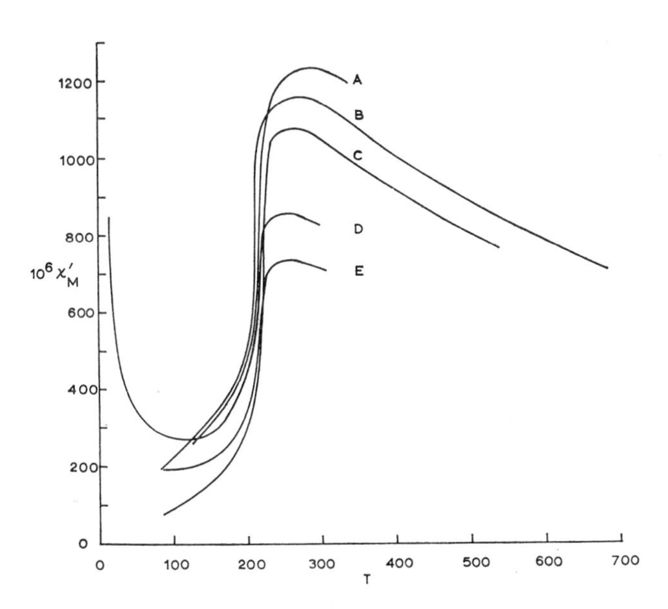

FIG. 77 Molar magnetic susceptibility (csgu) of α-$TiCl_3$ as a function of temperature. A: ref. 421. B: ref. 1422. C: ref. 1811. D: ref. 2206. E: ref. 1524.

therefore has one electron readily available for metal–metal bond formation in an analogous manner to the d_ε^1 trivalent titanium.

The zirconium trihalides also only show a weak paramagnetism, $10^6\chi'_M = 70$, 60 and 65 cgsu for the chloride, bromide and iodide respectively (1524).

There are no metal–metal bonds in α–$TiCl_3$ or α–$TiBr_3$ at room temperature, and although there is some discrepancy in the absolute magnitudes of the susceptibilities at room temperature (which are independent of field strength (2206)), there is qualitative agreement in that all workers have found a *sharp* decrease in susceptibility on cooling, occurring at about −58° C in the case of the chloride (421, 1422, 1524, 1811), and at about −103° C in the case of the bromide (1811). (Fig. 77).

There is a simultaneous sharp change in cell parameters (Fig. 78) and a contraction of the unit cell (1811).

The α- and γ-forms of titanium trichloride are violet, and the diffuse reflectance spectra shows strong broad bands at about 27 000 and

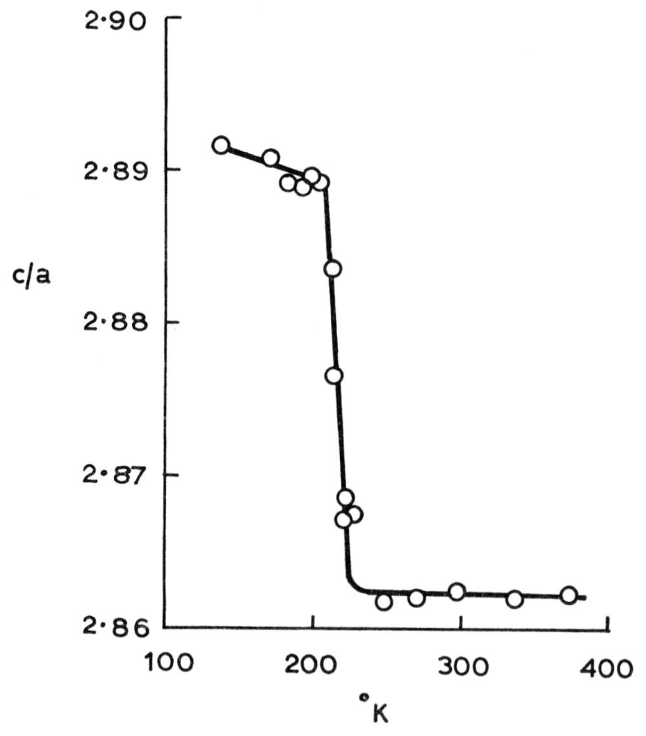

FIG. 78 Ratio of the hexagonal unit cell parameters, c/a, for α-$TiCl_3$ as a function of temperature.

19 000 cm^{-1} probably due to charge transfer, and a weaker band at about 14 000 cm^{-1} with a shoulder at about 12 000 cm^{-1} due to the expected $^2T_{2g} \rightarrow {}^2E_g$ transition. The spectrum of the brown β–$TiCl_3$ is completely different. The charge transfer bands move to higher energy (24 000 cm^{-1}) due to the increased positive charge on the metal atom, with a much weaker band in the region of the *d–d* transitions (494, 665, 667). The diffuse reflectance spectra of $ZrCl_3$ and $ZrBr_3$ have also been recorded (494).

These spectral differences between solid α–$TiCl_3$ and solid β–$TiCl_3$ are carried over into aqueous solution. The violet α–$TiCl_3$ forms a violet solution with bands at 19 800 ($\varepsilon = 3$) and 16 100 cm^{-1} (shoulder), whereas the brown β–$TiCl_3$ forms a brown solution with bands at 23 300 ($\varepsilon = 14$) and 17 800 cm^{-1} (shoulder). Even more remarkable is the observation that the latter spectrum survives partial oxidation followed by reduction to titanium(III) again, indicating the persistence of the polymeric structure. Distinctive spectra for the two forms are also obtained in ethanol and acetonitrile (974).

D. Adducts of the Trihalides

1. *Titanium*

A number of coordination compounds of the type $TiCl_3(ligand)_3$, $TiCl_3(ligand)_2$ and $TiCl_3(ligand)$ have been prepared from α–$TiCl_3$ and various ligands. Other ligands are nonreactive, and the reasons for these differences are not very well understood.

Other titanium(III) complexes have been obtained by reduction of the tetrachloride. For example with trimethylamine, $TiCl_3(Me_3N)_2$ was obtained, which can alternatively be obtained from α–$TiCl_3$. The organic oxidation product was not identified, but all the chlorine lost was found as $(Me_3NH)Cl$. The observed ease of reduction was

$Me_3N > Me_2NH > MeNH_2 > NH_3$ (82, 908, 2303).

Similarly reduction occurs with pyridine forming $TiCl_3(py)_3$ (908) where the pyridine oxidation product is presumably the same as that characterized for the reduction of $NbCl_5$ to $NbCl_4(py)_2$ (page 202).

On the other hand oxidation of titanium(III) to titanium(IV) can occur with arsine ligands. The reaction of *o*-triarsine or *v*-triarsine (page 85) with α–$TiCl_3$ in tetrahydrofuran leads to the isolation of $TiCl_3(THF)_3$ only (501). This was unexpected since vanadium and chromium trichlorides give VCl_3 (triarsine) and $CrCl_3$ (triarsine) respectively, and clearly demonstrates the more electropositive nature of titanium(III) and hence greater preference for ligands containing oxygen donor atoms. However refluxing of the trichloride with either of the triarsines in acetonitrile leads to the isolation of the $TiCl_4$(triarsine) compounds. Similarly the diarsine (*o*-$(C_6H_4)(AsMe_2)_2$) complex $TiCl_4(diars)_2$ is obtained by reaction of the trichloride in glacial acetic acid.

The only structurally characterised adduct appears to be $TiBr_3(Me_3N)_2$ in which the titanium atom is five coordinate, although the spectra and magnetism of a number of such adducts have been interpreted in terms of an octahedral environment.

In contrast to the tetravalent compounds, there appears to be no evidence for coordination numbers greater than six for any of these adducts.

Oxygen donor ligands form adducts of stoicheiometry $TiCl_3(ligand)_3$, $TiCl_3(ligand)_2$ and $TiCl_3$(ligand). Refluxing α–$TiCl_3$ with dioxane for a few hours forms the blue-green $TiCl_3(dioxane)_2$ (504, 909), with a room temperature effective magnetic moment of 1·73 B.M. The infrared spectrum shows the presence of both free and coordinated oxygen atoms. Prolonged refluxing of $TiCl_3$ with dioxane forms the blue-green antiferromagnetic $TiCl_3$(dioxane), which on the basis of infrared evidence has both oxygen atoms of the ligand coordinated (909, 1602). The third derivative, $TiCl_3(dioxane)_3$, is obtained from the monomeric $TiCl_3(Me_3N)_2$ and dioxane (909). This blue compound is soluble in dioxane in contrast to the two insoluble derivatives above. The infrared spectrum suggests the presence of both free and coordinated oxygen atoms. The dark green $TiBr_3(dioxane)_3$ is similarly obtained from $TiBr_3(Me_3N)_2$ (939).

With other oxygen donors, α–$TiCl_3$ forms $TiCl_3(THF)_3$ (504, 1363, 2049), $TiCl_3,1\frac{1}{2}(MeO.CH_2.CH_2.OMe)$ (909), $TiCl_3(Me_2CO)_3$ (504, 909), $TiCl_3(Et_2CO)_2$ (909) and $TiCl_3(Ph_2CO)_2$ (909), which are from green to blue, and have magnetic moments from 1·61 to 1·74 B.M. The structure of $TiBr_3,1\frac{1}{2}(MeO.CH_2.CH_2.OMe)$ shows that it should be formulated [*cis*-$TiBr_2(bidentate)_2$][$TiBr_4$(bidentate)] (917).

The products of the reaction between $TiCl_3$ and alcohols have been studied by a number of workers (996, 997, 2049, 2404), and include neutral complexes such as dark brown $TiCl(OMe)_2(MeOH)$, cationic species such as blue $[TiCl_2(MeOH)_4]^+$, and anionic species such as $[TiCl_4(MeOH)_2]^-$.

Dimethyl sulphide reacts with α–$TiCl_3$ to form the green $TiCl_3(Me_2S)_2$, which is less stable than the complexes with oxygen donor ligands, and disproportionates at about 50° C to $TiCl_4(Me_2S)_2$ and a titanium(II) compound (125, 918). The titanium(III) compound, and also its tetrahydrothiophen analogue, is strongly antiferromagnetic with a Néel temperature of about 320° K. Thioethers do not replace trimethylamine completely from $TiCl_3(Me_3N)_2$, although the amine is readily replaced by the more electronegative oxygen donor ligands (125). Thioxane, C_4H_8OS, forms $TiCl_3$(thioxane) with both oxygen and sulphur atoms apparently coordinated (909); however $TiCl_3(thioxane)_2$ and $TiBr_3(thioxane)_2$ are obtained from $TiX_3(Me_3N)_2$ (939).

Trimethylamine reacts with α–$TiCl_3$ and α–$TiBr_3$ to form $TiCl_3(Me_3N)_2$ and $TiBr_3(Me_3N)_2$ (82, 908). The structure (1972) of the latter shows it to be monomeric with a trigonal bipyramidal stereochemistry; the titanium atom is at the centre of a plane containing three bromine atoms (Ti–Br = 2·40, 2·44 and 2·44 Å, Br–Ti–Br = 118°, 121° and 121°) with the two donor atoms above and below the plane. The imperfect triangular shape was ascribed to a Jahn–Teller effect. The effective magnetic moment of $TiCl_3(Me_3N)_2$ is 1·69 B.M. at room temperature dropping to 1·50 at 85° K. It has been shown that this cannot be explained on the basis of D_{3h} ligand fields, and that a small additional component in the plane of the triangle must be present (2412). The visible spectra of these five coordinate complexes are different to that of octahedral titanium(III), and have been studied in some detail (611). The visible absorption spectrum in nitrogen-donor solvents appears significantly different to that in the solid state.

The alkyl cyanides form blue complexes $TiCl_3(RCN)_3$ (where R is Me, Et, Pr, $CH_2 = CH$) which have normal infrared spectra for coordinated nitrile groups (504, 701, 1364, 1597).

Pyridine and γ-picoline form similar compounds $TiCl_3(py)_3$ and $TiCl_3(pic)_3$ which are both green, whereas the sterically hindered α-picoline, and 2,6-lutidine form blue-green $TiCl_3(pic)_2$ and brown $TiCl_3(lut)$ respectively (908).

With excess of the bidentate dipyridyl in organic solvents $TiCl_3$, $1\frac{1}{2}$ dipy is formed, which although of unknown structure is probably best formulated as $[TiCl_2(dipy)_2][TiCl_4(dipy)]$ (910). If the reaction is carried out at 150° C in a sealed tube, $TiCl_3(dipy)$ is formed, which can subsequently add on a sixth donor atom forming, for example, $TiCl_3(dipy)(MeCN)$ (910). Both these compounds are dark blue. The reaction of $TiCl_3$ with dipyridyl in oxygen-free water initially forms a violet solution from which *colourless* crystals can eventually be isolated which were formulated as $[TiCl_2(dipy)_2]Cl$ (2133), although the colour suggests that oxidation to titanium(IV) may have occurred. Ethylenediamine forms $[Ti(en)_3]Cl_3$ and not $TiCl_3$, 4en as reported earlier (500, 749, 1597). Diethylenetriamine similarly forms $Ti(dien)_2Cl_3$. The magnetic moments of these compounds (μ_{eff} = 0·93 and 1·42 B.M. respectively) are abnormally low (1597). Hexamethylcyclotrisilazane behaves as a tridentate ligand forming $TiCl_3[(Me_2Si)_3(NH)_3]$ (2394).

The diphosphine $Et_2P.CH_2.CH_2.PEt_2$ forms the brown $TiCl_3$(diphosph) in benzene (463). The reaction of titanium trichloride with arsine ligands has been noted above.

Titanium tribromide and triiodide have been studied in less detail, but usually form octahedral complexes with normal magnetic moments. Examples of adducts analogous to those of the trichloride include $TiX_3(THF)_3$, $TiX_3(Diox)_3$, $TiX_3(Me_3N)_2$ (where X is Br or I), $TiI_3(py)_3$, $TiI_3(\gamma\text{-pic})_3$ and $TiI_3(\alpha\text{-pic})_2$ (906, 915, 916, 917). Oxygen abstraction from oxygen donors such as acetone and dimethoxyethane appears to occur more readily for the triiodide than for the other trihalides. However when diluted with benzene the latter forms $[TiI_2(MeO.CH_2.CH_2.OMe)_2]I$ (916). Another slight difference is that acetonitrile forms $TiCl_3(MeCN)_3$, $TiBr_3(MeCN)_3$ (701), but $[TiI_2(MeCN)_4]I$ (916).

2. *Zirconium*

The reaction of zirconium trihalides with neutral ligands unexpectedly sometimes forms complexes containing less ligand than with the corresponding titanium trihalides, although products with higher coordination numbers would be expected. For example pyridine forms $ZrX_3(py)_2$ ($X = Cl$, $\mu_{eff} = 1{\cdot}16$ B.M.; $X = Br$, $\mu_{eff} = 1{\cdot}24$ B.M.; $X = I$, $\mu_{eff} = 1{\cdot}10$ B.M.) in contrast to $TiX_3(py)_3$, acetonitrile forms ZrX_3, $2\frac{1}{2}MeCN$($X = Cl$, $\mu_{eff} = 0{\cdot}46$ B.M.; $X = Br$, $\mu_{eff} = 0{\cdot}29$ B.M.; $X = I$, $\mu_{eff} = 1{\cdot}20$ B.M.) in contrast to $TiCl_3,3MeCN$, $TiBr_3,3MeCN$ and $TiI_3,4$ MeCN (932), and tetrahydrofuran forms $ZrCl_3(THF)$ (ferromagnetic) in contrast to $TiCl_3(THF)_3$. These zirconium halides are isostructural with β–$TiCl_3$, and the differences in chemical behaviour between β–$TiCl_3$ and α–$TiCl_3$ have been referred to earlier (page 131).

E. Oxohalides

The preparation of the reddish yellow TiOCl has been achieved by the following methods (2039):

(i) The reaction of TiO_2 with excess $TiCl_3$ (or Ti–$TiCl_4$ mixtures) in a temperature gradient from 550° to 650° C:

$$2\,TiCl_3 + TiO_2 \rightarrow 2\,TiOCl + TiCl_4$$

(ii) The hydrolysis of $TiCl_3$ with the stoicheiometric quantity of water at 600° C:

$$TiCl_3 + H_2O \rightarrow TiOCl + 2\,HCl$$

(iii) The reaction of $TiCl_3$ with oxygen at 650° C according to the reaction:

$$6\,TiCl_3 + O_2 \rightarrow 2\,TiOCl + 4\,TiCl_4$$

The product is stable to air at room temperature. In the absence of air it decomposes when heated, although it can be sublimed in the presence of gaseous $TiCl_3$.

TiOCl has the FeOCl structure, which is based on close packing of anion layers with the metal atom in octahedral interstices. It exhibits a slight field independent paramagnetism, $10^6\chi'_M = 400$ cgsu at room temperature.

The tetravalent $TiOI_2$ loses iodine at 120° C in vacuum with the formation of TiOI (647). This behaviour is in contrast to $TiOF_2$, $TiOCl_2$ and $TiOBr_2$, which on heating form the tetrahalide and dioxide. The oxoiodide shows a small temperature independent paramagnetism, $10^6\chi'_M = 50$ cgsu.

Other trivalent Group IV oxohalides are apparently not known.

F. Complexes with Anionic Ligands

In this section attention will be largely concentrated on some halide, cyanide and acetylacetonate complexes of titanium(III).

The hexahaloanions $[TiX_6]^{3-}$ cannot be obtained from aqueous solution. Titanium trichloride and tribromide form hexahydrates in concentrated hydrochloric acid and hydrobromic acid respectively, which should be formulated as $[TiX_2(H_2O)_4]X,2H_2O$ and not $[Ti(H_2O)_6]X_3$ as commonly assumed (2047). These compounds are also commonly prepared by the electrolytic reduction of titanium(IV) in concentrated hydrohalic acid. Species claimed in aqueous solution include $[Ti(H_2O)_6]^{3+}$, $[TiCl(H_2O)_5]^{2+}$, $[TiCl_2(H_2O)_4]^+$ and $[TiCl_5(H_2O)]^{2-}$ depending upon conditions (1331).

The reaction of titanium trichloride with pyridinium chloride in acetonitrile forms the orange $(pyH)_3[TiCl_6]$, yellow-green $(pyH)_2[TiCl_5(MeCN)]$, or yellow $(pyH)_4HTi^{III}Cl_8$ depending upon conditions (996, 1971). The structure of the latter is not known, but is unlikely to contain eight coordinate titanium. The analogous reactions with tetraethylammonium halides form $(Et_4N)[TiCl_4(MeCN)_2]$, $(Et_4N)[TiCl_3Br(MeCN)_2]$ and $(Et_4N)[TiBr_4(MeCN)_2]$ (1971). Heating the first to 100° C forms the light brown $(Et_4N)[TiCl_4]$ which on the basis of spectral and magnetic properties was formulated as a polymeric octahedrally coordinated complex, in contrast to the tetrahedrally coordinated anion found in $(Ph_4As)[VCl_4]$ (931). In glacial acetic acid, $Cs_2[TiCl_5(HOAc)]$ can be obtained (1597).

The cubic violet, Na_3TiF_6, K_2NaTiF_6 and K_3TiF_6 are obtained by electrolytic reduction of the quadrivalent fluorides in molten potassium fluoride and/or sodium fluoride (326). Examination of the alkali metal (or diethylammonium) halide–titanium trihalide systems shows the existence of A_3TiX_6, $A_3Ti_2X_9$, A_2TiCl_5 and $ATiCl_4$ (where X is Cl or Br) (612, 753, 1457, 2116).

The reaction of titanium tribromide with potassium cyanide in liquid ammonia forms the dark green, air unstable, $Ti(CN)_3$, 5KCN (2048). This compound was originally formulated as the octahedral $K_3[Ti(CN)_6]$, 2KCN. The spectrum in liquid ammonia shows absorptions at 18 900 and 22 300 cm^{-1}, which is not very consistent with octahedral titanium(III). An alternative formula could be the eight coordinate $K_5[Ti(CN)_8]$. The magnetic properties (effective magnetic moments of 1·74, 1·70 and 1·65 B.M. at 294, 197 and 93° K respectively) do not greatly assist in deciding between these formulae. The esr spectrum of the solid shows a sharp line with a *g* value of 1·990 (1021), which is very similar to that observed for $K_3[W(CN)_8]$ and $K_3[Mo(CN)_8]$, but not to the lower values normally observed for octahedral titanium(III).

In aqueous solution in the absence of air, only the hexacyanide $K_3[Ti(CN)_6]$ is obtained (1166).

Titanium(III) with ammonium thiocyanate in water or acetone forms the violet $(NH_4)[Ti(SCN)_4(H_2O)_2]$ or orange $(NH_4)[Ti(SCN)_4]$ depending upon conditions. Both compounds have abnormally low magnetic moments (2250). Under anhydrous conditions, $(Bu_4N)_3[Ti(SCN)_6]$ can be obtained (1533).

The reaction of titanium trichloride with acetylacetone forms the dimeric $[(acac)_2Ti\langle{}^{Cl}_{Cl}\rangle Ti(acac)_2]$ (1866). The monomeric $[Ti(acac)_3]$ has been obtained from sodium acetylacetonate in oxygen-free water or from acetylacetone plus ammonia in dry benzene (148, 455, 598). Similar compounds have been obtained with other β-diketones (598). These compounds have room temperature effective magnetic moments of about 1·5 B.M., which have been studied in some detail as a function of temperature (847). All decompose slowly in air to form orange compounds of the type $Ti^{IV}O(acac)_2$ (page 104) (455, 598).

The oxalato complex $(NH_4)Ti(C_2O_4)_2,2H_2O$ can also be obtained from aqueous solutions, but readily forms the anhydrous salt at 110° C (800). An octahedral structure containing bidentate and bridging tetradentate oxalato groups was suggested. The structure of $Ti_2(C_2O_4)_3$, $10H_2O$ shows a dimeric molecule, one oxalato group bridging two seven coordinate, pentagonal bipyramidal, titanium atoms (693):

$$[(H_2O)_3(C_2O_4)Ti\langle{}^{O-C-O}_{O-C-O}\rangle Ti(C_2O_4)(H_2O)_3],\ 4H_2O.$$

G. Organometallic Compounds

The organometallic chemistry of titanium(III) has been studied in considerable detail, and cannot be adequately summarized here. This interest is partly because of the importance of the Ziegler–Natta catalytic process (page 118), and because of the capability of some complexes to fix nitrogen.

Titanium trialkyls are not stable at room temperature and have not been extensively studied (484). Cyclopentadienyl however forms $(C_5H_5)TiCl_2$, $(C_5H_5)_2TiCl$ and $(C_5H_5)_3Ti$.

The monocyclopentadienyl dichloride has been obtained from $(C_5H_5)_2TiCl_2$ and 1·3 moles of dibutylaluminium chloride (157). Both $(C_5H_5)TiCl_2$ and $(C_5H_5)_3Ti$ show normal paramagnetic behaviour for titanium(III) (421,865).

Most attention has been concentrated on $(C_5H_5)_2TiCl$ which can be readily prepared by the reaction of titanium trichloride with sodium cyclopentadienyl or magnesium dicyclopentadienyl, or by the reduction of $(C_5H_5)_2TiCl_2$ (for example with zinc dust, lithium aluminium hydride, or excess magnesium dicyclopentadienyl) (234, 1761, 1916). The green-brown product is stable in oxygen-free water to give a blue solution, and behaves as a 1 : 1 electrolyte. It is also soluble in benzene, in which it is dimeric (1761). The magnetic moment of the solid at room temperature is 1·6 B.M., but drops to 0·6 B.M. at 90° K due to direct titanium–titanium interactions (1656).

A large number of derivatives of $(C_5H_5)_2TiCl$ of the general type $(C_5H_5)_2$ TiX have been prepared, where X can be H^-, alkyl (329), π-allyl (329, 1651), CN^-, NCS^-, NCO^- (587), acac$^-$ (587), $R.COO^-$ (1916), $R.COS^-$ (586), BH_4^- (1799), PR_2^- (1350), $GePh_3^-$ and $SnPh_3^-$ (589) (the latter being isolated with one molecule of tetrahydrofuran). Some of these are paramagnetic monomers which follow the Curie–Weiss law (for example where X is π-allyl, NCS^-, NCO^-, acac$^-$), others dimeric (for example $[(C_5H_5)_2Ti(PR_2)]_2$), while the mass spectrum of $(C_5H_5)_2Ti(CN)$ shows the existence of a trimer. The $(C_5H_5)_2Ti^{III}\langle Cl \rangle AlEt_2$ (chloride-bridged, $Ti \leftarrow Cl$, $Cl \rightarrow Al$) is described on page 119.

The chloride ligand in $(C_5H_5)_2TiCl$ can also be replaced by neutral ligands forming $[(C_5H_5)_2Ti(MeCN)_2](Ph_4B)$, $[(C_5H_5)_2Ti(py)_2](Ph_4B)$ and $[(C_5H_5)_2Ti(H_2O)_3](Ph_4B)$. The first two cations are monomeric and the magnetic moments follow the Curie–Weiss law, whereas the last is dimeric and exhibits more complex magnetic behaviour (583). The behaviour of bidentate ligands has also been studied (588).

These complexes are unstable to air, and products such as $[(C_5H_5)_2Ti^{III}–O–Ti^{III}(C_5H_5)_2]$ and $[(C_5H_5)_2Ti^{IV}(Cl)–O–Ti^{IV}(Cl)(C_5H_5)_2]$ have been characterized (583, 993).

The reduction of $(C_5H_5)_2TiCl_2$ with ethyl magnesium chloride or sodium naphthalide in tetrahydrofuran forms the green-brown $[(C_5H_5)_2TiCl]_2$ which on further reduction forms a very dark brown solution thought to contain $(C_5H_5)_2Ti^{III}(\mu\text{-}H)_2Ti^{III}(C_5H_5)_2$. The latter reacts with nitrogen at room temperature and pressure to form ammonia, the suggested intermediates being of the type

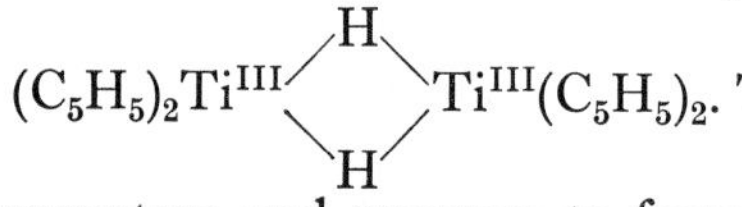

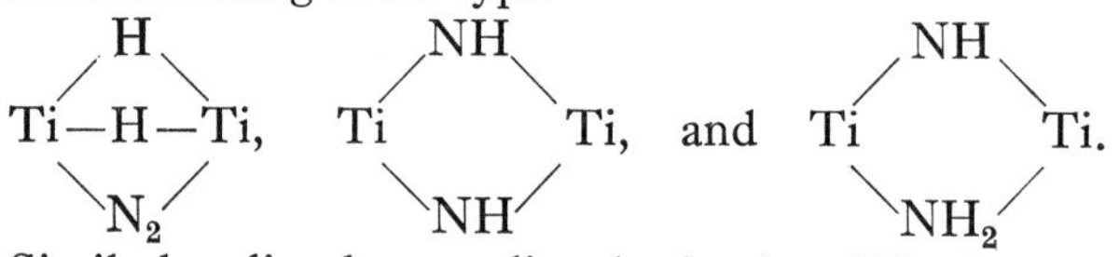

Ti–H–Ti (bridged by H and N_2), Ti(NH)(NH)Ti, and Ti(NH)(NH_2)Ti. (328, 1174, 1659)

Similarly dicyclopentadienyl titanium(II) forms a dimeric complex with molecular nitrogen at room temperature and pressure, which on reduction with six moles of sodium naphthalide and subsequent hydrolysis forms ammonia. Deuterium labelling of the solvents (diethyl ether, tetrahydrofuran, benzene) shows that the hydrogen of the ammonia originates from the solvents (2260).

Similar nitrogen complexes can be obtained in the absence of organometallic groups. The reduction of $TiCl_3(THF)_3$ in tetrahydrofuran with magnesium in the presence of molecular nitrogen forms a product of composition $TiNMg_2Cl_2(THF)$, which on hydrolysis also liberates ammonia (2416). A number of related complexes were also described.

These extremely air-unstable nitrogen complexes are not to be confused with the stable molecular nitrogen complexes formed by the heavier transition metals such as rhenium, osmium and iridium.

3. OXIDATION STATES $\leqslant$ II

A. Dihalides

1. *Titanium*

Titanium difluoride is unknown.

Titanium dichloride can be prepared by the disproportionation of the trichloride at 450° C (1967, 2079), or by reduction of the tetrachloride with titanium metal. The former method is not very convenient due to the formation of metallic titanium impurities. A convenient apparatus for the second method has been described, (752) in which the tetrachloride is passed down a titanium tube which is held at a temperature higher than 1035° C so that the dichloride formed is molten (M.P. = 1035° C, B.P. = 1500° C) and can be collected at the bottom of the tube. Reduction with titanium has also been carried out in sealed tubes (1419). The tetrachloride or trichloride can also be reduced with hydrogen above 700° C, but more conveniently at low temperatures in the presence of an electric discharge (1101).

Titanium dibromide (2427) and titanium diiodide (815) have also been prepared by the disproportionation of the higher halides at 400° and 480° C respectively (with again some contamination from metallic titanium), or by reaction of stoicheiometric quantities of metal and halogen in sealed tubes (1419). The dibromide has also been prepared by passing the tetrabromide over the metal as above (778).

All three halides are black and very unstable to air and moisture. Some preparations are pyrophoric in dry air, and they react violently with water liberating hydrogen.

The chloride (111, 971, 1419), bromide (751) and iodide (1419) have the cadmium iodide structure, in which the metal atoms occupy all the octahedral holes between every alternate layer of close packed halide ions.

The room temperature effective magnetic moment is about 1·1 B.M. The more recent results (1524) show a suggestion of antiferromagnetic behaviour with a Néel point of 85° K, which was not observed by earlier workers (1420, 2206), possibly due to the presence of paramagnetic impurities. The susceptibility is independent of field strength (2206). Higher values for the susceptibility of titanium dibromide and diiodide have been reported (1420).

2. *Zirconium and hafnium*

Zirconium difluoride has been prepared by the reduction of the tetrafluoride with atomic hydrogen at 350° C (1613). The compound is black, has an orthorhombic structure, is unstable to air, and disproportionates above 800° C to the tetrafluoride and metal.

The other zirconium dihalides are formed by thermal disproportionation of the trihalides (which again form metallic zirconium at higher temperatures), or by reduction of the tetrahalides with zirconium in a similar manner as described above for the titanium analogues (258, 680, 814, 1499, 1776, 2255, 2310, 2421). The compounds are black and are readily oxidized by oxygen and water, as was also noted for the titanium dihalides. The stability apparently depends upon the method of preparation. For example zirconium dibromide prepared by reduction of the tetrabromide has been reported to be only slowly decomposed by water (258), while the compound prepared by disproportionation reacts violently with water and is pyrophoric (2421).

The magnetic susceptibility of zirconium dichloride ($10^6\chi_M = 150$ cgsu at 300° K; $\mu_{eff} = 0 \cdot 6$ B.M.) indicates strong antiferromagnetic interactions (1524).

Hafnium tetrachloride is more difficult to reduce than zirconium tetrachloride, and this fact can be used to separate these elements (1776). The black hafnium dibromide has been formed by thermal disproportionation of the tribromide, and has similar properties to the zirconium compound (2078).

B. Inorganic Complexes

In liquid ammonia, $TiCl_2$ forms $TiCl_2(NH_3)_4$. In general however common ligands form $TiCl_2L_2$ (where L is dimethylformamide (760), acetonitrile, tetrahydrofuran, pentamethylene oxide, or pyridine (914)), $TiCl_2B$ (where B is dipyridyl or *o*-phenanthroline (914)), $TiBr_2L_2$ (where L is acetonitrile (914)), or $TiBr_2B$ (where B is dipyridyl (914)). These compounds all have low magnetic moments ($\mu_{eff} = 1 \cdot 0$–$1 \cdot 2$ B.M. at room temperature), and are readily oxidized by air or water.

The binary systems ACl–$TiCl_2$ show the formation of the compounds A_2TiCl_4 (where A is Rb or Cs) and $ATiCl_3$ (where A is K, Rb or Cs) (757). A fused mixture of $TiCl_2$ and NaCl corresponding to the stoicheiometry Na_2TiCl_4 has a magnetic moment of $2 \cdot 43$ B.M. at room temperature (1524).

Complexes of the type $[M(dipy)_3]$ are formed between 2,2′-dipyridyl and these Group IV metals. Whether they should be regarded as $[M^{\circ}(dipy)_3]$ or $[M^{III}(dipy^-)_3]$, or some intermediate formula, is probably a matter of opinion, since the filled *d* orbitals on the metal atoms undoubtedly participate in extensive back bonding to the antibonding π^* orbitals of the ligand. These compounds can be further reduced by one or even two electrons to form $[M^{-I}(dipy)_3]^-$ and $[M^{-II}(dipy)_3]^{2-}$ respectively (or $[M^{II}(dipy^-)_3]^-$ and $[M^{I}(dipy^-)_3]^{2-}$ respectively) (1186, 1187).

C. Organometallic Complexes

Dicyclopentadienyl titanium(II) was originally prepared by the reaction of titanium dichloride with sodium cyclopentadienyl (850), although a more convenient route appears to be the reduction of $(C_5H_5)_2TiCl_2$ with sodium naphthalide (2349). The dark green compound is pyrophoric. The diamagnetism and dimeric molecular weight indicate direct titanium–titanium bonding. Infrared and nmr evidence show that both σ-bonded and π-bonded cyclopentadienyl groups are present (1986).

In the presence of carbon monoxide at atmospheric pressure, the diamagnetic monomer $(C_5H_5)_2Ti(CO)_2$ is formed (413, 1746). The reaction of $(C_5H_5)_2Ti$ with molecular nitrogen is discussed together with the titanium(III)-nitrogen complexes on page 138 (2260).

The analogous zirconium dicyclopentadienyl has been prepared by similar methods (2350). It is isomorphous with the titanium compound, and is similarly pyrophoric and diamagnetic.

The reaction of titanium tetrachloride with sodium cyclooctatetraene, $Na_2(C_8H_8)$, forms $Ti(C_8H_8)_2$ and $Ti_2(C_8H_8)_3$ (322). The structure of the latter shows that each titanium atom is bonded to one planar eight electron donor cyclo-octatetraene molecule, while the third organic group is nonplanar and bridges both metal atoms (664).

Finally hexamethylbenzene forms $[Ti_3(C_6Me_6)_3Cl_6]Cl$ and $[Zr_3(C_6Me_6)_3Cl_6]Cl$, both of which are paramagnetic but of unknown structure (871).

CHAPTER 3

Group V—Vanadium, Niobium, Tantalum

I. OXIDATION STATE V

A. Oxides

1. *Vanadium*

In V_2O_5 the cation is considerably smaller than the octahedral hole formed by close packing of oxide ions, and whether the stereochemistry is considered to be a distorted trigonal bipyramid (five vanadium–oxygen bond lengths of 1·58–2·02 Å) or a distorted octahedron (sixth vanadium–oxygen distance of 2·79 Å) is largely a matter of choice. Two views of the structure are shown in Fig. 79. Fig. 79a shows strings of double octahedra, or double trigonal bipyramids, sharing edges. These form sheets by corner sharing with the double strings on either side. These sheets form a three-dimensional network by corner sharing with adjacent sheets (Fig. 79b). Fig. 79c is the same projection as Fig. 79b, only drawn as idealized trigonal bipyramids rather than as idealized octahedra, so that only the former sheet structure is created. These weak vanadium–oxygen bonds which complete the octahedra give rise to perfect cleavage between the sheets.

A comparison between the structures of V_2O_5 (Fig. 79a) and MoO_3 (Fig. 140, page 255) shows the two to be closely related, and it is therefore not unexpected that some of the vanadium atoms can be replaced by molybdenum atoms without change of structure (1386). In the structures of V_2MoO_8 (761, 1830) and Nb_3O_7F (59), the size of the *corner* shared blocks are increased from $2 \times \infty \times \infty$ to $3 \times \infty \times \infty$ (Fig. 80).

Vanadium pentoxide readily loses oxygen from one third of the (010) planes on heating to form the dark blue V_6O_{13} (20). Thus single crystals of V_6O_{13} may be prepared from single crystals of V_2O_5 (1000). The structure of V_6O_{13} (Fig. 81) shows that there are no metal–metal bonds at room temperature, the vanadium–vanadium distances for polyhedra sharing edges being 3·25 Å. The compound is paramagnetic (1073, 1462), but at about $-100°$ C the formation of metal–metal bonds is suggested by a sharp drop in magnetic susceptibility and electrical conductivity (1328, 1461).

A non-stoicheiometric sodium vanadium oxide can be prepared by

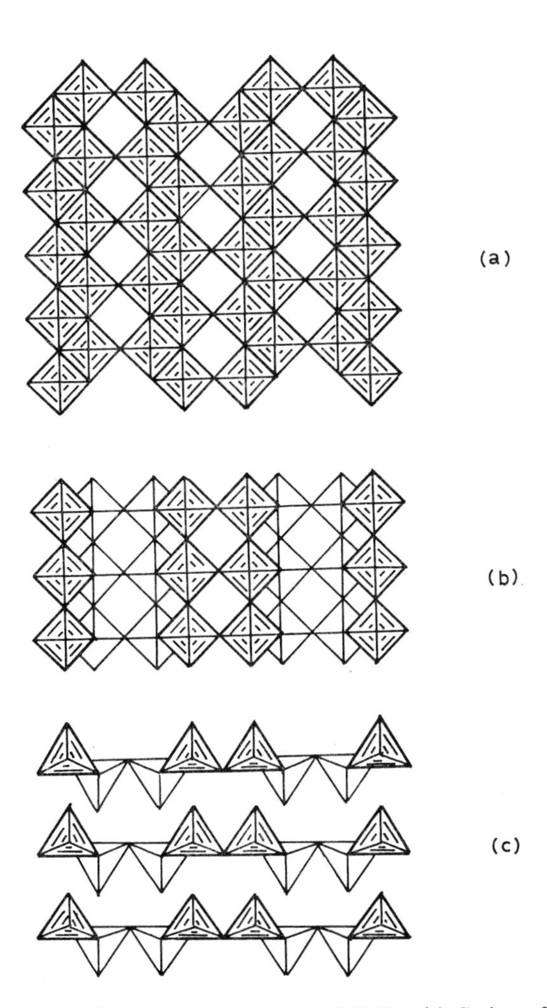

FIG. 79 Three views of the structure of V_2O_5. (a) Strings formed by VO_6 octahedra sharing edges. Each octahedron also shares corners with similar octahedra above and below the plane of the page as indicated in (b), forming sheets of corner shared octahedra $2 \times \infty \times \infty$, which share octahedral edges with each other. These sheets are not formed if the metal atom is considered to be five co-ordinate as indicated in (c).

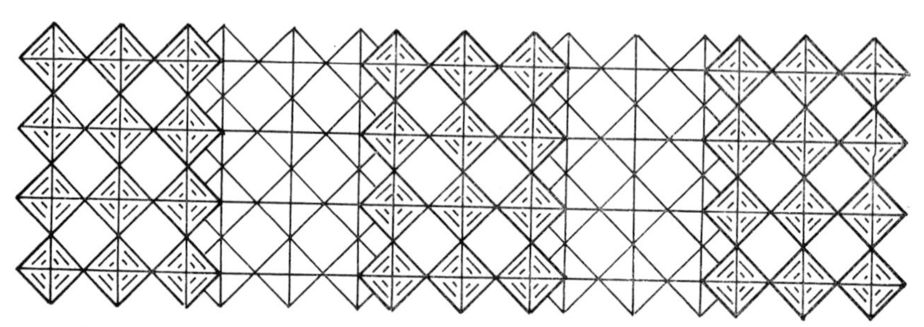

FIG. 80 The structure of V_2MoO_8 and Nb_3O_7F. (Compare with Fig. 79(b)).

heating vanadium pentoxide with, for example, Na_2O, Na_2CO_3 or $NaVO_3$, with the evolution of oxygen. The general formula is $Na_xV_2O_5$, with values of x ranging from about 0·15 to 0·33. Other double oxides, $NaV_3^VO_8$ and NaV^VO_3, are obtained from aqueous solution and are discussed with the other isopolyvanadates on page 184. The non-stoicheiometric $Na_xV_2O_5$ is almost black with a dark violet or dark green metallic lustre, is chemically inert, and is often referred to as a "vanadium bronze" by analogy with the tungsten bronzes. However unlike the tungsten bronzes, the vanadium bronzes do not exhibit metallic conduction, and it is probable that the electron released by the sodium atom is largely localized on one of the metal atoms.

The structure of $Na_{0 \cdot 33}V_2O_5$ shows double strings of VO_6 octahedra forming sheets by the sharing of corners, and these sheets are joined to-

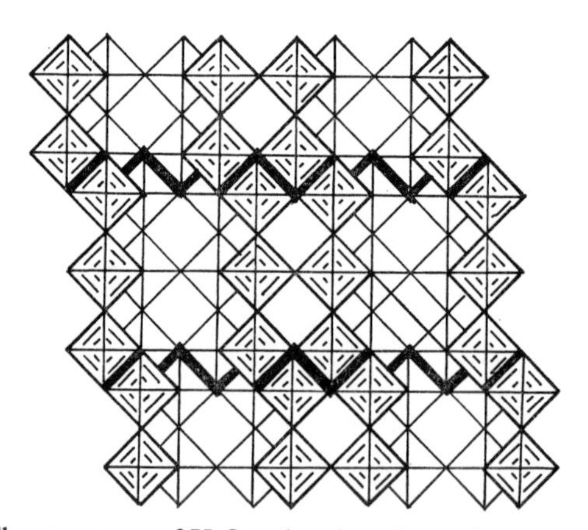

FIG. 81 The structure of V_6O_{13} showing shear planes running through the V_2O_5 structure (Fig. 79(b)) forming blocks of corner shared octahedra $2 \times 3 \times \infty$.

gether by strings of VO_5 trigonal bipyramids, resulting in open channels, parallel to the y-axis, which accommodate the sodium atoms (Fig. 82) (1826, 2327). The five coordinate vanadium atom can alternatively be considered to be octahedrally coordinated if the longer vanadium–oxygen bond shown in the Fig. is also included. This vanadium–oxygen distance

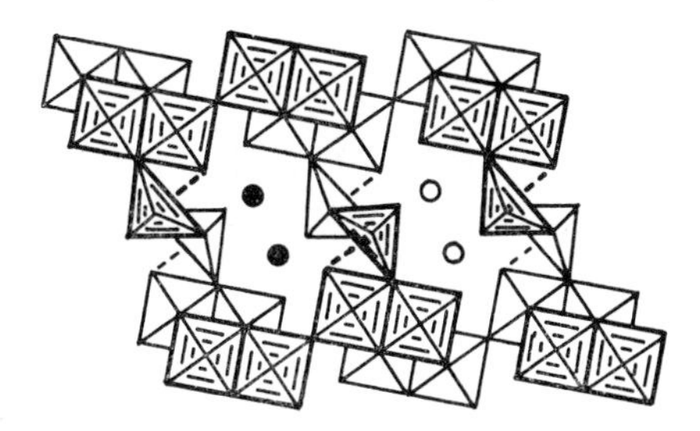

FIG. 82 The structure of $Na_{0\cdot3}V_2O_5$. Infinite ribbons formed by the edge sharing of double octahedra, and infinite strings formed by the edge sharing of trigonal bipyramids, are normal to the plane of the page. These share corners with each other leaving channels which accommodate the sodium atoms (circles). The sixth vanadium-oxygen bond which converts each trigonal bipyramid into an octahedron is shown by the broken line.

is 2·68 Å compared with the average for the other five of 1·83 Å, and for the average of 1·94 Å for the other VO_6 octahedra. The channels parallel to the y-axis are wide enough to accommodate pairs of sodium atoms, but if all these sites were occupied the formula would be $Na_4V_6O_{15}$ or $Na_{0\cdot67}V_2O_5$. Although the actual distribution of sodium atoms in these sites is not known, since the maximum sodium content experimentally available is only half this value, it appears that only one of any pair of sites is occupied.

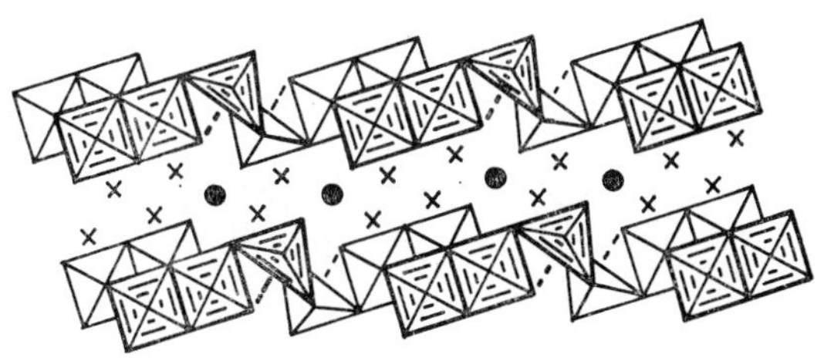

FIG. 83 The structure of $Li_{1+x}V_3O_8$. Infinite ribbons formed by the edge sharing of double octahedra, and infinite strings formed by the edge sharing of trigonal bipyramids, are normal to the plane of the page. These share corners with each other forming sheets between which the lithium atoms are situated in octahedral (circles) or tetrahedral (crosses) sites. The sixth vanadium-oxygen bond which converts each trigonal bipyramid into an octahedron is shown by the broken line.

Tunnels which contain the alkali metal cations are also found in the nonstoicheiometric semiconductors $Li_{1+x}V_3O_8$ and the isomorphous $Na_{1+x}V_3O_8$ (Fig. 83) (2328). Again the five coordinate vanadium atom could be considered to be octahedral if the next closest oxygen atom at 2·86 Å was considered bonded to the metal atom.

2. *Niobium and tantalum*

The structural chemistry of Ta_2O_5 (1478, 1515) and Nb_2O_5 is very complex, and only some general principles will be illustrated here.

Four of the structural isomers of Nb_2O_5 are illustrated in Figs. 84, 85, 86 and 79. In the first (Fig. 84), blocks of corner sharing octahedra, 3 octahedra × 3 octahedra, extend normal to the plane of the paper; these ReO_3-type blocks share octahedral edges with four neighbouring blocks as shown. The composition based on these octahedra is Nb_9O_{25}, and the

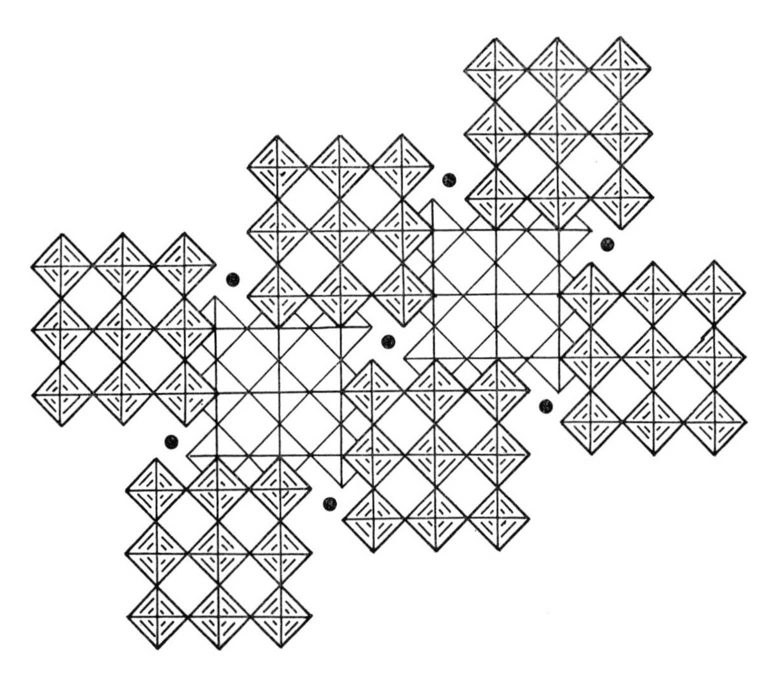

FIG. 84 The structure of Nb_2O_5. Blocks of corner shared NbO_6 octahedra $3 \times 3 \times \infty$ extend normal to the plane of the page. Each octahedron on the edge of each block shares an edge with an octahedron on a neighbouring block. Additional metal atoms in tetrahedral sites are indicated by circles.

composition $Nb_{10}O_{25}$ is attained by additional tetrahedrally coordinated niobium atoms near the corners of each block. In the second form, H-Nb_2O_5 (Fig. 85), the ReO_3-type blocks are larger, half being 4 octahedra $\times$ 3 octahedra and the other half being 5 octahedra $\times$ 3 octahedra. How-

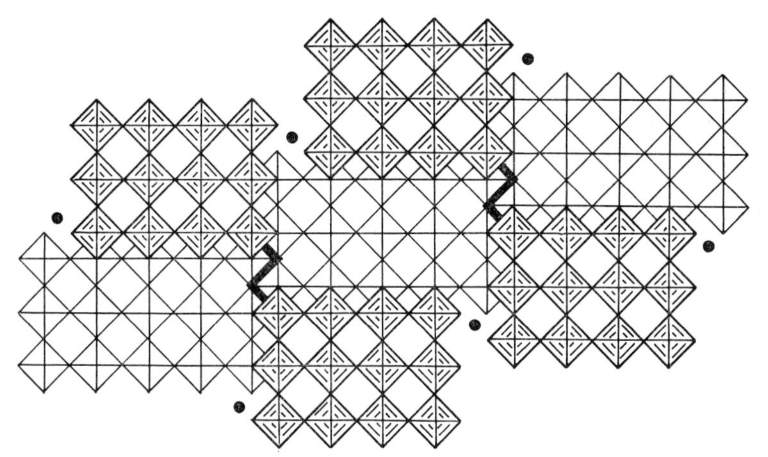

FIG. 85 The structure of H–Nb_2O_5.

ever the oxygen : metal ratio is reduced by edge sharing of octahedra between the 5 $\times$ 3 blocks. Tetrahedrally coordinated metal atoms are again situated between the blocks (983). In the third form, N–Nb_2O_5 (Fig. 86), all metal atoms are octahedrally coordinated, the corner shared ReO_3-type blocks being 4 octahedra $\times$ 4 octahedra $\times$ ∞ (62). In a fourth

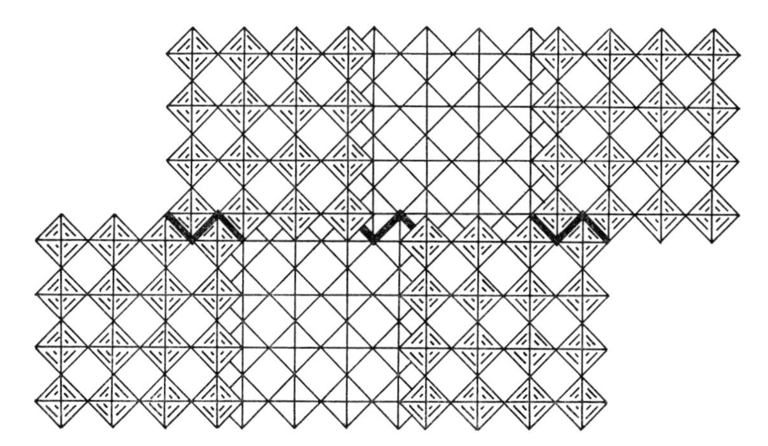

FIG. 86 The structure of N–Nb_2O_5. (Compare with Fig. 81).

form, R–Nb_2O_5 (Fig. 79), the ReO_3-type blocks are 2 octahedra thick and extend infinitely in the other two dimensions, which is the same as that found for V_2O_5 (1075).

Closely related oxides can be obtained with slightly lower oxygen : metal ratios by mild reduction of Nb_2O_5 or by incorporating metal ions of lower valency into the structure, and conversely higher oxygen : metal ratios can be obtained by replacing the niobium(V) atoms with ions of higher charge, or by replacing some of the oxide ions with fluoride ions.

For higher oxygen : metal ratios, the structures contain larger ReO_3-type blocks than in Nb_2O_5. For example $WNb_{12}O_{33}$ ($MO_{2\cdot54}$) has blocks (4 octahedra $\times$ 3 octahedra) with again isolated metal atoms in the tetrahedral holes between the blocks (Fig. 87). Larger blocks are found in

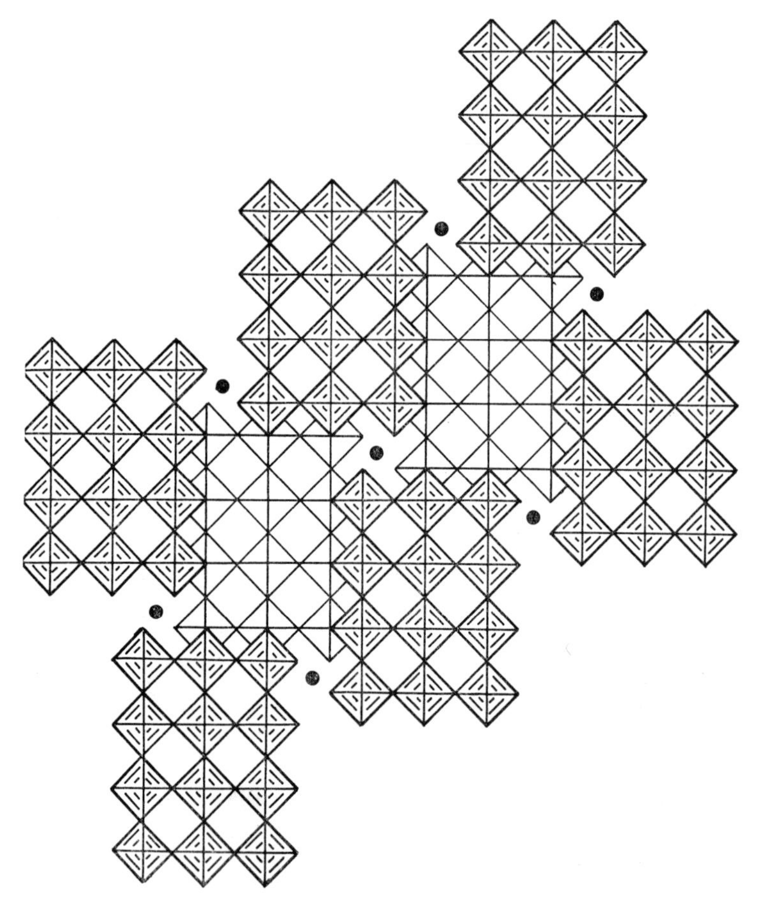

FIG. 87 The structure of $WNb_{12}O_{33}$.

$W_3Nb_{14}O_{44}$ ($MO_{2\cdot59}$) (4 octahedra × 4 octahedra) (Fig. 88), $W_5Nb_{16}O_{55}$ ($MO_{2\cdot62}$) (blocks 5 octahedra × 4 octahedra), $W_8Nb_{18}O_{69}$ ($MO_{2\cdot65}$) (blocks 5 octahedra × 5 octahedra), and so on (1953). Intermediate compositions such as $W_4Nb_{26}O_{77}$ ($MO_{2\cdot57}$), whose composition is the sum of $WNb_{12}O_{33}$ and $W_3Nb_{14}O_{44}$, has a mixture of (4 × 3) blocks and (4 × 4) blocks in each layer (Fig. 89) (66).

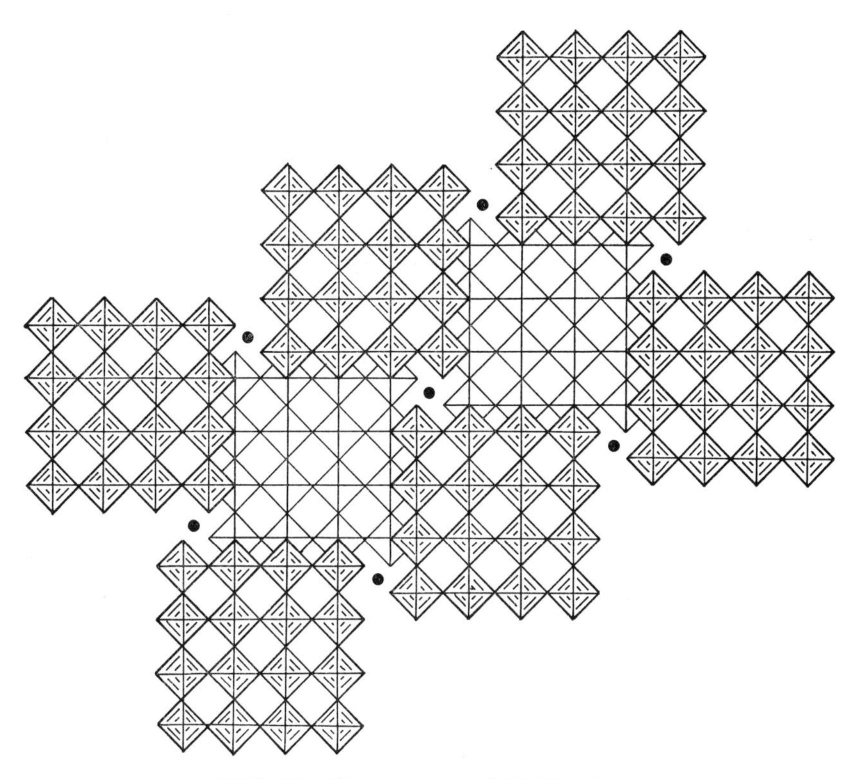

FIG. 88 The structure of $W_3Nb_{14}O_{44}$.

The stoicheiometry of the second form of Nb_2O_5 can also be changed by varying the size of the blocks from (4 × 3) and (5 × 3) in H–Nb_2O_5. Increased anion content is again attained by increasing the size of the blocks, for example to (5 × 3) and (5 × 3) in $Nb_{31}O_{77}F$ ($MO_{2\cdot51}$) (Fig. 90), or to (5 × 3) and (6 × 3) in $Nb_{17}O_{42}F$ ($MO_{2\cdot53}$) (93). Conversely a lower oxygen : metal stoicheiometry is attained by decreasing the size of the blocks, for example to (4 × 3) and (4 × 3) in $Nb_{25}O_{62}$ (1798) and $TiNb_{24}O_{62}$ ($MO_{2\cdot48}$) (1952).

These formidable looking formulae can be reduced to a general homo-

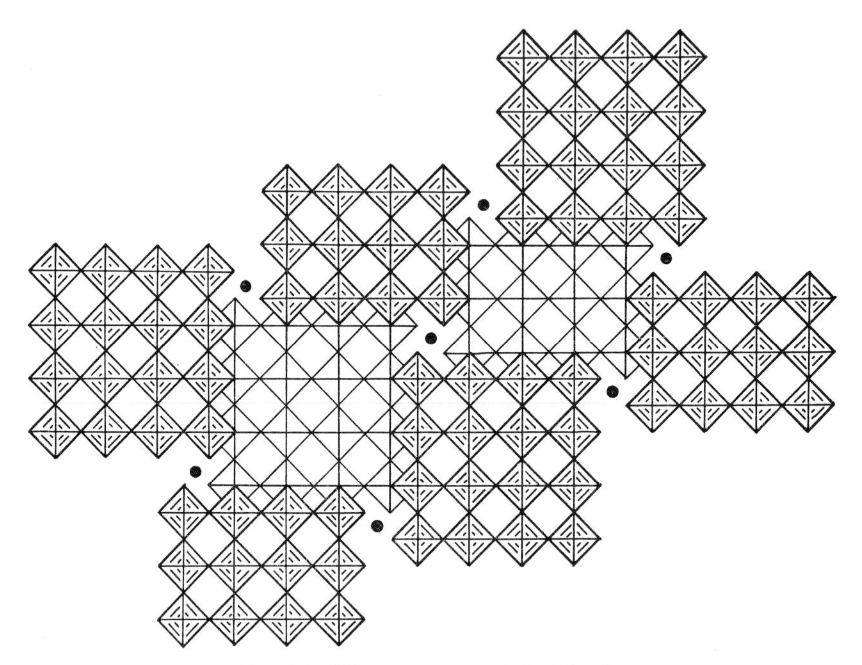

FIG. 89 The structure of $W_4Nb_{26}O_{77}$.

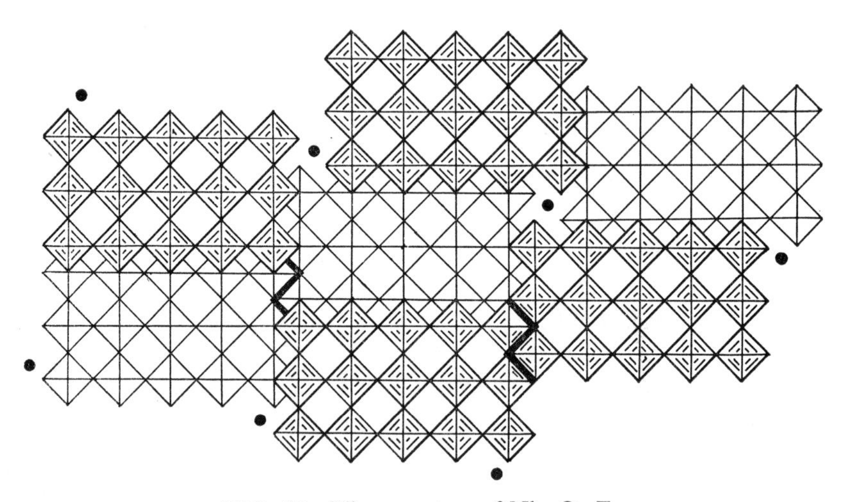

FIG. 90 The structure of $Nb_{31}O_{77}F$.

logous series $M_{3n+1}O_{8n-2}$, where $n = 7$ ($Nb_{22}O_{54}$) (983), $n = 8$ ($Nb_{25}O_{62}$ and $TiNb_{24}O_{62}$), $n = 9$ (Nb_2O_5), $n = 10$ ($Nb_{31}O_{77}F$), and $n = 11$ ($Nb_{17}O_{42}F$). Compositions between $NbO_{2\cdot48}$ ($n = 8$) and Nb_2O_5 ($n = 9$), for example $NbO_{2\cdot49}$ or $Nb_{53}O_{132}$, have elements of both the Nb_2O_5 (4×3 blocks) and $Nb_{25}O_{62}$ (4×3 blocks sharing edges) structures. Similarly $Nb_{47}O_{116}$ has elements of both the $Nb_{22}O_{54}$ ($n = 7$) and $Nb_{25}O_{62}$ ($n = 8$) structures (1078, 1079). The same feature is observed in the metal deficient phases; $Nb_{59}O_{147}F$ ($MO_{2\cdot51}$) can be considered to be a "mixture" of Nb_2O_5 ($n = 9$) and $Nb_{31}O_{77}F$ ($n = 10$), and $Nb_{65}O_{161}F_3$ ($MO_{2\cdot52}$) to be a "mixture" of $Nb_{31}O_{77}F$ ($n = 10$) and $Nb_{17}O_{42}F$ ($n = 11$) (1076).

Additional edge sharing of octahedra between the (4×3) blocks within the layers also leads to lower oxygen content, as in $Nb_{12}O_{29}$ (1797) and $Ti_2Nb_{10}O_{29}$ (2329), both of which exist in orthorhombic and monoclinic modifications (Fig. 91).

The stoicheiometry of H–Nb_2O_5 can also be varied by altering the proportion of layers containing the (4×3) blocks, to the layers containing the edge-shared (5×3) blocks. In any one crystal the composition is not always uniform, as the stacking sequence may vary from one part of the crystal to another.

Just as V_6O_{13} (Fig. 81, page 144) can be considered to be a shear structure based on V_2O_5 or R–Nb_2O_5 (Fig. 79, page 143), so can these structures containing blocks three octahedra wide be considered to be shear structures based on the V_2MoO_6 or Nb_3O_7F structures (Fig. 80, page 144). It may also be noted that a slight shift in this shear plane forms holes large enough to accommodate alkali metal cations, as in $NaNb_{13}O_{33}$ (60).

Other niobium and tantalum oxides are based on different structural principles, and in addition to octahedrally coordinated metal atoms, contain seven coordinate pentagonal bipyramidal metal atoms. The structures of $LiNb_6O_{15}F$ (1575) and $NaNb_6O_{15}F$ (61) are shown in Figs. 92 and 93 respectively. (The alkali metal atoms were not located). The relationship between these structures and those of tetragonal K_xWO_3, Mo_5O_{14} and $Nb_{16}W_{18}O_{94}$ is discussed on page 260. A number of other mixed oxides in the A_2O–M_2O_5 (where A is Na or K and M is Nb or Ta) and BO–Nb_2O_5 (where B is Ba or Pb) systems show X-ray powder patterns very similar to those of the tetragonal potassium tungsten bronzes, and those with oxygen : metal ratios less than 3 :1 presumably also contain pentagonal bipyramidal metal atoms (2163). Similarly Ta_3O_7F has been reported to be isomorphous with $LiNb_6O_{15}F$ (1271). The 3 :1 ratio however is retained in KNb_2O_5F, which has the normal tetragonal K_xWO_3 structure (1635).

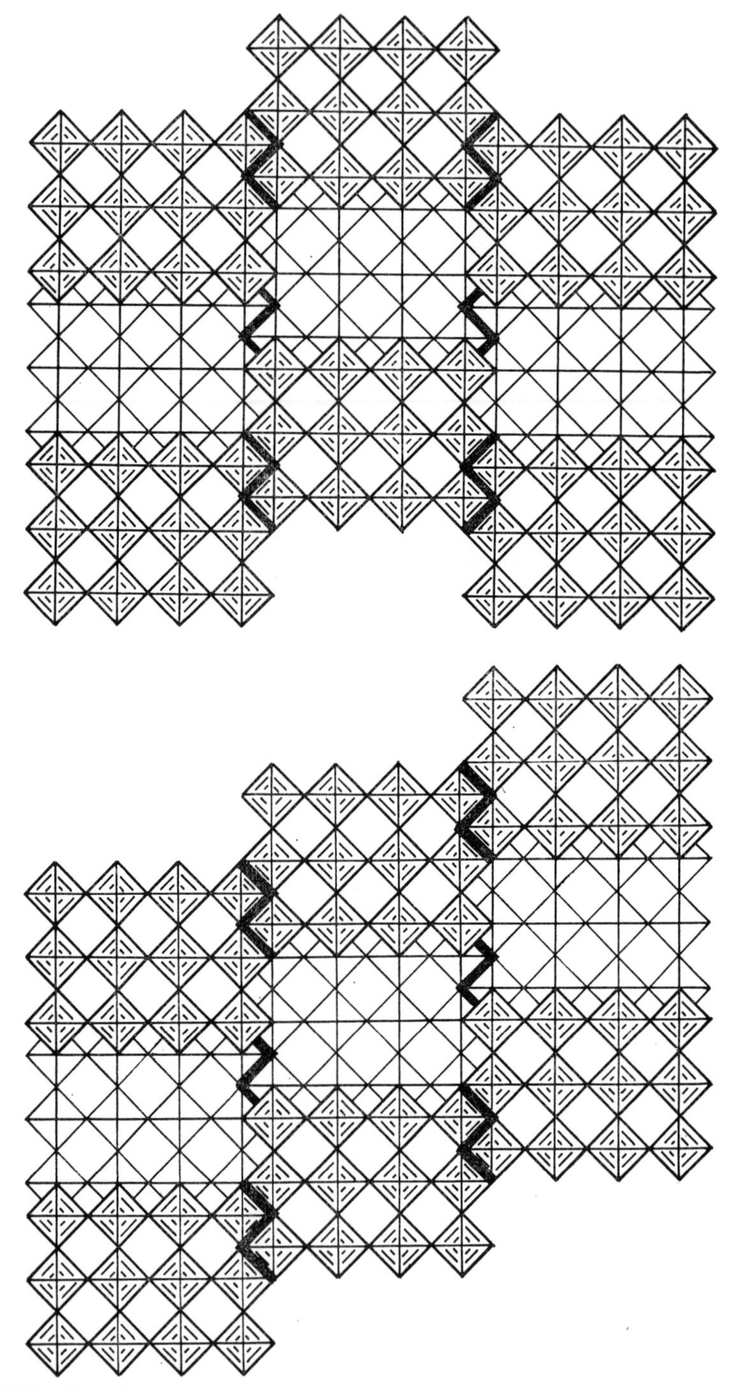

FIG. 91 The structure of the orthorhombic (a) and monoclinic (b) modifications of $Nb_{12}O_{29}$.

FIG. 92 The structure of $LiNb_6O_{15}F$. Each octahedron and pentagonal bipyramid shares corners with similar polyhedra above and below the plane of the page.

FIG. 93 The structure of $NaNb_6O_{15}F$. Each octahedron and pentagonal bipyramid shares corners with similar polyhedra above and below the plane of the page.

Niobium oxides related to the cubic tungsten bronzes Na_xWO_3 (page 265) include Sr_xNbO_3. As the value of x decreases fron 0·95 to 0·7, the colour changes from red to deep blue, and the cell constant decreases from 4·02 to 3·98 Å (1922). Niobium and tantalum can also be substituted into the hexagonal tungsten bronzes $Rb_{0.33}WO_3$ to form, for example $Rb_{0.33}Nb_{0.33}W_{0.67}O_3$; these compounds are white as expected for the formal d° electron configurations (1265).

B. Pentahalides

The known pentahalides are:

VF_5	NbF_5	TaF_5
—	$NbCl_5$	$TaCl_5$
—	$NbBr_5$	$TaBr_5$
—	NbI_5	TaI_5

All these pentahalides can be prepared by the direct halogenation of the metals or of the oxides, although the latter method can result in the formation of oxohalide impurities, particularly in the case of niobium. The niobium compounds are also more readily reduced than the tantalum compounds, and the halogen-to-metal charge transfer transitions occur at lower energies, leading to more yellow or orange colours. All compounds are unstable to moist air, water, and hydroxylic solvents, and these reactions are discussed later.

In all cases the coordination about the metal atom in the solid state is octahedral, and all structures could be considered to be based on close packing of halide ions with metal atoms in one fifth of the octahedral holes. However the relatively high volatility of these compounds, their solubility in nonpolar solvents, and the persistence of polymeric units in the liquid and solution state, indicate that they are better considered as covalent molecules. In all cases the MX_6 octahedra share two mutually *cis*-halogen atoms with a neighbouring octahedron or octahedra, to form dimers as in $(NbCl_5)_2$, tetramers as in $(NbF_5)_4$, or infinite polymers as in $(NbI_5)_{\infty}$. These structures are in sharp contrast to, for example, the structures of the phosphorus pentahalides.

1. *Pentafluorides*

Vanadium pentafluoride is prepared by the action of fluorine or bromine trifluoride on vanadium, vanadium tetrafluoride, or vanadium pentoxide at about 300° C (443, 774, 2167, 2302). The white solid melts at 190° C, boils at 48° C, and is monomeric in the vapour state with trigonal bipyramidal stereochemistry (445, 482, 486, 2302). The liquid state is highly associated, as shown by the very high Trouton constant of 33·1 (compared with, for example, 20·6 for MoF_6) (486), the high viscosity (444), and Raman spectra measurements (482).

Vanadium pentafluoride is extremely reactive, and reacts with water, air, grease, glass, etc. It often appears yellow, due to the attack on glass even at room temperature to form yellow VOF_3 and V_2O_5. It is a more powerful oxidizing agent and fluorinating agent than NbF_5, TaF_5, MoF_6, WF_5 and UF_6. It fluorinates carbon tetrachloride to CCl_2F_2, $CClF_3$ and CF_4, and oxidizes PF_3 to PF_5, I_2 to IF_5, S to SF_4, MoF_5 to MoF_6, and UF_4 to UF_6, in all cases being itself reduced to VF_4 (419).

Vanadium pentafluoride is isomorphous with molybdenum oxotetrafluoride, and with the pentafluorides of chromium, technetium and rhenium. Infinite chains of octahedrally coordinated vanadium atoms are formed by the sharing of fluorine atoms in the *cis*-positions (Fig. 94)

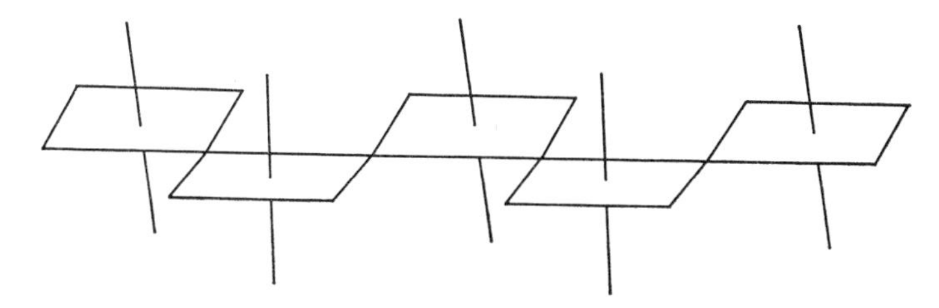

FIG. 94 The structure of VF_5.

(735, 736). The V–F–V angle of 150° compares with about 180° in TaF_5, MoF_5 and WF_5, and about 130° in RuF_5, OsF_5, RhF_5, IrF_5 and PtF_5. The vanadium-bridging fluorine bond length is 1·97 Å, and the vanadium-nonbridging fluorine bond length of 1·69 Å is the same as that in the vapour phase (1943).

The reaction between niobium and fluorine at 300° C, or other fluorinating agents such as chlorine trifluoride, bromine trifluoride, or hydrogen fluoride, results in the formation of niobium pentafluoride (775, 1970). Alternatively it may be more convenient to start with niobium(V) compounds such as $NbCl_5$ with ZnF_2 (2027) or HF (1968), or K_2NbF_7 (2027). Niobium pentoxide appears a less satisfactory material due to the formation of stable oxyfluorides. Glass is an unsatisfactory reaction vessel, and reacts even with the pure pentafluoride at 150° C; quartz, nickel and platinum vessels have been used, although quartz reacts above 400° C forming NbO_2F and SiF_4.

Tantalum pentafluoride is prepared by analogous methods (775, 805, 1968, 1970), although the reaction appears somewhat slower and more difficult than for the niobium compound (in spite of the higher standard enthalpy of formation (1054)).

The colourless pentafluorides of niobium and tantalum are more stable to air than the other pentahalides. The melting and boiling points are (805, 1322, 1968):

NbF_5:	M.P. = 79° C	B.P. = 234° C
TaF_5:	M.P. = 96° C	B.P. = 225° C

The structure of niobium and tantalum (and molybdenum) pentafluorides is shown in Fig. 95a as the tetrameric molecule, and in Fig. 95b as a cubic close packed fluoride lattice formed from these tetramers packing together (729). The metal-fluorine distances are 2·06 Å (bridging fluorine) and 1·77 Å (non-bridging fluorine). The linear metal-fluorine-metal bridges, in contrast to the bent bridges for other pentafluorides (see above), may be due to significant fluorine-to-metal p_π–d_π bonding (418). The infrared spectrum of gaseous niobium pentafluoride is consistent with a trigonal bipyramidal stereochemistry (240).

The mixed fluorochloro compounds $NbCl_4F$ (1442) and $TaCl_4F$ (1443) have been prepared by the reaction of the compounds $MCl_4.PCl_5$ with the stoicheiometric amount of AsF_3 in $AsCl_3$ as solvent:

$$MCl_5.PCl_5 + 2\,AsF_3 \rightarrow MCl_4F + PF_5 + 2\,AsCl_3$$

The compounds are tetrameric and the structures (1897, 1898) are closely related to those of the pentafluorides with bridging fluorine atoms but terminal chlorine atoms.

2. *Pentachlorides, bromides and iodides*

The reaction between dry chlorine and niobium metal at 200°–500° C forms niobium pentachloride as a yellow sublimate. Good yields can be obtained from the pentoxide by mixing the latter with carbon before chlorination, although if the temperature is too high there is considerable contamination with white $NbOCl_3$; however the two compounds can be separated by vacuum sublimation, the oxychloride being less volatile. A number of other chlorinating agents have been used. For example prolonged reluxing with thionyl chloride converts the oxide to the pentachloride (282, 961). The oxide can similarly be chlorinated with carbon tetrachloride in a suitable bomb with the simultaneous formation of carbon dioxide and phosgene (454, 961), although it is more convenient to use a higher boiling point chlorinated hydrocarbon at atmospheric pressure such as refluxing hexachlorobutadiene (100) or octachlorocyclopentene (144).

Probably because of the instability of niobium pentachloride to air, there is not good agreement concerning its physical properties, for example the more reliable determinations of the melting point have been given as 206° C (1833) and 210° C (29), and the boiling point as 247° C (23) and 254° C (29).

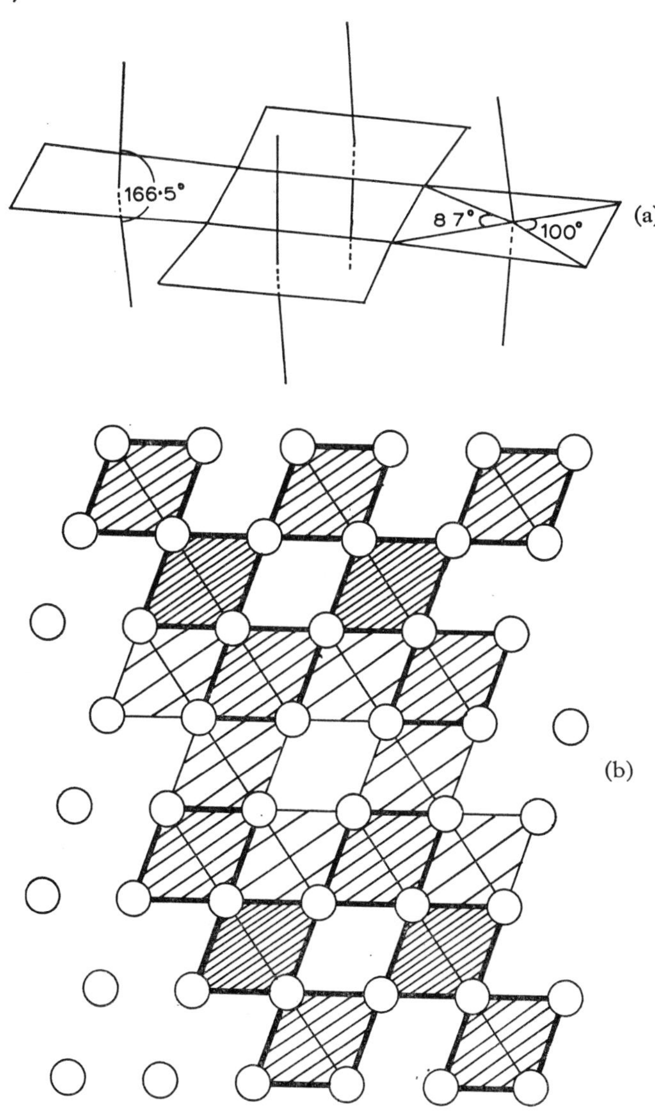

FIG. 95 The structures of NbF_5 and TaF_5. (a) The tetrameric molecule. (b) The packing of the tetrameric molecules viewed side on showing the relationship to a cubic close packed fluoride lattice.

Tantalum pentachloride is prepared in an analogous manner to niobium pentachloride, and although higher temperatures are normally required a purer product is usually obtained as there is less tendency to form tantalum oxychlorides. Thus tantalum, tantalum pentoxide, or tantalum pentoxide-carbon mixtures can be readily chlorinated with gaseous chlorine, thionyl chloride, aluminium trichloride (448), carbon tetrachloride in sealed tubes at 300° C, or higher halocarbons at atmospheric pressure (94). Niobium and tantalum can be partially separated in this manner. For example if the mixed pentoxides are heated to 270° C with carbon tetrachloride in a sealed tube, the Ta_2O_5 is hardly attacked whereas the Nb_2O_5 is almost completely transformed to the pentachloride, which can be separated by sublimation (454, 961).

Pure tantalum pentachloride is white, the pale yellow colour quoted in the old literature being due to niobium pentachloride impurities. The melting and boiling points are fairly close to 200° and 239° C respectively (29). The separation of niobium pentachloride from tantalum pentachloride by fractional sublimation is not very successful, and a better method is the fractional distillation of the liquids.

The orange yellow niobium pentabromide and pale yellow tantalum pentabromide are prepared in a completely analogous manner to the pentachlorides, that is by heating the metals, pentoxides or pentoxide-carbon mixtures with bromine at about 500° C, or other brominating agents such as aluminium tribromide (448) or carbon tetrabromide at about 200° C (454). Niobium pentabromide has also been prepared by halogen exchange between niobium pentachloride and boron tribromide at room temperature (694). It is again found that the niobium compound is more sensitive to oxygen impurities, the less volatile yellow $NbOBr_3$ being readily formed.

As was also noted for the pentachlorides, niobium pentabromide has a lower melting point but higher boiling point than tantalum pentabromide (29, 210, 2405):

$NbBr_5$:	M.P. = 254°–268° C	B.P. = 356°–363° C
$TaBr_5$:	M.P. = 280° C	B.P. = 345°–349° C

Niobium and tantalum pentaiodides can be prepared from the elements above about 300° C, as dark brown or black sublimable solids (28, 1458, 1568, 1700, 1938). The tantalum compound may be prepared directly from the pentoxide, but similar reactions with niobium were less successful (449, 451, 1792). The pentaiodides can also be prepared by halogen exchange from the pentachlorides or pentabromides by heating with aluminium triiodide, silicon tetraiodide, or hydrogen iodide (149, 1792, 2314).

Niobium pentaiodide dissociates below its melting point liberating iodine (28, 1458), and products prepared by the above techniques may often contain lower iodides. Decomposition to the metal ultimately occurs above 700° C. Tantalum pentaiodide appears stable up to about 500° C, and the melting point has been variously given as 365° and 496° C, and the boiling point as 400° and 543° C (28, 1458).

Reaction of mixtures of aluminium tribromide and aluminium triiodide with tantalum pentoxide have been reported to yield the red brown $TaBr_4I$ and dark brown $TaBrI_4$ (453).

The structure of niobium pentachloride in the solid state shows a dimeric molecule Nb_2Cl_{10}, consisting of two octahedra sharing a common edge (Fig. 96a) (2435). The octahedra are distorted due to mutual repulsion of the niobium atoms from their octahedral centres. The packing of these dimeric molecules is such that the chlorine atoms form a hexagonally close

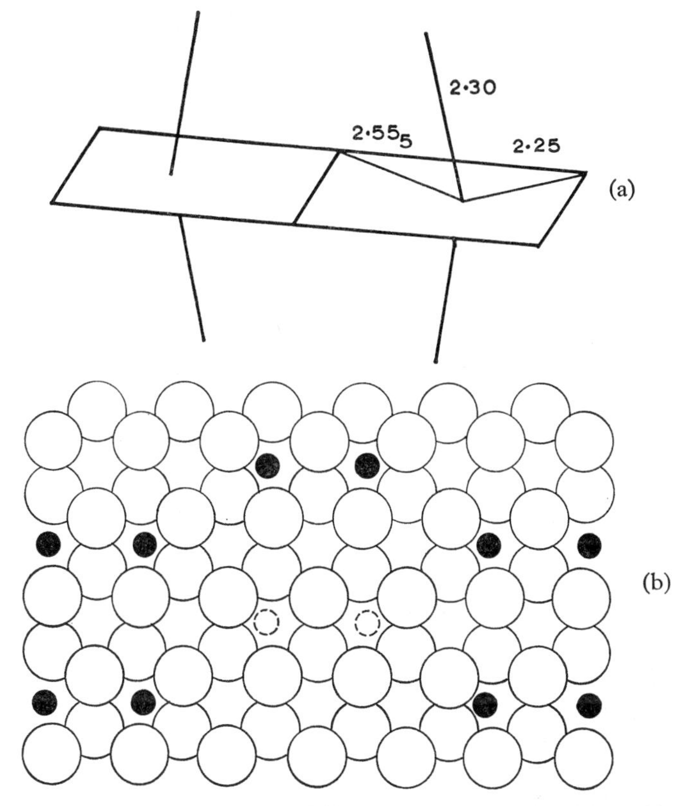

FIG. 96 The structure of $NbCl_5$. (a) The dimeric molecule. (b) Packing of the dimeric molecules showing relationship to a close packed chloride lattice.

packed structure with niobium atoms occupying one fifth of the octahedral holes (Fig. 96b). Tantalum pentachloride and pentabromide, and niobium pentabromide, are isomorphous with niobium pentachloride (212, 1938, 1939, 2435). Dimeric molecules M_2Cl_{10} are also found in UCl_5 (2175), $PaCl_5$ and β-$PaBr_5$ (360), but now the dimers are stacked so that there is a cubic close packing of halide ions. Another isomer of $PaCl_5$ however has seven coordinate metal atoms, linked by edge sharing of polyhedra into infinite strings (675).

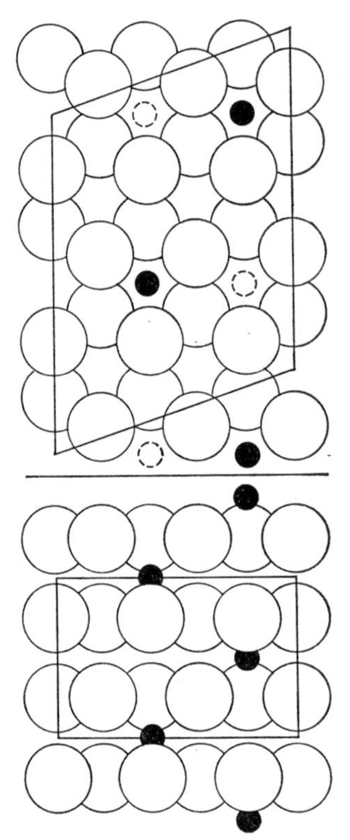

FIG. 97 The structure of NbI_5, showing relationship to a close packed iodide lattice. (Compare with Fig. 94).

Niobium pentaiodide has a different structure to the other pentahalides, although it is again found that the metal atom is octahedrally coordinated with two *cis*-halogen atoms shared with adjacent octahedra. In this case however a linear polymer is formed similar to that in vanadium pentafluoride (Fig. 94) (1568). Once again the structure can be related to hexagonal close packing of iodide ions (Fig. 97). Tantalum pentaiodide does not appear to be isomorphous with the niobium compound.

In the gaseous state $NbCl_5$, $TaCl_5$ and $TaBr_5$ are monomeric (1793, 2405), and the decrease in molecular complexity can also be inferred from the high entropy of sublimation of these pentahalides (39–44 cal deg.$^{-1}$) (29, 210, 805). The Raman spectra of the gases are consistent with a trigonal bipyramidal structure (176). Early work on the interpretation of electron diffraction data indicated trigonal bipyramidal stereochemistries, with the axial and planar bond lengths equal within experimental error. The bond lengths obtained were $2{\cdot}30 \pm 0{\cdot}03$ Å for the pentachlorides and $2{\cdot}45 \pm 0{\cdot}03$ Å for the pentabromides (2161).

Niobium and tantalum pentachlorides, and to a lesser extent the other pentahalides, are readily soluble in dry non-hydroxylic solvents such as carbon tetrachloride, chloroform (and other halo-hydrocarbons), carbon disulphide, benzene or acetonitrile. For example at 20° C, tantalum pentachloride is 0·8% and 3·1% soluble in carbon tetrachloride and carbon disulphide respectively (810). In hydroxylic solvents hydrogen chloride is readily evolved, and the resultant compounds are discussed later (pages 175 and 178).

In the liquid state the pentachlorides (230, 231) and pentafluorides (806, 807) are virtually nonconducting. The viscosities of the pentafluorides are considerably greater than the pentachlorides, indicating the persistence of polymeric molecules in the former case (807).

The infrared and Raman spectra of these pentahalides have been investigated by a number of workers. In the solid and liquid states, and as solutions in carbon disulphide and carbon tetrachloride, the pentachlorides show similar spectra which were assigned to Jahn–Teller distorted trigonal bipyramidal structures with varying success (109, 425, 985, 1106, 1728), in spite of the known structures in the solid state. However more recent work on the solids has been interpreted in terms of the known dimeric octahedrally coordinated structures (174, 2331). A major difficulty with these compounds is the preparation and manipulation of dry, oxygen-free solutions, and additional spectral bands in the solution spectra are probably due to decomposition products. For example niobium pentachloride formed significantly conducting solutions when it was 10^{-3} M in solvents such as nitromethane and acetonitrile, but it was shown that the conducting species were decomposition products arising from water impurities, even although stringent precautions were taken (1360). Indeed water can be conductometrically titrated against niobium pentachloride to give an end point at $H_2O/NbCl_5 = 5/1$ (1360). Nevertheless the molecular weights in carbon tetrachloride and nitromethane have been determined, and the dimeric form is retained in solution (1360). In donor solvents such as acetonitrile however monomeric compounds are formed (page 162). It is of relevance to the infrared and Raman results above to point out that addition compounds are also probably formed in carbon disulphide. Niobium pentachloride is also a non-electrolyte in propylene carbonate (1140).

Niobium pentaiodide dissociates in carbon tetrachloride liberating iodine (471), whereas tantalum pentaiodide remains unchanged.

C. Adducts of the Pentahalides

Niobium and tantalum pentafluorides, chlorides and bromides are Friedel–Crafts catalysts, and also readily form addition compounds. Tantalum pentaiodide on the other hand does not appear to have significant Friedel–Crafts activity, and nor does it appear to form addition compounds with, for example, ethers (591).

The compounds generally obtained are monomeric, octahedral MX_5 (ligand), although there are also indications of seven coordinate compounds, the best characterized probably being MX_5(diarsine). (In contrast antimony pentachloride forms only six coordinate complexes (2353)). The existence of $NbCl_5(ligand)_{\frac{1}{2}}$ with ligands such as triphenylamine should also be noted.

Vanadium pentafluoride does not appear to form simple adducts, but forms vanadium(IV) with ligands such as pyridine, ammonia and ethylenediamine (page 200) (442). Reduction does not occur with reactants such as NOF or NO_2F (488), but here the products should probably be described as $(NO^+)(VF_6^-)$ and $(NO_6^+)(VF_6^-)$ (page 169).

1. *Oxygen, sulphur and selenium donors*

Niobium pentafluoride and tantalum pentafluoride react with dimethyl ether and dimethyl sulphide forming adducts of the type $MF_5(Me_2O)$ and $MF_5(Me_2S)$, with evidence of further solvation at room temperature. These compounds are volatile solids which melt below 100° C, while the four corresponding ethyl derivatives are all liquids (807). Dimethylsulphoxide forms the bis-adducts $NbF_5(Me_2SO)_2$ and $TaF_5(Me_2SO)_2$, which also melt below 100° C (808).

The pentachlorides and pentabromides form adducts at room temperature of the type MX_5(Ether) with dimethyl ether, diethyl ether, dipropyl ether, pentamethylene oxide, and 1,4-dioxane (548, 591, 819). Further solvation may occur at 0° C. Polymeric organic products are obtained with tetrahydrofuran. At about 100° C oxygen abstraction from the ethers occurs to produce the oxohalide and alkyl halide. Oxygen abstraction to give adducts of the oxotrihalides also occurs with excess Ph_3AsO, Ph_3PO, Me_2SO and Ph_2SO which are discussed on page 169

(353, 547, 549). However with no excess ligand, $NbCl_5(Ph_3PO)$ and $TaCl_5(Ph_3PO)$ may be obtained, while $NbCl_5[(Me_2N)_3PO]$, $TaCl_5[(Me_2N)_3PO]$ and $TaCl_5(Ph_2BzPO)$ are obtained even with excess ligand (353).

Phosphorus oxychloride similarly forms $NbCl_5(POCl_3)$ and $TaCl_5(POCl_3)$ (1092, 1790, 2119). These compounds are monomeric in benzene but dissociate in nitrobenzene, the degree of dissociation increasing in the order Ta $<$ Nb $<$ Sb (2117, 2320). They fully dissociate in the vapour (1092). The structure of the niobium compound confirms octahedral coordination, the important parameters being (314):

$$\begin{aligned} \text{Nb–O} &= 2\cdot16\ \text{Å} \\ \text{Nb–Cl} &= 2\cdot25\text{–}2\cdot35\ (\text{av.} = 2\cdot30)\ \text{Å} \\ \text{Nb–O–P} &= 149°\ \text{C}. \end{aligned}$$

Niobium and tantalum pentachlorides and pentabromides dissolve in dimethyl sulphide or diethyl sulphide to give red solutions from which yellow compounds can be obtained of composition $MX_5(R_2S)$ (809). These compounds have unusually low melting points, for example $NbCl_5(Et_2S)$ melts at 27° C (548). Further solvation may occur at lower temperatures, for example $TaCl_5(Et_2S)$ forms $TaCl_5(Et_2S)_2$ below 15° C. Pentamethylene sulphide forms 1 :1 adducts, whereas tetrahydrothiophene has been reported to yield both 1 : 1 and 1 : 2 adducts with all four pentahalides (809, 819).

It has been reported that dimethyl sulphide will not readily displace diethyl ether from $TaCl_5(Me_2O)$, nor will dimethyl ether readily displace dimethyl sulphide from $TaCl_5(Me_2S)$, although in both cases products of composition $TaCl_5(Me_2O)(Me_2S)$ may be isolated at about −30° C, but on warming to room temperature the second ligand which was added is evolved. On the other hand it is interesting that diethyl sulphide and dipropylsulphide displace the ether from $NbCl_5(Et_2O)$ and $NbCl_5(Pr_2O)$ respectively, suggesting that the sulphur ligand is more strongly bonded to the metal than the oxygen ligand (548).

Similarly it should be noted that thiodioxane and selenodioxane form $NbCl_5(C_4H_8OS)$ and $NbCl_5(C_4H_8OSe)$, which on the basis of infrared and n.m.r. data were suggested to be sulphur- and selenium-bonded, and not oxygen-bonded, to the metal atom (127, 819).

2. *Nitrogen, phosphorus and arsenic donors*

In contrast to the behaviour with vanadium pentafluoride, niobium and tantalum pentafluorides form simple compounds with nitrogen-donor ligands such as $NbF_5(NH_3)_2$, $NbF_5(EtNH_2)_2$, $NbF_5(Et_2NH)$, $NbF_5(Et_3N)$, $NbF_5(py)_2$, and $NbF_5(en)_{1\cdot6}$ (395, 442, 488).

Niobium and tantalum pentachlorides and pentabromides react with trimethylamine and appear to give compounds of the type $MX_5(Me_3N)_2$ (429, 927). Triphenylamine forms either $NbCl_5,Ph_3N$ or $NbCl_5,\frac{1}{2} Ph_3N$ depending on conditions (652).

The reaction of niobium pentachloride and pentabromide with pyridine leads initially to the formation of $NbX_5(py)$, which are rapidly reduced to $NbX_4(py)_2$ with formation of 1-(4-pyridyl) pyridinium dichloride (33, 1585). With the tantalum pentahalides there have been reports of the formation of $TaX_5(py)$ (where X is Cl or Br) (1585) and $TaCl_5(py)_2$ (1543), as well as reduction to $TaX_4(py)_2$ (where X is Cl or Br) (33). Reduction of the pentaiodides with pyridine has also been noted (1581, 1585). In acetonitrile at 0° C, dipyridyl forms $NbCl_5$(dipy)(MeCN) and TaX_5(dipy)(MeCN) (where X is Cl or Br), for which infrared evidence suggests that the acetonitrile is not coordinated. However niobium pentabromide and tantalum pentaiodide under the same conditions form MX_5,2 dipy, which were formulated as the eight coordinate $[MX_4(dipy)_2]X$ (934). Reduction occurs above 0° C, as has also been observed with other heterocyclic ligands (33). These products have also been formed directly from the tetrahalides (page 202).

Niobium and tantalum pentachlorides, bromides and iodides with alkylnitriles readily form the well-characterized MX_5(RCN) (where R is Me, Et, Pr) (818, 934, 1360). The molecular weight of $NbCl_5$(MeCN) in boiling acetonitrile shows it to be monomeric, but to be partly dissociated in boiling nitromethane (1360).

The influence of solvent has been demonstrated with the Group V triphenyls, which form $NbCl_5(Ph_3X)$ (where X is N, P, As, Sb or Bi) from carbon tetrachloride, *n*-hexane or cyclohexane, but $NbCl_5,\frac{1}{2} Ph_3X$ from benzene (652).

The bidentate *o*-phenylenebisdimethylarsine forms MX_5(diars) (where M is Nb or Ta and X is Cl or Br) (502). These compounds are isomorphous, diamagnetic, monomeric, and non-conducting, and are therefore seven coordinate.

D. Oxohalides and Adducts

1. *Oxohalides*

The known oxotrihalides and dioxohalides are given below:

VOF_3	—	—
$VOCl_3$	$NbOCl_3$	$TaOCl_3$
$VOBr_3$	$NbOBr_3$	$TaOBr_3$
—	$NbOI_3$	—

VO_2F	NbO_2F	TaO_2F
VO_2Cl	NbO_2Cl	TaO_2Cl
—	NbO_2Br	TaO_2Br
—	NbO_2I	TaO_2I

In addition niobium forms Nb_3O_7F and Nb_3O_7Cl. The structure of Nb_3O_7Cl (2066) consists of octahedra sharing edges and corners in an infinite three dimensional structure, and as such is more closely related to the oxides than to the halides. Similarly a number of oxofluorides with complex formulae such as $Nb_{31}O_{77}F$ and $Nb_{65}O_{161}F_3$ have been mentioned on page 148.

Niobium and tantalum also form some analogous gaseous sulphur and selenium compounds which have been established by mass spectrometry, namely $NbSCl_3$, $TaSCl_3$, $NbSBr_3$, $NbSeCl_3$, $NbSeBr_3$ and $TaSeBr_3$.

(a) *Vanadium.* The oxofluoride VOF_3 has been formed by the action of oxygen on the trifluoride at red heat (1966), and from fluorine and the pentoxide at 475° C (2301). It has been briefly noted (644) that a convenient preparation is from $VOCl_3$ and ClF_3 at room temperature. The oxotrifluoride is a pale yellow solid which has been little studied. The melting point has been reported as 300° C (1966) and 110° C (2301), the latter value undoubtedly being closer to the true value. The infrared spectrum of the vapour has been interpreted as showing a trigonally distorted tetrahedron, with a vanadium-oxygen stretching frequency at 1058 cm^{-1} (239, 2104).

Vanadium dioxomonofluoride has been prepared by the action of fluorine on the corresponding dioxomonochloride (2355).

The oxides V_2O_3 or V_2O_5 can be chlorinated with dry chlorine at 600°–800° C (1966), or with aluminium trichloride (1299), to yield $VOCl_3$. Mixtures of the pentoxide and carbon are chlorinated at 300° C (1817). More recently the chlorination with aluminium trichloride has been carried out in LiCl–NaCl–KCl mixtures; $VOCl_3$ evolution begins at about 100° C and the reaction is essentially complete when the temperature reaches 250° C, although the reaction mixture is only partially molten at this temperature (683).

The physical properties of $VOCl_3$ are somewhat similar to VCl_4. For example it is a pale yellow liquid, M.P. = −80° C and B.P. −127° C. The vapour pressure has been determined from 15° to 125° C; at 20° C it is 14 mm of mercury (1817, 2257). It is extremely unstable to moist air. It is completely miscible with hydrocarbons and carbon tetrachloride. The visible and ultravoilet spectrum (666), electron diffraction data (1834), and infrared spectrum (1687) of the vapour have been interpreted as indicating a trigonally distorted tetrahedron. The high vanadium-oxygen stretching frequency of 1035 cm^{-1} indicates the occurrence of strong oxygen-metal π bonding (2100).

The corresponding oxobromide has been prepared by heating V_2O_3 with bromine and carbon according to the reaction (1884):

$$V_2O_3 + 3\ Br_2 + C \rightarrow 2\ VOBr_3 + CO$$

The analogous reaction with V_2O_5 has also been studied (1685, 1802). $VOBr_3$ is a deep red hygroscopic liquid, M.P. = −59° C and B.P. = 130° C at 100 mm of mercury. It slowly decomposes at room temperature to $VOBr_2$ and Br_2, and rapidly on gentle warming. The infrared spectrum has again been interpreted as showing a trigonally distorted tetrahedral structure, with a lower vanadium-oxygen stretching frequency of 1025 cm^{-1} (1685).

The orange-yellow vanadium dioxochloride has been prepared from the oxotrichloride and a number of oxygen-containing molecules (642):

$$VOCl_3 + Cl_2O \rightarrow VO_2Cl + 2\ Cl_2$$
$$VOCl_3 + O_3 \rightarrow VO_2Cl + O_2 + Cl_2$$
$$3\ VOCl_3 + As_2O_3 \rightarrow 3\ VO_2Cl + 2\ AsCl_3$$

An infrared band at 990 cm^{-1} shows the presence of non-bridging oxygen atoms (642, 651). The compound disproportionates above 180° C into V_2O_5 and $VOCl_3$ (1817).

(b) *Niobium and tantalum.* Niobium oxotrichloride is often produced as an impurity during the preparation of the pentachloride, particularly if prepared from the pentoxide. The yellow niobium pentachloride can be conveniently sublimed off in vacuo at about 100° C, followed by the white niobium oxytrichloride at about 200° C. The reaction with thionyl chloride can be made to yield pure $NbOCl_3$ if the stoicheiometric quantities are heated to 200° C according to the reaction:

$$Nb_2O_5 + 3\ SOCl_2 \rightarrow 2\ NbOCl_3 + 3\ SO_2$$

The oxytrichloride is also conveniently prepared by the action of dry oxygen on the pentachloride at about 300° C. Yet another convenient method is simply to heat the stoicheiometric quantity of pentachloride and pentoxide (1683, 2021).

Vapour pressure data from 200° to 400° C are consistent with monomeric $NbOCl_3$, although there is often some thermal dissociation to $NbCl_5$ and Nb_2O_5, and also possibly to Nb_3O_7Cl, particularly at the higher temperatures. The vapour pressure reaches one atmosphere at 335° C. (1016, 1141, 1233, 2021). The Raman spectrum in the gaseous state resembles that of $VOCl_3$, with a niobium-oxygen stretching frequency at 997 cm^{-1}, but the highest niobium-oxygen mode in the solid is at 769 cm^{-1} which is characteristic of –Nb–O–Nb– bonding (1828).

The solid $NbOCl_3$ obtained from sublimation appears as long fibrous needles. The structure (Fig. 98) shows the existence of infinite –Nb–O–Nb–O– chains, joining together dimeric units similar to those which occur in $NbCl_5$ (1988). The structure and properties are clearly in sharp contrast to those of $VOCl_3$.

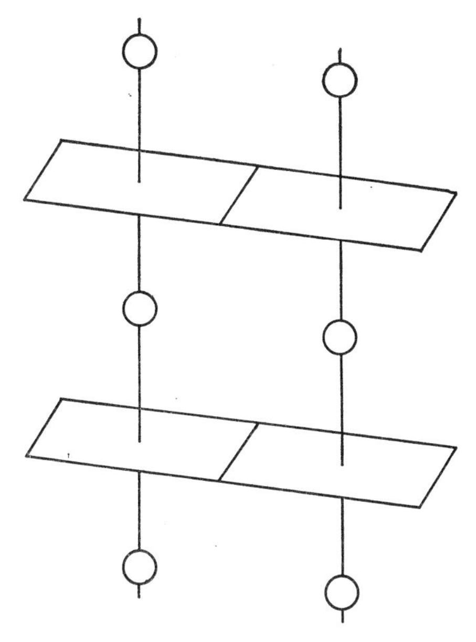

FIG. 98 The structure of $NbOCl_3$. Dimeric Nb_2Cl_6 units are linked into infinite ribbons through bridging oxygen atoms.

The analogous tantalum oxytrichloride has been much less extensively studied (640, 804, 1722, 2031). It can be prepared from the action of Ta_2O_5, Sb_2O_3 or Cl_2O on $TaCl_5$. Unlike $NbOCl_3$, it apparently cannot be prepared simply by partial hydrolysis of $TaCl_5$ under normal conditions, and it has been suggested that this reaction could be used to separate these two elements.

Niobium oxytribromide can be prepared by the action of bromine on niobium pentoxide-carbon mixtures, or by the reaction of the pentabromide with oxygen. The reaction between niobium pentoxide and carbon tetrabromide at 200° C also yields $NbOBr_3$, in contrast to similar reactions which produce the pure halides $NbCl_5$, $TaCl_5$ and $TaBr_5$ (454, 804). The infrared spectrum of the yellow solid shows the niobium-oxygen stretching vibration at 734 cm^{-1}, indicating a similar polymeric structure to that observed in $NbOCl_3$.

Tantalum oxotribromide has not been studied in any detail (804).

Niobium oxotriiodide has been prepared by the following reaction in a 400°–275° C temperature gradient (2013):

$$6\,Nb + 15\,I_2 + 2\,Nb_2O_5 \rightarrow 10\,NbOI_3$$

It decomposes when heated into $NbOI_2$ and iodine.

In the gaseous state monomeric MOX_3 appear to be formed in all cases. The decrease in metal-oxygen stretching frequency observed as the size of the halogen atom increases is relatively small (2076).

	POX_3	VOX_3	$NbOX_3$
MOF_3	1418	1058	—
$MOCl_3$	1290	1035	999
$MOBr_3$	1261	1025	994
MOI_3	—	1012	986

The dioxyfluorides of niobium and tantalum, NbO_2F and TaO_2F, are obtained by evaporation of hydrofluoric acid (48%) solutions to dryness, and drying at 250° C. The compounds have the ReO_3 structure with the fluorine and oxygen atoms randomly distributed in the octahedron about the metal atom (947). These compounds are discussed further on page 203.

Disproportionation of $NbOCl_3$ and $TaOCl_3$ at 300°–400° C produces NbO_2Cl and TaO_2Cl respectively, together with the corresponding pentachlorides (1142, 1233, 1722). Another compound, Nb_3O_7Cl, has been prepared from Nb_2O_5 and $NbOCl_3$ at 600 °C (1233, 2032). It is colourless in the presence of oxygen or chlorine, but blue in the absence of oxidizing agents. The structure contains octahedrally coordinated metal atoms (2066). The corresponding tantalum compound has also been mentioned (2041).

The dioxobromides NbO_2Br and TaO_2Br have been prepared, but no details appear at present to be available (2041).

Niobium dioxoiodide has been obtained as red needles by heating a 1·0 : 2·0 : 6·1 molar ratio of $Nb : Nb_2O_5 : I_2$ in a temperature gradient of 500°–475° C (2041). The isomorphous tantalum analogue was similarly obtained using a 500°–600° C temperature gradient (2041).

2. *Oxohalide adducts*

Vanadium oxotrichloride, like titanium tetrachloride, forms coloured charge transfer complexes with many aromatic hydrocarbons, but these will not be discussed further. Ligands can form adducts of the type $VOCl_3L_2$ (where L is Et_3N, py, MeCN or PhCN) under mild conditions, but reduction to $VOCl_2$ complexes is more common with many amines, nitriles, dialkyl sulphides, etc. (124, 964).

The reaction between oxygen containing ligands and niobium and tantalum pentachlorides and pentabromides under mild conditions can lead to adducts of the pentahalides, but with excess ligand or under more vigorous conditions oxygen abstraction occurs with the formation of adducts of the oxotrihalides. For example dimethylsulphoxide forms $NbOX_3(Me_2SO)$ together with Me_2SX_2 or $X.CH_2.S.CH_3$ depending upon conditions. Similarly diphenylsulphoxide forms $NbOCl_3(Ph_2SO)_2$ and p-$Br.C_6H_4.S.C_6H_5$ (547). Similarly excess triphenylphosphine oxide and excess triphenylarsine oxide form compounds of the type $NbOCl_3(Ph_3PO)_2$ and $NbOCl_3(Ph_3AsO)_2$, together with Ph_3PCl_2 and Ph_3AsCl_2 respectively (353, 549). Oxygen abstraction however does not occur with hexamethylphosphoramide.

Similar compounds have also been obtained by oxygen abstraction from the solvent. For example $NbCl_5[o\text{-}C_6H_4(AsMe_2)_2]$ in nitromethane for several days formed $NbOCl_3[o\text{-}C_6H_4(AsMe_2)_2]$ (502).

Other compounds have been prepared directly from the oxotrihalides, and this has also allowed the preparation of compounds containing other than oxygen-donor ligands, for example $NbOCl_3L_2$ (where L is diethylaniline, 2-methylquinoline, 2-ethylquinoline) and $NbOCl_3$(dipyridyl) (340, 547, 1518).

All these compounds appear to be monomeric and nonelectrolytes, and the niobium-oxygen stretching frequencies in the range 935–970 cm^{-1} confirm the absence of bridging oxygen atoms.

E. Anionic Halo and Oxohalo Complexes

1. *Fluoro and oxofluoro complexes*

(a) *Vanadium.* The direct reaction of vanadium pentafluoride with potassium fluoride (774), nitrosyl fluoride or nitryl fluoride (488) forms KVF_6, $(NO)VF_6$ and $(NO_2)VF_6$ respectively. Alternatively compounds containing the VF_6^- anion have been prepared by the fluorination of a mixture of VCl_3 or V_2O_5 and a salt containing the appropriate cation, with bromine trifluoride (594, 2112). The products are white solids which fume in moist air and are immediately hydrolysed by water.

The HF–V_2O_5–H_2O ternary system shows complex hydrated oxofluorides with F/V ratios of 1·33, 3·75 and 4·5 (391).

(b) *Niobium and tantalum.* In contrast to vanadium, both oxofluoro and fluoro complexes of niobium and tantalum may be obtained from aqueous solution. The coordination number of the metal atom in those which have been structurally characterized include six, seven and eight.

The particular solid salt which is obtained when alkali metal (or other) fluorides are added to aqueous hydrofluoric acid solutions of niobium or tantalum is dependent upon the cation, its concentration, and the hydrogen fluoride concentration. For example $KTaF_6$ is obtained from hydrofluoric acid solutions greater than 45% HF, while K_2TaF_7 is obtained from less concentrated acid. Alternatively if ammonium is the cation present, the compounds isolated in order of decreasing acid concentration are $(NH_4)TaF_6$, $(NH_4)_2TaF_7$ and $(NH_4)_3TaF_8$. The influence of cation is further shown by the salts $CsTaF_6$, K_2TaF_7 and $(NH_4)_3TaF_8$, which are obtained from solutions of the same hydrogen fluoride concentration (1997). These trends are probably not those which would have been predicted. The conditions for preparation of the different sodium salts have also been examined (725). Analogous salts containing NbF_6^-, NbF_7^{2-} and NbF_8^{3-} have also been obtained.

The hexafluoroniobate and hexafluorotantalate complexes have also been prepared through the action of bromine trifluoride and bromine at low temperatures, followed by double decomposition of the resultant $(BrF_2)(MF_6)$, for example with $KBrF_4$ (1098). The existence of K_3NbF_8 (M.P = 760° C) has also been noted during an examination of the KF (M.P. = 856° C)–K_2NbF_7 (M.P. = 733° C) phase diagram (1738).

The structures of the alkali metal, silver and thallium hexafluorovanadates, niobates and tantalates all contain octahedral MF_6^- anions. The particular crystal structure obtained depends upon the relative size of the cation (254, 594, 1347).

The structure of K_2NbF_7 shows discrete NbF_7^{2-} groups with slightly distorted C_{2v} symmetry, that is the capped trigonal prismatic stereochemistry (page 8) (367, 1198). The niobium-fluorine bond lengths lie in the range 1·94–1·98 Å (average 1·96 Å), the fluorine–fluorine distances lie in the range 2·40–2·91 Å, and the F–Nb–F angles are

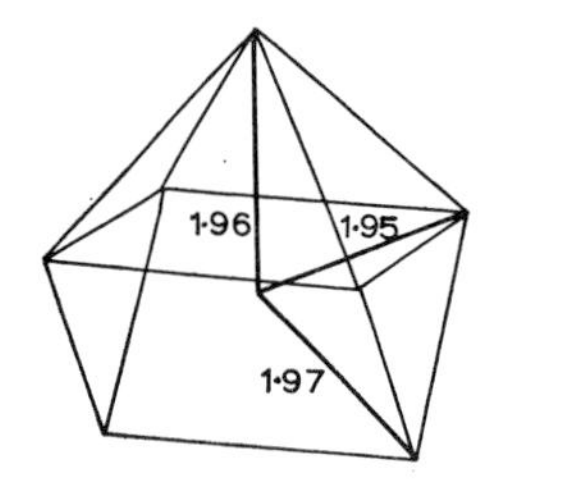

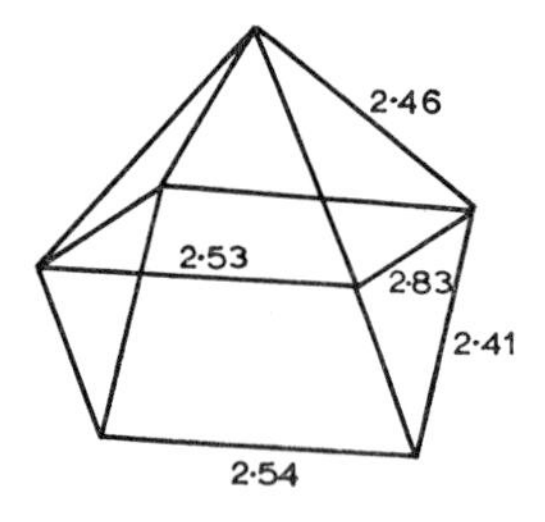

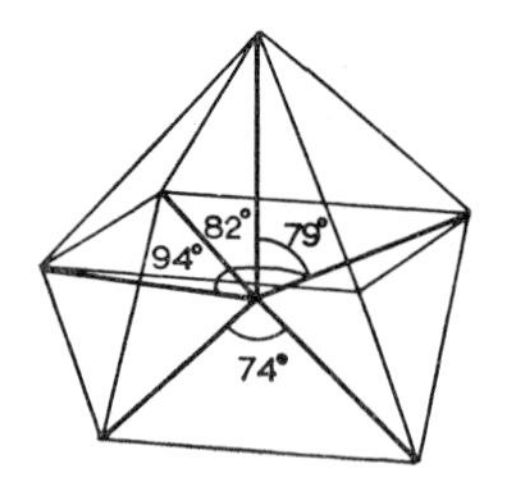

FIG. 99 Nb–F distances, F–F distances, and F–Nb–F angles in K_2NbF_7.

between 74° and 98° The parameters in Fig. 99 have been averaged assuming ideal C_{2v} symmetry. The tantalum analogue is isomorphous. In contrast the metal atom in K_2PaF_7 is nine coordinate (metal–fluorine distances in the range 2·13 to 2·46 Å), and each polyhedron shares two edges with adjacent polyhedra forming infinite strings (361).

The infrared reflection spectrum of TaF_5 in 1 : 1 LiF : KF melts at approximately 700° C is similar to that of solid K_2TaF_7, but considerably different to that of solid $CsTaF_6$, and hence TaF_7^{2-} appears to be the predominant species present. In 1 : 1 LiF : NaF melts however, the spectrum of TaF_6^- is also apparent (896).

The anion in Na_3TaF_8 is eight coordinate with a square antiprismatic stereochemistry. The detailed stereochemistry is discussed on page 17 (1205). It may again be noted that the protactinium analogue, Na_3PaF_8 has a different structure (354) in which eight fluorine atoms around the protactinium are at the corners of a cube, the structure being related to that of fluorite (page 12).

Aqueous hydrofluoric acid solutions of niobium differ from those of tantalum in that oxyfluoro complexes may be obtained much more readily. This difference is used as the basis for the classical method for the separation of niobium and tantalum. For example when a 2–3% hydrofluoric acid solution is concentrated, K_2TaF_7 precipitates first, followed by $K_2(NbOF_5)H_2O$. Under appropriate conditions salts containing $NbOF_6^{3-}$ and $TaOF_6^{3-}$ can also be obtained (1345).

$K_3(NbOF_6)$ has a seven coordinate structure and is isomorphous with $K_3(ZrF_7)$ (2395). A comparison of the Raman spectra of $K_3(NbOF_6)$ and $(NH_4)_3(NbOF_6)$ shows the latter also contains the $(NbOF_6)^{3-}$ ion (1345); $(NH_4)_3TaOF_6$ is isomorphous with $(NH_4)_3NbOF_6$.

K_2NbOF_5,H_2O contains octahedral $NbOF_5^{2-}$ ions (1204, 1869), and the Raman spectrum of $NbOF_5^{2-}$ is similar to that of NbF_6^-, but not to that of NbF_7^{2-}. The $NbOF_5^{2-}$ ion is also present in K_3HNbOF_7, which also contains the linear HF_2^- ion and should be written $K_3(HF_2)(NbOF_5)$ (1204).

The major species present in aqueous solution has been determined by a comparison of Raman spectra of these solutions with the Raman spectra of solids containing the MF_8^{3-}, MF_7^{2-}, MF_6^-, MOF_6^{3-} and MOF_5^{2-} ions. In 23 *M* hydrofluoric acid, niobium(V) was present as NbF_6^-, while at 11 *M*, 5 *M* and approximately zero molar, $NbOF_5^{2-}$ was the only species detected (1067, 1344). The ^{93}Nb–n.m.r. and ^{19}F–n.m.r. of niobium(V) in concentrated hydrofluoric acid confirms that NbF_7^{2-} is not present in significant quantities, and indeed there is little exchange between NbF_6^- and added fluoride (1829). Tantalum(V) in 24 *M* hydrofluoric acid was found to be present largely as TaF_6^-, although TaF_7^{2-}

could also be detected. However, the latter species was dominant at 10 *M*, 3 *M* and approximately zero molar, and its relative concentration further increased on addition of ammonium fluoride (1067, 1345). In 0 to 0·5 *M* hydrofluoric acid a third species was detected, which was possibly $TaOF_6^{3-}$ or $TaOF_5^{2-}$. The TaF_8^{3-} ion was not detected. These results are not consistent with those obtained by potentiometric and anion exchange techniques (2315).

The reaction between selenium tetrafluoride and niobium pentafluoride forms $(Se^{IV}F_3)^+(Nb_2^VF_{11})^-$. The structure of the the dimeric anion shows two NbF_6 octahedra sharing a common corner (732). Niobium and tantalum pentafluorides readily react with xenon tetrafluoride to form $XeF_2.MF_5$ and $XeF_2.2\,MF_5$ (521, 731, 1213). The structure of the latter is undoubtedly very similar to that of the antimony analogue, $(XeF)^+(Sb_2F_{11})^-$, the anion of which also consists of two octahedra with a linear Sb–F–Sb bridge (1612). It may be relevant to note that platinum forms compounds with the same stoicheiometries, that is $Xe(PtF_6)_2$ and $Xe(PtF_6)$, the latter being the first compound prepared of xenon. In contrast vanadium pentafluoride is unreactive to xenon difluoride and tetrafluoride, but reacts with the hexafluoride and oxotetrafluoride to form $VF_5.2\,XeF_6$ and $VF_5.2\,XeOF_4$ respectively (1702).

The solid state reaction between equimolar amounts of potassium carbonate and niobium pentoxide in a large excess of potassium fluoride at 750° C forms K_2NbO_3F according to:

$$K_2CO_3 + Nb_2O_5 + 2\,KF \rightarrow 2\,K_2NbO_3F + CO_2$$

The compound has the K_2NiF_4 structure. It was suggested that the fluoride ions were preferentially situated at the unshared *trans* positions of the octahedra, while the oxygen atoms were preferentially situated at the four corners which are shared with neighbouring octahedra (970).

2. *Chloro, bromo, iodo, oxochloro and oxobromo complexes*

The addition of ionic chlorides to $VOCl_3$ dissolved in hydrochloric acid saturated with hydrogen chloride at 0° C forms salts of $[VOCl_4]^-$ or $[VOCl_5]^{2-}$ depending upon the cation and its concentration (692, 1374). Reduction to vanadium(IV) may occur under more vigorous conditions.

Potentiometric and conductometric titration of chloride against niobium or tantalum pentachlorides in nitromethane (1360), benzyl chloride (1104), phosphorus oxychloride (1096, 1100) and iodine monochloride (1093) show the formation of MCl_6^-. Continuation of the titrations up to a mole ratio of 3 :1 showed no further end points. The salts may be conveniently obtained from solutions in thionyl chloride (112, 966). These hexachloro anions are undoubtedly monomeric and octahedral (9).

Analogous salts of $NbBr_6^-$, $TaBr_6^-$, NbI_6^- and TaI_6^- have been obtained from acetonitrile solution or acetone/chloroform mixtures (359, 1827). Mixed haloanions MX_5Y^- (where M is Nb or Ta, X is Cl or Br, Y is Cl, Br or I) have also been studied (1827).

The binary systems of niobium pentachloride or tantalum pentachloride with alkali metal and ammonium chloride show either congruent or incongruent melting points corresponding to the formation of $A^IM^VCl_6$ (with the exception of the LiCl–$NbCl_5$ system) (543, 1093, 1234, 1718, 1833, 2445). There is no evidence for the formation of other than simple 1 : 1 compounds. Heating the white $(NH_4)(TaCl_6)$ forms the yellow-green nitride dichloride $TaNCl_2$ (1019, 1020). In contrast to the cyclic phosphonitrilic compounds formed from $(NH_4)(PCl_6)$, the tantalum compound appears to contain infinite –Ta–N–Ta– chains (1018).

Many reactions of niobium pentachloride and tantalum pentachloride with chloro compounds are probably best written as forming the hexachloroanions. For example phosphorus pentachloride forms $NbCl_5.PCl_5$ and $TaCl_5.PCl_5$ (1092, 1442). The products are probably $[PCl_4^+][MCl_6^-]$ rather than $[MCl_4^+][PCl_6^-]$, as conductometric titration of silver perchlorate against phosphorus pentachloride in nitromethane shows the existence of PCl_4^+, whereas no $NbCl_4^+$ was noted in the titration against niobium pentachloride (1360). Arsenic trichloride, chlorine and tantalum pentachloride form $(AsCl_4)(TaCl_6)$ (1445), while sulphur dichloride and chlorine form $NbSCl_9$ and $TaSCl_9$ of unknown structure (961).

The heptachloroniobate ion $(NbCl_7)^{2-}$ has also been claimed (963).

Niobium pentachloride, niobium oxotrichloride or niobium pentoxide dissolve in concentrated hydrochloric acid, from which the pale yellow salts $A_2(NbOCl_5)$ (where A is NH_4, K, Rb or Cs, but not Na) may be obtained by the addition of excess alkali metal chloride and saturation with hydrogen chloride at 0° C (963, 1558, 2358). These compounds are unstable in more dilute acid, and to moist air. The anion sites in the K_2PtCl_6-type structure appear to be randomly occupied by oxygen and chlorine atoms (352, 1559). The niobium-oxygen stretching frequency occurs in the range 922–930 cm^{-1} (352, 1974).

However pyridinium hydrochloride and quinolinium hydrochloride form $(pyH)(NbOCl_4)$ and $(quinH)(NbOCl_4)$ from aqueous hydrogen chloride solutions. The niobium-oxygen stretching frequencies drop to 850 and 800 cm^{-1} respectively indicating Nb–O–Nb bridging groups. The $(pyH)_2(NbOCl_5)$ and $(quinH)_2(NbOCl_5)$ are obtained from ethanolic hydrogen chloride solutions (1721).

The binary systems $NbOCl_3$–NaCl and $NbOCl_3$–KCl show the formation of $NaNbOCl_4$, $KNbOCl_4$ and K_2NbOCl_5 (2114).

Niobium forms oxobromo complexes such as the dark red $Rb_2(NbOBr_5)$

and $Cs_2(NbOBr_5)$, and the orange (pyH)($NbOBr_4$) and (quinH)($NbOBr_4$), from aqueous solutions of concentrated hydrobromic acid (2358). The niobium–oxygen stretching frequency at 928 cm^{-1} in Cs_2NbOCl_5 is shifted to 977 cm^{-1} in Cs_2NbOBr_5 (1974). Both $KNbOBr_4$ and K_2NbOBr_5 are observed in theKBr–$NbOBr_3$ binary system (2169).

The charge transfer spectrum in the visible region of these oxohaloanions is very similar to the haloanions themselves, showing that oxygen is not involved in the ligand–to–metal charge transfer bands (966).

Attempts to prepare analogous tantalum salts have been less successful. However, the reaction of hexachlorotantalates with antimony(III) oxide and chlorine at 500° C yields the oxotetrachlorotantalates (1722).

F. Alkoxides

1. *Vanadium*

Vanadium forms oxo alkoxides $VO(OR)_3$ which are sometimes referred to as alkylvanadates, R_3VO_4. They may be prepared by the action of sodium alkoxides on $VOCl_3$, or by the direct action of alcohol on V_2O_5, or preferably, NH_4VO_3 (434, 1885). The compounds are moisture-sensitive liquids or low melting point solids; for example $VO(OEt)_3$ melts at about 0° C, $VO(OPr)_3$ and $VO(Bu^s)_3$ melt from −5° to −10° C, and $VO(OBu^t)_3$ melts at about 45° C.

Intermediate chloro-alkoxides of the types $VOCl_2(OR)$ and $VOCl(OR)_2$, and aryl compounds such as $VOCl(OPh)_2$, are also known (964, 1697).

The vanadium oxo-trialkoxides have been reported to be monomeric in benzene (434), but a more recent structural determination shows

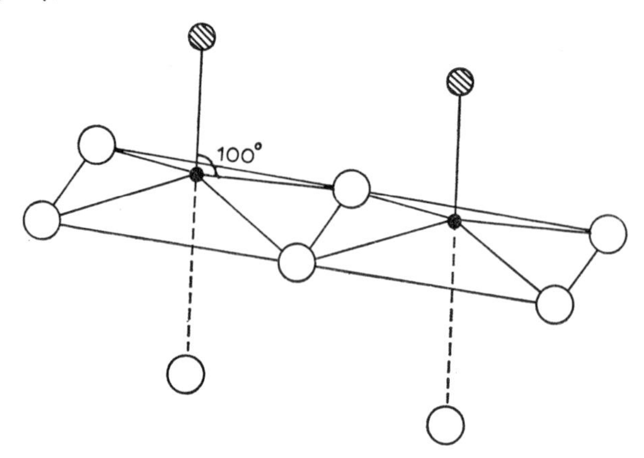

FIG. 100 The structure of $VO(OMe)_3$. The open circles represent the oxygen atoms of methoxy groups, and the hatched circles represent the remaining oxygen atoms.

$VO(OMe)_3$ to be dimeric (Fig. 100) (441). The metal atoms are five co-ordinate with methoxy groups forming the square base and the oxo group the apex of a square pyramid. Pairs of metal atoms share two methoxy groups, the relevant vanadium–oxygen bond lengths being:

$$\begin{aligned} \text{V–OMe(bridge)} &= 2\cdot 02\text{Å} \\ \text{V–OMe(non bridge)} &= 1\cdot 80\ \text{Å (av.)} \\ \text{V} = \text{O} &= 1\cdot 54\ \text{Å} \end{aligned}$$

These dimers can be considered to be linked into infinite linear polymers by completion of octahedral coordination about the metal atoms with methoxy groups from adjacent dimeric molecules (V. . . OMe = 2·28 Å).

2. *Niobium and tantalum*

The niobium and tantalum penta-alkoxides are prepared in a similar manner to the titanium tetra-alkoxides (page 94). The addition of the appropriate anhydrous alcohol to the pentachloride results in vigorous evolution of hydrogen chloride with the formation of mixed chloro alkoxides of the type $MCl_2(OR)_3$. The reaction is driven to the penta-alkoxide by the addition of anhydrous ammonia. The products are colourless oils (for example $Ta(OEt)_5$, $Ta(OPr)_5$ and $Ta(OBu)_5$) or low melting solids (for example $Ta(OMe)_5$, M.P. = 50° C), although frequently yellow when impure. They can also be conveniently prepared as before by alcohol exchange.

The simple pentaalkoxides are dimeric in benzene, and this has been confirmed by a structure determination of $Ta(OEt)(OBu)_4$, in which two octahedrally coordinated metal atoms share a common edge (1255).

The n.m.r. of the dimeric $Ta(OMe)_5$ in *n*-octane at −58° C shows three peaks with relative intensities 1 : 2·0 : 2·0 due to bridging methoxy groups, and non-bridging groups *cis* and *trans* to the bridging groups respectively (292). Similar results were obtained with other tetraalkoxides.

The detailed study of the molecular weight of these compounds has lead to the following trends, which may be compared with those observed for the Group IV alkoxides (page 97):

1. The compounds are dimeric in boiling benzene, although some dissociation occurs in higher boiling point solvents (282, 309). For example the molecular complexity of the tantalum alkoxides is 1·8 in boiling toluene.

2. The molecular weight decreases as the donor ability of the solvent increases. For example the tantalum alkoxides have a molecular complexity of 1·5 in acetonitrile although it has approximately the same boiling point

as benzene. The isolation of adducts however is rare. For example tantalum alkoxides are monomeric in pyridine, but evaporation of the solvent merely leaves unchanged pentaalkoxide (309).

Both these trends are shown in the molecular complexity of the alkoxides measured in the same alcohol as used to form the alkoxy group (Table 19) (282, 308, 309). The observed variations are due to the increase in boiling point along the series ethanol, propanol, butanol, while methanol has a compensating higher donor power.

TABLE 19

Molecular complexities of $Nb(OR)_5$ and $Ta(OR)_5$ in the parent alcohol.

	Molecular complexity		Molecular complexity
$Nb(OMe)_5$	1·3	$Ta(OMe)_5$	1·2
$Nb(OEt)_5$	1·5	$Ta(OEt)_5$	1·8
$Nb(OPr^n)_5$	1·3	$Ta(OPr^n)_5$	1·7
$Nb(OBu^n)_5$	1·1	$Ta(OBu^n)_5$	1·4

3. The molecular weight decreases as the degree of branching of the alkyl group increases (Table 20) (283, 310).

TABLE 20

Molecular complexities of $Nb(OR)_5$ and $Ta(OR)_5$ in boiling benzene

	Molecular complexity	
R	$Nb(OR)_5$	$Ta(OR)_5$
$CH_3.CH_2.CH_2.CH_2$	2·01	2·02
$(CH_3)_2CH.CH_2$	1·83	2·04
$(C_2H_5)(CH_3)CH$	1·14	1·06
$(CH_3)_3C$	—	1·00
$CH_3.CH_2.CH_2.CH_2.CH_2$	2·00	2·01
$(CH_3)_2CH.CH_2.CH_2$	1·81	1·98
$(C_2H_5)(CH_3)CH.CH_2$	1·81	1·98
$(CH_3)_3C.CH_2$	1·52	1·35
$(C_2H_5)_2CH$	1·16	1·02
$(C_3H_7)(CH_3)CH$	1·04	1·00
$(CH_3)_2CH.CH(CH_3)$	1·04	1·00

The steric limitations imposed by bulky alkoxy groups are also reflected in the preparation of these alkoxides by alcohol exchange (281). Thus the reaction of $Ta(OR)_5$, where R is Me, Et or Pr^i with alcohols R′OH containing bulky alkyl groups such as Pr^i, Bu^t or Am^t, leads to $Ta(OR)(OR')_4$.

Attempts to prepare the penta(*tert*-butoxide) compounds by conventional means resulted in the formation of oxo-alkoxides such as $Nb_2O(OBu^t)_8$ (281, 283). The $Nb(OBu^t)_5$ however has been prepared from butanol and $Nb(NR_2)_4$ (2276), or from butylacetate and $Nb(OEt)_5$ (1676). The n.m.r. spectrum indicates a monomeric five coordinate complex (292).

The hydrolysis of these pentaalkoxides forms oxo alkoxides (289, 296). The structure of $Nb_8O_{10}(OEt)_{20}$ shows corner sharing and edge sharing of octahedrally coordinated metal atoms (Fig. 101) (296). The relation to the structure of $(H_2W_{12}O_{42})^{10-}$ may be noted (page 295).

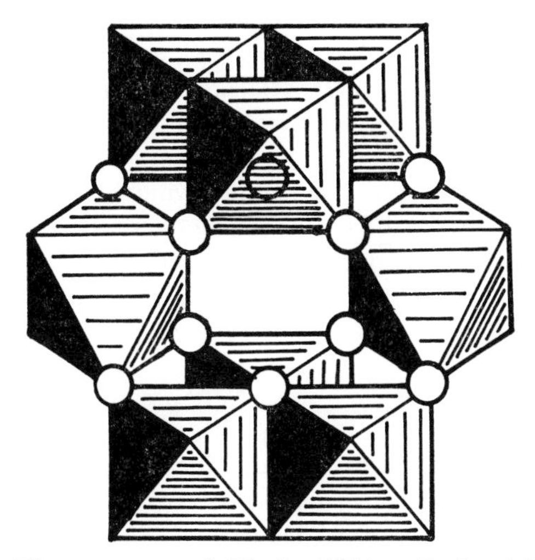

FIG. 101 The structure of $Nb_8O_{10}(OEt)_{20}$. Each niobium atom is octahedrally coordinated by oxygen atoms (open circles) and ethoxy groups.

As was also noted for the Group IV alkoxides, a number of derivatives closely related to the alkoxides may be prepared. These include alkoxy anions of the type $[M(OR)_6]^-$ (where M is Nb or Ta and R is Me, Et, Pr or Bu) (1675), aryloxides such as $Nb(OPh)_5$ (950, 963, 1676), siloxanes of the type $M(OSiR)_5$ (2276), and β-diketonate derivatives of the types $MX_2(OR)_2(\beta\text{-diketonate})$, $M(OR)_2(\beta\text{-diketonate})_3$, $NbOCl_2(acac)$ and $NbOBr(acac)_2$ (340, 672, 952, 964). The last two contain bridging oxygen atoms (niobium–oxygen stretching frequencies at 835 and 825 cm^{-1} respectively.)

G. Complexes with other Anionic Ligands

In alkaline or slightly acid solutions, vanadium(V), niobium(V) and tantalum(V) are present as the polymerised isopolyanions (see page 181), but in moderately acid solutions (0·5–4 *M*) colloidal suspensions are observed. At higher acid concentrations the solubility again increases and chloro (page 172), nitrato or sulphato complexes are obtained depending upon the particular acid used. Alternatively complexes may be obtained from solutions at pH 7–0 by the addition of carboxylic acids, hydroxy-acids, hydroxyamines or amines (107, 811, 812, 1032, 1235, 1749). Most of the these complexes have not yet been adequately characterized, which is experimentally difficult due to the formation of polymeric species containing hydroxo and oxo bridges. The anion : metal ratio may be relatively high, as in the structurally characterized seven coordinate oxalato complex $[NbO(C_2O_4)_3]^{2-}$ (341, 1662), or it may be much less than unity with polymeric units of metal atoms. The best characterized species are the pseudohalageno complexes such as the thiocyanates, and the peroxo complexes.

Compounds containing anionic ligands which have been well characterized have been obtained from nonaqueous solution.

Attention here will be concentrated on amido complexes, pseudo-halogeno complexes, peroxo complexes, and nitrato complexes.

1. *Amido complexes*

The reaction of vanadium tetrachloride with chlorine azide forms $V^VCl_3(NCl)$ through the intermediate azide $VCl_4(N_3)$ (2236). A more convenient synthesis is the chlorination of vanadium nitride (2235). The related $VCl_3(NMe)$ is also known (2162).

Niobium pentachloride reacts with dimethylamine to form a deep red solution, which may be extracted with benzene yielding orange $NbCl_3(NMe_2)(Me_2NH)$ and a residue of the amine hydrochloride. The compound is monomeric in benzene (429, 430) (but also see reference 927). A number of other compounds of general formula $MX_3(NR_2)(R_2NH)$ (where M is Nb or Ta, X is Cl or Br, and R is Me, Et or Pr), have been obtained by similar methods (429, 430, 927).

There is thus less solvolysis of the metal–chlorine bond with secondary amines than with alcohols.

Complete solvolysis can be achieved with lithium dialkylamides, for example (304, 306):

$$TaCl_5 + 5\,Li(NMe_2) \rightarrow Ta(NMe_2)_5 + 5\,LiCl$$

This methyl derivative is stable to heat and sublimes at about 100° C, but the ethyl and higher homologues decompose at 60° C to form the amido–imide $Ta(NEt)(NEt_2)_3$, which is thought to contain the Ta = NEt group rather than –Ta–N(Et)–Ta– group (304, 307). Reaction of lithium dialkylamides with niobium pentachloride normally leads to reduction to, for example, $Nb(NEt_2)_4$. The pentavalent $Nb(NR_2)_5$ however was obtained for $(NMe_2)^-$ and $(NMeBu)^-$; the latter decomposed on heating to a mixture of $Nb(NMeBu)_4$ and $Nb(NBu)(NMeBu)_3$. The pentavalent ethyl compound $Nb(NEt_2)_5$ was successfully obtained by amide exchange from $Nb(NMe_2)_5$ under mild conditions (306).

Solvolysis similarly occurs with primary amines, but the products have not been as well characterized (429, 430).

The reaction of anhydrous liquid ammonia with niobium or tantalum pentachlorides again results in the solvolysis of two metal–chlorine bonds, but the compounds are of uncertain composition as they contain both ammonia and ammonium chloride. Removal of these substances at about 160° C under vacuum leaves the air unstable, yellow $NbCl_3(NH_2)_2$ and $TaCl_3(NH_2)_2$ (928, 1727, 2191).

2. *Pseudo-halogeno complexes*

Hydrogen cyanide and niobium pentachloride yield $NbCl_4(CN)(Et_2O)$ from diethyl ether, and $NbCl_5.HCN$ from carbon tetrachloride. Addition of triethylamine to the latter formed $(Et_3NH)[NbCl_5(CN)]$. The corresponding bromo complex was also obtained (321).

The addition of potassium thiocyanate to niobium and tantalum pentachlorides in acetonitrile formed $M(NCS)_5(MeCN)$ or $[M(NCS)_6]^-$ (where M is Nb or Ta) depending upon conditions (264, 368, 1426). Infrared spectra indicate bonding through the nitrogen atom as expected. Similar reactions have been carried out in pyridine, dimethylformamide, and methanol (264, 1020).

3. *Peroxo complexes*

The addition of ethanol to cold aqueous solutions of pentavalent vanadium (837), niobium or tantalum (1082) containing hydrogen peroxide precipitate the tetraperoxo complexes, for example $(NH_4)_3[Nb(O_2)_4]$. The salts are isomorphous with the eight coordinate chromium analogues (262, 837) showing the presence of four bidentate O_2^{2-} groups. The structure is discussed more fully on pages 13 and 318.

The salts $A_3[Nb(O_2)_4]$ (where A is K, NH_4, Rb or Cs) decompose rapidly in 12 *M* hydrochloric acid and turn pale yellow due to the presence of $NbOCl_5^{2-}$. However, if the solution is kept at 0° to −10° C and saturated with hydrogen chloride, dark red crystals of $A_2[Nb(O_2)Cl_5]$ may be obtained (1085). The analogous peroxopentafluoro complexes, and the tantalum complexes, have also been obtained (638, 2369). The ^{19}F-n.m.r. spectrum of aqueous solutions of $[Ta(O_2)F_5]^{2-}$ consists of a doublet and a quintuplet of relative intensities 4 : 1 consistent with the structure shown in Fig. 102 (791).

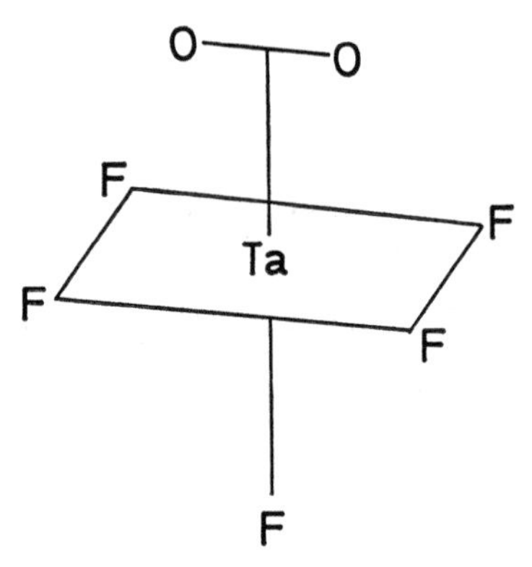

FIG. 102 The proposed structure of $Ta(O_2)F_5^{2-}$.

Peroxo-oxalato complexes of niobium and tantalum containing $[M(O_2)(C_2O_4)_3]^{3-}$ and $[M(O_2)_2(C_2O_4)_2]^{3-}$ ions can also be obtained from aqueous solution (674, 1066, 1084, 1662). The structure of $(NH_4)_3[Nb(O_2)_2(C_2O_4)_2],H_2O$ shows an eight coordinate structure as expected, but perhaps unexpectedly both the peroxo groups are on the same side of the molecule with both oxalato groups on the other side (1662). This "*cis*" arrangement would seem to suggest ligand-to-metal π bonding.

Other apparently eight coordinate complexes include $K[M(O_2)_3(\text{dipyridyl})]$ and $K[M(O_2)_3(o\text{-phen})]$ (where M is Nb or Ta) (674).

The peroxo-hydroxo complexes $K[Nb(O_2)(OH)_4],xH_2O$ and $K[Nb(O_2)_2(OH)_2],H_2O$ appear to contain polymeric anions (1061).

4. *Nitrato complexes*

Vanadium, niobium and tantalum form oxo complexes containing covalently bound nitrato groups with N_2O_5 or N_2O_4, the product depending upon the conditions used.

Freshly precipitated and dried niobium(V) or tantalum(V) hydroxides (113), $VOCl_3$ (2055, 2057), $NbCl_5$ (840) or $TaCl_5$ (113) react with liquid N_2O_5 to form the mono-oxo complexes $M^{V}O(NO_3)_3$. The vanadium complex is a liquid at room temperature, freezing at 2° C. The infrared spectrum shows that the nitrate groups are bidentate (16). A band at 906 cm^{-1} for the niobium compound is typical of a non-bridging oxygen atom, although a corresponding band could not be detected in the analogous tantalum compound.

Liquid N_2O_4 is fairly unreactive by itself, but in the presence of ionizing solvents it readily reacts with metallic vanadium to form the dioxo complex $VO_2(NO_3)$ (1837). Under similar conditions niobium pentachloride forms solvates of $NbO_2(NO_3)$, for example $NbO_2(NO_3),0\cdot67$ MeCN from acetonitrile (113). This compound is involatile and amorphous, and presumably contains bridging oxo groups.

Mono-oxo nitrato complex anions may similarly be formed with liquid N_2O_5. For example the reaction of $(Me_4N)(MCl_6)$, where M is Nb or Ta, with liquid N_2O_5 forms the corresponding $(Me_4N)[M^VO(NO_3)_4]$ complexes (358). The protactinium complex $(Me_4N)(PaCl_6)$ however forms $(Me_4N)[Pa(NO_3)_6]$.

H. Polyvanadates, Polyniobates and Polytantalates

1. *Isopolyvanadates*

(a) *Introduction.* In strongly alkaline solutions, pH > 14, the Raman spectrum of vanadium(V) ($\sim$2 molar) shows the existence of VO_4^{3-} (1065).

On acidification the solutions turn orange-yellow due to the formation of polymerized ions with the vanadium atom having a coordination number greater than four. A schematic representation of the ions as a function of

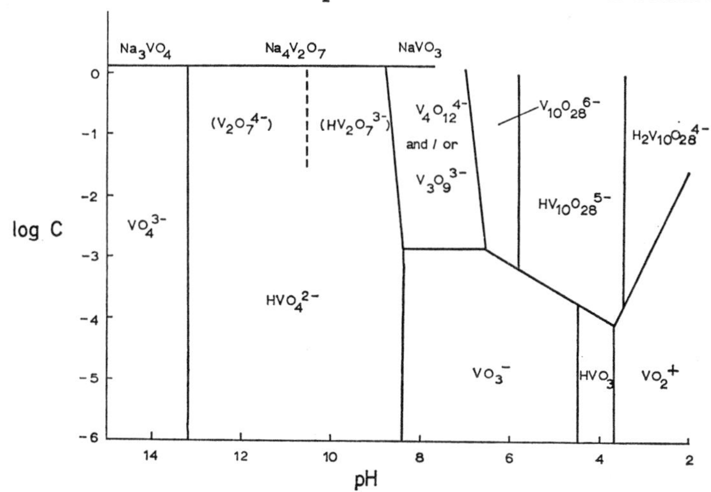

FIG. 103 Concentration and pH ranges of stability of isopolyvanadate anions.

vanadium concentration and pH is shown in Fig. 103, where each area corresponds to the presence of more than 50% of the vanadium atoms in the stated form.

It is only in very dilute solutions that the vanadium ion remains monomeric and forms HVO_4^{2-} (pH ~ 13–9), $H_2VO_4^-$ (which is commonly written as VO_3^- although it may have a coordination number even greater than four) (pH ~ 9–5), hydrated vanadium oxide (pH ~ 4) and finally the vanadyl cation VO_2^+ below pH 3.

Figure 103 shows that the chemistry is more complex in more concentrated solutions. A small amount of dimeric $V_2O_7^{4-}$ and $HV_2O_7^{3-}$ exist in equilibrium with HVO_4^-; the so-called metavanadates which are trimeric and/or tetrameric exist in equilibrium with $H_2VO_4^-$;while the decavanadates $V_{10}O_{28}^{6-}$, $HV_{10}O_{28}^{5-}$ and $H_2V_{10}O_{28}^{4-}$ are formed over the pH range 6–2.

These isopolyvanadates will be discussed in the following pages under three headings:

(i) Equilibria between species present in alkaline solution, namely VO_4^{3-}, HVO_4^{2-}, $H_2VO_4^-$, $V_2O_7^{4-}$, $HV_2O_7^{3-}$, $V_3O_9^{3-}$ and/or $V_4O_{12}^{4-}$.

(ii) Equilibria involving the decavanadates in acid solution.

(iii) Equilibria involving cationic species in strongly acid solution.

(b) *Alkaline solutions.* Potentiometric results at an ionic strength of 3·0 show that the hydrolysis of VO_4^{3-} to HVO_4^{2-} is given by (1775):

$$K_h = \frac{[HVO_4^{2-}][OH^-]}{[VO_4^{3-}]} = 10^{-1.0}$$

or

$$K_3 = \frac{[VO_4^{3-}][H^+]}{[HVO_4^{2-}]} = 10^{-13.0}$$

Similar studies by ultra violet spectrophotometry yield (2043):

$$K_h = 10^{-0.8} \text{ and } K_3 = 10^{-13.2}$$

Above about 0·1 *M* some dimerisation, to $V_2O_7^{4-}$ and $HV_2O_7^{3-}$ occurs, which has been studied potentiometrically (338, 1257, 1775):

$$K_h = \frac{[HV_2O_7^{3-}][OH^-]}{[HVO_4^{2-}]^2} = 10^{-3.2}$$

$$K = \frac{[HV_2O_7^{3-}]}{[HVO_4^{2-}]^2[H^+]} = 10^{10.8}$$

$$K = \frac{[V_2O_7^{4-}]}{[HVO_4^{2-}]^2} = 10^{0.4}$$

These anionic species VO_4^{3-}, HVO_4^{2-}, $V_2O_7^{4-}$ and $HV_2O_7^{3-}$ can be determined independently in aqueous solution by ^{51}V-n.m.r. spectroscopy (1229), and Raman spectroscopy (1065).

Temperature-jump relaxation studies have shown that the rate of dimerization is very fast ($k = 3{\cdot}1 \times 10^4 M^{-1} sec^{-1}$, compared with $k = 1{\cdot}8 M^{-1} sec^{-1}$ for the dimerization of chromate to dichromate (2383)).

Potentiometric and spectrophotometric results also agree in the evaluation of the second hydrolysis constant for the vanadate ion (338, 1257, 2043, 2083, 2183):

$$K_h = \frac{[H_2VO_4^-][OH^-]}{[HVO_4^{2-}]} = 10^{-5.6}–10^{-5.9}$$

$$K_2 = \frac{[HVO_4^{2-}][H^+]}{[H_2VO_4^-]} = 10^{-8.1}–10^{-8.4}$$

In more concentrated solutions at about pH 8 polymerization to trimeric $V_3O_9^{3-}$ and/or tetrameric $V_4O_{12}^{4-}$ occurs. This polymerization is accompanied by a very rapid increase in pH due to the polymers being anions of weaker acids than the monomer. There is general agreement from molecular weight measurements (546, 1011, 1274, 1278, 1764, 2084, 2152, 2224) that above $0{\cdot}01 M$ (with respect to $H_2VO_4^-$) the anionic species present is a tetramer. There appears to be only one species present under these conditions (1062). However spectrophotometric data on dilute solutions are best interpreted as showing an equilibrium between the monomer and a trimer (2043):

$$K = \frac{[VO_3^-]^3}{[V_3O_9^{3-}]} = 10^{-5.6}$$

The most recent and careful potentiometric work suggests that the trimeric ion predominates in very dilute solutions, with the tetrameric ion predominating in more concentrated solutions (338, 1257).

$$K = \frac{[VO_3^-]^3}{[V_3O_9^{3-}]} = 10^{-6.2}$$

$$K = \frac{[VO_3^-]^4}{[V_4O_{12}^{4-}]} = 10^{-8.8}$$

The ^{51}V-n.m.r. spectrum in this region shows two just resolvable absorptions, also suggesting the presence of two closely related species (1229).

The crystalline phases which are obtained from these solutions have infinite polymeric structures. The vanadium atoms in anyhydrous KVO_3

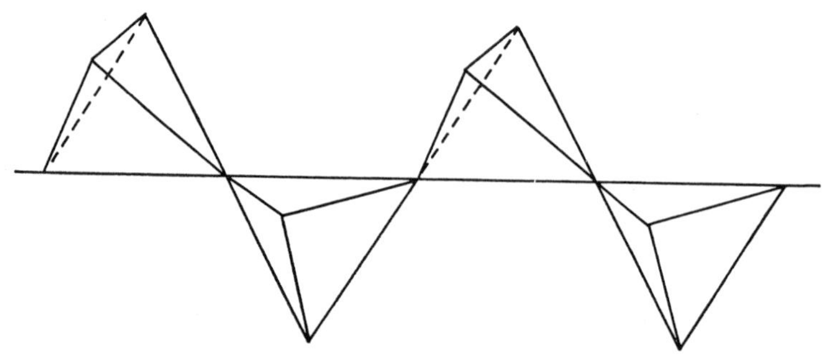

FIG. 104 The structure of KVO_3.

are tetrahedrally coordinated, and share oxygen atoms to form infinite linear strings (Fig. 104) (793, 796). The vanadium atoms are five co-ordinate in the hydrate $KVO_3.H_2O$ (Fig. 105) (473).

(c) *Acid solutions.* Hydrated V_2O_5 is precipitated from acid solutions, and often appears in a finely dispersed colloidal form. Quantitative work is not

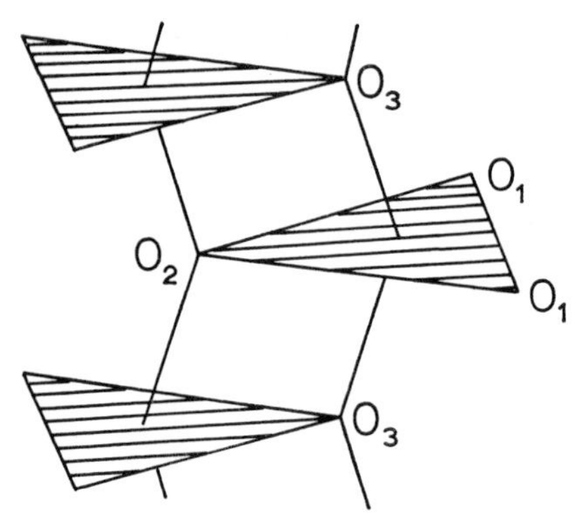

FIG. 105 The structure of $KVO_3.H_2O$ showing linking of VO_5 trigonal bipyramids.
$V–O_1$ and $V–O_2$ = 1·66–1·68 Å; $V–O_3$ = 1·91 and 2·01 Å;
$O_1–V–O_2$ = 105°; $O_1–V–O_3$ = 129°; $O_2–V–O_3$ = 126°;
$O_3–V–O_3$ = 150°.

very reproducible due to the slow approach to equilibrium. Values based on spectral results extrapolated to infinite dilutation are (2043):

$$K_1 = \frac{[H_2VO_4^-][H^+]}{[H_3VO_4]} = 10^{-4.3}–10^{-4.8}$$

Potentiometric work in a flow apparatus yields $K_1 = 10^{-3.4}$, which is significantly different to the result obtained for equilibrated solutions (2083).

Above about 10^{-3} molar, addition of approximately 2·5 moles of acid to each mole of VO_4^{3-} yields a solution of pH6 to pH 3 containing the orange decavanadates, $V_{10}O_{28}^{6-}$ $HV_{10}O_{28}^{5-}$ and $H_2V_{10}O_{28}^{4-}$, which have been studied in some detail (338, 468, 1011, 1257, 1276, 1277, 1617, 1896, 1950, 2043, 2083, 2084). The equilibrium between these three decavanadates is strongly dependent upon cation (Table 21) (1273, 1950, 2083).

TABLE 21

Acid dissociation constants of $H_2V_{10}O_{28}^{4-}$ as a function of cation present.

	Li^+	Na^+	K^+	Rb^+	Cs^+
$K_1 = \dfrac{[H^+][HV_{10}O_{28}^{5-}]}{[H_2V_{10}O_{28}^{4-}]}$	4·4	4·3	3·9	3·7	3·3
$K_2 = \dfrac{[H^+][V_{10}O_{28}^{6-}]}{[HV_{10}O_{28}^{5-}]}$	6·9	6·9	6·4	6·2	6·0

The structure of the $V_{10}O_{28}^{6-}$ anion has been determined in the minerals pascoite $Ca_3V_{10}O_{28},17H_2O$ (799, 2252) and hummerite $K_2Mg_2V_{10}O_{28},16H_2O$ (794, 799). The structure consists of ten VO_6

FIG. 106 The structure of $V_{10}O_{28}^{6-}$.

octahedra sharing edges (Fig. 106). Each of the octahedra is considerably distorted due to the mutual repulsion of the vanadium atoms.

Solutions of $HV_{10}O_{28}^{5-}$ appear to disproportionate over a period of some months into $H_2V_{10}O_{28}^{4-}$ and more basic vanadates. Trivanadates AV_3O_8

(where A is NH_4, K, Rb or Cs) or $K_3V_5O_{14}$ are precipitated on warming or ageing depending upon conditions. The structure of the trivanadates consist of infinite two dimensional layers of VO_6 octahedra sharing edges, which are separated from similar layers by layers of cations (Fig. 107) (248,

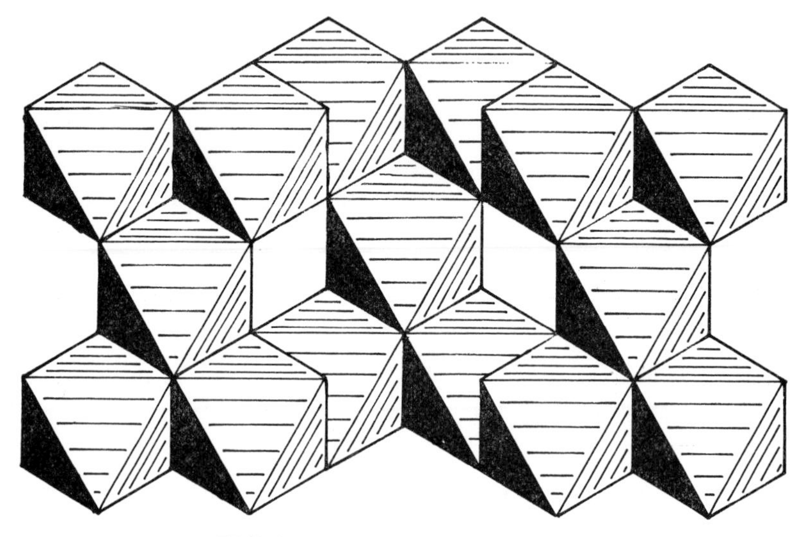

FIG. 107 The structure of KV_3O_8.

797, 1346). The octahedra are extremely distorted, the vanadium-oxygen distances ranging from 1·58 to 2·97 Å. An infinite two dimensional structure is also found in $K_3V_5O_{14}$ (Fig. 108) (399). In this case the oxygen atoms required to complete the octahedral coordination and to form a three dimensional structure are about 3·4 Å from the vanadium atoms, which is much too long to be considered as bonding.

(*d*) *Strongly acid solutions.* The colloidal hydrated vanadium oxide dissolves in excess acid to form the cation VO_2^+ (1229, 1950, 2043). Estimates for the equilibrium constant vary widely (714, 2043, 2432):

$$K = \frac{[HVO_3][H^+]}{[VO_2^+]} = 10^{-3}\text{–}10^{-5}$$

(*e*) *Reduced isopolyvanadates.* Reduction of vanadate in weakly acid solution produces decavanadates with vanadium(V) : vanadium(IV) ratios of 8 : 2, 7 : 3, 6 : 4, 5 : 5, 4 : 6 and 3 : 7, the colours progressively changing from green through red to brown (1823). The 7 : 3 and 3 : 7 complexes, $[H_5V_7^{V}V_3^{IV}O_{28}]^{4-}$ and $[H_9V_3^{V}V_7^{IV}O_{28}]^{4-}$, appear particularly stable.

A compound which is probably similar is obtained by reducing a suspension of V_2O_5, and has been variously formulated as $V_3O_5(OH)_4$ or V_6O_{14}, 6 H_2O. Its properties have been reviewed in some detail (1931).

Reaction of vanadylacetylacetonate with tetraethylammonium chloride in the presence of a metal complex such as copper acetylacetonate yields $(Et_4N)_4[H_4V^{V}_8V^{IV}_2O_{28}]$ (1168). The compound is deep purple due to a

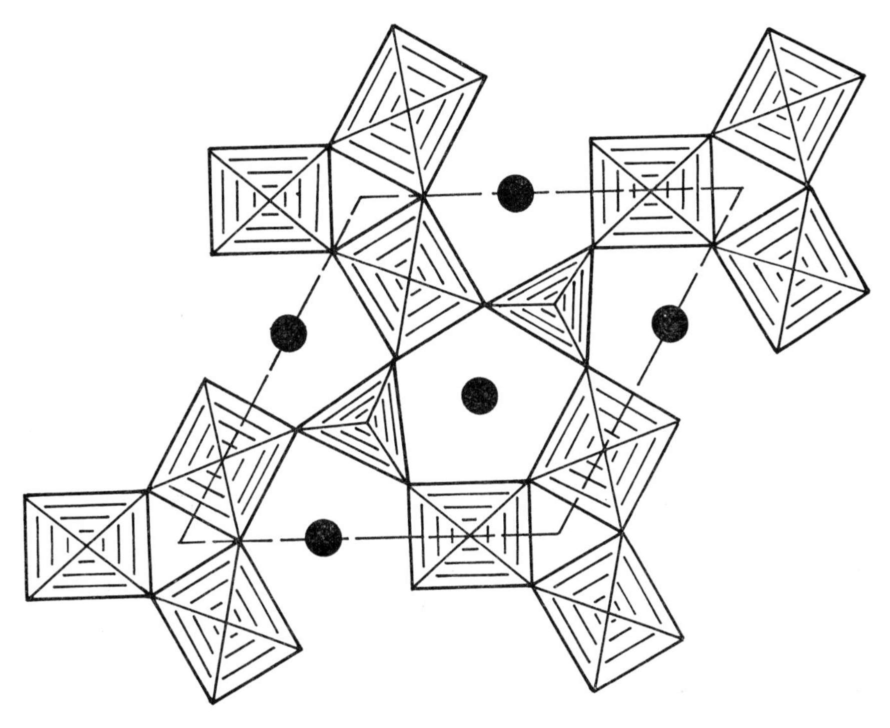

FIG. 108 The structure of $K_3V_5O_{14}$. The VO_6 octahedra and VO_5 trigonal bipyramids also share corners with similar polyhedra above and below the plane of the page forming infinite tunnels which accommodate the potassium ions.

very broad and intense band at 20 000 cm^{-1}, and another very broad but less intense band at 12 000 cm^{-1}. It is soluble in acetonitrile, nitromethane and dimethylsulphoxide to give a purple solution, and in water to give a green solution. The effective magnetic moment of 1·9 B.M. appears somewhat high in view of the e.s.r. results which show a 15 line spectrum due to strong interaction of two equivalent ^{51}V nuclei (nuclear spin 7/2).

Similar reactions of vanadylacetylacetonate in the presence of fluoride leads to the closely related $(Et_4N)_4[H_2V^{V}_8V^{IV}_2O_{26}F_2]$ (1169).

2. *Isopolyniobates and isopolytantalates*

Isopolyniobate and isopolytantalate anions are present in aqueous alkaline solution, from which salts of the type $Na_8Nb_6O_{19},xH_2O$, $Na_7HNb_6O_{19},xH_2O$, and $Na_6H_2Nb_6O_{19},xH_2O$ can be isolated (626, 1349, 1742). In highly alkaline solution there is evidence of the monomeric anions NbO_4^{3-} and TaO_4^{3-} (637, 1083), 1514, 1741).

The structures of $Na_7HNb_6O_{19},15H_2O$ (1551) and $K_8Ta_6O_{19},16H_2O$ (1552) show the presence of $M_6O_{19}^{8-}$ units (Fig. 109). Each metal atom is

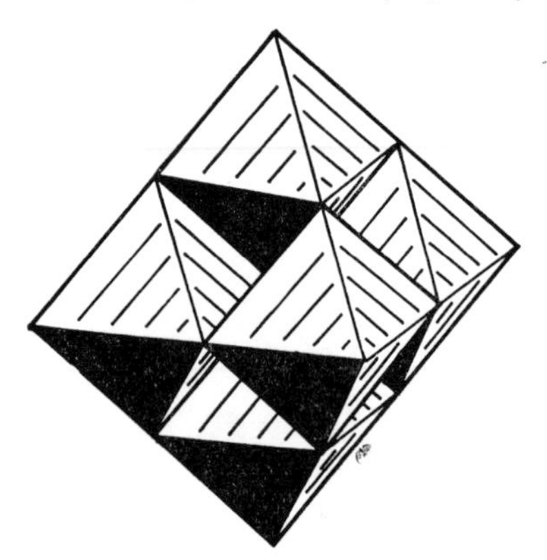

FIG. 109 The structure of $Nb_6O_{19}^{8-}$ and $Ta_6O_{19}^{8-}$.

octahedrally coordinated by six oxygen atoms, the metal atoms themselves, also being at the corners of an octahedron, with niobium–niobium distances of approximately 3·3 Å. It has been pointed out that this is the most favourable way to arrange six NbO_6 octahedra to allow the distortion necessary to accommodate the niobium(V)–niobium(V) Coulombic repulsions (1354).

A variety of physical techniques show that these six-fold condensed ions persist in solution (103, 538, 1287, 1288, 1289, 1767, 1768, 1769, 2284). The acid dissociation constants of $H_2Nb_6O_{19}^{6-}$ have been estimated as $pK_1 = 10{\cdot}9$ and $pK_2 = 13{\cdot}8$ (1773).

3. *Heteropolyvanadates and heteropolyniobates*

Vanadate forms the 1 : 12 phosphovanadates which can be isolated, for example, as $Na_7[PV_{12}O_{36}]$, 38 H_2O (1274, 1764, 2126). This anion is less stable towards acid, base and heat than the corresponding 12-phosphotungstates and 12-phosphomolybdates. However more stable products can be obtained by replacing some of the molybdenum or tungsten atoms in heteropolymolybdates or heteropolytungstates by vanadium atoms

(579, 2184, 2305). Examples of analogues of $[PMo_{12}O_{40}]^{3-}$ include $[PVMo_{11}O_{40}]^{4-}$, $[PV_2Mo_{10}O_{40}]^{5-}$ and $[PV_3Mo_9O_{40}]^{6-}$.

The hexaniobate ion $Nb_6O_{19}{}^{8-}$ can behave as a tridentate ligand forming, for example, $[Co^{III}(Nb_6O_{19})(H_2O)(en)]^{5-}$ (893), $[Cr^{III}(Nb_6O_{19})(H_2O)(en)]^{5-}$ (893), $[Mn^{IV}(Nb_6O_{19})_2]^{12-}$ (620, 621, 894), and $[Ni^{IV}(Nb_6O_{19})_2]^{12-}$ (894). The structure of $Na_{12}[Mn^{IV}(Nb_6O_{19})_2],50H_2O$ confirms that the $Nb_6O_{19}{}^{8-}$ anion remains essentially unchanged (Fig. 110)

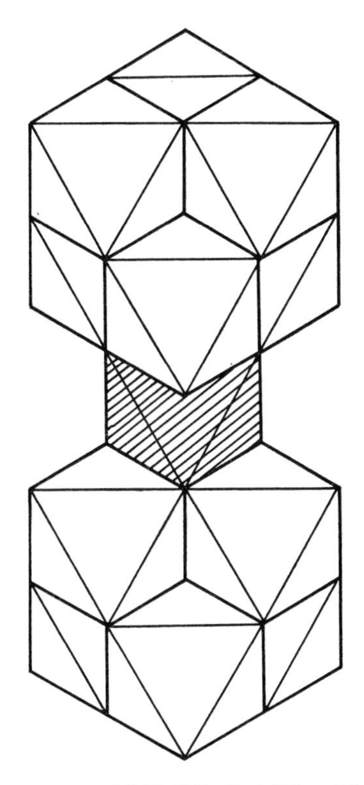

FIG. 110 The structure of $[Mn(Nb_6O_{19})_2]^{12-}$. The MnO_6 octahedron is shaded to distinguish it from the NbO_6 octahedra.

(895). The molecule has approximately D_{3d} symmetry. The MnO_6 octahedron (Mn-O = 1·87 Å) is stretched along the three-fold axis so that the O-Mn-O angle across the plane perpendicular to the three-fold axis is 97°, which is attributed to mutual repulsion of the two $Nb_6O_{19}{}^{8-}$ groups. The three niobium–niobium distances facing the manganese atom are longer than the other niobium–niobium distances (3·53 Å compared with 3·34 Å) due to electrostatic repulsion of the manganese(IV). The variations in niobium–distances can also be considered to be a consequence of metal–metal Coulombic repulsions.

I. Organometallic Complexes

Apparently the only examples of pentavalent Group V σ-bonded alkyls are $(CH_3)_3NbCl_2$ and $(CH_3)_3TaCl_2$ prepared from dimethyl zinc and the metal pentachloride in pentane at low temperatures (1323). These yellow compounds decompose spontaneously at room temperature with the evolution of methane, and are also unstable to oxygen and water. The solids are readily volatile and the vapours are monomeric.

The cyclopentadienyl $(C_5H_5)VOCl_2$ has been obtained by the action of oxygen and hydrogen chloride on $(C_5H_5)V(CO)_4$ (876).

Niobium and tantalum form the bis-cyclopentadienyls $(C_5H_5)_2MCl_3$ and $(C_5H_5)_2MBr_3$. They are obtained from the pentahalides and sodium cyclopentadienyl in tetrahydrofuran (2340). These trichlorides and tribromides are unstable to water, whereas the corresponding $(C_5H_5)_2NbI_3$ is stable to water (2299). Reduction with sodium borohydride forms the air unstable trihydride $(C_5H_5)_2TaH_3$ (1052, 1594).

2. OXIDATION STATE IV

A. Oxides

Vanadium and niobium dioxides have been studied in considerable detail, but there is some doubt concerning the tantalum analogue (1478). The easier reduction of Nb_2O_5 compared with Ta_2O_5, and the subsequent solubility of NbO_2 in hot concentrated sulphuric acid, can be used to separate these elements (2018).

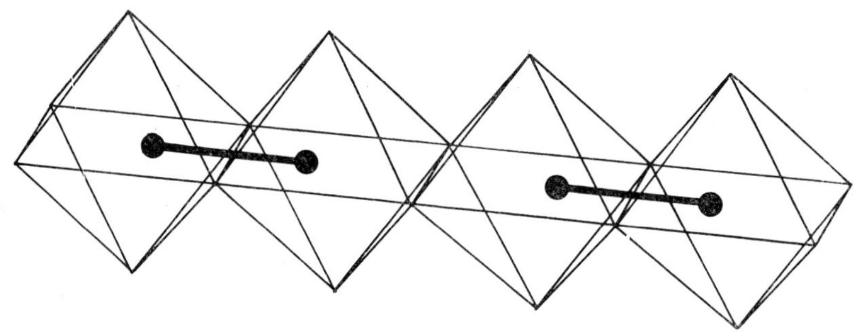

FIG. 111 The formation of metal–metal bonds between octahedrally coordinated metal atoms in VO_2 and NbO_2. (Compare with Figs. 17 and 42).

The vanadium and niobium compounds have the rutile structure, but the metal–metal distances along the *c*-axis are not equal as in rutile itself, but are alternately long and short due to the formation of V_2 and Nb_2 doublets (Fig. 111).

Those metals which have a dioxide with the rutile structure are shown in Table 22. The oxides enclosed in the box have distorted rutile structures with metal–metal bonds.

TABLE 22

Oxides with the rutile structure. Those enclosed in the box have distorted structures due to the formation of metal–metal bonds. The numbers refer to the *c/a* axial ratio of the unit cell

d^0	d^1	d^2	d^3	d^4	d^5	d^6	d^{10}	$d^{10}s^2$
TiO_2 0·644	VO_2 0·625	CrO_2 0·660	MnO_2 0·653				GeO_2 0·651	
	NbO_2 0·618	MoO_2 0·577	TcO_2 0·586	RuO_2 0·692	RhO_2 0·688		SnO_2 0·673	TeO_2 0·787
	TaO_2 0·651	WO_2 0·567	ReO_2 0·575	OsO_2 0·708	IrO_2 0·701	PtO_2 0·694	PbO_2 0·683	

The metal–metal distances along the rutile *c*-axis are shown in Table 23.

TABLE 23

Metal–metal distances along the *c*-axis of dioxides with rutile or distorted rutile structures

		M–M (Å)			$M{-}M_{long}—M{-}M_{short}$	References
d^0	TiO_2		2·959		0	1630
d^1	VO_2	2·65		3·12	0·47	56, 1630
	NbO_2	2·80		3·20	0·40	1646, 1934
	TaO_2		3·07		0	2071
d^2	CrO_2		2·92		0	2387
	MoO_2	2·51		3·11	0·60	316, 1630
	WO_2	2·49		3·08	0·59	1934
d^3	MnO_2		2·86		0	1934, 2240
	TcO_2	2·48		3·06	0·58	1934
	ReO_2	2·49		3·08	0·59	1630
d^4	RuO_2		3·11		0	2025
	OsO_2		3·19		0	1934
d^5	RhO_2		3·09		0	2110
	IrO_2		3·15		0	1934
d^6	PtO_2		3·14		0	2110

Two types of distortion have been observed for these metal–metal bonded dioxides:

(i) in VO_2, MoO_2, WO_2, TcO_2 and ReO_2 the metal atom doublets are tilted in the (100) direction of the rutile substructure, to form a large monoclinic cell (56).

(ii) In NbO_2 the metal atom doublets are tilted in the (110) or $(1\bar{1}0)$ direction of the rutile substructure, the unit cell being tetragonal (1631, 1646, 2269).

The existence of metal–metal bonding can be seen in the absence of a complete structural determination by a relative shortening of the c-axis of the unit cell (1648). The c/a axial ratios are given in Table 22, and it can be seen that if $c/a > 0{\cdot}64$, there is no metal–metal bond. (The figures for the rutile substructure of the distorted structures have been calculated from:

For MoO_2-type: $a_{\text{rutile}} = b_{\text{monoclinic}}$, $c_{\text{rutile}} = \frac{1}{2}\, a_{\text{monoclinic}}$.

For NbO_2 : $a_{\text{rutile}} = \dfrac{2^{\frac{1}{2}}}{4}\, a_{\text{tetragonal}}$, $c_{\text{rutile}} = \frac{1}{2}\, c_{\text{tetragonal}}$.)

It can be seen that the values quoted for TaO_2 are not consistent with the formation of the expected Ta_2 pairs.

Mixed oxides (218, 1647, 1648, 1649, 1708, 1956, 1957, 2248, 2318) of these elements normally have a c/a axial ratio greater than that obtained by averaging the axial ratios of the components, indicating a less pronounced bonding when the metals are dissimilar. This effect is particularly noticeable for the VO_2–NbO_2 mixture, where $V_{\frac{1}{2}}Nb_{\frac{1}{2}}O_2$ has a higher axial ratio ($c/a = 0{\cdot}646$) then either VO_2 or NbO_2. In the case of TiO_2–TaO_2 mixtures, the rutile structure is only maintained as the tantalum concentration is increased up to $Ti_{\frac{1}{2}}Ta_{\frac{1}{2}}O_2$, which is best written $Ti^{III}Ta^{V}O_4$.

Vanadium dioxide is particularly interesting, as on heating to about 70° C the V–V pairs are broken. The phase change has been studied by structural, magnetic, electrical, ^{51}V-n.m.r. and spectral methods. The high temperature phase has an undistorted rutile structure, the vanadium–vanadium distance increasing from 2·65 Å to 2·87 Å, although the c/a axial ratio remains unchanged (0·636 and 0·634 respectively) (2380). At approximately 70° C (the exact temperature varies by 2°–3° C between different workers, and there is also a hysteresis of a few degrees on cooling), the magnetic susceptibility increases sharply from a small temperature independent value (approximately 100 cgsu mol^{-1}) to a temperature dependent value corresponding to $\mu_{\text{eff}} \sim 1{\cdot}5$ B.M. (1073, 1224, 1270, 1462, 1857, 1958, 1958). The magnetic susceptibility at room temperature is found to increase with decreasing particle size (2259); as the particle size decreases from above 1000 Å to approximately 90 Å, the magnetic susceptibility increases by about 45%. The freeing of the bonding electrons on warming is also shown by a sharp 100-fold increase in the electrical conductivity (259, 1270, 1714, 1958). Below 68° C the ^{51}V-n.m.r. spectrum is complex due to the asymmetry introduced by the vanadium–vanadium bond, but a simple spectrum is obtained above this temperature (2312). Finally the diffuse reflectance spectrum of VO_2 shows a minimum at 41 000 cm^{-1}, which is lost on heating past 70° C (1958).

The vanadium–vanadium bonding is also disrupted by dilution with TiO_2 as shown by magnetic susceptibility, electrical conductivity, and spectral measurements (1958). The susceptibility per vanadium atom is a linear function of composition, and extrapolation to TiO_2, that is breaking of all the vanadium–vanadium bonds, corresponds to an effective magnetic moment of 1·73 B.M. per vanadium atom. Similarly addition of TiO_2 increases the electrical conductivity and leads to a diffuse reflectance spectrum characteristic of the high temperature form of VO_2.

Niobium and tantalum dioxides have only small paramagnetic susceptibilities which are almost independent of temperature (319, 1473).

Vanadium dioxide loses oxygen to form a number of phases of composition V_nO_{2n-1} (where $n = 3,4,5,6,7$ or 8) with structures analogous to Ti_nO_{2n-1} which are discussed in more detail on page 65 (55, 64, 89, 1334, 1461). The structure shows that periodic dislocations or shear planes run through a crystal of VO_2, resulting in slabs of VO_2-type structure infinitely extended in two dimensions, but only n octahedra thick, which are linked to similar slabs by the face sharing of the octahedra (65). These compounds are paramagnetic at room temperature, with the susceptibility decreasing linearly as the composition approaches VO_2 (1073). However vanadium–vanadium pairs are again formed on cooling through the following transition temperatures (259, 1073, 1461, 1462):

V_3O_5	V_4O_7	V_5O_9	V_6O_{11}	VO_2
−140° C	−23° C	−134° C	−96° C	+68° C

B. Tetrahalides

1. *Tetrafluorides*

Vanadium tetrafluoride was first prepared by the direct action of anhydrous hydrogen fluoride on vanadium tetrachloride at −28° C (1966), although a purer product can be obtained if the reaction is carried out in a cold inert solvent (CCl_3F at −78° C) (443). It has also been prepared by the direct fluorination of vanadium metal followed by sublimation from the resultant mixture with VF_5 and VF_3, and by reduction of the pentafluoride (419).

Vanadium tetrafluoride is a bright green hygroscopic solid with an effective magnetic moment at room temperature of 2·17 B.M., which follows the Curie–Weiss law with $\theta = 198°$ It disproportionates rapidly above 100° C to VF_3 and VF_5, and even at room temperature there is significant disproportionation after a few days (443).

The mixed fluorochlorides $VClF_3$ and $V_2Cl_3F_5$ have been obtained by the action of fluorides (for example PF_3, AsF_3, AsF_5, SbF_3, SF_4 or VF_5) on vanadium tetrachloride (420, 1444).

Reduction of niobium pentafluoride with niobium or silicon forms the dark blue or black NbF_4 (1037, 2007, 2027). The compound is involatile and disproportionates above 350° C to NbF_5 and NbF_3, which is in contrast

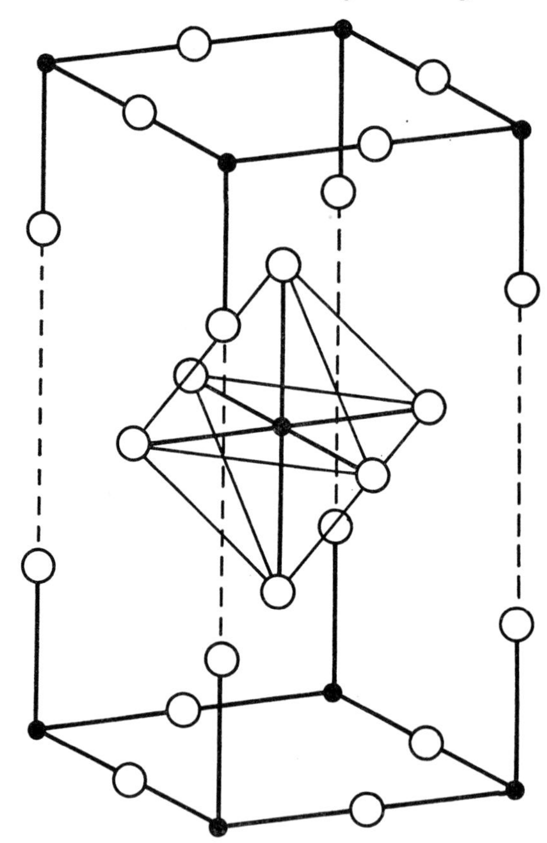

FIG. 112 The structure of NbF_4. The NbF_6 octahedra share four corners with neighbouring octahedra to form an infinite sheet structure. The filled circles represent niobium atoms and the open circles fluorine atoms.

to TiF_4, VF_4 and CrF_4 which sublime at 100°–200° C, and to ZrF_4 and HfF_4 which sublime only above 700° C. The structure (Fig. 112) shows octahedral coordination of niobium, with the NbF_6 octahedra sharing four corners with adjacent octahedra to form a two dimensional layer structure (1037, 2026, 2027). The magnetic susceptibility is dependent upon field strength; extrapolation to infinite field strength gives
$10^6\chi_M$ (297° K) = 98 c.g.s.u. and $10^6\chi_M$ (80° K) = 115 c.g.s.u., (1037) or $10^6\chi_M$ (90–295° K) = 175–200 c.g.s.u. (2027) depending upon the worker.

Tantalum tetrafluoride has not been reported.

2. *Tetrachlorides, bromides and iodides*

(a) *Vanadium.* Vanadium tetrachloride is produced when chlorine is passed over vanadium metal at about 400° C. Other chlorinating agents such as $COCl_2$, $SOCl_2$, S_2Cl_2 or CCl_4 can be used on V_2O_5 or the metal itself (1433). Vanadium tetrachloride is a red brown viscous liquid, M.P. = −26° C, B.P. = 153° C. It is readily soluble in solvents such as carbon tetrachloride, and forms adducts with a number of other solvents (page 200). It hydrolyses rapidly in air, and reacts readily with hydroxylic solvents.

Decomposition to vanadium trichloride and chlorine occurs slowly at room temperature, and rapidly on boiling, the equilibrium constants being given below (2155):

$$2\,VCl_{3(s)} + Cl_{2(g)} \rightleftarrows 2\,VCl_{4(g)} \quad \Delta H = 13\cdot 8 \text{ Kcal/mole.}$$

° C	$Kp = \frac{p^2_{VCl_4}}{p_{Cl_2}}$
160	1480
170	2070
180	3000

Vapour density measurements and mass spectrometry show that vanadium tetrachloride is essentially monomeric in the gas phase (1074, 1944). The vapour pressure (in mm of mercury) is given by:

$$\log p = -\frac{1998}{T} + 7\cdot 58 \quad (310°\text{–}350° \text{ K}) \quad \text{(Reference 2155)}$$

$$\log p = -\frac{2020}{T} + 7\cdot 62 \quad (300°\text{–}420° \text{ K}) \quad \text{(Reference 2257)}$$

The second expression has the lower vapour pressure at higher temperatures, and may therefore be more correct in view of the known thermal instability of this compound. It is also derived from the most recent determinations, examples of which are shown:

°C	
29·5	p = 9 mm Hg
41	p = 15
104	p = 180
153	p = 746

Electron diffraction of the vapour shows the molecule to be tetrahedral as expected, with bond lengths of 2·14 Å (1715)

The effective magnetic moment of VCl_4 has been reported as 1·79 B.M. (1421), 1·67 B.M. (682, 2382), and 1·61 B.M. (507). The lower value suggesting less contamination with VCl_3 may be most reliable. The moment is independent of field strength, and the Curie law is obeyed.

The infrared spectrum of liquid vanadium tetrachloride, and vanadium tetrachloride dissolved in carbon tetrachloride, show some similarities to the spectrum of the gas phase, and to the spectrum of titanium tetrachloride, although all show significant differences (682). Earlier work (2155) on the freezing point depression of VCl_4 in CCl_4 were interpreted as showing the presence of large amounts of dimeric V_2Cl_8, but were later (2382) shown to be due to the fact that the solid phase crystallizing out was a solid solution of VCl_4 and CCl_4, and not pure CCl_4 as assumed earlier. The magnetic susceptibility of VCl_4 in VCl_4–CCl_4 mixtures was found to increase uniformly from $\mu_{eff} = 1{\cdot}63$ B.M. (extrapolated to zero mole fraction of VCl_4) to $\mu_{eff} = 1{\cdot}6$ B.M. for pure VCl_4, and was thought to be experimentally significant (2382).

Vanadium tetrachloride has a single unpaired electron, and may therefore be expected to exhibit a Jahn–Teller effect. Electron spin resonance of solutions in carbon terachloride and of the solid indicate that such an effect is indeed observed, the ground state corresponding to 58% flattened tetrahedra and 42% elongated tetrahedra (1301). The results are intermediate between those predicted earlier by perturbation methods (135), and by a crystal field approach (136). The behaviour in the gas phase has not been resolved from the available Raman and infrared spectra (241, 682, 1074), or the electron diffraction data. (1656, 1715).

The tetrabromide and apparently the tetraiodide are found in equilibrium mixtures of the trihalides and halogens in the gas phase (219, 1324, 1587, 2285). The magenta tetrabromide has been isolated by quenching the vapour 78° C. It decomposes above —45° C (1587).

(*b*) *Niobium and tantalum.* Halogenation of the metals at low temperatures results in the direct formation of the pentahalides, while reaction at higher temperatures causes increasing decomposition to the lower halides. For example:

$$NbCl_5 \rightarrow NbCl_{3\cdot13} \rightarrow NbCl_{2\cdot67} \rightarrow NbCl_{2\cdot33}$$

The tetrachlorides and tetrabromides of niobium and tantalum are more difficult to prepare, as at moderate temperatures they disproportionate:

$$MX_{4(s)} \rightleftarrows MX_{5(g)} + MX_{3(s)}$$

The experimental technique which has been developed is to carry out the reduction under a high pressure of pentahalide, so as to drive the above equilibrium to the left. This pressure is maintained inside sealed tubes using a temperature gradient, one end being sufficiently hot for reaction to occur, while the other end is sufficiently hot to maintain the necessary saturated vapour pressure of the pentahalide.

Niobium pentachloride is conveniently reduced to the tetrachloride with aluminium or niobium metals, or niobium trichloride, in a temperature gradient 150°–230° C (2016). The compound disproportionates when heated, the approximate conditions for stability being indicated in Fig. 113 (2008, 2009). The compound is formed as black crystals, but is

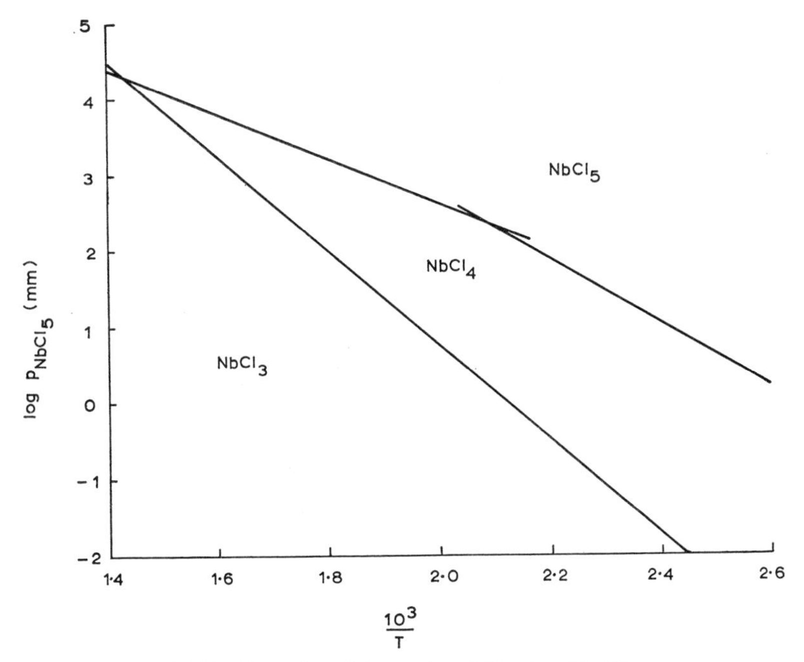

FIG. 113 Conditions of stability for $NbCl_4$.

red-brown when finely ground. The structure (Fig. 17, page 26) (2026, 2068) shows rows of metal atoms between close packed layers of chloride ions, but the niobium–niobium distances are alternately short and long (3·06 and 3·76 Å) due to metal–metal bonding by pairs of niobium atoms.

Niobium pentabromide has similarly been reduced with niobium in temperature gradients to form the dark brown $NbBr_4$, which is isomorphous with $NbCl_4$ (211, 1590, 2012).

The grey niobium tetraiodide is formed directly from the elements, there being no need to maintain a pressure of the pentaiodide in this case (471, 550, 2086). The structure (Fig. 114) (619) again shows metal–metal

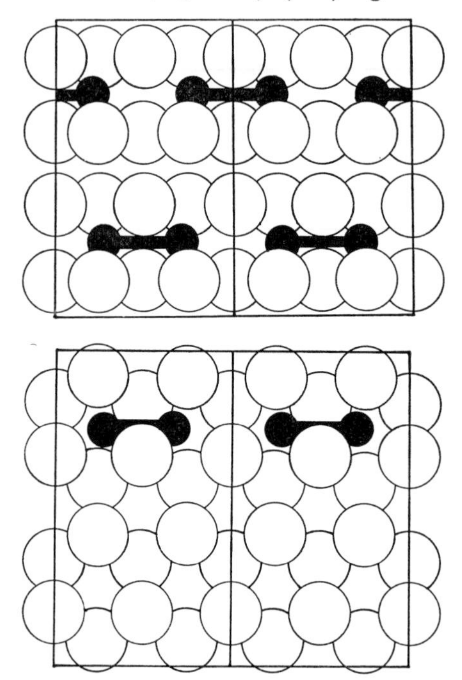

FIG. 114 The structure of α–NbI_4 showing pairs of niobium atoms in a close packed iodide lattice.

bonded pairs of niobium atoms in a close packed halide lattice, but the pairs are arranged slightly differently to those in the chloride and bromide. Alternatively the structure can be considered to consist of chains of NbI_6 octahedra sharing opposite edges, but with alternate niobium–niobium distances being short and long (3·31 and 4·36 Å respectively). Although niobium–niobium bonding undoubtedly occurs, the bond length is fairly long, and is discussed further on page 42. There is a weakly endothermic phase change from this α-NbI_4 to β-NbI_4 at 348° C, but there is no apparent change in the X-ray powder data. A second and more profound change occurs at 417° C to form γ-NbI_4, which finally melts incongruently at 503° C to form an iodine-rich liquid and Nb_3I_8 (1590, 2086). The structure of γ-NbI_4 is remarkable in that it is derived from the β-$TiCl_3$ structure but with $\frac{1}{4}$ of the cation sites vacant in a disordered manner (2086). Distances between neighbouring cation sites appear to be somewhat shorter than in α-NbI_4.

Niobium tetrachloride, tetrabromide, and all three forms of the tetraiodide have been reported to be diamagnetic or to show a small temperature-independent paramagnetism (1475, 1590, 2026, 2032, 2086).

Tantalum tetrachloride can be conveniently prepared by the reduction of the pentachloride with aluminium in a temperature gradient (200°–400° C) in an analogous manner to niobium tetrachloride (2017). A more difficult technique is to use tantalum metal as reductant in a non-uniform temperature gradient of 280°–630° C, the higher temperature requiring silica tubes; at 280° C the tantalum pentachloride pressure is approximately three atmospheres (946, 2030). Under these conditions prolonged heating causes attack on the silica tube with the formation of $TaOCl_2$ and $SiCl_4$ or Ta_2Si and Ta_2O_5 (2024, 2030). The pentachloride has also been reduced at lower temperatures with hydrogen in the presence of an electric discharge (1103).

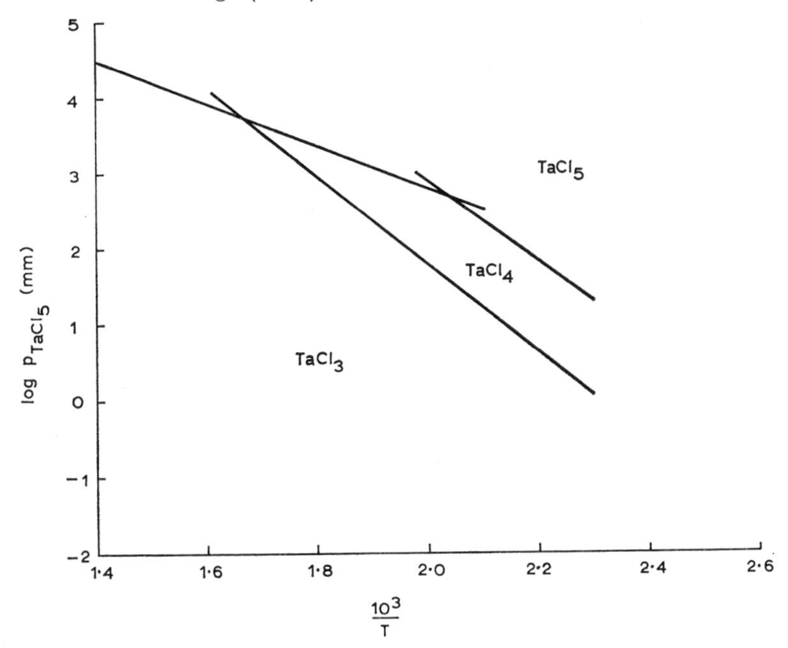

FIG. 115 Conditions of stability for $TaCl_4$.

Disporportionation of $TaCl_4$ into $TaCl_3$ and $TaCl_5$ occurs at 200°–300° C, the approximate conditions for stability being shown in Fig. 115 (2017). The compound is black in mass, but green when finely ground. It is isomorphous with $NbCl_4$, and the presence of tantalum–tantalum bonds is confirmed by the diamagnetism (2030).

Tantalum tetrabromide can likewise be formed by the reduction of the pentabromide with aluminium (temperature gradient 250°–500° C) or tantalum (temperature gradient 300°–630° C), (211, 1581, 2014) or at lower temperatures with hydrogen in the presence of an electric discharge (1103). It is a dark blue diamagnetic substance, isomorphous with $NbCl_4$, $NbBr_4$, and $TaCl_4$ (211, 1581, 2014, 2032).

The tetraiodide is similarly obtained from tantalum pentaiodide and aluminium (temperature gradient 180°–400° C), or more conveniently by decomposition of the pyridine complex $TaI_4(py)_2$ (page 202) (1581). These two preparations are not isomorphous with each other or with α-NbI_4, although a form which is isomorphous with α-NbI_4 has also been reported (619). A structurally uncharacterized tetraiodide has been reported to be diamagnetic (1938, 2032).

C. Adducts of the Tetrahalides

1. *Vanadium*

Vanadium tetrafluoride reacts readily with pyridine, ammonia and selenium tetrafluoride forming $VF_4(py)$, $VF_4(NH_3)$, and $VF_4.SeF_4$ respectively (443). Similar compounds had been obtained earlier by the reaction of vanadium pentafluoride with these ligands (442).

Adducts of vanadium tetrachloride are unstable to air, forming adducts of $VOCl_2$. In many instances they are also unstable with respect to reduction, and it is probably this tendency which sometimes leads to observed magnetic moments greater than the spin-only value of 1·73 B.M. Complexes with Group V donor atoms are discussed after those with Group VI donor atoms.

(a) *Oxygen and sulphur donors.* Oxygen-donor ligands form brown complexes of the general formula VCl_4L_2, where L is diethyl ether, tetrahydrofuran, tetrahydropyran, benzophenone and pyridine-N-oxides, and VCl_4B with the bidentates 1,4-dioxan and ethyleneglycol dimethyl ether (324, 497, 2088).

Reduction to vanadium(III) however occurs with the analogous sulphur ligands pentamethylene sulphide, tetrahydrothiophen, 1,4-dithian and 1,4-thioxan (324). However the bidentate ligands $RS.CH_2.CH_2.SR$ (where R is Me, Et or Ph) form the violet or black VCl_4B (497). *cis*-

Dimethylthiomaleodinitrile also forms $VCl_4[(MeS)_2C_2(CN)_2]$, although it is uncertain whether the ligand is bonded to the metal through the sulphur or nitrogen atoms (497).

(b) *Nitrogen, phosphorus and arsenic donors.* Vanadium tetrachloride with acetonitrile or ethyl cyanide forms the purple $VCl_4(RCN)_2$, which can be isolated before reduction to vanadium(III) if the reaction is carried out in an inert solvent such as carbon tetrachloride (701, 957). Molecular weight determinations in benzene indicate that one of the nitrile ligands is lost in solution, which is also consistent with the increase in V-Cl stretching frequencies from below 400 cm^{-1} in a nujol mull to 432 and 438 cm^{-1} in benzene. The mononitrile adduct however has not been isolated.

Similarly vanadium tetrachloride reacts with pyridine in toluene at $-20°$ C to form the dark red $VCl_4(py)_2$ ($\mu_{eff} = 1{\cdot}80$ B.M.); reduction to vanadium(III) occurs on warming (1389). Similar compounds have been obtained with pyrazine and dimethylpyrazine (324). The visible spectra of the analogous brown compounds VCl_4(dipy) and VCl_4(*o*-phen) show broad asymmetric bands with components at approximately 18 000 and 21 500 cm^{-1} which were assigned to the transitions $^2B_{2g} \rightarrow {}^2A_{1g}$ and $^2B_{2g} \rightarrow {}^2B_{1g}$ respectively (493). The room temperature effective magnetic moments of 1·76 B.M. fall to about 1·60 B.M. at 80° K, which was interpreted as indicating fairly large separations ($\sim$1000 cm^{-1}) between the orbitals derived from the octahedral $^2T_{2g}$ orbitals, due to fairly large distortions from regular octahedral stereochemistries.

Trimethylamine appears to form only the 1 : 1 adduct at low amine concentration, but products of varying composition apparently containing more amine and/or vanadium(III) were obtained at higher amine concentration (926). Similar compounds $VCl_4(amine)_2$ and VCl_4(diamine) have been obtained with a number of aromatic amines (1892, 1893).

Reduction of vanadium tetrachloride is observed with triphenylarsine, triphenylphosphine or bis(diphenylphosphino)ethane in carbon tetrachloride or benzene (324). The bidentate *o*-phenylenebisdimethylarsine forms the eight coordinate dodecahedral $VCl_4(MeD)_2$ (505, 506), which is isomorphous with $TiCl_4(MeD)_2$, whereas the ethyl analogue *o*-phenylenebisdiethylarsine forms only the six coordinate VCl_4(EtD) (495). These compounds have room temperature effective magnetic moments of 1·74 and 1·76 B.M. respectively. Vanadium tetrachloride with these ligands thus behaves more like titanium tetrachloride than like niobium or tantalum tetrachlorides. The tridentate arsines $MeAs(o\text{-}C_6H_4.AsMe_2)_2$ and $MeC(CH_2.AsMe_2)_3$ form the seven coordinate VCl_4(triars) compounds, which are again similar to the titanium analogues (501).

2. *Niobium and tantalum*

(a) *Oxygen and sulphur donors.* The adducts of tantalum tetrahalides are in general less well characterized than those of the niobium analogues. For example oxygen donors form the yellow or brown solids NbX_4L_2 (where X is Cl or Br and L is tetrahydrofuran, tetrahydropyran or dioxan), whereas only oils were isolated from the tantalum tetrahalides (933).

Tetrahydrothiophen forms the 1 : 2 adducts $MX_4(C_4H_8S)_2$ (where M is Nb or Ta and X is Cl or Br). Niobium tetrachloride and dimethylsulphide however form $NbCl_4(Me_2S)$ which is dimeric in benzene and has a low room temperature effective magnetic moment of 0·44 B.M. compared with 1·30–1·42 B.M. for other niobium(IV) complexes, suggesting considerable niobium–niobium interaction. It has been tentatively suggested that some reduction to tantalum(III) occurs during the reaction of tantalum tetrabromide with dimethylsulphide (935).

(b) *Nitrogen, phosphorus and arsenic donors.* Niobium pentachloride and pentabromide react with pyridine to initially form complexes of niobium(V), but which are slowly reduced to $NbCl_4(py)_2$ and $NbBr_4(py)_2$, the oxidation product being the 1-(4-pyridyl)pyridinium ion (33,1585, 1590). For example:

$$2\,NbBr_5 + 7\,C_5H_5N \rightarrow 2\,NbBr_4(C_5H_5N)_2 + (C_5H_5N.C_5H_5N)Br + (C_5H_5NH)Br.$$

Tantalum pentachloride and pentabromide form $TaCl_4(py)_2$ and $TaBr_4(py)_2$ more slowly, and in this case the pentavalent adducts can also be characterized (33, 1543, 1585). The pentaiodides form $MI_4(py)_2$ with the evolution of iodine, which forms an adduct with excess pyridine (1585):

$$2\,NbI_5 + 5\,C_5H_5N \rightarrow 2\,NbI_4(C_5H_5N)_2 + C_5H_5NI_2$$

The compounds are all light brown to red, but in the case of niobium tetrabromide a red compound is obtained only above 60° C, which reverts to a green compound on cooling. Reduction to tetrachloro and tetrabromo niobium adducts similarly occurs with γ-picoline (brown compounds except for the green $NbBr_4(\gamma\text{-pic})_2$), dipyridyl and *o*-phenanthroline (green or purple compounds) (33).

Purer products are obtained directly from the tetrahalides and pyridine, as the difficult separation of pyridine oxidation products is avoided (933, 1581, 1590). There are considerable differences in the magnetic behaviour depending upon the method of preparation. For example reduction of the tantalum pentahalides with pyridine yields $TaCl_4(py)_2$ and $TaBr_4(py)_2$ ($\mu_{eff} = 1·31$ and 1·02 B.M. respectively) whereas the same compounds prepared from the tetrahalides have room temperature effective magnetic moments of 0·69 and 0·43 B.M. respectively.

As has been noted above, the tantalum compounds are more difficult to characterize than the niobium ones. For example although $NbX_4(CH_3CN)_2$ (where X = Cl, Br) are well characterized, the tantalum analogues analyse perhaps less satisfactorily, and have infrared spectra much more complex than is normal for coordinated acetonitrile (933).

The bidentate *o*-phenylenebisdimethylarsine precipitates the eight coordinate $NbX_4(diars)_2$, where X is Cl, Br, or I, from acetonitrile solutions of the corresponding tetrahalide (656). The same compounds are obtained from the pentahalides under sealed tube conditions at higher temperatures. The chloro and bromo complexes are isomorphous with the titanium analogues, so the stereochemistry is dodecahedral, with the diarsine ligand spanning the two A–A edges of the dodecahedron and the halide ions occupying the four B vertices (page 13). From an examination of the *d–d* bands in the visible spectrum, and by varying the halogen atoms in the series Cl, Br, I, and by varying the bidentate ligand in the series
o-$C_6H_4(AsMe_2)_2$, 1,2,4-$C_6H_3(Me)(AsMe_2)_2$, *o*-$C_6H_4(AsEt_2)_2$, it was deduced that the orbital splitting was in the order
$d_{xy} < d_{z^2} < d_{xz}, d_{yz} < d_{x^2-y^2}$ (page 17).
The ethyldiarsine complexes dissolve unchanged in toluene, but partially dissociate in acetronitrile and completely dissociate in pyridine. The room temperature effective magnetic moments of these eight coordinate complexes are in the range 1·60–1·67 B.M. and follow the Curie–Weiss law, which is not unexpected for a single *d*-electron in an orbitally non-degenerate ground state.

The ligands *o*-$C_6H_4(AsPh_2)_2$ and Ph_3As did not react with $NbCl_4$ under these conditions (656).

Tantalum tetrachloride under the same conditions formed only $TaCl_4(diars)$ with *o*-$C_6H_4(AsMe_2)_2$. This relatively large difference in the chemistry of the electropositive niobium and tantalum brought about by a ligand of low electronegativity can be compared to the differences in zirconium and hafnium brought about by the same ligand. The elusive $TaCl_4(diars)_2$ could only be obtained in an impure state by the reaction of tantalum tetrachloride and diarsine in a sealed tube (656).

D. Oxohalides

The vanadium(IV) oxohalides VOX_2 (where X is F, Cl, Br or I) are described separately in Section G of this Chapter, page 209.

Attempts to prepare $NbOF_2$ resulted in the discovery of a single phase between the stoicheiometric limits Nb^VO_2F and $NbO_{1\cdot25}F_{1\cdot75}$ (average niobium oxidation state of 4·25), which has the ReO_3 structure with octahedral coordination about the metal atom. It was concluded that $NbOF_2$, does not exist (2007).

However apart from the fluorides, all six MOX_2 compounds, where M is Nb or Ta and X is Cl, Br or I, have been reported. All probably exhibit metal–metal bonding, which is to be expected by analogy with the metal–metal bonded dioxides and tetrahalides.

These compounds are prepared by reduction of the pentavalent metal. For example (2013, 2032):

$$NbOCl_3 + H_2 \xrightarrow{400°\text{–}500° C} NbOCl_2$$

$$Nb + NbCl_5 + Nb_2O_5 \xrightarrow{350°\text{–}370° C} NbOCl_2$$

$$Ta + TaCl_5 + Ta_2O_5 \xrightarrow{400°\text{–}500° C} TaOCl_2$$

$$Nb + I_2 + Nb_2O_5 \longrightarrow NbOI_2$$

These compounds can also be grown as single crystals by transport in the gaseous phase in an analogous manner to the halides and oxides. For example (2013, 2032):

$$NbOCl_{2(s)} + NbCl_{5(g)} \rightleftarrows NbOCl_{3(g)} + NbCl_{4(g)}$$

$$NbOI_{2(s)} + I_{2(g)} \rightleftarrows 2\ NbOI_{3(g)}$$

The structure of $NbOCl_2$ shows that each metal atom is octahedrally surrounded by four chlorine and two oxygen atoms (Fig. 116d). The four chlorine atoms and the metal atom are in the one plane, and the octahedra share these edges to form infinite one dimensional strings; the niobium atoms are then grouped in pairs by a displacement from their octahedral centres towards each other, in an analogous manner to that occurring in $NbCl_4$. The two oxygen atoms are *trans* to each other, and are shared by octahedra in adjacent strings in an analogous manner to that occurring in $NbOCl_3$, to form a two dimensional layer structure. There is thus a close structural relationship between $NbCl_5$ (Fig. 116a), $NbOCl_3$ (Fig. 116b), $NbCl_4$ (Fig. 116c) and $NbOCl_2$ (Fig. 116d).

The same structure is found for $NbOBr_2$, $NbOI_2$, and $TaOCl_2$, with the interatomic distances shown (2026, 2032, 2068):

	Nb–Nb (Å)	Nb–O (Å)
$NbOCl_2$	3·14	1·83 and 2·11
$NbOBr_2$	3·12	1·83 and 2·10
$NbOI_2$	3·16	1·82 and 2·11

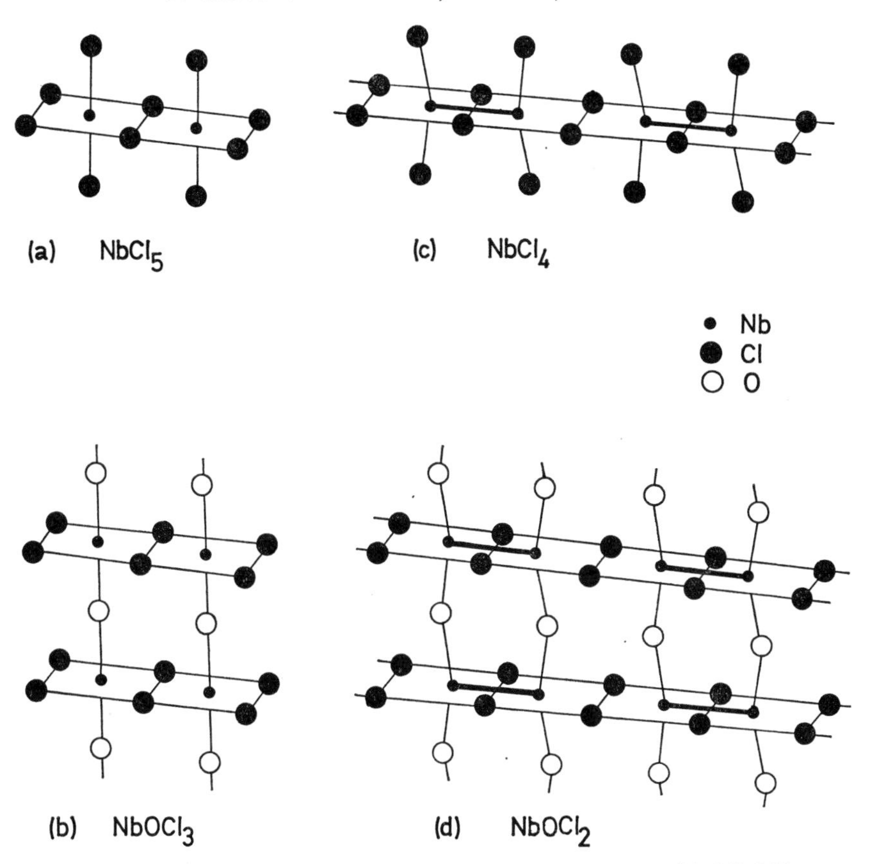

FIG. 116 Structural relationship between (a) $NbCl_5$, (b) $NbOCl_3$, (c) $NbCl_4$, and (d) $NbOCl_2$.

The existence of metal–metal bonding is confirmed by the diamagnetism of $NbOCl_2$, $NbOBr_2$, $NbOI_2$ and $TaOCl_2$ (2013, 2026, 2032, 2068).

The structure of NbS_2Cl_2 (2064) is related to $NbOCl_2$, with niobium atoms octahedrally surrounded by four Cl^- and two S_2^{2-} ions (Fig. 117). The niobium–niobium distance spanning the S_2^{2-}–S_2^{2-} edge is only 2·90 Å, indicating strong metal–metal bonding, while the next shortest niobium–niobium distance is 4·12 Å. This is at least partly due to the lack of bridging atoms in the plane of the niobium–niobium bond, allowing a closer approach of the metal atoms. It could be considered that the niobium atoms are eight coordinate, being surrounded by four chlorine and four sulphur atoms in a distorted square antiprismatic stereochemistry. NbS_2Br_2, $NbSe_2Br_2$ and $NbSe_2I_2$ are isomorphous with NbS_2Cl_2, but not with NbS_2I_2 or $NbSe_2Cl_2$ (2064), although the diamagnestism of all six compounds suggests similar niobium–niobium bonding (2010).

E. Anionic Halo and Oxohalo Complexes

The salts $M_2V^{IV}F_6$ (where M is K, Rb or Cs) are formed by the fluorination of $M_2V^{III}F_5,H_2O$ (1535), or by the addition of KF to VF_4 dissolved in SeF_4 (443). These hexafluoro salts have normal structures with octahedral coordination of the vanadium atom. The high room temperature magnetic moments of 1·97, 1·99 and 2·03 B.M., and the high Curie–Weiss constants of 78°, 100° and 103° for the above three salts should be noted. The transition $^2T_{2g} \rightarrow {}^2E_g$ occurs at 20 120 cm^{-1} in solid K_2VF_6 (183).

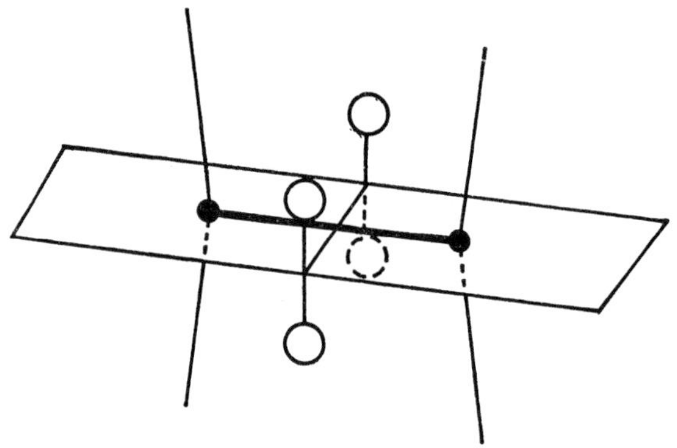

FIG. 117 The structure of NbS_2Cl_2.

The salts A_2VCl_6 have been obtained by the addition of $(Et_2NH_2)Cl$ or $(Et_3NH)Cl$ to a solution of VCl_4 (or $VCl_4(MeCN)_2$) in chloroform (940), or by the chlorination of the green A_2VOCl_4 (where A is pyH, quinH, *iso*-quinH or Cs) (see page 213) with thionyl chloride or with thionyl chloride/nitromethane mixtures at room temperature (1389):

$$A_2VOCl_4 + SOCl_2 \rightarrow A_2VCl_6 + SO_2$$

All these dark red compounds have room temperature effective magnetic moments in the range 1·73–1·75 B.M. The expected $^2T_{2g} \rightarrow {}^2E_g$ transition for a d^1 octahedral complex is at 15 400 cm^{-1}; the absorption peak is somewhat broad and asymmetric in a similar manner to the spectra of octahedral d^1 titanium(III) compounds, and can also be attributed to a Jahn–Teller distortion.

The conductometric titration of potassium chloride against vanadium tetrachloride in iodine monochloride suggests the formation of K_2VCl_6,

(1903), whereas the conductometric titration against tetramethylammonium chloride in arsenic trichloride suggests the formation of $(Me_4N)(VCl_5)$ (1094) (or $(Me_4N)[VCl_5(AsCl_3)]$).

The phase diagrams of niobium or tantalum tetrachlorides with alkali metal chlorides in some instances have shown the formation of the hexachloroanion, for example Na_2NbCl_6 (M.P. = 582° C), K_2NbCl_6 (M.P. = 782° C), K_2TaCl_6 (M.P. = 732° C), Rb_2TaCl_6 (M.P. = 756° C) and Cs_2TaCl_6 (M.P. = 796° C) (1460, 1977, 1978). The thermal stability of the tetravalent state under these conditions is in marked contrast to the tetrachlorides themselves.

The compounds prepared in this way show complex magnetic behaviour as a function of temperature (1249). Hexachloroniobates(IV) have also been obtained by reduction of hexachloroniobates(V) with iodide ion (1217), or from the tetrachloride in chloroform/acetonitrile mixtures (933) or in aqueous solutions saturated with hydrogen chloride (599, 1720). Compounds such as Cs_2NbCl_6 apparently have the K_2PtCl_6 structure (1720) (as do also compounds such as Cs_2NbOCl_5 also obtained from aqueous hydrochloric acid). The room temperature effective magnetic moments are in the range 0·8–1·2 B.M. for the niobium(IV) compounds. In addition to charge transfer bands above 30 000 cm^{-1}, $NbCl_6^{2-}$ shows two bands at about 20 000 cm^{-1} which are separated by about 5000 cm^{-1}, which have been assigned to the $^2T_{2g} \rightarrow {}^2E_g$ transition split by a Jahn–Teller distortion, although it should again be noted that values vary from worker to worker. In addition there is another band at about 10 000 cm^{-1} which cannot be assigned (933, 1217).

The chemistry of oxovanadium(IV) complexes is so extensive that it is dealt with separately in Section G, page 209.

The reaction of alkali metal fluorides with niobium dioxide at about 800° C leads to the formation of $MNbO_2F$ (where M is Li, Na or K) (2081). $KNbO_2F$ has a perovskite structure with a unit cell slightly larger than NbO_2F itself (a = 4·01 and 3·90 Å respectively). $LiNbO_2F$ and $NaNbO_2F$ (and also $LiNbO_3$) have rhombahedrally distorted perovskite structures. These compounds are electrical insulators and exhibit only temperature independent paramagnetism.

F. Complexes with other Anionic Ligands

In this section we shall deal with amido, alkoxy, dithiocarbamato, β-diketonato and related complexes of vanadium(IV) (excluding VO^{2+} complexes), niobium(IV) and tantalum(IV). There is also a brief mention in the literature of $K_2[Nb(NCS)_6]$ (368).

Vanadium tetrachloride reacts with liquid ammonia with the formation of $VCl(NH_2)_3$ (921). However only two vanadium–chlorine bonds are solvolysed with primary or secondary aliphatic amines, with the formation of compounds of the general type $VCl_2(RNH)_2,xRNH_2$ (699, 926). The tetraamido complexes $V(NMe_2)_4$ and $V(NEt_2)_4$ are obtained from the tetrachloride using the corresponding lithium dialkylamide (304).

Vanadium tetrachloride similarly reacts vigorously with alcohols to form $VCl_2(OR)_2(ROH)$, but further replacement of chloride could not be obtained, even in the presence of ammonia or sodium alkoxide (301); at 150° C these compounds decompose to $V_2^{IV}OCl_3(OR)_3$. The tetraalkoxides however can be obtained from $V(NEt_2)_4$ and the alcohol (299, 2276). The tetramethoxide and tetraethoxide are brown solids, whereas the higher molecular weight tetrapropoxides, tetrabutoxides and tetraamyloxides are brown liquids. This behaviour reflects the decreasing molecular weight with increasing length of the alkyl group; the degree of polymerization of $V(OMe)_4$ and $V(OEt)_4$ in benzene are 2·8 and 2·0 respectively, whereas the other compounds have molecular complexities in the range 1·4–1·1 This may be compared with the analogous titanium tetraalkoxides (page 97).

Magnetic and e.s.r. studies on the monomeric $V(OBu)_4$, $V(NMe_2)_4$ and $V(NEt_2)_4$ imply considerable distortion from a regular tetrahedral structure ($\mu_{eff} = 1{\cdot}69$–$1{\cdot}70$ B.M., $g = 1{\cdot}962$–$1{\cdot}976$), which is also consistent with the observation that the ligand field bands in the visible spectra are split into two components separated by 3000–4600 cm^{-1} (46, 352).

Reaction of the tetraamides with carbondisulphide forms the dithiocarbamato derivatives $V^{IV}(S_2C.NMe_2)_4$ and $V^{IV}(S_2C.NEt_2)_4$ (286). Magnetic and e.s.r. measurements on $V(S_2C.NMe_2)_4$ showed $\mu_{eff} = 1{\cdot}72$ B.M. and gave a g-value of 1·969 which is consistent with eight coordination. The e.s.r. spectrum of $V(S_2C.NEt_2)_4$ in benzene, carbon disulphide, methylene dichloride or chloroform showed an equilibrium between two isomers, one apparently containing bidentate ligands and the other monodentate ligands, the latter being favoured at higher temperatures (46, 300).

The electrolytic reduction of niobium pentachloride in alcohols saturated with anyhydrous hydrogen chloride, followed by the addition of organic cations precipitates compounds of the type $(BH)_2[Nb(OR)Cl_5]$, where B is pyridine or quinoline, and R is Me, Et or Pr (2371). The visible spectra, e.s.r. spectra, and magnetism ($\mu_{eff} \sim 1{\cdot}73$ B.M.) have been studied (910, 2371). Reduction in alcoholic solutions in the absence of hydrogen chloride precipitates the red diamagnetic dimer $[NbCl(OEt)_3py]_2$ (2372). Reaction with sodium ethoxide further forms the diamagnetic red-brown oil, $Nb(OEt)_4$.

The tetrakis-amido compounds $Nb(NR_2)_4$ (where R_2NH is Et_2NH,

Me(Bu)NH or $(CH_2)_5NH$) have also been prepared, but no magnetic behaviour was reported (306). Dimethylamine forms $Nb(NMe_2)_5$. Reaction of the tetrakisamido compounds with alcohols forms $Nb(OR)_5$ (where R is Bu, Pr, Et_3Si) with reduction of the alcohol, even at 0° C (2276). Carbon disulphide again forms the niobium(IV) dithiocarbamates $Nb(S_2C.NMe_2)_4$ ($\mu_{eff} = 1 \cdot 32$ B.M.) and $Nb(S_2C.NEt_2)_4$ ($\mu_{eff} = 0 \cdot 55$ B.M.), but tantalum forms the pentavalent $Ta(S_2C.NMe_2)_5$ (286).

Eight coordinate complexes of the type $Nb^{IV}(chelate)_4$ have been obtained by the reaction of niobium tetrachloride with tropolone, 8-hydroxyquinoline, and the β-diketones acetylacetone, dibenzoylmethane, benzoyltrifluoroacetone and thenoyltrifluoroacetone, in the presence of base (657). In the presence of dioxane, the acetylacetonate complex forms $Nb(acac)_4(dioxane)$ which is probably nine coordinate. Chlorine-containing products of the type $Nb^{IV}Cl_2(chelate)_2$ and $Nb^{IV}Cl(chelate)_3$ were obtained in the absence of base. Similar reactions were studied with tantalum tetrachloride, but these were complicated by oxygen-abstraction reactions forming, for example, $[Ta^{V}Cl_3(chelate)]_2O$. The effective magnetic moments of the eight coordinate complexes are in the range $1 \cdot 43$–$1 \cdot 66$ B.M., and the Curie–Weiss law is obeyed.

G. Vanadyl Complexes

1. *Introduction*

Vanadium forms oxodihalides VOX_2, where X is F, Cl or Br. The VO^{2+} group is retained in the extensive aqueous chemistry of vanadium(IV).

In general these vanadyl complexes are green and monomeric, the oxygen atom being strongly bonded to the vanadium atom as shown by the short vanadium–oxygen bond of $1 \cdot 60$ Å, the magnetic moments in the range $1 \cdot 6$–$1 \cdot 8$ B.M., and the infrared stretching vibration at 940–1040 cm^{-1} (2102). The mere existence of the oxo–aquo complex $[VO(H_2O)_4]^{2+}$ rather than a hydroxo complex such as $[V(OH)_2(H_2O)_4]^{2+}$ implies unusually strong vanadium–oxygen bonding (see page 46).

It is generally found that the vanadium atom is five coordinate with square pyramidal stereochemistry, the oxygen atom occupying the unique axial position. In some cases a sixth very weakly bound ligand is *trans* to the oxygen atom completing a distorted octahedral stereochemistry. (For example, V. . .$OH_2 = 2 \cdot 0$ to $2 \cdot 5$ Å). Similarly ^{17}O-n.m.r. spectroscopy in aqueous solution reveals two kinds of water molecule; there are four (equatorial) water molecules with a residence time of about 10^{-3} s, and other more rapidly exchanging water molecules giving rise to a shift in the solvent water peak, with a residence time estimated from the band width of the order of 10^{-11} s (1919, 2413). The study of the addition of ligands to the acetylacetonate complex $VO(acac)_2$ is referred to later (page 215).

The five coordinate square pyramid is however not the only known stereochemistry. For example the stereochemically forcing trimethylamine forms $VOCl_2(Me_3N)_2$ where the oxygen atom and two chlorine atoms occupy the three equivalent equatorial positions of a trigonal bipyramid, the complete molecule having trigonal symmetry (page 212).

It is also very interesting to note that whereas the β-diketone derivatives $[VO(acac)_2]$ (676, 1216) and $[VO(bzac)_2]$ (1215, 1216) have five coordinate square pyramidal stereochemistry, the related sulphur ligand monothiothenyltrifluoroacetone forms the yellow brown $VO(C_4H_3S.C(S)=C.C(O).CF_3)_2$ which is dimeric in nitrobenzene, does not show a vanadium–oxygen band at 950–1050 cm^{-1}, and was therefore formulated as having an oxygen bridged V(O)(O)V structure with octahedrally coordinated metal atoms (1197). Like the other oxovanadium(IV) compounds, it has a magnetic moment close to the spin-only value of 1·73 B.M.

The e.s.r. and visible spectra of the square pyramidal molecules have been studied in considerable detail, but the interpretation is not completely resolved due to the complex nature of the vanadium–oxygen bond. (For example see references 134, 274, 462, 1871, 2103 and reference therein).

TABLE 24

Vanadium–oxygen bond lengths and O–V–Ligand angles in square pyramidal vanadyl complexes

	V = O (Å)	O = V–L	Reference
$[VO(SO_4)(H_2O)_3],2H_2O$	1·59	98°	134
$(NH_4)_2[VO(NCS)_4],5H_2O$	1·62	97°	1155
$[VO(acac)_2]$	1·57	106°	676, 1216
$[VO(bzac)_2]$	1·61	106°	1215, 1216
$Na(Et_4N)[VO(Ph_2CO.CO_2)_2],2PrOH$	1·58	106°	462
$Na_4[VO(dl\text{-tartrate})]_2,2H_2O$	1·62	106°	2263
$(NH_4)_4[VO(d\text{-tartrate})]_2,2H_2O$	1·60	109°	2263

The three bands in the visible region are assigned as originating from the single electron in the d_{xy} orbital:

$$d_{xy} \rightarrow d_{xz}d_{yz} \qquad {}^2B_2 \rightarrow {}^2E(I)$$
$$d_{xy} \rightarrow d_{x^2-y^2} \qquad {}^2B_2 \rightarrow {}^2B_1$$
$$d_{xy} \rightarrow d_{z^2} \qquad {}^2B_2 \rightarrow {}^2A_1$$

The relative energies of the $d_{xz}d_{yz}$ and $d_{x^2-y^2}$ orbitals is very sensitive to the distance the vanadium atom has moved out of the plane of the four ligands towards the oxygen atom, that is the O–V–Ligand angle. A summary of the known crystal structures is given in Table 24.

In the following section only some of vanadyl chemistry will be summarized. In particular, little emphasis is placed on compounds with multidentate ligands obtained from aqueous solution; they have been recently reviewed elsewhere (2099).

2. *Halides*

The yellow anhydrous VOF_2 is obtained from concentrated aqeuous hydrofluoric acid (above 69% HF). The blue and green hydrates which are apparently $VOF_2,4H_2O$ and $VOF_2,2H_2O$ respectively are obtained from more dilute acid (396, 2101).

The bright green, anhydrous $VOCl_2$ has been prepared by the reduction of $VOCl_3$ or $VOCl_3$–V_2O_5 mixtures at about 500° C with hydrogen, zinc, VCl_3, or VOCl (961, 1817, 2099). The compound is very deliquescent to give blue solutions. The hydrated material is commercially available and is formed by hydrolysis of the tetrachloride or by the action of concentrated hydrochloric acid on the pentoxide. The structure of the anhydrous compound appears related to that of $NbOCl_2$ and $TaOCl_2$ (page 204), but with no distortion of the VO_2Cl_4 octahedra caused by metal–metal bonding. The V–O–V distance is 3·38 Å (2146).

The oxodibromide $VOBr_2$ has been prepared by the action of bromine and sulphur monobromide on V_2O_5 at 500°–600° C (1965).

3. *Adducts of the halides*

Neutral adducts of the type $VOCl_2L_2$, where L represents a unidentate ligand, have been obtained by the action of reducing ligands (Me_3N, Me_2NH, $MeNH_2$, py, MeCN, Me_2S, Et_2S) on $VOCl_3$ under non-aqeuous conditions (124, 964).

In spite of the general implicit assumption that all these five coordinate $VOCl_2$ adducts are square pyramidal, the only one whose structure is known is trigonal bipyramidal. The structure of $VOCl_2(Me_3N)_2$ (Fig. 118) shows that the oxygen atom (V–O = 1·59 Å) and two chlorine atoms are coplanar (Cl–V–Cl = Cl–V–O = 120°) (691). In this instance the vanadium–oxygen multiple bonding is insufficient to stabilize the tetragonal pyramid, and the shape of the polyhedron is apparently the result of ligand–ligand repulsion, the oxygen atom and two chlorine atoms being staggered with respect to the relatively bulky trimethylamine ligands, the complete molecule having trigonal symmetry. Comparison of the diffuse reflectance spectra and the spectra in benzene indicate that this stereochemistry is retained in solution. Similar structures are found for $TiCl_3(Me_3N)_2$, $VCl_3(Me_3N)_2$ and $CrCl_3(Me_3N)_2$. The vanadium–chlorine bond length in $VOCl_2(Me_3N)_2$ (2·25 Å) is similar to that in $VCl_3(Me_3N)_2$ (2·24 Å), whereas the vanadium–nitrogen bond length is slightly shorter (2·18 and 2·21 Å respectively) (1055).

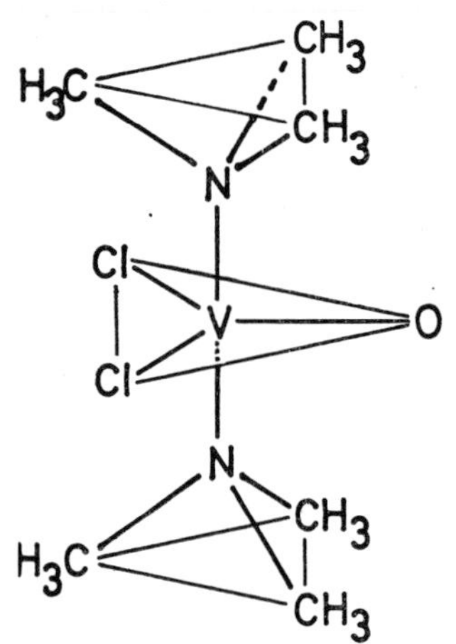

FIG. 118 The structure of $VOCl_2(Me_3N)_2$.

A large number of complexes containing neutral monodentate ligands (mostly with oxygen or nitrogen donor atoms) have been obtained by the partial hydrolysis of vanadium tetrachloride adducts (830), or even from aqueous, ethanolic or benzene solutions of vanadium(III) (1223, 1364) or vanadium(IV) (957, 964, 1363, 1638, 1639, 2101). In addition to complexes of the type $[VOCl_2(ligand)_n]$, solvolysis of the vanadium–chlorine bond can also occur to produce cationic complexes, although in many cases this has not been confirmed by the appropriate physical measurements.

Under similar conditions the neutral bidentates dipyridyl, *o*-phenanthroline and dipyridyl–N,N′-dioxide form $[VOCl_2(bidentate)]$ and $[VO(bidentate)_2]^{2+}$ (493, 1616, 2101). It is interesting to note that the adducts $VOCl_2(triars)$ formed by the tridentate arsines $CH_3.As(o\text{-}C_6H_4.AsMe_2)_2$ and $CH_3.C(CH_2.AsMe_2)_3$ are again similar, being monomeric and non-electrolytes (501).

4. *Anionic halo complexes*

Fluoride forms reasonably stable complexes with VO^{2+}, and $MVOF_3$, M_2VOF_4 and M_3VOF_5 can be isolated from solution, where M represents a univalent cation (21, 2099).

The species present in aqueous solution have been determined by a potentiometric technique (21), which shows that VOF_3^- is the predominant species when the fluoride ion concentration is 0·1 *M*, VOF_2 at 0·01 *M*, VOF^+ at 0·001 *M*, and VO^{2+} below 0·0001 *M*. These results appear to be independent of vanadium concentration indicating only the presence of monomeric vanadium species. Chloride on the other hand forms only very weak complexes with VO^{2+} in aqueous solution, because of the strongly electropositive character of vanadium(IV). These results are shown graphically in Fig. 119.

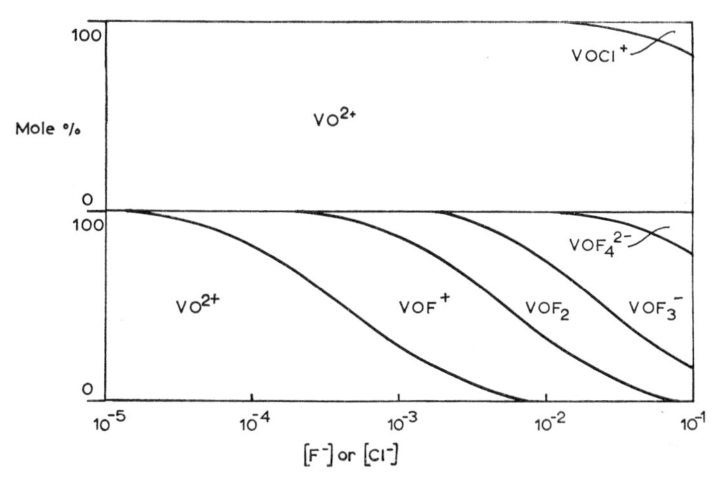

FIG. 119 Concentration ranges of stability for chloro and fluoro complexes of vanadyl (reference 21).

However chloro complexes can be obtained from aqeuous solutions saturated with hydrogen chloride at 0° C, or from non-aqueous solvents such as ethanol saturated with hydrogen chloride (1375, 1390, 1456, 2099). Salts of the type A^IVOCl_3,xH_2O or $A^I_2VOCl_4,xH_2O$ are obtained depending upon cation and conditions. In acetonitrile solutions the complexes obtained are of the type $(pyH)[VOCl_3(MeCN)_2]$ or $(NH_4)_2[VOCl_4(MeCN)]$, again depending upon cation, which decompose on warming to the solvent-free $(pyH)[VOCl_3]$ and $(NH_4)_2[VOCl_4]$ respectively (831). Similarly in acetonitrile/dioxane mixtures,

$(pyH)_2[(MeCN)Cl_3OV.(C_4H_8O_2).VOCl_3(MeCN)]$ is formed, in which the dioxane bridges two vanadium atoms (831).

5. *Complexes with other anionic ligands*

The action of a large number of multidentate anionic ligands on oxovanadium(IV) has been studied. The most common complexes are the nonionic $[VO(\beta\text{-diketonate})_2]$ and $[VO(\text{Schiff base})_2]$ which will be discussed in some detail, and [VO(phthalocyanine)] (92, 154, 1256, 2107). In this section attention will be concentrated on some pseudo-halide, β-diketonate, Schiff base and oxalate complexes. Other complexes have been reviewed elsewhere (2099).

Spectrophotometric studies in aqueous solution have shown that thiocyanate forms $VO(NCS)^+$ (967). With thiocyanate or cyanide, complexes can be isolated from aqueous solution of composition $M_2[VO(NCS)_4],xH_2O$, $M_3[VO(NCS)_5],xH_2O$ and $M_3[VO(CN)_5],xH_2O$, where M represents a univalent cation (1455, 2101). The structure of $(NH_4)_2VO(NCS)_4,5H_2O$ shows that the vanadium atom is above the plane of the four nitrogen-bonded thiocyanate groups (1155). The sixth coordination position required to complete the octahedron is occupied by a water molecule, with a relatively large vanadium–oxygen distance of 2·22 Å.

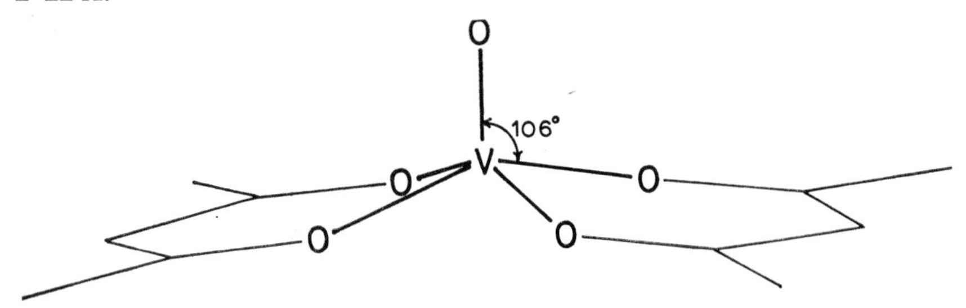

FIG. 120 The structures of $VO(acac)_2$ and $VO(bzac)_2$.

The blue green $VO(acac)_2$ is easily prepared from aqueous solution (1711, 1954). Similar complexes are obtained with other β-diketones (802, 1227, 1315, 1711). The structures of the square pyramidal $VO(acac)_2$ (676, 1216) and $VO(bzac)_2$ (1215, 1216) are shown in Fig. 120. The shorter vanadium–oxygen bond in the acetylacetonate complex compared with the benzoylacetonate complex (1·57 and 1·61 Å respectively) is thought to be significant. The unsymmetrical benzoylacetone ligands are *cis* to each other.

The different stereochemistry with bridging oxygen atoms obtained with a monothio-β-diketonate has been noted earlier (page 210) (1197).

Both $VO(acac)_2$ and $VO(bzac)_2$ are monomeric in boiling benzene (1315), although it has also been noted that the former is dimeric in triphenylmethane, biphenyl and diphenylaniline, but monomeric in camphor (1039).

These two vanadyl compounds form 1 :1 adducts with amines such as methylamine, isoquinoline and γ-picoline, but not with other donor atoms (1315). Some β-diketonates contain a molecule of water of crystallization, for example $VO(Me.C(O).C(CN) = C(O).Me)_2,H_2O$ but it is probably not coordinated to the vanadium atom (802). (It was originally thought that $VO(acac)_2$ itself was a monohydrate (1949)).

The $^2B_2 \rightarrow {}^2E(I)$ band in the vanadyl acetylacetonate and trifluoroacetylacetonate complexes shifts by about 1000 cm^{-1} as bases are added to the complex in noncoordinating solvents such as benzene nitrobenzene. This has allowed the equilibrium constant for the formation of these adducts to be determined over a temperature range (Table 25), and also to be compared with a calorimetrically determined ΔH (424). The following results were obtained:

TABLE 25

Stability constants for the formation of complexes between vanadylacetyacetonate and different ligands

Ligand	$K = \dfrac{[Complex]}{[VO(acac)_2][Ligand]}$
py	58
14 other Amines	2 to $\sim$ 1000
MeOH	0·6
Ph_3PO	32
$(Me_2N)_3PO$	105

(i) The same values were obtained in benzene and nitrobenzene.

(ii) The largest changes were due to amines. Changes due to dioxane, thiourea, acetone, acetonitrile and triphenylphosphine could not be detected.

(iii) There appeared to be no correlation between the magnitude of the spectral shift and the strength of the interaction. Thus methanol and hexamethylphosphoramide caused the largest spectral shifts, but have the smallest heats of reaction (5·8 and 5·9 kcal mol^{-1} respectively, compared, for example with 7·4 kcal mol^{-1} for pyridine). This discrepancy is not expected from the molecular orbital description of the bonding in these vanadyl complexes. There also appears to be no correlation between the spectral data and the nature of the ligand, or with the nature of the β-diketone (424, 1250).

The tetradentate Schiff bases formed by the condensation of β-diketones or salicylaldehyde with primary diamines readily form stable

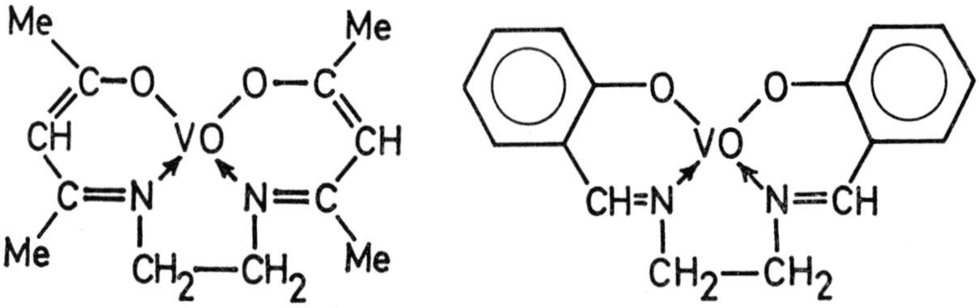

FIG. 121 Tetradentate Schiff base complexes with vanadyl.

plexes with VO^{2+} (Fig. 121) (1228, 1865, 1906). These may again form addition compounds with bases such as pyridine.

Tridentate Schiff bases form $[VO(ligand)]_2$, a possible structure being shown in Fig. 122 (1003). The magnetic properties have been interpreted

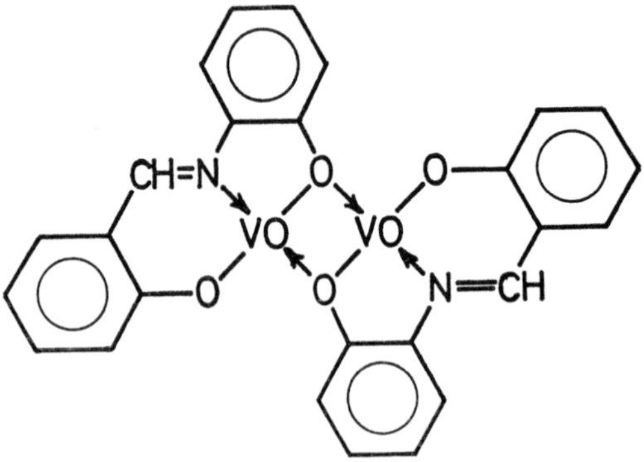

FIG. 122 Tridentate Schiff base complex with vanadyl.

in terms of a direct σ vanadium–vanadium interaction between unpaired spins in the $3d_{xy}$ orbitals of the vanadium atoms. Normal monomeric paramagnetic compounds are obtained upon addition of pyridine, [VO(ligand)(py)].

Salts of two common oxalato complexes can be readily obtained from aqueous solution, those containing the simple $[VO(C_2O_4)_2]^{2-}$ ion, and those containing the more complex $[(VO)_2(C_2O_4)_3]^{2-}$ ion of unknown structure (1455, 1996, 2101).

In aqueous solution at pH ~8, the predominant species is $[VO(C_2O_4)_2]^{2-}$, for which the dissociation constant is (703):

$$K = \frac{[VO^{2+}][C_2O_4^{2-}]^2}{[VO(C_2O_4)_2^{2-}]} = 9 \times 10^{-13}$$

At higher pH, the less stable $[VO(OH)(C_2O_4)]^-$ is formed:

$$K = \frac{[VO^{2+}][OH^-][C_2O_4^{2-}]}{[VO(OH)(C_2O_4)^-]} = 3 \times 10^{-7}$$

The non-ionic compounds $[VO(C_2O_4)(o\text{-phen})]$ and $[VO(C_2O_4)(dipy)]$ can be obtained with the appropriate ligand (2101).

When a suspension of V_2O_5 in boiling water is reacted with three moles of oxalic acid, the blue $[VO(C_2O_4)(H_2O)_2]$ and $[VO(C_2O_4)(H_2O)_2], 2H_2O$ are obtained (1995):

$$V_2O_5 + 3\ H_2C_2O_4 \rightarrow 2\ VO(C_2O_4) + 3\ H_2O + 2\ CO_2$$

H. Organometallic Complexes

The reaction of vanadium tetrachloride with sodium cyclopentadienyl or cyclopentadienyl magnesium bromide forms the light green $(\pi\text{-}C_5H_5)_2VCl_2$ (2389, 2391). Similar compounds of the type $(\pi\text{-}C_5H_5)_2VX_2$ are known for X = Br, N_3, NCO, CN, NCS and NCSe, and of the type $[(\pi\text{-}C_5H_5)_2VB]^+$ where B is the anion of acetylacetone, benzoylacetone, dibenzoylmethane, hexafluoroacetylacetone, tropolone and ethylacetate (686).

The analogous $(\pi\text{-}C_5H_5)_2NbX_2$ are obtained by reduction of biscyclopentadienyl niobium(V) compounds with iodide and benzyl mercaptan (2299). Reaction of $(\pi\text{-}C_5H_5)_2NbCl_2$ with sodium phenyl in benzene forms the σ-bonded organometallic $(\pi\text{-}C_5H_5)_2Nb(Ph)_2$. (2136). The apparently similar monomeric $Nb(C_5H_5)_4$ and $Ta(C_5H_5)_4$ are obtained from the pentachlorides and sodium cyclopentadienyl in diethyl ether or benzene, and appear to contain two π-bonded and two σ-bonded cyclopentadienyl rings (874). The effective magnetic moments of $(\pi\text{-}C_5H_5)_2NbI_2$, $(\pi\text{-}C_5H_5)_2Nb(\sigma\text{-}C_5H_5)_2$ and $(\pi\text{-}C_5H_5)_2Ta(\sigma\text{-}C_5H_5)_2$ are 1·80, 1·41 and 1·73 B.M. respectively (874, 2299).

3. OXIDATION STATE III

A. Introduction

There is very little resemblance between the chemistry of vanadium(III) and that of niobium(III) and tantalum(III). For example vanadium forms V_2O_3 and VOCl, whereas sesquioxides and oxochlorides of niobium and tantalum do not exist. Vanadium forms the stoicheiometric trihalides VX_3 whereas niobium and tantalum form the nonstoicheiometric $M_{3-x}Cl_8$. Vanadium(III) in solution forms adducts $VX_3(ligand)_n$ and haloanions such as $(VCl_4)^-$ under nonaqueous conditions, and complexes such as $[VCl_2(H_2O)_4]^+$ under aqueous conditions; analogues with niobium and tantalum are not known. The chemistry of vanadium(III) has therefore been dealt with more or less separately.

There are a number of important polynuclear compounds of niobium and tantalum which contain average oxidation numbers between +II and +III. The chlorides of the higher valency states are characterized by structures based on close packing of chloride ions with the niobium atoms in octahedral holes, whereas those of the lower valency states are characterized by the presence of octahedral clusters of niobium atoms. There is some overlap between these two types in the region between the two oxidation numbers. For example:

Average Oxidation Number:	2·33	2·50	2·67	3·00
Close Packed Structures:	—	$CsNb_4Cl_{11}$	Nb_3Cl_8	$Cs_3Nb_2Cl_9$
Cluster Structures:	$(Nb_6Cl_{12})^{2+}$	$(Nb_6Cl_{12})^{3+}$	$(Nb_6Cl_{12})^{4+}$	—

A rather arbitrary division has been made between these two structural types: Nb_3Cl_8 and $CsNb_4Cl_{11}$ are discussed in this section, whereas the cluster compounds are discussed under divalent compounds (page 236).

Vanadium(III) may have a coordination number of four (as in tetrahedral VCl_4^-), five (as in trigonal bipyramidal $VCl_3(Me_3N)_2$) or six (as in octahedral VF_6^{3-}). The spectra due to *d–d* transitions arising from the splitting of the ^3F and ^3P terms by the appropriate ligand fields have been studied for all three stereochemistries (Fig. 123a, b and c), and in all cases it is relatively simple to evaluate the ligand field splitting 10Dq and the Racah parameter B.

The visible spectrum of octahedral vanadium(III) is characterized by two bands at about 16 000 and 24 000 cm^{-1} ($\varepsilon \sim 20$) (for oxygen ligands) which are assigned to $^3T_{1g}(F) \rightarrow {}^3T_{2g}$ and $^3T_{1g}(F) \rightarrow {}^3T_{1g}(P)$ respectively (494, 504, 671, 701, 1221, 1601). These bands are at about 13 000 and 18 000 cm^{-1} respectively for sulphur ligands (965). The effective magnetic moments fall slightly from about 2·7 B.M. at room temperature to 2·3–2·7 B.M. at liquid nitrogen temperatures (1601).

The visible spectrum of vanadium(III) in a tetrahedral crystal field shows bands at 9000 and 15 000 cm^{-1} (for VCl_4^-) which are assigned to the transitions $^3A_2 \rightarrow {}^3T_1(F)$ and $^3A_2 \rightarrow {}^3T_1(P)$ respectively (438, 511).

The spectrum in a trigonal bipyramidal crystal field is more complex,

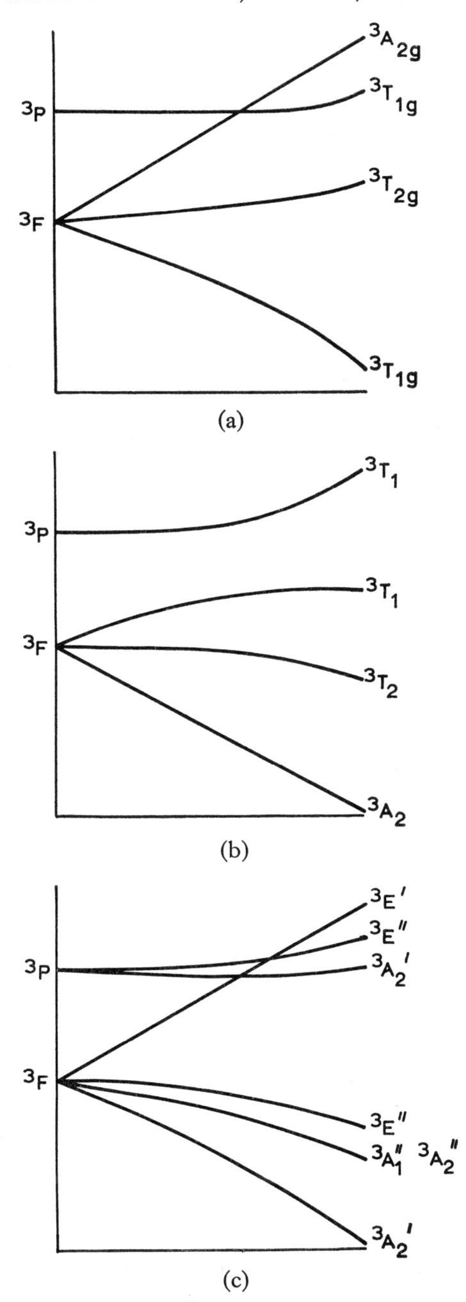

FIG. 123 Energy level diagram for vanadium(III) in (a) octahedral, (b) tetrahedral, and (c) trigonal bipyramidal ligand fields.

and in particular is characterized by two bands in the near infrared region at 5000 and 7000 cm^{-1} which are assigned to the transitions $^3A_2' \rightarrow {}^3A_1''$ and $^3A_2' \rightarrow {}^3E''$ (611, 2411).

B. Oxide

Niobium and tantalum sesquioxides are not known.

Vanadium sesquioxide has the same structure at Ti_2O_3 (page 122), and its physical properties have likewise been studied in some detail as a function of temperature. The metal-metal distances along the *c*-axis are again comparable to those in the metal, and vanadium–vanadium bonding in pairs along the *c*-axis has again been postulated to explain the physical properties. In addition, the extra electron results in another rapid lower temperature transition at $-105°$, and it has been suggested this is due to bonding in the plane at right angles to the *c*-axis. Each vanadium atom is bonded to only one of the three surrounding vanadium atoms in this plane, resulting in a contraction of the *a*-axis, and a slight expansion and tilting of the *c*-axis (2339). Similar bonding between close packed sheets is observed in $MoCl_3$ (page 336).

Cooling from room temperature to this low temperature form results in a decrease in magnetic susceptibility (432, 1462), a change in the ^{51}V–n.m.r. spectrum (1750), and a decrease in the electrical conductivity by a factor of about 10^8 with a change from metallic to semiconductor properties (821, 1029, 1714).

C. Trihalides

1. *Vanadium*

Vanadium trifluoride has been prepared by the action of gaseous hydrogen fluoride on VCl_3 (1966) or VCl_2 (774) at red heat. The compound is dark green and sublimes at about 800° C. The structure shows octahedrally coordinated metal atoms (1266). The effective magnetic moment of 2·55 B.M. at room temperature (1803), and the diffuse reflectance spectra (494), are normal for octahedral vanadium(III).

Vanadium tetrachloride slowly decomposes at room temperature to vanadium trichloride and chlorine. The decomposition is more rapid at higher temperatures, and vanadium trichloride can be conveniently prepared by refluxing VCl_4 for several days either by itself or in an inert solvent such as carbon disulphide. Removal of the unreacted tetrachloride by distillation leaves the trichloride as a residue. The equilibrium constant for the reaction:

$$2\,VCl_{3(s)} + Cl_{2(g)} \rightleftarrows 2\,VCl_{4(g)}$$

has been measured above the boiling point of the tetrachloride (152° C) (1818, 2115):

$$\log Kp_{(at)} = 8 \cdot 8513 - \frac{3758}{T} \pm 0 \cdot 093$$

Vanadium trichloride is an involatile, violet, deliquescent solid. It is relatively stable to air, although can appear to fume in air due to VCl_4 impurities. It starts to disproportionate into the gaseous tetrachloride and solid dichloride at about 600° C (1588, 1818):

$$2\,VCl_{3(s)} \rightleftarrows VCl_{2(s)} + VCl_{4(g)}$$

$$\log Kp_{(mm\,Hg)} = 11 \cdot 449 - \frac{8237}{T} \pm 0 \cdot 025$$

The structure (1422) is similar to that of α-$TiCl_3$, and is discussed in more detail on page 124. The room temperature effective magnetic moment is 2·78 B.M., and the susceptibility is independent of field strength (1421, 2206).

Vanadium tribromide is formed directly from the elements above 400° C, or from vanadium pentoxide and carbon tetrabromide in a sealed tube at 350° C (2115). It can be purified by sublimation at 350° C only in the presence of bromine vapour, for which the following equilibria apply (1587):

$$2\,VBr_{3(s)} \rightleftarrows 2\,VBr_{2(s)} + Br_{2(g)}$$

$$\log p_{(mm\,Hg)} = 5 \cdot 20 - \frac{5070}{T}$$

$$2\,VBr_{3(s)} \rightleftarrows VBr_{2(s)} + VBr_{4(g)}$$

$$\log p_{(mm\,Hg)} = 10 \cdot 47 - \frac{8240}{T}$$

For example:

$$370^\circ\,C: \quad p_{Br_2} = 0 \cdot 002\ \text{mm Hg}$$

$$p_{VBr_4} = 0 \cdot 005\ \text{mm Hg}$$

$$530^\circ\,C: \quad p_{Br_2} = 0 \cdot 08\ \text{mm Hg}$$

$$p_{VBr_4} = 1 \cdot 70\ \text{mm Hg}$$

The structure, magnetism (1421) and diffuse reflectance spectra (494) are normal.

Vanadium triiodide is prepared from vanadium and iodine at 150°–280° C, and can be purified in a 320°–400° C temperature gradient

(1324, 1709). It decomposes when heated to the diiodide and iodine (2285):

$$2\ VI_{2(s)} + I_{2(g)} \rightleftarrows 2\ VI_{3(g)}$$

°C	$K = \frac{p_{VI_3}^2}{p_{I_2}}$
750	0·01
800	0·08
850	0·5

Vanadium trichloride and vanadium tribromide form a continuous solid solution over the whole composition range. Similarly VBr_2I has been prepared. All phases have a hexagonal unit cell of the BiI_3 type (1589).

Green aqueous solutions of vanadium(III) can be obtained by electrolytically reducing V_2O_5 or VO_2 in hydrochloric acid, or simply by dissolving V_2O_3 or VCl_3 in hydrochloric acid. On cooling to −10° C and saturating with hydrogen chloride, the green $VCl_3,6H_2O$ is obtained. Dehydration of $VCl_3,6H_2O$ at 85°–90° C in a stream of dry hydrogen chloride forms the green $VCl_3,4H_2O$. If it is assumed that it is possible to linearly interpolate the energy of the bands in the visible spectrum between the extreme cases of VCl_6^{3-} (in fused salts) and $V(H_2O)_6^{3+}$ (in vanadium alum or vanadium(III) perchlorate solutions), then the spectra of the above hydrates indicates an environment containing four water molecules and two chloride ions, that is the formulae should be written $[VCl_2(H_2O)_4]Cl,2H_2O$ and $[VCl_2(H_2O)_4]Cl$ respectively (1221).

The dissolution of vanadium pentoxide in saturated aqueous hydrobromic acid forms the corresponding dark green $VBr_3,6H_2O$, which again on the basis of its visible spectrum should be similarly formulated $[VBr_2(H_2O)_4]Br,2H_2O$ (1784). Thus bromide will reduce vanadium(V) to vanadium(III), whereas chloride reduces to the vanadium(IV) ions $VOCl_4^{2-}$ and $VOCl_3^-$.

The corresponding iodide has also been prepared.

2. *Niobium and tantalum*

The preparation of niobium trifluoride and tantalum trifluoride have been claimed by the action of hydrogen/hydrogen fluoride mixtures on the metals. These blue compounds were reported to be stable to air, water and heat, and to be paramagnetic ($\mu_{eff} = 0{\cdot}7$ and $1{\cdot}4$ B.M. respectively) with the ReO_3 structure (756, 775, 1099, 1732, 1803). The ReO_3 structure was also reported for Nb^VO_2F and Ta^VO_2F, with random distribution of oxygen and fluorine atoms about the octahedral metal atom (947). This

ReO_3 phase has most recently (1037, 2007, 2026, 2027) been found to be homogeneous between NbO_2F and $NbO_{1\cdot25}F_{1\cdot75}$, that is between oxidation states 5·00 and 4·25, and it appears that the earlier claims to the trifluorides were incorrect. It may be observed that the ReO_3 structure was also claimed for MoF_3, but again this structure is only found if oxygen is present (page 335).

Niobium trichloride was first prepared by the thermal decomposition of the pentachloride (1945). The compound is also prepared by reduction of the pentachloride with hydrogen (374, 1103, 2246) or stannous chloride

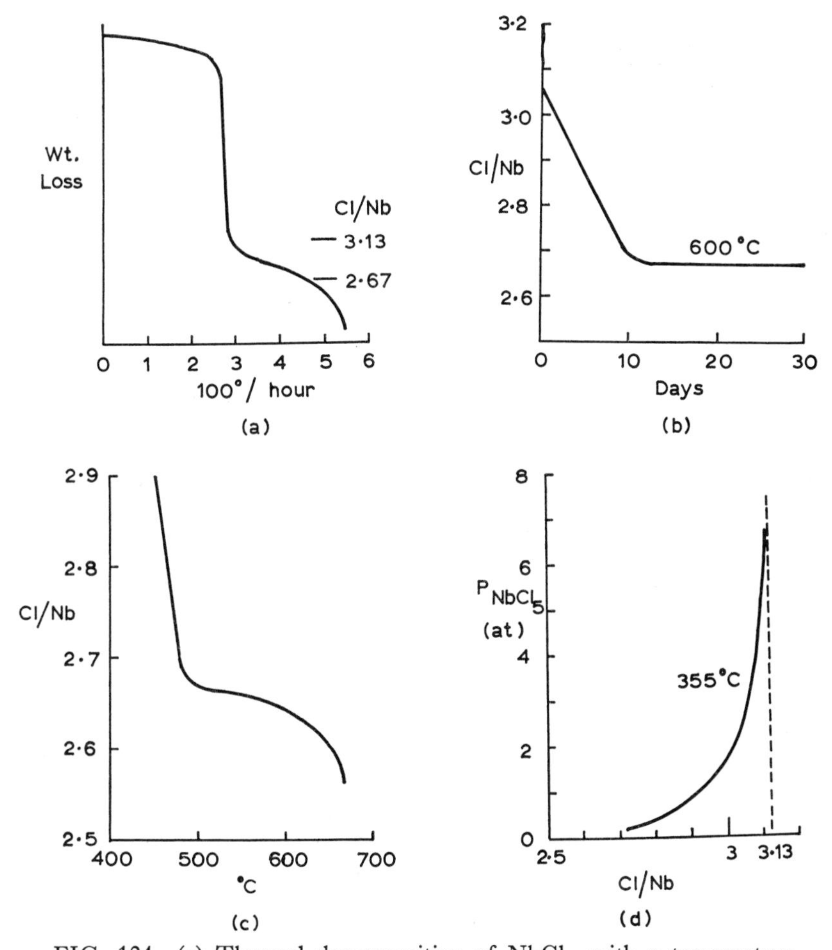

FIG. 124 (a) Thermal decomposition of $NbCl_4$, with a temperature increase of 100°C h^{-1}. (b) Thermal decomposition of $NbCl_4$ at 600°C. (c) Reduction of $NbCl_5$ with hydrogen. (d) Isotherm at 355° C for the composition range of $NbCl_3$ as a function of $NbCl_5$ pressure.

(1719). The early workers found that the Cl/Nb ratio was 2·7–3·0/1·00. More detailed work (2005, 2011) on the preparation by the thermal decomposition of $NbCl_4$ (Fig. 124a and b), and by the reduction of $NbCl_5$ with hydrogen (Fig. 124c) show that at low temperatures (~ 300° C) the phase corresponds to a limit of $NbCl_{3\cdot13}$ (or $Nb_{2\cdot55}Cl_8$), and at higher temperatures (~ 500° C) corresponds to a limit of $NbCl_{2\cdot67}$ (or Nb_3Cl_8). Similarly the preparation by transport in a temperature gradient (355°–390° C) (Fig. 124d) shows a smooth trend from $NbCl_{3\cdot13}$ (under eight atmospheres pressure of $NbCl_5$) to $NbCl_{2\cdot67}$ (with less than one atmosphere of $NbCl_5$.) The colour is brown for Cl/Nb $\gg$ 3·0, changing to light green for Cl/Nb = 2·67. The powder pattern also shows a smooth change between these limits.

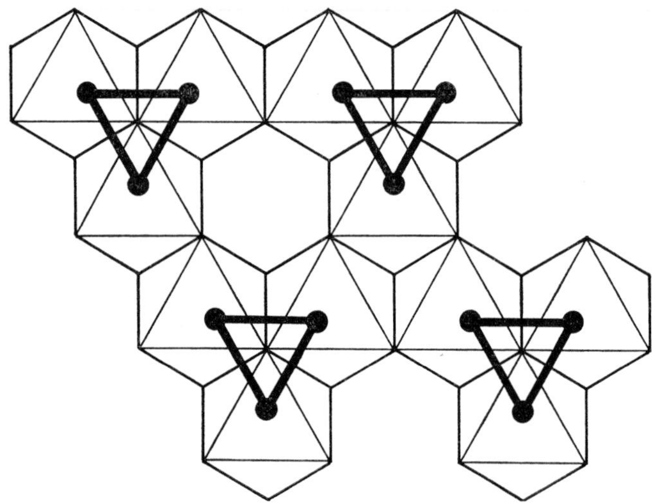

FIG. 125 The sturcture of Nb_3Cl_8. (Compare with Fig. 19).

The heats of formation of these and other niobium chlorides have been obtained from chemical equilibria (2019) and from direct thermochemical measurements (2023), and show a substantial enthalpy increase from $NbCl_{2\cdot67}$ to the non-stoicheiometric $NbCl_{3\cdot18}$:

	Nb → $NbCl_{2\cdot67}$	→ $NbCl_{3\cdot18}$	→ $NbCl_4$	→ $NbCl_5$
$-\Delta H^0_{298}$ (kcal per mole)	128·6	143·6	166·0	190·5
$-\Delta H^0_{298}$ (kcal per Cl)	48·1	45·2	41·5	38·1

The crystal structure (2068, 2069) shows a hexagonally close packed CdI_2 structure with one quarter of the metal sites vacant (Fig. 125). The metal atoms are arranged in triangular groups of three with niobium–

niobium distances of 2·81 Å within the group, which is considerably shorter than the distance between the centres of the octahedral holes (3·37 Å), or than in $NbCl_4$ (3·06 Å), or than in the metal (2·94 Å). Based on the formula Nb_3Cl_8, there are seven electrons available for metal–metal bonding within the triangle, and if the same molecular orbital scheme is used as has been described for the structurally related $Zn_2Mo_3O_8$ (554), then six electrons are accommodated in bonding orbitals and one in a non-bonding orbital. The effective magnetic moment of 1·83 B.M. confirms that this metal–metal bonded compound is unusual in being paramagnetic (2068).

In Nb_3Cl_8, every second row of octahedral sites between the layers of close packed chloride ions is completely occupied, but every alternate row is only half filled. The structure is therefore closely related to that of $NbCl_4$, as removal of the metal atoms from these half filled rows together with the distortion caused by the bonding of these atoms, leaves the structure of $NbCl_4$. Although there is no continuous phase between Nb_3Cl_8 and $NbCl_4$, the gross non-stoicheiometry observed is very unusual in halide chemistry.

Niobium tribromide has also been prepared by the reduction of the pentahalide with hydrogen at 500° C (374), or by the reduction of the tetrabromide with niobium in a temperature gradient (400°–500° C) (209, 2012). This compound again shows non-stoicheiometric behaviour, the lower limit corresponding to Nb_3Br_8 which is isomorphous with Nb_3Cl_8, but the upper limit which has been reached is $NbBr_{3\cdot04}$. The compound is dark brown, but the colour again depends upon stoicheiometry. The breaking of the triangular cluster of niobium atoms as the stoicheiometry is changed from Nb_3Br_8 to $NbBr_{3\cdot04}$ (or $Nb_{2\cdot63}Br_8$) results in a significant increase in the *a*-axis of the unit cell (209):

	Nb_3Br_8	$NbBr_{3\cdot04}$
a	7·227	7·258 Å
c	12·93	12·94 Å

After correction for the diamagnetic Nb_2 clusters, the magnetic susceptibility of $NbBr_{3\cdot08}$ corresponds to an effective magnetic moment of approximately 1·8 B.M. per Nb_3 cluster (1475).

A slightly different structure of exact stoicheiometry Nb_3Br_8 is obtained by transport at higher temperatures (800°–760° C) (2148). The structure of this β-Nb_3Br_8 again has Nb_3 groups, but the sequence of anion packing is more complex, being a mixed hexagonal and cubic type. The unit cell is rhombahedral, but the pseudo-hexagonal cell corresponding to α-Nb_3Br_8 has the dimensions $a = 7{\cdot}08$ and $c = 12{\cdot}99$ Å suggesting stronger niobium–niobium bonding within the close packed layers. The niobium–niobium bond length is 2·88 Å, compared with the distance between the centres of the octahedral holes of 3·54 Å. Thus although the metal–metal bond is longer than in Nb_3Cl_8, the relative shortening with respect to the lattice is greater. The low effective magnetic moment (1·95 B.M. at 573° K falling to 0·5 B.M. at 90° K) indicates some magnetic exchange between different Nb_3 clusters.

Pure NbI_3 can be obtained as a residue from the thermal decomposition of the higher iodides, or as a sublimate from Nb_3I_8 (see below) (2086). It also appears to be the principle product of the reaction between the metal and iodine in a sealed tube up to about 500° C (449, 451, 471, 550, 2086). The structure is hexagonal with $a = 6{\cdot}61$, $c = 6{\cdot}82$ Å and $c/a = 1{\cdot}03$, which is greater than γ-NbI_4 or the Group IV iodides ($c/a = 0{\cdot}90$–$0{\cdot}92$) (2086). However this NbI_3 is considered to be a metastable phase and decomposes irreversibly at 513° C to γ-NbI_4 and Nb_3I_8.

There are again two forms of Nb_3I_8 corresponding to the two forms of Nb_3Br_8 (2148). The low temperature α-form is again non-stoicheiometric $NbI_{2{\cdot}89-3{\cdot}05}$, while the high temperature β-form is stoicheiometric Nb_3I_8. The dimensions of β-Nb_3I_8 using the pseudo-hexagonal cell above are $a = 7{\cdot}60$ and $c = 13{\cdot}90_5$ Å. The niobium–niobium bond lengths are 3·00 Å, compared with the distance between the octahedral centres of 3·80 Å. The magnetic behaviour is similar to β-Nb_3Br_8 (effective magnetic moments of 1·3 and 0·4 at 573 and 90° K respectively).

The tantalum trihalides have not been studied so extensively. Attempts to prepare the trichloride by reduction of the tetrachloride with hydrogen have not been successful; however it has been prepared by reduction with aluminium in a temperature gradient. The reduction to the trichloride can be used to separate tantalum from niobium (2021, 2022, 2023). Tantalum trichloride forms black crystals but is green when finely ground. It is isomorphous with niobium trichloride with a homogeneity range of $TaCl_{2{\cdot}9}$–$TaCl_{3{\cdot}1}$ (2030), although earlier measurements indicated $TaCl_{3{\cdot}3}$ (2424). The magnetic susceptibility corresponds to an effective magnetic moment of approximately 1·0 B.M. based on the Ta_3 cluster (2030). The isomorphous Ta_3Br_8 (1475, 2014, 2314, 2425) and TaI_3 (1458) show similar behaviour.

D. Adducts of the Trihalides

Adducts of niobium and tantalum trihalides do not appear to be known.

Vanadium trichloride reacts with neutral monodentate ligands forming monomeric five coordinate $VCl_3(ligand)_2$, polymeric six coordinate

$VCl_3(ligand)_2$, or monomeric six coordinate $VCl_3(ligand)_3$ depending upon the particular ligand.

Trimethylamine, dimethylsulphide and diethylsulphide form $VCl_3(ligand)_2$, which have room temperature effective magnetic moments in the range 2·50–2·70 B.M. (416, 700, 702, 926). These compounds are monomeric (in nitrobenzene and benzene) and non-conducting (in nitrobenzene) and therefore five coordinate in solution, but the visible spectra of the sulphide compounds indicate dimeric octahedral structures in the solid state (700, 2411). The red $VCl_3(Me_3N)_2$ can also be prepared from vanadium tetrachloride and trimethylamine, but if a dilute solution of trimethylamine in a hydrocarbon solvent is used, the buff-coloured $VCl_4(Me_3N)$ is obtained (926). The trimethylamine adduct $VCl_3(Me_3N)_2$ has a trigonal bipyramidal structure with the three chlorine atoms in the equatorial plane, and is isomorphous with $TiCl_3(Me_3N)_2$ (page 133), and $CrCl_3(Me_3N)_2$. The d^1 titanium(III) and d^3 chromium(III) complexes should be distorted by the Jahn–Teller effect, but not the d^2 vanadium(III) complex. The structure shows three equal vanadium–chlorine bond lengths of 2·24 Å, and Cl–V–Cl angles of 118·2, 121·0 and 121·0°, and is significantly less distorted than the corresponding chromium(III) complex (1055). Similar five coordinate adducts are formed with a number of aliphatic and aromatic phosphine oxides (1262). Ethyl acetate forms VCl_3,2EtOAc (957). The bidentate dipyridyl forms $VCl_3(dipy)$, which with acetonitrile forms $VCl_3(dipy)(MeCN)$ (905).

Alkyl cyanides form the green $VCl_3(RCN)_3$ and the brown $VBr_3(RCN)_3$ (where R is Me, Et or $CH_2 = CH$) (511, 701, 1147, 1364). These compounds are nonelectrolytes, and the appearance of only two vanadium–chlorine stretching vibrations suggest *cis* octahedral structures (511). The effective magnetic moments vary from 2·74 B.M. for $VCl_3(MeCN)_3$ to 2·50 for $VBr_3(MeCN)_3$. Reduction of VCl_4 to $VCl_3(MeCN)_3$ occurs with excess acetonitrile, with the simultaneous evolution of hydrogen chloride (701). Further solvolysis to $[VCl_2(MeCN)_4]Cl$ can also occur (1147).

Dioxan (906, 1363, 1485) and aliphatic alcohols (439) also form the six coordinate 1 : 3 adducts $VCl_3(diox)_3$, $VBr_3(diox)_3$, and $VCl_3(ROH)_3$. Further solvolysis to $[VCl_2(ROH)_4]Cl$ can occur depending upon the alcohol and conditions. Similarly $VX_3,1\frac{1}{2}(MeO.CH_2.CH_2.OMe)$ (where X is Cl or Br) are obtained with the bidentate ether, which lose ligand at 106° C to form $VX_3(MeO.CH_2.CH_2.OMe)$ (906). These red compounds have room temperature effective magnetic moments in the range 2·68–2·85 B.M., which fall to 2·4–2·5 B.M. at 90° K.

Tridentate arsine ligands (501) and hexamethyltrisilazane (2394) form $[VCl_3(tridentate)]$.

E. Oxohalides

Vanadium oxychloride may be prepared by the thermal decomposition of $VCl_3,6H_2O$ into VOCl, $VOCl_2$ and oxides (1854), by the disproportionation of $VOCl_2$ into VOCl and gaseous $VOCl_3$ at about 300° C (758, 1854), by the reduction of $VOCl_3$ with hydrogen at 580°–600° C (1818), or by the reaction between V_2O_3 and VCl_3 in a 620°–720° C temperature gradient (2037). It decomposes in vacuo above about 700° C into $V_2O_{3(s)}$, $VCl_{2(s)}$, and $VOCl_{3(g)}$. It is brown, isomorphous with TiOCl, but in contrast to the antiferromagnetic titanium compound has a magnetic susceptibility which falls only slightly with decreasing temperature (758, 2037).

The violet or black oxybromide has been prepared by the thermal decomposition of $VOBr_2$ (1965) or $VBr_3,6H_2O$ (1784).

F. Anionic Halo Complexes

Salts of the fluorovanadate(III) anions are the only purely halo compounds which can be obtained from aqueous solution. Complexes such as $[VCl_4(MeCN)_2]^-$ and $[VBr_4(MeCN)_2]^-$ are obtained from non-aqueous solvents, which on heating form, for example, VCl_4^-. Vanadium(III) in fused salts froms VCl_4^-, VCl_6^{3-} and $V_2Cl_9^{3-}$. None of these complexes appear to have analogues in the chemistry of niobium and tantalum. Solid state reactions form salts such as $Cs_2Nb^{III}Cl_5$ and $CsNb_4Cl_{11}$, the niobium atoms in the latter case having an average oxidation state of 2·5.

These complexes are now discussed in more detail.

Salts of the general types A_3VF_6 (where A is K or NH_4) (91, 1803), A_2VF_5,H_2O (where A is K, Rb or Cs) (91, 1535) and $AVF_4,2H_2O$ (where A is Rb or NH_4) (91, 1535) have been obtained from aqueous hydrofluoric acid solutions. All have room temperature effective magnetic moments of about 2·8 B.M., and the susceptibility follows the Curie–Weiss law.

Reaction of VCl_3 or $VCl_3(MeCN)_3$ with excess pyHCl in a sealed tube at 150° C or in chloroform solutions forms the pink $(pyH)_3(VCl_6)$ for which $\mu_{eff} = 2{\cdot}71$ B.M. (931).

However, the addition of tetraethylammonium chloride to VCl_3 in thionyl chloride yields the maroon $(Et_4N)_3V_2Cl_9$ (9).

Vanadium trichloride and tetraethylammonium chloride or tetraphenylarsonium chloride in acetonitrile yields the yellow $(Et_4N)[VCl_4(MeCN)_2]$ and $(Ph_4As)[VCl_4(MeCN)_2]$. Acetonitrile is lost at 80° C forming the green $(Et_4N)(VCl_4)$ and $(Ph_4As)(VCl_4)$ which are conclusively tetrahedral since the latter is isomorphous with $(Ph_4As)[FeCl_4]$

of known crystal structure. The magnetic moments of 2·8 B.M. are as expected for the tetrahedral d^2 configuration. The $(Et_4N)[VCl_4(MeCN)_2]$ reacts with other nitrogen donor ligands forming $(Et_4N)[VCl_4(py)_2]$, $(Et_4N)[VCl_4(dipy)]$ and $(Et_4N)[VCl_4(o\text{-phen})]$ (438, 511).

Vanadium tribromide similarly forms $(pyH)_3[VBr_6]$ only from non-coordinating solvents such as chloroform (931), since $(Et_4N)[VBr_4(HOAc)_2]$ (1784) and $(Et_4N)[VBr_4(MeCN)_2]$ (511) are obtained from acetic acid and acetonitrile respectively. The latter again lost solvent at 100° C to form the green tetrahedral $(Et_4N)(VBr_4)$. The mixed $(Et_4N)[VCl_3Br(MeCN)_2]$ is stable with respect to disproportionation in boiling acetonitrile, whereas attempts to prepare other mixed haloanions were unsuccessful. The tetraiodo complexes could not be obtained (511).

The addition of KCl, RbCl or CsCl to the green aqueous solutions of VCl_3 or $VCl_3,6H_2O$, cooling to $-10°$ to $-30°$ C and saturating with hydrogen chloride precipitates the green $KVCl_4,6H_2O$ (1221), $RbVCl_4,6H_2O$ (2096) and $Cs_2VCl_5,4H_2O$ (2096) respectively. Earlier reports (603, 2203) of other green compounds such as $K_2VCl_5,4H_2O$ could not be repeated. The same reaction under warm or hot conditions precipitates the red compounds $KVCl_4,1\frac{1}{2}\ H_2O$ (1221) or K_2VCl_5,H_2O (2096), Rb_2VCl_5,H_2O (2096) and Cs_2VCl_5,H_2O (2096). These red and green compounds are interconvertible.

The corresponding bromides A_2VBr_5,xH_2O (where A = Rb, Cs or pyH, and x = 0, 1, 3 or 5) are similarly obtained from saturated aqueous hydrobromic acid (1784). The anhydrous compounds appear to have significantly lower magnetic moments (μ_{eff} = 2·35–2·47 B.M. at room temperature) than the hydrated compounds (μ_{eff} = 2·55–2·82 B.M. at room temperature), which are presumably monomeric and octahedral.

The binary systems VCl_3–ACl (where A is Na, K, Rb or Cs) show the formation of A_3VCl_6 (where A is Na, K, Rb), $A_3V_2Cl_9$ (where A is K, Rb or Cs) and $RbVCl_4$ (2113, 2316). The effective magnetic moment of 2·74 B.M. for $Cs_3V_2Cl_9$ and the similarity of its visible spectrum to that of VCl_6^{3-}, shows the absence of metal–metal bonding, in contrast to compounds such as $Cs_2W_2Cl_9$ (page 341) (1983). A spectrophotometric study of vanadium(III) in a LiCl–KCl eutectic mixture has indicated the existence of the following equilibrium (1077):

$$VCl_6^{3-} \rightleftarrows VCl_4^- + 2\,Cl^-$$

A study of the binary systems between alkali metal chlorides and niobium and tantalum "trichlorides" shows the formation of A_2TaCl_5 (where A is K, Rb, or Cs) (1976, 1980, 1981).

Reduction of Nb_3Cl_8 with niobium in the presence of caesium chloride according to the following stoicheiometry leads to the formation of

$CsNb_4Cl_{11}$, where each niobium atom has a formal oxidation number of 2·5 (347):

$$Nb + 5\,Nb_3Cl_8 + 4\,CsCl \rightarrow 4\,CsNb_4Cl_{11}$$

The structure is shown in Fig. 126, considered as a hexagonally close packed array of caesium and chloride ions with niobium atoms occupying

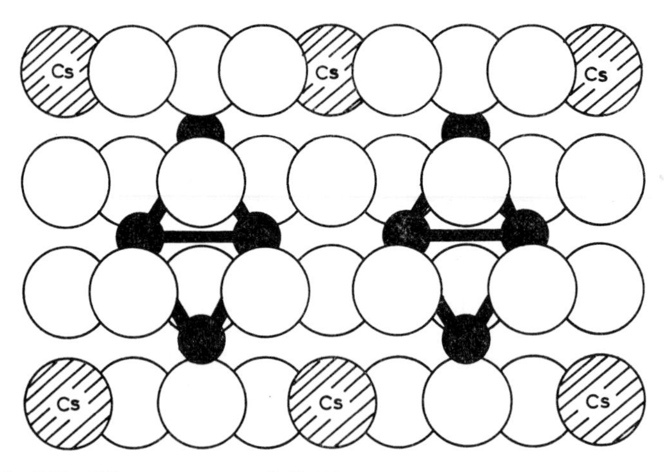

FIG. 126 The structure of $CsNb_4Cl_{11}$ showing the niobium atoms (black) between hexagonally close packed chloride (white) and caesium (hatched) ions.

one third of the octahedral holes, or two thirds of those holes formed by six chlorine atoms. The ten electrons formally available can be used for the formation of five niobium–niobium bonds of a *planar* Nb_4 group:

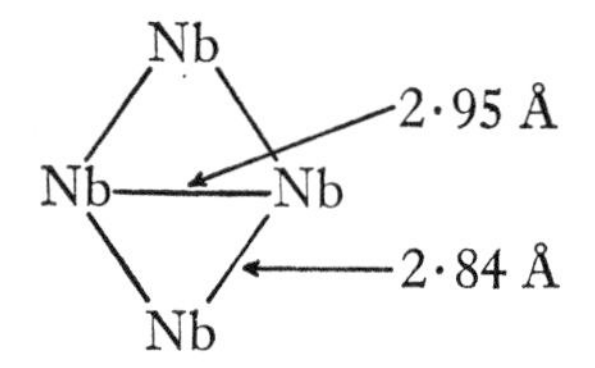

The niobium–niobium bond distances of $4 \times 2{\cdot}84 + 2{\cdot}95$ Å are slightly *longer* than the 2·81 Å found in Nb_3Cl_8. The compound is black, but dark brown when finely ground, the absorption spectrum being somewhat similar to Nb_3Cl_8. It is insoluble in water, concentrated hydrochloric acid, dimethylsulphoxide, acetonitrile, etc. A rather high temperature independent paramagnetism of 283 cgsu has been recorded. Similar compounds are $RbNb_4Cl_{11}$ and $CsNb_4Br_{11}$ (Nb–Nb $= 4 \times 2{\cdot}96 + 3{\cdot}05$ Å, compared with 2·88 Å for β-Nb_3Br_8).

G. Complexes with other Anionic Ligands

Some vanadium pseudohalo complexes, amides, and carboxylates, and some niobium sulphato complexes, will be briefly dealt with in this section.

The product resulting from the reaction of vanadium(III) with cyanide in dilute aqueous acid appears to be $K_4[V(CN)_7], 2H_2O$, with a magnetic moment of 2·8 B.M. (447).

Thiocyanate ion reacts with vanadium trichloride in acetonitrile to progressively form $[VCl_2(NCS)(MeCN)_3]$, $[VCl(NCS)_2(MeCN)_3]$, $[V(NCS)_3(MeCN)_3],2MeCN$ and $A_3[V(NCS)_6],2MeCN$, where A^+ is a univalent cation such as K^+, NH_4^+, pyH^+ or Ph_4P^+ (261). Similar reactions formed adducts of $V(NCS)_3$ with diethyl ether (260), pyridine, tetrahydrofuran and acetone (261). All compounds had room temperature effective magnetic moments in the range 2·31–2·56 B.M.

One vanadium–halogen bond is readily solvolysed in liquid ammonia or primary aliphatic amines to form compounds of the type $VCl_2(NH_2),xNH_3$ (912, 1917), $VBr_2(NH_2),xNH_3$ (1782), and $VCl_2(RNH),xRNH_2$ (913).

Vanadium triacetate could not be prepared by the action of acetic acid on vanadium, or acetic anhydride on vanadium tribromide, but could be obtained using the diboride (1056):

$$2\,VB_2 + 6\,HOAc \rightarrow V_2(OAc)_6 + 3\,H_2 + 4\,B$$

On the basis of the dimeric and nonconducting nature in freezing acetic acid, and the low magnetic moment ($\mu_{eff} = 0.77$ B.M. per metal atom at room temperature falling to 0·6 B.M. at 90° K), a structure similar to $Mo_2(OAc)_4$ (page 31) was suggested with four acetate groups bridging the two metal atoms, and one attached to each metal atom. The non-bridging groups can be replaced by pyridine forming $V_2(OAc)_4(py)_2$. Benzoic acid similarly forms $V_2(Ph.COO)_6$.

Vanadium(III) is stable in aqueous solution, and in addition to the hexahydrates $VX_3,6H_2O$ discussed above, a very large number of complexes have been obtained, particularly with oxygen donor and/or multidentate ligands such as ethylenediaminetetraacetic acid (2085), oxalic acid (1519), picolinic acid (1682) and ethylenediamine (500, 749, 920). The hydrolysis of hexaaquovanadium(III) to the dimeric V–O–V and/or V(OH)₂V [V⟨OH/OH⟩V] has also been studied (1780, 1831, 1832). These complexes do not appear to possess the properties which are usually associated with early transition metals such as metal–metal bonding, high coordination number, and polyanion formation, and will therefore be described in no further detail. They have been well treated elsewhere (496).

The electrolytic reduction of niobium(V) in 75% sulphuric acid produces red-brown solutions from which an acidic red-brown crystalline material can be obtained, or alternatively the corresponding salts can be precipitated by the addition of suitable cations. The analytical results of various workers lead to the following stoicheiometric ratios:

Cation	Cation : Nb : SO_4^{2-}	Reference
NH_4^+	1·6 : 1·0 : 2·5	1824
K^+	1·3 : 1·0 : 2·0–2·2	1379
H^+	— : 1·00 : 2·00	1023
NH_4^+	1·06 : 1·00 : 2·00	1023
K^+	1·33 : 1·00 : 2·00	1023
Na^+	1·47 : 1·00 : 2·00	1023
K^+	— : 1·00 : 2·00	1474
NH_4^+	0·50 : 1·00 : 2·00	1033
K^+	1·33 : 1·00 : 2·00	1033

On the basis of titrations against permanganate, the niobium oxidation state has been reported to be +3 or +3·67 (1023, 1033, 1379). The potassium salt has been formulated as $K_8[Nb_6(OH)_{12}(SO_4)_{12}],18H_2O$ or $K_8[Nb_6O_3(SO_4)_{12}],21H_2O$. It exhibits a slight paramagnetism (1474).

Electrolysis in more dilute sulphuric acid (41%) results in an air-sensitive green solution, from which green crystals could be obtained in which the niobium appears to have an average oxidation number of 3·33 or 3·67 (1023, 1933).

Similar electrolytic reduction of niobium(V) in concentrated hydrochloric acid produces niobium(IV) and niobium(III) species, and in the presence of 20% ethyleneglycol, niobium(II) species (599). The visible spectra of these solutions do not resemble those of $(Nb_6Cl_{12})^{x+}$ clusters (160).

Tantalum is not reduced under the same conditions.

H. Organometallic Complexes

The reaction between biscyclopentadienyl vanadium(IV) dichloride and biscyclopentadienyl vanadium(II) forms biscyclopentadienyl vanadium(III) chloride (1536):

$$(C_5H_5)_2VCl_2 + (C_5H_5)_2V \rightarrow 2\,(C_5H_5)_2VCl$$

The corresponding bromide and iodide can also be prepared. The sigma-bonded organometallic compounds $(C_5H_5)_2VR$ (where R is CH_3, σ-C_3H_5, σ-C_5H_5, C_6F_5, C_6H_5 and other aromatic groups), are obtained from $(C_5H_5)_2VCl$ and the appropriate alkali metal organometallic reagent or Grignard reagent (2137).

Treatment of $(C_5H_5)V(CO)_4$ with dimethylsulphide (1214), bis(trifluoromethyl)dithietene (1395) or acetic acid (1396) forms the formally vanadium(III) $[(C_5H_5)V(MeS)_2]_2$, $[(C_5H_5)V(S_2C_2(CF_3)_2)]$ and $(C_5H_5)V(OAc)_2$ respectively. The first two were formulated as having four sulphur atoms bridging the two metal atoms, and the magnetic moments of 1·04 B.M. at 345° K falling to 0·70 B.M. at 195° K, and 0·6 B.M. at room temperature, indicate some degree of vanadium–vanadium bonding. Similarly although the mass spectrum of the acetate complex shows that the vapour is essentially monomeric, the room temperature effective magnetic moment for the solid of 1·49 B.M. indicates some vanadium–vanadium interaction.

4. OXIDATION STATES $\leqslant$ II

A. Introduction

The chemistry of the lower valence states of vanadium bear very little resemblance to that of niobium and tantalum. Vanadium(II) forms a number of relatively simple compounds such as the dihalides which have "normal" dihalide structures based on close packing of halide ions, tetrahedral complexes such as VCl_4^-, and octahedral complexes such as $[V(H_2O)_6]^{2+}$, $[VCl_2(H_2O)_4]$, $[VCl_2(py)_4]$ and $[V(en)_3]^{2+}$ which have the expected magnetic and spectral properties for a d^3 system.

The halide chemistry of niobium and tantalum on the other hand is dominated by $(M_6X_8)^{3+}$, $(M_6X_{12})^{2+}$, $(M_6X_{12})^{3+}$ and $(M_6X_{12})^{4+}$ clusters which are based on octahedral groups of metal atoms, and which have the metal atoms in the average oxidation states of +1·83, +2·33, +2·50 and +2·67 respectively. These are therefore treated in some detail in a separate section.

Other compounds intermediate between niobium(II) and niobium(III) and which are based on the close packing of halide ions are discussed on page 218.

A number of organometallic compounds (for example carbonyls, cyclopentadienyls and arenes) have formal oxidation states in the range +II to −I, and these are all briefly discussed in this section. In addition vanadium is present in the −I oxidation state in the nitrosyls

$K_5[V(CN)_5(NO)]$ (1064) and $VCl_2(NO)_3$ (179). Finally if dipyridyl is assumed to remain as an uncharged bidentate ligand, reduction of $[V(dipy)_3]^{2+}$ to $[V(dipy)_3]^+$, $[V(dipy)_3]$ and $[V(dipy)_3]^-$ involves progressive reduction in the formal oxidation number from +II to −I (1178). Similar behaviour is noted with the closely related *o*-phenanthroline and terpyridyl ligands (194), and also with niobium analogues (1185).

B. Inorganic Vanadium Compounds

Whereas reduction of vanadium in aqueous hydrochloric, hydrobromic or hydroiodic acid readily proceeds to vanadium(II), the electrolytic reduction of V_2O_5 in 1 *M* hydrofluoric acid precipitates grey $V_2F_5,7H_2O$, which forms the brown anhydrous compound when heated in vacuum (2097). The hydrate has an effective magnetic moment of 3·35 B.M. which is independent of temperature, whereas the anhydrous compound has $\mu_{eff} = 2{\cdot}89$ B.M. at room temperature dropping to 2·30 B.M. at liquid nitrogen temperatures. However the blue vanadium(II) $(NH_4)_2VF_4,2H_2O$ and KVF_3 can be obtained by halogen exchange from $VCl_2,4H_2O$ in aqueous solution. The former again shows a temperature independent magnetic moment of 3·84 B.M., while the anyhydrous KVF_3, which has the cubic perovskite structure, has an effective magnetic moment of 3·16 B.M. at room temperature falling to 2·16 B.M. at liquid nitrogen temperatures, indicating considerable magnetic exchange (2097).

Vanadium dichloride is generally prepared by the disproportionation of the trichloride above 600° C with evolution of the volatile tetrachloride, or by reduction of the trichloride with hydrogen at 500° C (1854). The green product has the cadmium iodide layer structure (758, 2319), and the magnetic properties indicate considerable exchange between the metal atoms (1421, 2206).

The yellow dibromide is isomorphous with the dichloride and is similarly prepared by reduction of the tribromide with hydrogen (782, 1421). The vapour pressures of the chloride and bromide have been measured (1588).

Vanadium diiodide is formed by the disproportionation of the triiodide or by reacting stoicheiometric quantities of the elements. Two forms may be obtained, depending upon the reaction temperature (1324, 1420, 1709). The red form again has the cadmium iodide structure, whereas the black form is closely related but with mixed hexagonal and cubic close packing of the iodide ions (1324). The effective magnetic moment of 3·3 B.M. is significantly higher than that of the chloride or bromide (1324, 1420).

Vanadium dichloride is not readily soluble in many solvents, but will dissolve in pyridine in sealed tubes at 160° C to form $VCl_2(py)_4$. Alternatively electrolytic reduction of vanadium trichloride in methanol saturated with hydrogen chloride forms $VCl_2(MeOH)_4$. These adducts can then be used to form complexes with other ligands. Those of stoicheiometry $VCl_2(ligand)_4$ (where the ligand is H_2O, MeOH, MeCN or py), and also $VCl_2,6MeNH_2$, show normal paramagnetic behaviour for octahedral vanadium(III), that is $\mu_{eff} = 3{\cdot}78\text{–}3{\cdot}87$ B.M. and independent of temperature, the slight reduction from the spin-only value being due to spin-orbit coupling. In addition a number of $VCl_2(ligand)_2$ complexes are known (where the ligand is H_2O, MeOH, py, tetrahydrofuran or dioxan) which have considerably lower effective magnetic moments, falling from 3·23–3·56 B.M. at room temperature to 2·09–3·22 B.M. at liquid nitrogen temperatures. Similar adducts of VBr_2 and VI_2 have been obtained (913, 941, 2090, 2091).

The chloro complexes A_2VCl_4 (where A is Me_4N, Et_4N or Ph_3MeAs) and $AVCl_3$ where A is K, Rb, Cs or Me_4N, can be obtained by using the stoicheiometric amount of cation (2093), by the careful dehydration of $AVCl_3,6H_2O$ at 100° C under vacuum (1058), or by solid state reactions between the dichloride and alkali metal chloride (2092). The rubidium salt is isomorphous with $CsNiCl_3$, indicating a linear polymeric structure with VCl_6 octahedra sharing opposite faces (1058). The polymer is antiferromagnetic with a room temperature effective magnetic moment of 1·85 B.M. rising to a maximum at about 330° K, which is consistent with interaction between *pairs* of vanadium atoms. The effective magnetic moment of $CsVCl_3$ falls from 1·70 B.M. at room temperature to 0·94 B.M. at liquid nitrogen temperature, whereas that of $(Me_4N)VCl_3$ falls from 2·13 to 1·30 B.M. (2093). In $RbVCl_3$ each of the three bands normally present in the visible spectrum is split into three distinct components (1058). The change in the spectrum of vanadium(II) in a lithium chloride–potassium chloride eutectic as the temperature is raised above 400° C has been interpreted as being due to a change from VCl_6^{4-} to VCl_4^{2-} (1077).

In addition to the above hydrates $VCl_2,4H_2O$ and $VCl_2,2H_2O$, two series of hydrated double salts can be obtained, the red-violet $AVCl_3,6H_2O$ (where A is NH_4 or Rb), and the green $CsVCl_3,2H_2O$ (2094). The visible spectra of the former are completely consistent with the hexaaquo ion $[V(H_2O)_6]^{2+}$, whereas that of latter suggests that the vanadium atom is surrounded by six chlorine atoms (1058).

Aqueous blue-purple solutions of vanadium(II) may be obtained directly from the dichloride, or by reduction of the higher valence states electrolytically or by means of amalgamated zinc in acid solution. Any intermediate brown colour is due to vanadium(III) which can be prevented by increasing the acid concentration. The resulting hexaaquo ion can be readily isolated as the sulphate $[V(H_2O)_6](SO_4)$ or as the Tutton salts

$A_2[V(H_2O)_6](SO_4)_2$, where A is NH_4, K, Rb or Cs. The magnetic moments of 3·77 B.M. are again slightly lower than the spin only value of 3·87 B.M. due to spin-orbit coupling, and do not change with decreasing temperature (1495). There are the normal three spin-allowed transitions for a d^3 system: ${}^4A_{2g} \rightarrow {}^4T_{2g}$ at 12 000 cm^{-1}, ${}^4A_{2g} \rightarrow {}^4T_{1g}(F)$ at 18 000 cm^{-1}, and ${}^4A_{2g} \rightarrow {}^4T_{1g}(P)$ at 28 000 cm^{-1} (1495). The structure of $(NH_4)_2[V(H_2O)_6](SO_4)_2$ shows a vanadium–oxygen bond length of 2·15 Å (1701).

The water molecules from the hexaquo complex can be displaced in aqueous solution by cyanide to form $[V(CN)_6]^{4-}$ (1862), or by bidentate ligands such as picolinic acid (1682), ethylenediamine (1496) and acetylacetone (2004).

C. Niobium and Tantalum Clusters

1. *Halides*

The lowest halides formed by the reduction of niobium and tantalum pentahalides under fairly vigorous conditions are:

	Nb_6F_{15}	Nb_6Cl_{14}	(Nb_6Br_{14})		Nb_6I_{11}	
Average oxidation No.	2·50	2·33	(2·33)		1·83	
	—	Ta_6Cl_{15}	Ta_6Br_{14}	Ta_6Br_{15}	Ta_6I_{14}	Ta_6I_{15}
Average oxidation No.	—	2·50	2·33	2·50	2·33	2·50

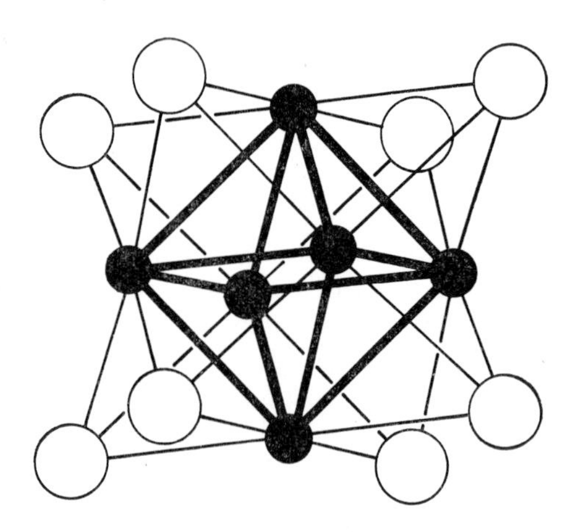

FIG. 127 The $(M_6X_8)^{x+}$ group. The metal atoms in the octahedral cluster are shown as shaded circles, while the bridging halogen atoms above the octahedral faces are unshaded.

Claims to the existence of dihalides such as Nb_6Cl_{12} (946, 2011), Nb_6Br_{12} (1103), Nb_6I_{12} (452) and Ta_6Cl_{12} (946, 2424) have not been confirmed by later workers. The average oxidation numbers of these lowest oxidation states follow the expected trends Ta > Nb and F > Cl > Br > I.

All compounds are based on octahedral clusters of metal atoms with halogen atoms above each octahedral face to form $(M_6X_8)^{x+}$ groups (Fig. 127) as in Nb_6I_{11}, or above each octahedral edge to form $(M_6X_{12})^{x+}$ groups (Fig. 128) as in all other cases. These $(Nb_6I_8)^{3+}$ and $(M_6X_{12})^{x+}$ units are then linked into polymeric structures by the sharing of the additional halogen atoms between different clusters.

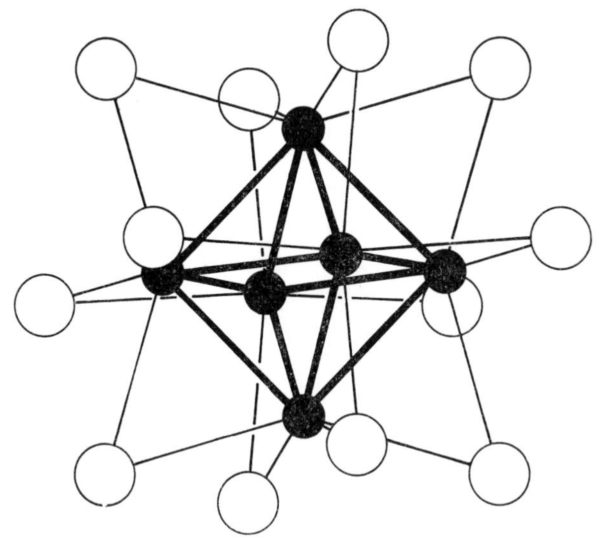

FIG. 128 The $(M_6X_{12})^{x+}$ group. The metal atoms in the octahedral cluster are shown as shaded circles, while the bridging halogen atoms above the octahedral edges are unshaded.

The niobium fluoride Nb_6F_{15} can be prepared by the thermal disproportionation of the tetrafluoride at 500° C, or by reduction of the pentafluoride with niobium in a nickel tube using a 400°–900° C temperature gradient, followed by sublimation of the excess pentafluoride at 200° C (2027). The brown-black powder is stable to air, water, concentrated acids and alkalis. It is only decomposed by fusing with sodium hydroxide (with evolution of hydrogen), or by heating to 700° C in vacuo (to the metal and the pentafluoride) or in air (to $Nb(O,F)_3$). The structure (2027) shows a regular octahedral cluster of niobium atoms with quite short niobium–niobium distances of 2·80 Å. Each edge of the octahedron is bridged by fluorine atoms forming $(Nb_6F_{12})^{3+}$ groups such that each niobium–bridging fluorine bond length is 2·05 Å (Fig. 128). Each niobium atom is then bonded to an additional fluorine atom in a "centrifugal" position at a slightly greater distance of 2·11 Å (Fig. 129). Each of these six centrifugal fluorine atoms acts as a bridge to one of the six neighbouring clusters forming a cubic unit cell with $a = 8{\cdot}19_0$ Å. The formula may be written $(Nb_6F_{12})F_{6/2}$. This sharing is reminiscent of the linking of ReO_6

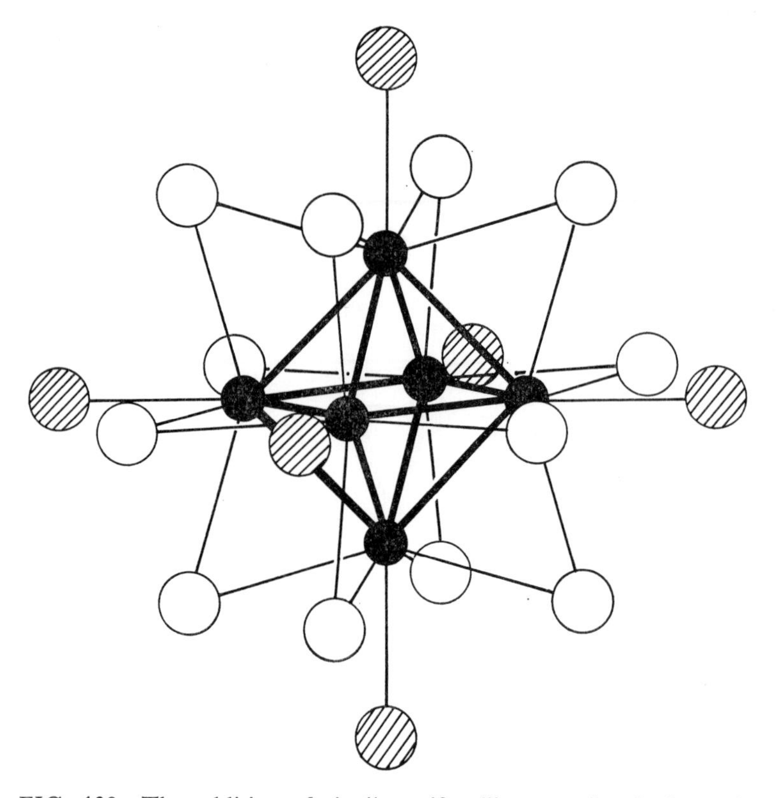

FIG. 129 The addition of six "centrifugal" atoms (hatched) to the $(M_6X_{12})^{x+}$ group shown in Fig. 128.

octahedra to form the ReO_3 structure, but an important difference is that the fairly large holes so formed are occupied by a different set of $(Nb_6F_{12})^{3+}$ clusters, so that the structure consists of two distinct but interpenetrating infinite three dimensional molecules. Figure 130 shows one quarter of the unit cell and the relationship to the sodium chloride structure the vacancies corresponding to $Nb_{3/8}F_{15/16}$.

The compounds of composition M_6X_{14} have been prepared in a pure

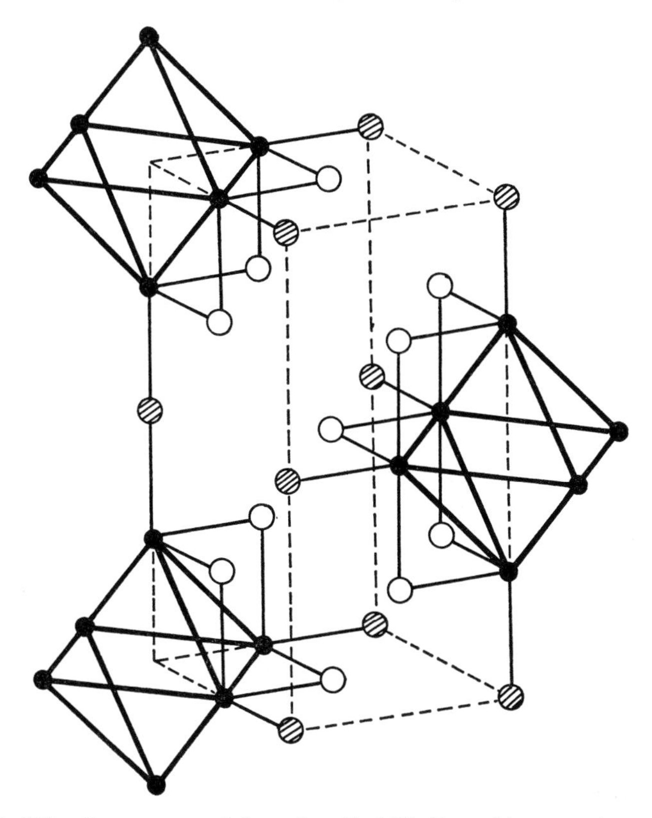

FIG. 130 One quarter of the unit cell of Nb_6F_{15}, with atoms denoted as in Fig. 129. The niobium atoms are displaced towards each other compared with a regular defect sodium chloride structure, so that they are at (0·242,0,0) etc. rather than (0·25,0,0) etc.

state by the reduction of the higher halides with metal in an appropriate temperature gradient:

	Temperature gradient	Reference
$Nb_3Cl_8 + Nb \rightarrow Nb_6Cl_{14}$	700–800 (in quartz)	2151
$TaBr_5 + Al \rightarrow Ta_6Br_{14}$	250–450	1479
$TaI_5 + Al \rightarrow Ta_6I_{14}$	300–475	1479
$TaI_5 + Ta \rightarrow Ta_6I_{14}$	510–660	164

The tantalum–iodine phase diagram (Fig. 131) shows that Ta_6I_{14} is the only tantalum cluster iodide, whereas the tantalum–bromine phase diagram (Fig. 132) shows the formation of both Ta_6Br_{14} and Ta_6Br_{15} (1582).

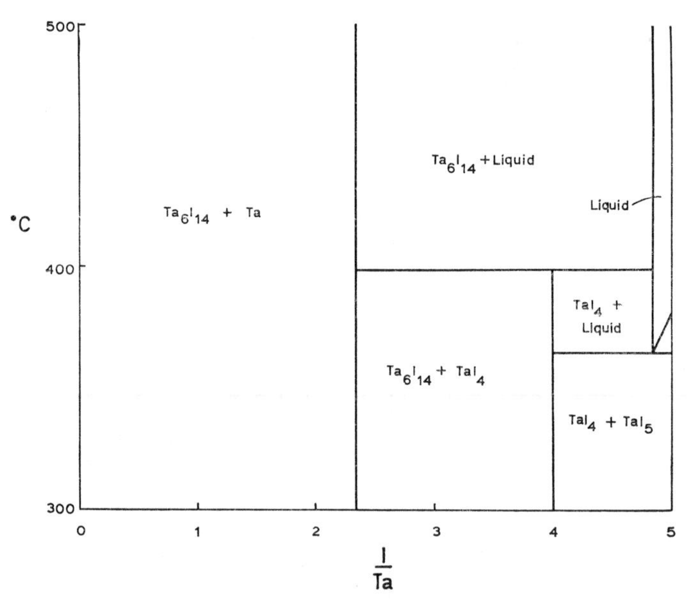

FIG. 131 Tantalum-iodine phase diagram.

The dark green-brown Nb_6Cl_{14} contains $(Nb_6Cl_{12})^{2+}$ clusters with niobium–bridging chlorine distances of 2·40 Å, analogous to the $(Nb_6F_{12})^{3+}$ clusters above. The structure (2151) (Fig. 133) clearly

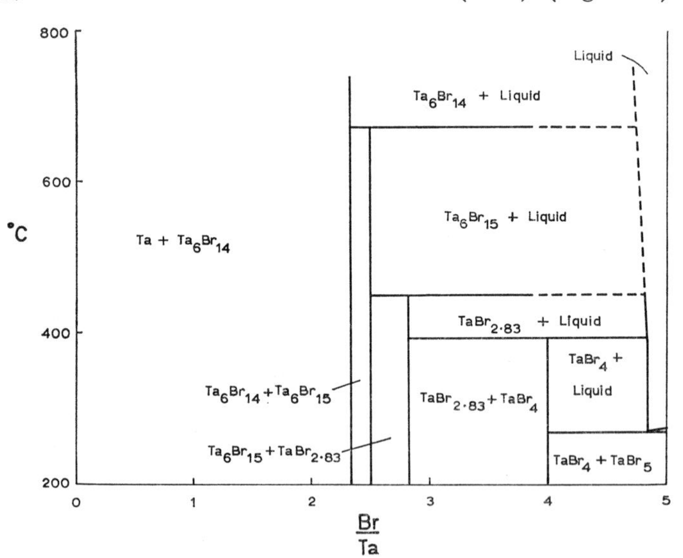

FIG. 132 Tantalum-bromine phase diagram.

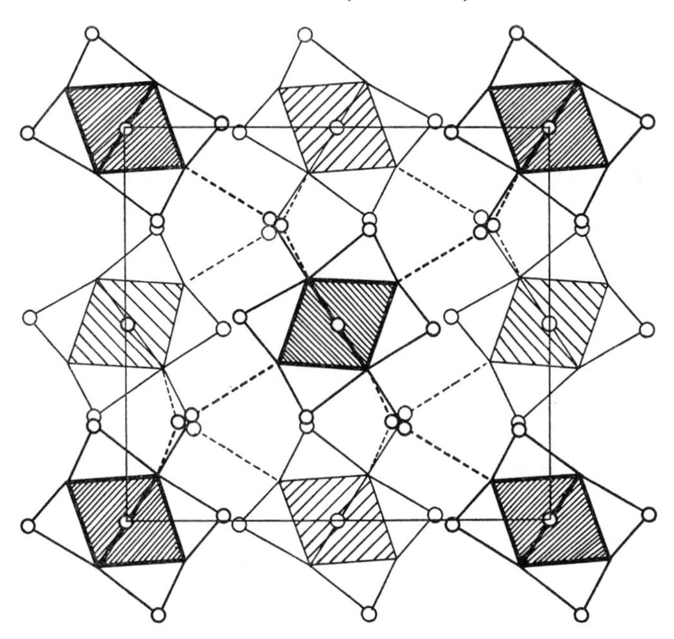

FIG. 133 The structure of Nb_6Cl_{14}. The cluster-centrifugal chlorine bonds are shown by broken lines.

shows that these clusters are linked together by four centrifugal chlorine atoms, with niobium–chlorine distances of 2·58 Å, into two dimensional sheets so that the formula can be considered as $(Nb_6Cl_{12})Cl_{4/2}$. The other two *trans*-centrifugal sites are occupied by chlorine atoms of (Nb_6Cl_{12}) groups in adjacent sheets at the somewhat greater distance of 3·01 Å. This non-equivalence of centrifugal sites results in a squashing of the Nb_6 octahedron along the resulting four-fold axis (Fig. 134), due to stronger niobium–niobium bonding for the two metal atoms participating in weaker niobium–centrifugal chlorine bonding. The relevant interatomic distances are:

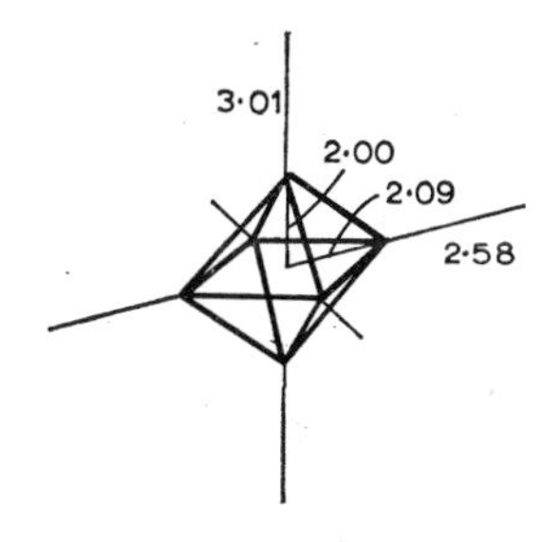

FIG. 134 Distortion of the Nb_6 cluster in Nb_6Cl_{14}.

$$\left.\begin{aligned} Nb_I\text{–}Nb_I &= 2\cdot95_5 \\ Nb_I\text{–}Nb_{II} &= 2\cdot89_5 \end{aligned}\right\} \text{Weighted average} = 2\cdot91_5\ \text{Å}$$

The structure of the dark green Ta_6I_{14} is closely related (164), except that the tantalum–centrifugal iodine bonds are considerably longer, for example compared with the tantalum–iodine distances within the $(Ta_6I_{12})^{2+}$ group. This is presumably due to the steric difficulty of surrounding a tantalum atom by four other tantalum atoms and five iodide ions. These long tantalum–iodine bonds result in shorter tantalum–tantalum bonds. Indeed the two equivalent *trans*-tanatalum-centrifugal iodine distances are so long that they can be scarcely regarded as chemical bonds, and the resulting stronger tantalum–tantalum bonds result in a greater distortion of the Ta_6 octahedron than observed above for the Nb_6 octahedron in Nb_6Cl_{14}. The two dimensional structure has layers parallel to (100) and it is found that the crystals readily cleave along these planes confirming the weakness of these longer tantalum–iodine bonds. The relevant bond distances are shown:

$$\left.\begin{aligned} Ta_I\text{-}Ta_I &= 3\cdot08 \\ Ta_I\text{-}Ta_{II} &= 2\cdot80_5 \end{aligned}\right\} \text{Weighted average} = 2\cdot90\ \text{Å}$$

$$\begin{aligned} Ta\text{-}I_{bridge} &= 2\cdot76 \quad \text{(average for } Ta_6I_{12} \text{ group)} \\ Ta_I\text{-}I_{centrifugal} &= 3\cdot11 \\ Ta_{II}\text{-}I_{centrifugal} &= 4\cdot31 \end{aligned}$$

The green Ta_6Br_{14} is isomorphous with Ta_6I_{14} (1479).

Reduction of tantalum pentachloride with aluminium in a 200°–400° C temperature gradient (1479) or with tantalum at 800° C (in a tantalum capsule) (163) forms the green brown Ta_6Cl_{15}. The corresponding black Ta_6Br_{15} can be prepared by the similar reduction of higher bromides (163, 2014); probably the best method uses a three temperature–temperature gradient of 330°–450°–620° C. Oxidation of Ta_6I_{14} with iodine at 240° C apparently forms the polyiodide $(Ta_6I_{12})^{2+}I_{18}$ which on heating in vacuo forms Ta_6I_{15} (162). These three Ta_6X_{15} compounds form very large cubic unit cells and contain undistorted $(Ta_6X_{12})^{3+}$ clusters linked by the sharing of bridging centrifugal halogen atoms (162, 163, 1479). The relevant bond distances in Ta_6Cl_{15} are (163):

$$\begin{aligned} Ta\text{-}Ta &= 2\cdot92_5\ \text{Å} \\ Ta\text{-}Cl_{bridge} &= 2\cdot43 \\ Ta\text{-}Cl_{centrifugal} &= 2\cdot56 \end{aligned}$$

The magnetic susceptibility of these $(M_6X_{12})^{3+}$ compounds show the presence of one unpaired electron per cluster (for example $\mu_{eff} = 2\cdot1$ B.M. at 630° K dropping to 1·3 B.M. at 90° K for Ta_6Cl_{15}) in contrast to the diamagnetic $(M_6X_{12})^{2+}$ compounds.

A particularly interesting compound is Nb_6I_{11}, formed by the thermal decomposition of Nb_3I_8, or its reduction with niobium (161, 1487, 2149). Under similar conditions Nb_3Br_8 is not reduced. The structure (161, 2149) is again found to contain octahedral clusters of niobium atoms, but now the average niobium–niobium bond length is significantly shorter at 2·85 Å. The cluster is slightly distorted, the niobium–niobium bond lengths ranging from 2·72 to 2·94 Å due to a tilting of one pair of *trans*-metal atoms relative to the other four coplanar metal atoms, and also a shortening of the metal–metal bonds to these tilted metal atoms. However in contrast to the above halides there are now only eight bridging halogen

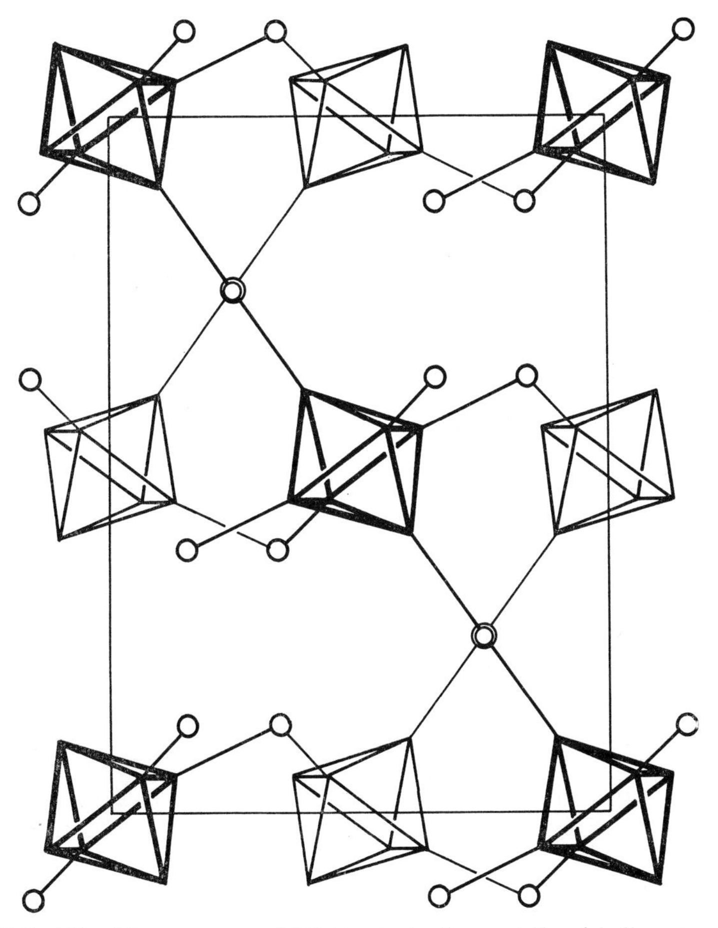

FIG. 135 The structure of Nb_6I_{11}. Only the centrifugal iodine atoms are shown; those above each face of the Nb_6 octahedron are omitted for clarity.

atoms, one above each face of the octahedron forming $(Nb_6I_8)^{3+}$ (Fig. 127). As in the $(M_6X_{12})X_3$ structures, each of these $(Nb_6I_8)^{3+}$ clusters is joined to six neighbouring clusters by the sharing of centrifugal iodine atoms forming $(Nb_6I_8)I_{6/2}$ or $(Nb_6I_8)I_3$ with average niobium–iodine bond lengths of 2·93 Å (Fig. 135).

The magnetic susceptibility of Nb_6I_{11} has been studied over the very large temperature range of 5–863° K (2149). After correction for both the diamagnetism of the iodide ions and the temperature independent paramagnetism of the cluster, the results show a magnetic moment of 2·24 B.M. ($\theta = -53°$) from 20° to 90° K, and 4·05 B.M. ($\theta = -470°$) above 300° K, with more complex behaviour between 300° and 90° K, and below 20° K. This suggests the presence of one unpaired electron per cluster at low temperatures, and three unpaired electrons per cluster at high temperatures.

Heating Nb_6I_{11} with hydrogen at 300° C under atmospheric pressure forms the remarkable HNb_6I_{11} (2147). Neutron diffraction investigations on HNb_6I_{11} and DNb_6I_{11} has located the hydrogen atom at the centre of the (Nb_6I_8) group. The volume of the unit cell (2352 Å^3) is slightly larger than that of Nb_6I_{11} itself (2345 Å^3). The hydride is diamagnetic below 200° K, but approaches the magnetic susceptibility of Nb_6I_{11} from 400° to 800° K.

2. *Chemical reactions*

For the preparation of derivatives of these clusters it is not generally necessary to isolate the pure halides themselves, and normally the reduced mass of chlorides or bromides is extracted and recrystallized from aqueous hydrochloric or hydrobromic acid respectively, yielding the compounds $Nb_6Cl_{14},7H_2O$, $Nb_6Br_{14},7H_2O$, $Ta_6Cl_{14},7H_2O$ and $Ta_6Br_{14},7H_2O$. Metallic cadmium is a particularly suitable reducing agent, and about 30% yields have been obtained (1138, 2006). Such compounds have been known since 1907 when it was reported (446) that reduction of tantalum pentachloride with sodium amalgam at red heat, followed by extraction with hydrochloric acid formed "hydrated tantalum dichloride". Later workers formulated this product as $Ta_6Cl_{14},7H_2O$ (459, 1137), $H(Ta_3^{II}Cl_7,H_2O),3H_2O$ (1544) and $(Ta_3^{III}Cl_7O),3H_2O$ (1969). The $(M_6X_{12})^{2+}$ structure was first correctly deduced from measurements of the low angle X-ray scattering from alcoholic solutions of these compounds (2317).

The crystal structure of $Ta_6Cl_{14},7H_2O$ (380) shows that the $(Ta_6Cl_{12})^{2+}$ group in the anhydrous chloride is retained. The six centrifugal sites are now occupied by four water molecules and two *trans*-chloride ions, so the structure can be written $[(Ta_6Cl_{12})Cl_2(H_2O)_4],3H_2O$. The $Ta\text{-}Cl_{centrifugal}$

distance is fairly short at 2·35 Å (compared with $2 \cdot 50_5$ Å for the tantalum–bridging chlorine bond length), whereas the Ta–$OH_{2\ centrifugal}$ distance appears relatively long by comparison at 2·25 Å. This stronger *trans*-tantalum–centrifugal ligand bond results in weaker metal–metal bonding, causing an elongation of the Ta_6 octahedron along this axis. The tantalum–tantalum bond lengths are 2·78 and 3·15 Å, with a weighted average of 3·03 Å.

A number of similar complexes $[(Nb_6Cl_{12})Cl_2(ligand)_4]$ have been obtained with the oxygen–donor ligands dimethylsulphoxide, dimethylformamide, triphenylarsine oxide, triphenylphosphine oxide, and pyridine–l–oxide (844).

These complexes show three characteristic intense bands ($\epsilon \sim 10^3$) in the visible and ultra violet spectra at 10 000, 25 000 and 35 000 cm^{-1} which can be assigned to allowed transitions between metal–metal bonding molecular orbitals and metal–metal antibonding molecular orbitals, with a number of additional peaks and shoulders of lower intensity (see later). They do not shift in energy very much in comparing $(Nb_6Cl_{12})^{2+}$ complexes with $(Nb_6Br_{12})^{2+}$ complexes (1361).

A study of the electronic transitions at 10 000 and 25 000 cm^{-1} in the complexes $[(Nb_6Cl_{12})Cl_2(4\text{–}Z.C_5H_4NO)_4]$ (where the ligand is a pyridine–1–oxide substituted in the 4 position with either electron withdrawing ($Z = NO_2$, Cl) or electron releasing ($Z =$ Me, MeO, Me_2N) groups) shows a linear relation between the energy of these transitions and the base strength of the ligand. That is the energy difference between the bonding orbitals and the antibonding orbitals, which is expected to be a function of the strength of the metal–metal bonding, increases as the strength of the metal–ligand bond decreases (845). The other intense band at 35 000 cm^{-1} is obscured by ligand bands.

Complexes of the type $[(Nb_6Cl_{12})Cl_2(ligand)_4]$ have also been isolated with pyridine and substituted pyridines, while the bidentate 2,2′-dipyridyl N,N′-dioxide forms $[(N_6Cl_{12})Cl_2(dipyO_2)_2]$ (845). The latter is particularly interesting since models show that it is capable of occupying two *cis*-centrifugal positions of the cluster and the cluster would be expected to show a more complex distortion.

The potassium salt of the hexachloroanion, $K_4[(Nb_6Cl_{12})Cl_6]$, has been prepared by the solid state reaction between Nb_6Cl_{14} and KCl (2150), or by the solid state reaction between Nb_3Cl_8 and KCl at high temperatures where disproportionation occurs to form also K_2NbCl_6 (886). However it has also been claimed that a study of the binary systems between "$NbCl_2$" and alkali metal chlorides show the formation of K_2NbCl_4, Rb_2NbCl_4 and Cs_2NbCl_4 (1979). The $(Et_4N)_4[(Nb_6Cl_{12})Cl_6]$ can be obtained from Nb_6Cl_{14}, $8H_2O$ in ethanol saturated with hydrogen chloride (1603). In line with the above discussion, it is found that the Nb_6 octahedron in $K_4[(Nb_6Cl_{12})Cl_6]$ is undistorted, the relevant bond lengths being given below (2150):

$$\text{Nb–Nb} = 2{\cdot}91_5\ \text{Å}$$
$$\text{Nb–Cl}_{\text{bridging}} = 2{\cdot}49$$
$$\text{Nb–Cl}_{\text{centrifugal}} = 2{\cdot}60$$

The preparation of pure $(Nb_6Cl_{12})^{2+}$ complexes from aqueous solution is not simple, as oxidation to $(Nb_6Cl_{12})^{3+}$clusters readily occurs. These oxidized species may be detected by the emergence of a band in the near infrared spectrum at about 8000 cm^{-1}, but otherwise the visible spectrum is not greatly altered (see later). The presence of this oxidation product may explain the so-called brown enneahydrate $Nb_6Cl_{14},9H_2O$, in contrast to the green $Nb_6Cl_{14},8H_2O$, both forms being noted as long ago as 1913 (1137). One alternative idea is that in the brown form some of the bridging chloride ions have been replaced by hydroxyl groups (39). It was also noted that the enneahydrate is more readily obtained at low temperatures (2193).

There is no exchange between $(Ta_6Cl_{12})^{2+}$ and $*Cl^-$ within 13 hours ($\pm$20%) (787).

The infrared spectra of these complexes below 400 cm^{-1} have been studied by a number of workers (271, 844, 845, 1603, 1604, 1667). For $(Nb_6Cl_{12})^{2+}$, two strong niobium–chlorine vibrations occur at 340 and 290 cm^{-1}. However there is little agreement about the assignments of many of the other bands between different authors (or in one case even by the same author in the same paper). Niobium–niobium stretching vibrations have been assigned to bands between 90 and 200 cm^{-1}. It must also be remembered that the symmetrical niobium–niobium stretch is not expected to be infrared active.

3. *Oxidized clusters*

The green $(M_6X_{12})^{2+}$ can be oxidised by air or iodine in aqueous solution to $(M_6X_{12})^{3+}$, and by two or more moles of stronger oxidizing agents such as bromate, ferric, hydrogen peroxide or chlorine, to the red brown $(M_6X_{12})^{4+}$ (where M is Nb or Ta, and X is Cl or Br) (788, 885, 1586, 1603, 2193, 2194, 2195, 2196, 2197). The products which can be isolated include the paramagnetic $(M_6X_{12})^{3+}$ compounds $(Et_4N)_3[(Nb_6Cl_{12})Cl_6]$, $Nb_6Cl_{15},7H_2O$, $HNb_6Cl_{16},6H_2O,EtOH$, $Ta_6Cl_{15},7H_2O$ and $Ta_6Br_{15},8H_2O$, and the diamagnetic $(M_6X_{12})^{4+}$ compounds $(Et_4N)_2[(Nb_6Cl_{12})Cl_6]$, $(pyH)_2[(Nb_6Br_{12})Cl_6)$ and $Ta_6Br_{16},7H_2O$. The pyridinium salts $(pyH)_2[(Nb_6Cl_{12})Cl_6]$, $(pyH)_2[(Nb_6Br_{12})Cl_6]$, $(pyH)_2[(Ta_6Cl_{12})Cl_6]$ and $(pyH)_2[(Ta_6Br_{12})Cl_6]$ form isomorphous hexagonal crystals.

The effective magnetic moment of $(Et_4N)_3[(Nb_6Cl_{12})Cl_6]$ is 1·62 B.M., and the e.s.r. spectrum shows a g value of 1·95 (1603). The complex splitting (49 detectable components) due to the nuclear spin (9/2) of ^{93}Nb (100% natural abundance) shows that the electron is delocalized over all six niobium atoms.

4. *Bonding and visible spectra*

Of the 30 electrons potentially available to the six metal atoms in clusters of the type $(Nb_6Cl_{12})^{2+}$, twelve will be used for bonding the twelve halogen atoms and two will be lost to provide the net charge, leaving eight pairs of electrons available for metal–metal bonding. However if electron pair metal–metal bonds are considered to lie along each of the octahedral edges, 12 electron pairs are required. A further and related difficulty is that the four metal atoms and four bridging halogen atoms around each metal atom form a distorted cubic stereochemistry (C_{4v} symmetry), and appropriate hybrid orbitals can only be constructed if f atomic orbitals are utilized.

Both these difficulties can be overcome using either simple valence bond or molecular orbital approaches. An "electron in box" approach has also been given (1534, 1739).

Before dealing with the metal–metal bonding within these clusters in some detail, it should be emphasized that the bridging halogen atoms also play an important role in holding the cluster together. For example the bridging halogen atoms must play the dominant role in the stabilization of the Pt_6Cl_{12} cluster, and any bonding in the niobium and tantalum clusters must include some contribution of this type.

A satisfactory valence bond description is to consider the hybrid orbitals of each metal atom to be directed down the four *faces* of the octahedron, so that eight three-centre orbitals are formed which accommodate the eight

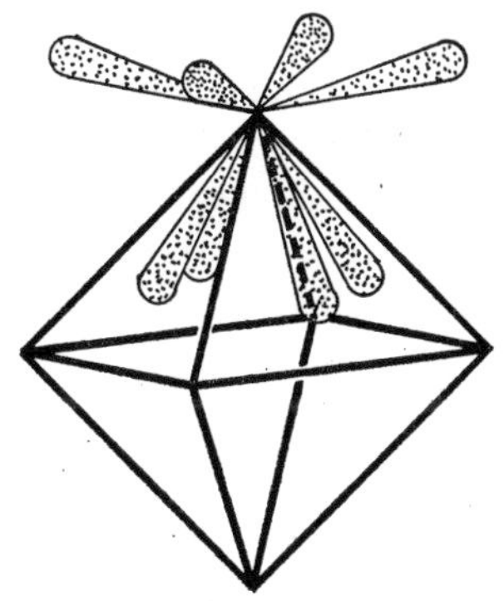

FIG. 136 Square antiprismatic hybridization shown on one of the metal atoms in $(M_6X_{12})^{x+}$.

electron pairs (Fig. 136) (706). The arrangement of electron pairs about each metal atom is in the form of a distorted square antiprism.

A fairly simple molecular orbital approach also leads to a very plausible bonding scheme (563, 607). Three simplifying assumptions can be made in the initial approach:

(i) Each metal atom uses planar $sp_xp_yd_{x^2-y^2}$ hybrids to bond to the four (non-planar) bridging halogen atoms (with bent σ bonds).

(ii) The centrifugal ligand is bonded purely through the p_z orbital, rather than a $p_zd_{z^2}$ hybrid.

(iii) Ligand–metal and halogen–metal π bonding are neglected.

The remaining four atomic d orbitals on each of the six metal atoms are then used to construct purely metal–metal molecular orbitals (Fig. 137).

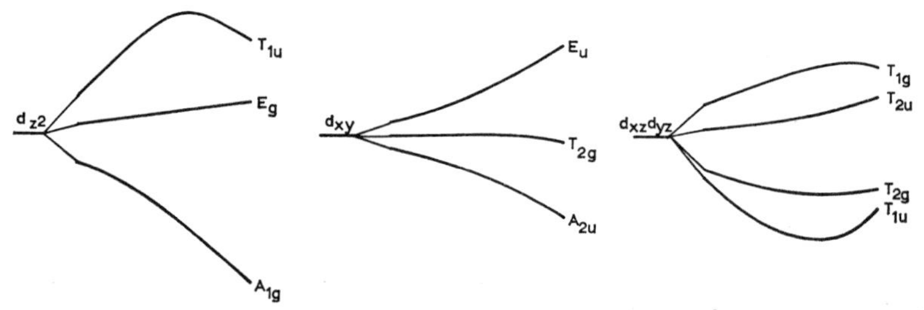

FIG. 137 Molecular orbital energy level diagram for an M_6X_{12} system.

The six d_{z^2} orbitals overlap at the centre of the octahedron forming a stable $A_{1g}(d_{z^2})$ orbital, the energy of which is strongly dependent upon the calculated overlap due to the relatively long distance between the metal atoms and the centre of the octahedron. The six d_{xy} orbitals overlap over the faces of the octahedron, while the twelve degenerate d_{xz} and d_{yz} orbitals overlap along the edges of the octahedron. One unusual feature of the resulting molecular orbital diagram is that there are only eight bonding orbitals compared with 16 antibonding orbitals (the $T_{2g}(d_{xy})$ is expected to be more unstable than shown due to interaction with the bonding $T_{2g}(d_{xz}, d_{yz})$). The 16 available electrons are then accommodated in the eight bonding orbitals, which is in agreement with the observed diamagnetism.

A number of observations may be made in the light of the above assumptions and subsequent experimental results.

Firstly it must be remembered that the whole d_{z^2} manifold should be both raised and the splitting increased by an unknown amount due to interaction with the centrifugal ligand and bridging halogen atoms.

Secondly the observations that both the $(Nb_6Cl_{12})^{2+}$ (16 electrons) and $(Nb_6Cl_{12})^{4+}$ (14 electrons) clusters are diamagnetic, and the high magnetic moment and g-value for the $(Nb_6Cl_{12})^{3+}$ (15 electron) cluster, show that the fifteenth and sixteenth electrons occupy a non-degenerate molecular orbital. This is presumably the $A_{2u}(d_{xy})$ in agreement with Fig. 137.

Thirdly all the bands in the visible spectrum move to lower energies upon oxidation $(M_6X_{12})^{2+} \rightarrow (M_6X_{12})^{3+} \rightarrow (M_6X_{12})^{4+}$ due to weaker metal–metal bonding (788, 885, 2193, 2197). Of particular interest is the very intense low energy band in the niobium complex at 10 000 cm^{-1}, which disappears on oxidation to $(Nb_6Cl_{12})^{4+}$ indicating that this band originates from electrons in the $A_{2u}(d_{xy})$ orbital. The $A_{2u}(d_{xy}) \rightarrow T_{2g}(d_{xy})$ is the only allowed band. The second very intense band at 25 000 cm^{-1} is probably the allowed $T_{2g}(d_{xz}, d_{yz}) \rightarrow T_{2u}(d_{xz}, d_{yz})$. It might also be relevant that although irradiation with a mercury vapour source (20 000–52 000 cm^{-1}) markedly accelerates the base hydrolysis of $(Nb_6Cl_{12})^{2+}$ and $(Ta_6Cl_{12})^{2+}$, irradiation with a xenon source (10 000–32 000 cm^{-1}) has no effect (348). It was therefore deduced that no chlorine–metal charge transfer bands occurred in this region, assuming that the rate is determined by an initial substitution of a bridging chloro group by a bridging hydroxo group. Similarly the spectral shifts as chloride is replaced by bromide do not appear large enough for halogen-to-metal charge transfer spectra.

Fourthly an alternative scheme has been proposed (2065), in which from a consideration of the effects of chlorine–metal π-bonding the order of the molecular orbitals derived from the d_{xy} atomic orbitals is reversed, that is the order of increasing energy is $E_u < T_{2g} < A_{2u}$. In order to explain the diamagnetism of both $(Nb_6Cl_{12})^{2+}$ and $(Nb_6Cl_{12})^{4+}$ it is then necessary to split this $E_u(d_{xy})$ orbital so as spin pairing can occur, which could only be achieved by a large tetragonal distortion to the octahedron.

The spectra of these clusters can be more or less satisfactorily assigned using the molecular orbital scheme outlined above (2062). Compilations of detailed visible spectra are available (see references 886, 1479, 1603 and 2194 in addition to those cited above).

Another spectral result of some interest is the observation that the solution spectra of mixed clusters having the average compositions $(Nb_{2\cdot4}Ta_{3\cdot6}Br_{12})^{2+}$ and $(Nb_{1\cdot6}Ta_{4\cdot4}Br_{12})^{2+}$ are different to spectra of solutions prepared from the appropriate amounts of $(Nb_6Br_{12})^{2+}$ and $(Ta_6Br_{12})^{2+}$ (2035).

D. Organometallic Complexes

The subject of organometallic chemistry of these low oxidation states is divided into carbonyl compounds (which may also contain similar ligands such as phosphines), and cyclopentadienyl and other π bonded complexes (which may also contain carbonyl groups).

1. *Carbonyl compounds*

The reductive carbonylation of common higher valence vanadium compounds such as VCl_3, $VOCl_3$, or $V(acac)_3$, with zinc and/or magnesium under a carbon monoxide pressure of 200 atmospheres at 135° C, and using pyridine as solvent, forms $[V^{II}(py)_6][V^{-I}(CO)_6]_2$. Acidification with aqueous acid and extraction with petroleum ether forms the hexacarbonyl $V(CO)_6$ (786). An alternative and perhaps preferable method is to reduce the trichloride with sodium in diglyme as solvent under 200–350 atmospheres pressure of carbon monoxide at 90°–120° C, to yield $[Na(diglyme)_2][V^{-I}(CO)_6]$ (2377). A good yield is obtained if an oxygen scavenger (for example benzophenone) is added to the bomb and the contents stirred for 24 hours before admission of the carbon monoxide and heating. This diglyme salt is also commercially available, from which the hexacarbonyl can be generated by treatment with phosphoric acid.

Niobium and tantalum heaxcarbonyls are not known, although salts such as $[Na(diglyme)_2][Ta^{-I}(CO)_6]$ are obtained from niobium and tantalum pentachlorides in an analogous manner to the vanadium complex. However in these cases the diglyme solution should not be left standing long before reaction with carbon monoxide due to the formation of $MCl_4(OR)$ which does not readily react (2376, 2377).

Vanadium hexacarbonyl is a volatile blue-green compound soluble in organic solvents, and in contrast to the Group VI hexacarbonyls is extremely sensitive to air. Unexpectedly single crystal X-ray work shows that it is isomorphous with the Group VI hexacarbonyls, that is in contrast to the other formally odd-electron transition metal carbonyls, the inert gas configuration is not achieved by the formation of metal–metal bonds (409). This is confirmed by the paramagnetism, both in the solid state and in solution (benzene, toluene), which indicates a low spin d_ε^5 configuration ($\mu_{eff} \sim 1 \cdot 8$ B.M.) (409, 410). The e.s.r. spectra at $1 \cdot 3$° K of the pure hexacarbonyl or in n-pentane gives a g-value of $2 \cdot 06$ (1894). This increase over the free electron value of $2 \cdot 002$ is consistent with a tetragonally distorted octahedral structure with an elongation along the tetragonal axis, but not with a trigonally distorted octahedral structure. This distortion can be ascribed either to the Jahn–Teller theorem, or to less vanadium-to-carbonyl π bonding from the t_{2g} orbital carrying one electron to the four square planar carbonyl groups. Such distortion is also consistent with the infrared spectrum of the vapour (1108), which shows a broad band centred at 1986 cm^{-1} due to the carbon–oxygen stretching frequency, in contrast to the sharp band for $Cr(CO)_6$ with P, Q and R branches of the rotational structure.

Although $V(CO)_6$ itself does not exhibit metal–metal bonding, substitution with tricyclohexylphosphine or *o*-phenylenebisdimethylarsine forms the diamagnetic dimers $[V(CO)_4\{(C_6H_{11})_3P\}_2]_2$ (1160) and $[V(CO)_4\{o\text{-}C_6H_4(AsMe_2)_2\}]_2$ (1332) respectively. The lowest bands in the carbonyl stretching region are at 1757 and 1870 cm^{-1} respectively, indicating that the metal–metal bond is also bridged by carbonyl groups, at least for the phosphine derivative. However the weaker bases Ph_3P, Et_3P and Pr_3P form the paramagnetic ($\mu_{eff} = 1{\cdot}80$ B.M.) $[V(CO)_4(R_3P)]_2$ (1160). Phosphine itself forms a phosphorus bridged dimer (1160):

$$2\,V(CO)_6 + 2\,PH_3 \rightarrow (OC)_4V\begin{matrix} \nearrow PH_2 \searrow \\ \nwarrow PH_2 \swarrow \end{matrix}V(CO)_4 + 4\,CO + H_2$$

A similar compound was obtained with diphenylphosphine. The diphosphine $Ph_2P.CH_2.CH_2.PPh_2$ causes disproportionation to $[V^{I}(Ph_2P.CH_2.CH_2.PPh_2)_3][V^{-I}(CO)_6]$ at room temperature, but substitution occurs at 120° C to form $[V(CO)_4(Ph_2P.CH_2.CH_2.PPh_2)]$ or $[V(CO)_2(Ph_2P.CH_2.CH_2.PPh_2)_2]$ depending upon reactant stoicheiometry (194). The triphosphine $Me.C(CH_2.PPh_2)_3$ similarly forms $[V(CO)_3(triphosph)]$ (194) and also the iodine oxidation product $[VI(CO)_3(triphosph)]$ (2304).

Oxygen and nitrogen bases (Bu_2O, dioxan, MeOH, Me_2CO, Me_2SO, Ph_3PO, C_5H_5N etc.) cause disproportionation to $[V^{-I}(CO)_6]^-$ and V^{I} or V^{II}, to form, for example $[V(Ph_3PO)_4][V(CO)_6]_2$ and $[V(py)_6][V(CO)_6]$ (786, 1158, 1161). Disproportionation also occurs with phosphine substituted carbonyls to form complexes such as $[V^{II}(dipy)_3][V^{-I}(CO)_4(Ph_2P.CH_2.CH_2.PPh_2)]_2$ (194) and $[V^{II}(Et_2O)_6][V^{-I}(CO)_5(Ph_3P)]_2$ (1158). Complete loss of carbon monoxide occurs under more vigorous conditions to form, for example, $[V(dipy)_3]$, which has been referred to above.

Reaction of $[Na(diglyme)_2][V(CO)_6]$ with $(Ph_3P)AuCl$ or (triars)CuBr (where triars is $Me.As(C_6H_4.AsMe_2)_2$) forms $(Ph_3P)Au\text{–}V(CO)_6$ and $(triars)Cu\text{–}V(CO)_6$, in which the vanadium atom attains a coordination number of seven with vanadium–gold and vanadium–copper bonds respectively (1332).

The similar reaction of $[Na(MeO.CH_2.CH_2.OMe)_3][Ta(CO)_6]$ with alkylmercury halides forms the diamagnetic $RHg\text{–}Ta(CO)_6$, where R is Me, Et or Ph. The tantalum–mercury bond is not bridged by carbonyl groups (1340).

2. *π Bonded compounds*

Bis-cyclopentadienyl vanadium(II), or "vanadocene" $(C_5H_5)_2V$, can be formed by the reduction of $(C_5H_5)_2VCl_2$ with lithium aluminium hydride in tetrahydrofuran, or by the action of cyclopentadienyl magnesium bromide on vanadium tetrachloride (234, 855, 877). The dark violet paramagnetic compound is extremely unstable to air, forming $(C_5H_5)_2V_2(OH)_2O_3$ (1537).

The mixed cyclopentadienyl carbonyl $(C_5H_5)V(CO)_4$ is prepared from $(C_5H_5)_2V$ and carbon monoxide (877), or from $[V(CO)_6]^-$ and sodium cyclopentadienyl in the presence of mercury(II) and using iron, ruthenium or osmium catalyts (2376). The structure (2385) shows that the cyclopentadienyl ring is disordered, having two possible orientations with respect to the carbonyl groups (Fig. 138).

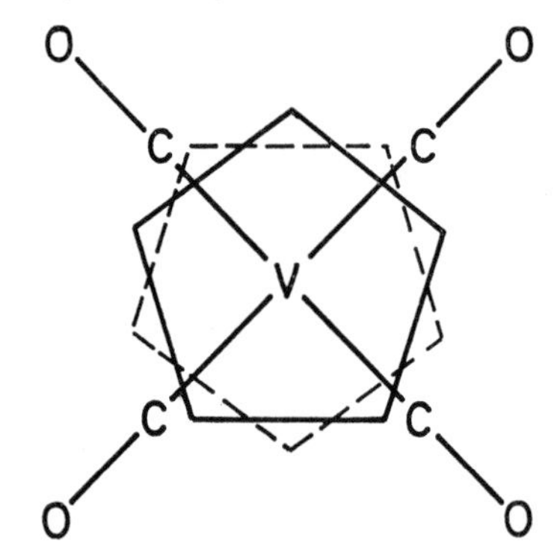

FIG. 138 The structure of $(C_5H_5)V(CO)_4$.

The niobium and tantalum analogues $(C_5H_5)M(CO)_4$ are similarly formed from the $[M(CO)_6]^-$ anions (1392, 2376).

One carbonyl group can be displaced from $(C_5H_5)V(CO)_4$ by phosphines forming $(C_5H_5)V(CO)_3(R_3P)$, where R is H (866), Bu (873) or Ph (2307).

Two molecules of carbon monoxide can be replaced by butadiene, dimethylbutadiene or cyclohexyl-1,3-diene forming the diamagnetic $(C_5H_5)V(CO)_2$(diene) (862). Cycloheptatriene forms paramagnetic $(\pi\text{–}C_5H_5)V(\pi\text{–}C_7H_7)$ ($\mu_{eff} = 1{\cdot}69$ B.M.) (1412), in which all twelve vanadium–carbon bond lengths are approximately equal (2·24 Å) (877).

Diphenylacetylene and $(C_5H_5)M(CO)_4$ form $(C_5H_5)M(CO)_2(Ph_2C_2)$ (where M is V or Nb) and $(C_5H_5)M(CO)(Ph_2C_2)_2$ (where M is V, Nb or Ta) (1771, 1772). Similar compounds of the type $(C_5H_5)V(CO)_2(RC_2H)$ are obtained with monoalkyl substituted alkynes (2307). Since two two-electron donor carbonyl groups have been replaced by one acetylenic group, the metal atom is electron deficient unless the acetylene is donating four electrons, that is it can be considered as an olefinic σ bonded bidentate with simultaneous donation of the π electrons. The structure (1172) of $(C_5H_5)Nb(CO)(Ph_2C_2)_2$, and some of its products, are shown in Fig. 139. The PhC–CPh bond is fairly long (1·35 Å) as expected, and similarly the Nb–C bond length of 2·19 Å is unusually short (for example compared with 2·46 Å for the niobium–carbon distances to the cyclopentadienyl ring).

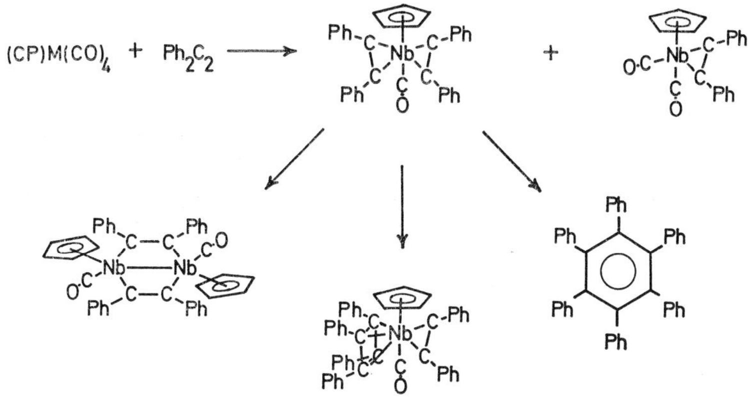

FIG. 139 Structures of the products from the reaction between $(C_5H_5)Nb(CO)_4$ and Ph_2C_2.

Refluxing $(C_5H_5)Nb(CO)(Ph_2C_2)_2$ in toluene forms the diamagnetic dimer $[(C_5H_5)Nb(CO)(Ph_2C_2)]_2$. The mass spectrum shows a peak due to $(C_5H_5)_2Nb_2$ but none to $(C_5H_5)Nb$, suggesting a very stable niobium–niobium bond (1771). This is confirmed by the structure which shows a short niobium–niobium bond length of 2·74 Å indicating multiple niobium–niobium bonding (Fig. 139). Each acetylene molecule bridges two metal atoms as shown, and have even longer carbon–carbon bonds (1·39 Å) than in the above molecule. Refluxing the same compound, $(C_5H_5)Nb(CO)(Ph_2C_2)_2$ in benzene forms $(C_5H_5)Nb(CO)(Ph_2C_2)(\pi\text{-}Ph_4C_4)$ (Fig. 139) (1091). On further heating fusion of the acetylenic units increases further with the formation of hexaphenylbenzene.

Other alkyne compounds include the paramagnetic $(C_5H_5)_2V(MeOOC.C{\equiv}C.COOMe)$ and $(C_5H_5)_2V(F_3C.C{\equiv}C.CF_3)$ obtained from dicyclopentadienyl vanadium and the alkyne (2306).

Reduction of $(C_5H_5)V(CO)_4$ with sodium amalgam in tetrahydrofuran forms $Na[(C_5H_5)V^{-I}(CO)_3]$, which with aqueous acid forms

$(CO)_2(C_5H_5)V\overset{CO}{\frown}V(C_5H_5)(CO)_2$ (873, 877). One of the terminal carbonyl groups may be replaced by triphenylphosphine forming

$(CO)_2(C_5H_5)V\overset{CO}{\frown}V(C_5H_5)(CO)(Ph_3P)$. However tributylphosphine and diethylphenylphosphine form equimolar mixtures of $(C_5H_5)V(CO)_3(R_3P)$ and $(C_5H_5)V(CO)_2(R_3P)_2$.

Vanadium dibenzene, $[V^{\circ}(\pi\text{-}C_6H_6)_2]$ and related arene compounds can be prepared by the action of aluminium/aluminium trihalide on vanadium tetrachloride or trichloride in benzene (404, 860), or by the action of phenylmagnesium bromide on vanadium tetrachloride or trichloride in tetrahydrofuran (1486). The red-brown compound is monomeric in benzene, and has an effective magnetic moment of 1·73 B.M. at room temperature. Vanadium hexacarbonyl reacts with benzene, and other arenes, forming $[V^{I}(CO)_4(\text{arene})][V^{-I}(CO)_6]$ (405, 406), whereas vanadium diarenes and vanadium hexacarbonyl or carbon monoxide form $[V^{I}(\text{arene})_2][V^{-I}(CO)_6]$ (404, 408). Reduction of $[V(CO)_4(C_6H_6)]^+$ with borohydride forms $[V(CO)_4(C_6H_7)]$, where C_6H_7 represents the π-cyclohexadienyl ring (407).

Cyclooctatetraene forms $[V(C_8H_8)_2]$ (322).

The treatment of niobium chloride with hexamethylbenzene, aluminium trichloride and aluminium in the melt forms the brown, diamagnetic $[Nb(C_6Me_6)Cl_2]_x$ and the green diamagnetic $[Nb_3(C_6Me_6)Cl_6]_xCl_x$ (871). These compounds clearly contain strong niobium–niobium bonding.

A preliminary report describes the reduction of niobium and tantalum pentabromides with phenyl lithium to form $Li_4[M(Ph)_6]n\,Et_2O$ ($n = 3 \cdot 5$–4) (1989). These compounds are extremely unstable to air and water. The ether is retained on recrystallization of the complexes from benzene.

CHAPTER 4

Group VI—Chromium, Molybdenum, Tungsten

1. OXIDATION STATE VI

A. Oxides

1. *Trioxides*

An early determination of the crystal structure of CrO_3 shows the formation of infinite chains formed from CrO_4 tetrahedra sharing corners, with Cr–O $\sim 1 \cdot 8$ Å (400).

The stereochemistry of the metal atom in MoO_3 can be considered to be intermediate between that of a tetrahedron and that of an octahedron, the molybdenum-oxygen distances being approximately $2(1 \cdot 7) + 2(2 \cdot 0) + 2(2 \cdot 3)$ Å. The structure can thus be considered to consist of infinite strings of corner sharing tetrahedra (Fig. 140a), or of sheets

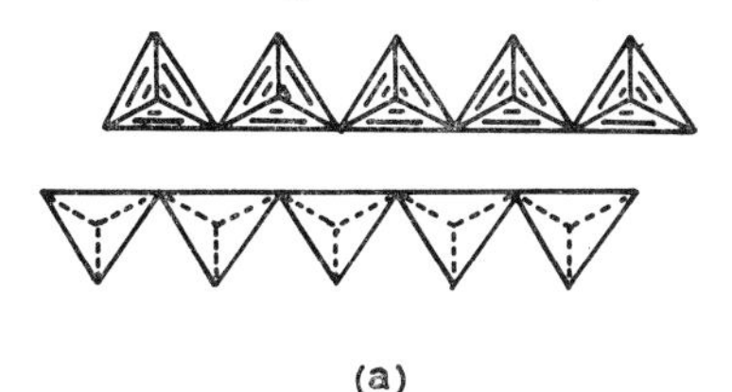

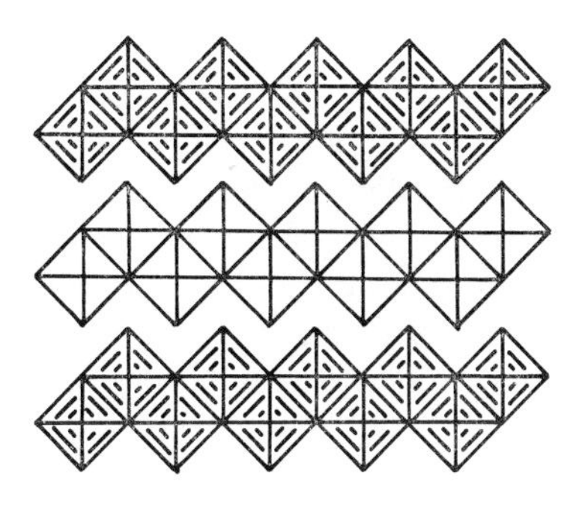

FIG. 140 The structure of MoO_3. (a) Viewed as infinite strings of corner shared tetrahedra. (b) Viewed as complete octahedra sharing edges and corners to form continuous sheets. Three such sheets are shown.

formed from octahedra sharing edges with two adjacent octahedra (Fig. 140b), and corners with two other octahedra (in and out of the page of Fig. 140b) (1381). Such a layer structure is very unusual for an oxide. The oxygen atoms can be considered to be in a distorted cubic close packing.

The structure of WO_3 is composed of more regular octahedra, tungsten-oxygen bond distances being $2\,(1{\cdot}78) + 2\,(1{\cdot}89) + 2\,(2{\cdot}11)$ Å (1573). The octahedra share all corners with neighbouring octahedra to form an ReO_3-type structure, but the antiparallel displacement of the tungsten atoms to the edges of the octahedral cages of oxygen atoms results in a

FIG. 141 The structure of WO_3.

puckering of the close packed layers of oxygen atoms (Fig. 141). At about 750° C there is a phase change corresponding to the antiparallel displacement of the tungsten atoms to the octahedral corners (1628). The undistorted ReO_3 structure is stabilized in the bronzes $WO_{3-x}F_x$ (2164) and Na_xWO_3 (see later).

Just as WO_3 can be considered to be the tungsten-oxygen skeleton remaining after the removal of sodium from the cubic bronze Na_xWO_3, so mixed molybdenum-tungsten oxides can be considered to be the skeleton remaining after removal of the potassium from the hexagonal bronze K_xWO_3 (Fig. 153, page 268). The structures of $MoW_{11}O_{36}$ and $MoW_{14}O_{45}$, prepared by sublimation from the more involatile WO_3 containing molybdenum impurities, are shown in Fig. 142 (1042). It is tempting to speculate that WO_3 grown from the vapour may also have this structure.

2. *Reduced molybdenum and tungsten oxides*

Reduction of MoO_3 with molybdenum at about 700° C, or WO_3 with tungsten at about 1000°C, forms intensely coloured (generally blue) mixed valence compounds, and finally the dioxides. The pentoxides M_2O_5 apparently do not exist. The same compounds are formed simply by heating the trioxides *in vacuo*, molybdenum oxide having the higher oxygen partial pressure.

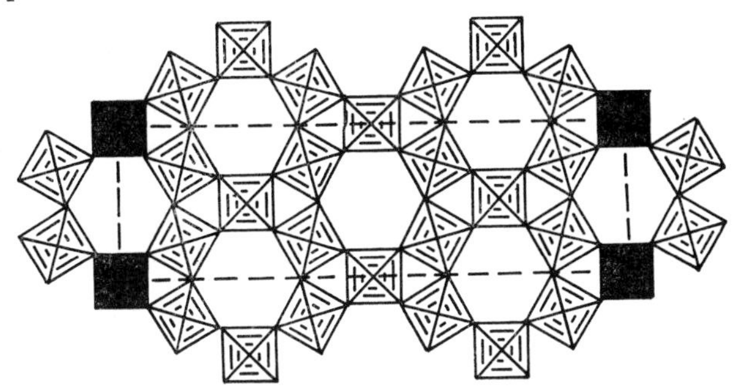

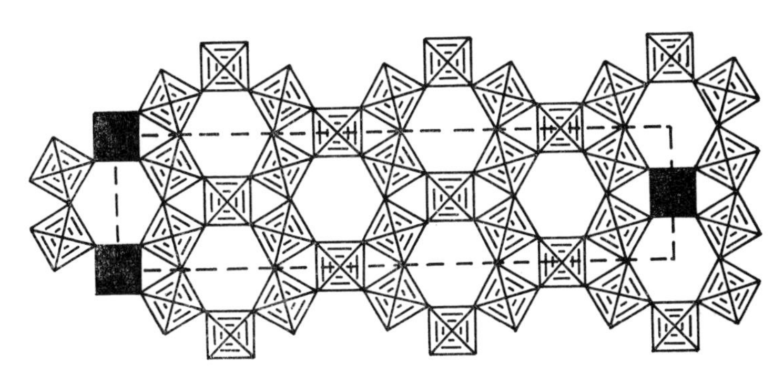

FIG. 142 The structure of $MoW_{11}O_{36}$ and $MoW_{14}O_{45}$ (half unit cell only) showing the corner sharing of WO_6 octahedra (hatched) and MoO_6 octahedra (black). Each layer shares octahedral corners with identical layers above and below the plane of the page.

The diversity of structures of these intermediate oxides is quite remarkable, and can be classified into three distinct types:

(a) Structures where blocks of corner sharing octahedra share some octahedral edges with similar blocks. These will be referred to as "shear" structures.

(b) Structures containing both six and seven coordinate metal atoms.

(c) Structures containing both four and six coordinate metal atoms.

(a) *Shear structures* (1627). A typical example of this type is Mo_8O_{23}, the (idealized) structure of which is shown in Fig. 143 (1619). The crystal can be considered to be composed of boards of MO_6 octahedra sharing corners, with the length parallel to the *b*-axis (normal to the plane of the page of Fig. 143). If the two side octahedra share edges with the two side octahedra of the next board, a characteristic block of four edge sharing octahedra is formed (for example, see the centre of the unit cell in Fig. 143). A homologous series of the general formula M_nO_{3n-1} is

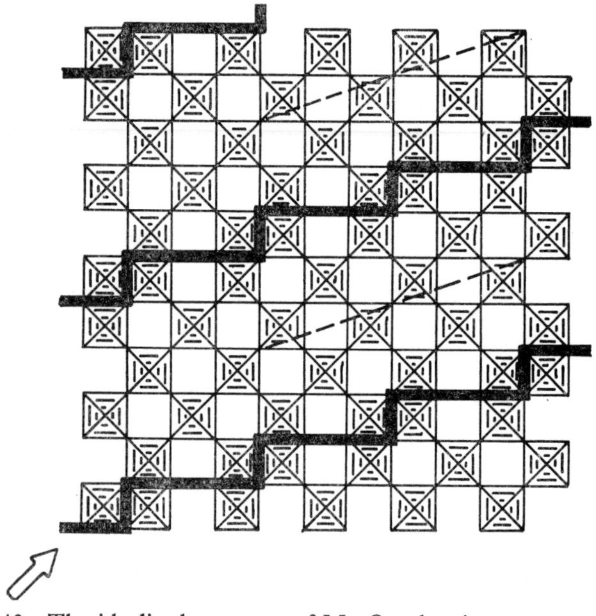

FIG. 143 The idealized structure of Mo_8O_{23} showing corner and edge sharing of MoO_6 octahedra. The shear planes which interrupt the perfect ReO_3 structure are shown by the heavy lines.

thus possible, where n is the width of the board in the direction of the arrow shown in the figure. Other homologues in this series can be made using mixed molybdenum-tungsten oxides $(Mo_xW_{1-x})_nO_{3n-1}$, where the value of n decreases and the number of edge shared octahedra increases, as the fraction of molybdenum increases relative to tungsten (249, 1633):

x	0–0·1	0·1–0·3	0·3–0·45	0·45–0·52	>0·52
n	8, 9	10	12	14	>14
b(Å)	4·05, 4·03	4·00	3·96	3·94	3·94–3·77

The unit cell dimension corresponding to the corner sharing of octahedra normal to the plane of Fig. 143, which is a measure of the puckering of these layers, decreases as the molybdenum is progressively replaced by tungsten.

A second homologous series of the type M_nO_{3n-2} is formed if the three side octahedra of each board share edges with the adjacent board forming a characteristic block of six edge sharing octahedra. Figure 144 shows the idealized structure of $W_{20}O_{58}$, where the boards are 20 octahedra wide in the direction of the arrow (1624).

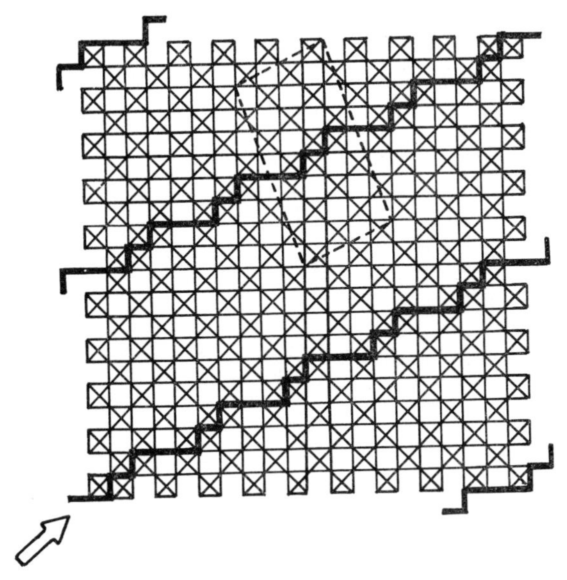

FIG. 144 The idealized structure of $W_{20}O_{58}$.

Shear structures can also be obtained by the replacement of some of the tungsten(VI) ions with tantalum(V) ions, rather than reduction to tungsten(V) ions as above. Thus $W_{35}Ta_4O_{115}$ has the structure expected for $W_{39}O_{115}$ (969).

An alternative way of viewing these structures is to consider that periodic dislocations run through a crystal of ReO_3 type structure, caused by sliding one block relative to another. The term "shear structures" however should not presuppose this mechanism of formation.

At the shear plane the structural elements are the same as in MoO_2 and WO_2, but there are no metal–metal bonds. For example metal–metal distances are 3·23–3·28 Å in the molybdenum oxides and 3·28–3·33 Å in the tungsten oxides, compared with 2·51 and 2·49 Å in MoO_2 and WO_2.

Shear structures can also be based on the edge sharing of MoO_3 structural elements rather than of WO_3 structural elements. The structures are less easy to represent in two dimensions, but Fig. 145 shows the linking of MoO_3 units by the edge sharing of the last three octahedra in each unit. The length of each unit is again related to the stoicheiometry, for example in $Mo_{18}O_{52}$ the MoO_3 units are 18 octahedra long (1384). The real structure at the shear plane is not exactly as represented in the idealized Fig. 145, as the second last metal atom of each unit is located in a tetrahedral site rather than an octahedral site, reducing the number of shared polyhedron edges.

(b) *Structures containing both six and seven coordinate metal atoms.* The structures of $Mo_{17}O_{47}$ ($MoO_{2\cdot77}$) (1380), $W_{18}O_{49}$ ($WO_{2\cdot72}$) (1623), and $Nb^{V}_{16}W^{VI}_{18}O_{94}$ ($MO_{2.77}$) (2163) are very unusual.

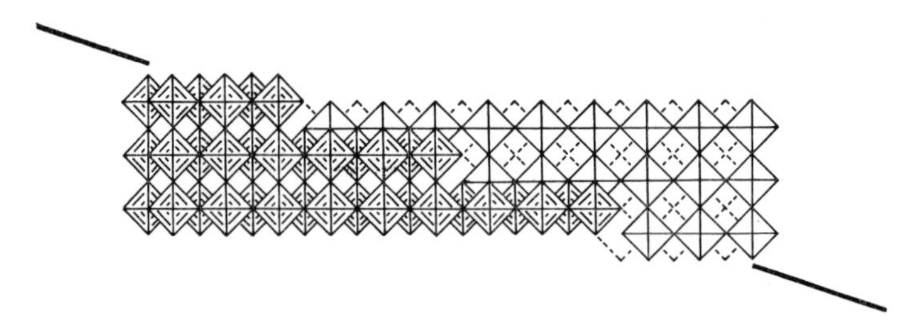

FIG. 145 The structure of the shear plane in $Mo_{18}O_{52}$ shown running through one of the sheets (in the plane of the page) of the MoO_3 structure. The portion of the sheet on the left hand side of the shear plane (the direction of which is shown by the dark line) is shaded, while that on the right hand side of the shear plane is unshaded. In $Mo_{18}O_{52}$ each zig-zag row contains 36 octahedra between consecutive shear planes, the last six of which share edges with similar octahedra across the shear plane.

Reference must first be made to the structure of the tetragonal potassium tungsten bronze, K_xWO_3 (Fig. 152, page 267), which is built up of WO_6 octahedra sharing all corners. Groups of five octahedra are grouped around a hole, and the stacking of these on top of each other forms channels which accommodate the potassium ions.

In $Nb_{16}W_{18}O_{94}$ one third of these channels are occupied by MO groups. The oxygen atoms occupy the potassium sites of K_xWO_3, so that the stereochemistry about these additional metal atoms is pentagonal bipyramidal, each pentagonal bipyramid sharing the five equatorial edges with the five neighbouring octahedra (Fig. 146). The reduced $Ta^{V}_{8}W^{V}_{8}W^{VI}_{18}O_{94}$ appears identical.

These compounds could be considered to be members of a homologous series $(MO)_x(MO_3)$, the value of x depending upon the degree of occupancy

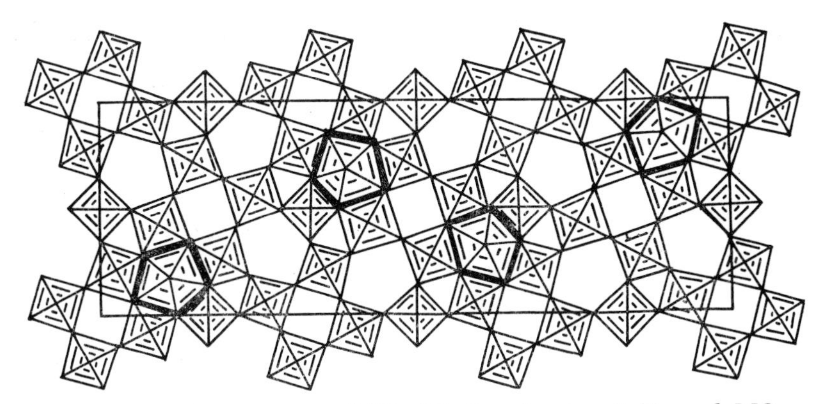

FIG. 146 The structure of $Nb_{16}W_{18}O_{94}$ showing linking of MO_6 octahedra and MO_7 pentagonal bipyramids. Each layer shares polyhedral corners with identical layers above and below the plane of the page. The unit cell contains three unit cells of tetragonal K_xWO_3 shown in Fig. 152.

FIG. 147 The structure of Mo_5O_{14} showing linking of MoO_6 octahedra and MoO_7 pentagonal bipyramids. Each layer shares polyhedral corners with identical layers above and below the plane of the page.

of the pentagonal channels. For low degrees of occupancy, the filled tunnels appear to be randomly distributed.

There are a number of other ways MO_6 octahedra can link up to form pentagonal channels running through the crystal, some or all of which can accommodate metal atoms with pentagonal bipyramidal stereochemistry. The structures of Mo_5O_{14} (1383), $Mo_{17}O_{47}$ (1380) and $W_{18}O_{49}$ (1623) are shown in Figs. 147, 148 and 149 respectively.

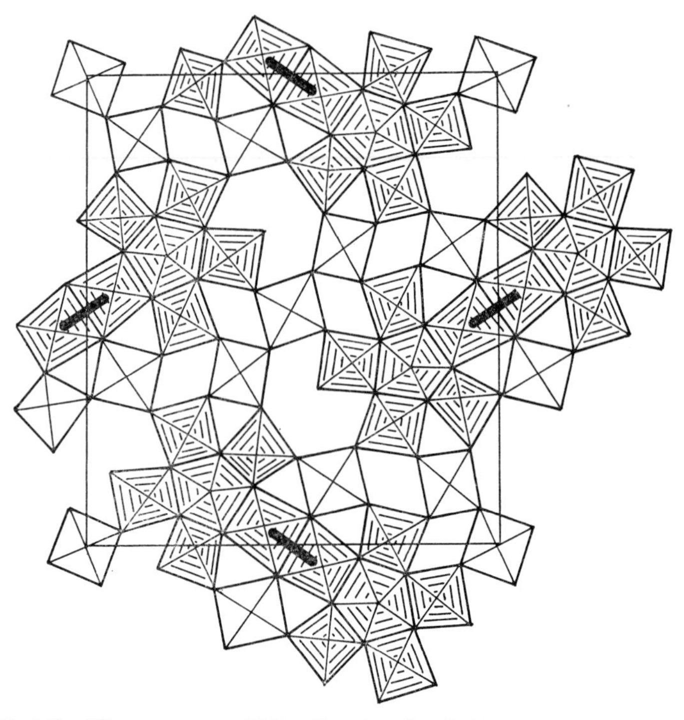

FIG. 148 The structure of $Mo_{17}O_{47}$ showing linking of MoO_6 octahedra and MoO_7 pentagonal bipyramids. The short molybdenum–molybdenum distances are shown by the bold lines. Each layer shares polyhedral corners with identical layers above and below the plane of the page.

The Mo_5O_{14} structure is also found for mixed vanadium-molybdenum and molybdenum-tungsten oxides (1387).

In $Mo_{17}O_{47}$ and $W_{18}O_{49}$ these units formed by the six edge shared polyhedra (shaded in the figures), are further linked into pairs by the edge sharing of octahedra. These two octahedra are strongly distorted, and the metal-metal distances (shown by the bold lines) are exceptionally short, Mo–Mo = 2·63 Å and W–W = 2·60 Å. These are only slightly longer than the metal–metal bond length in the dioxides, and 0·7–0·8 Å shorter than the other metal–metal distances in these structures.

The distortion of the MoO_6 octahedra observed in the higher molybdenum oxides is again found in Mo_5O_{14} and $Mo_{17}O_{47}$, where the layers are puckered by $\pm 0 \cdot 3$ Å. This is again reflected in a comparison of the relevant cell dimension with that of the undistorted tungsten compound:

Mo_5O_{14}	3·94 Å
$Mo_{17}O_{47}$	3·95 Å
$W_{18}O_{49}$	3·77 Å

(c) *Structures containing both four and six coordinate metal atoms.* The other possible way an MO_3 structure composed of corner shared octahedra can become oxygen deficient, apart from edge sharing as described above, is to replace some MO_6 octahedra with MO_4 tetrahedra.

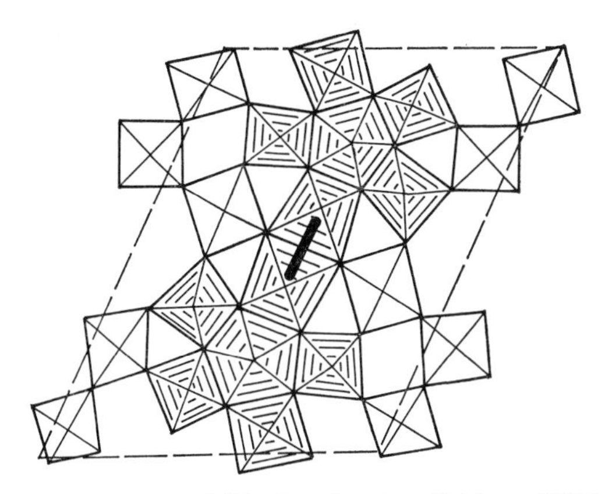

FIG. 149 The structure of $W_{18}O_{49}$ showing linking of WO_6 octahedra and WO_7 pentagonal bipyramids. The short tungsten-tungsten distance is shown by the bold line. Each layer shares polyhedral corners with identical layers above and below the plane of the page.

This is observed in both the orthorhombic and monoclinic modifications of the red-violet Mo_4O_{11} (Figs. 150a and 150b respectively). In the former infinite zig-zag strings of three octahedra and one tetrahedron run along the x-axis, complete corner sharing being attained by linking with parallel strings. In the monoclinic modification these strings are almost linear rather than zig-zag (1382, 1620).

These structures are very spacious, as is illustrated by comparing the average volume occupied by an oxygen atom in these structures (20·2 Å^3) with that in other molybdenum oxides (16–19 Å^3) (neglecting the size of the molybdenum atoms) (1591):

	Average volume per oxygen atom (Å³)
MoO_3	16·9
$Mo_{18}O_{52}$	17·1
Mo_8O_{23}	19·1
Mo_5O_{14}	18·6
$Mo_{17}O_{47}$	16·8
MoO_2	16·3
Mo_4O_{11}	20·2

3. *Bronzes*

(a) *Introduction.* In 1824 it was discovered that reduction of sodium polytungstate with hydrogen at red heat yielded a chemically inert substance, which because of its metallic, bronze-like appearance, was called a bronze (2409). Other properties of these compounds which are characteristic of metals include the high electrical conductivity which decreases with increasing temperature.

The structure of the sodium tungsten bronze of high sodium content

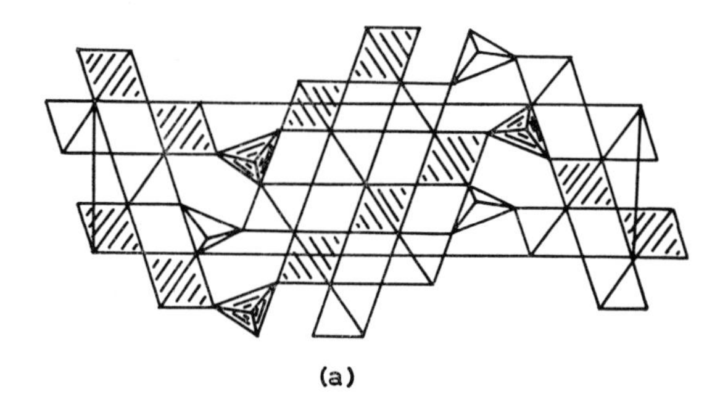

(a)

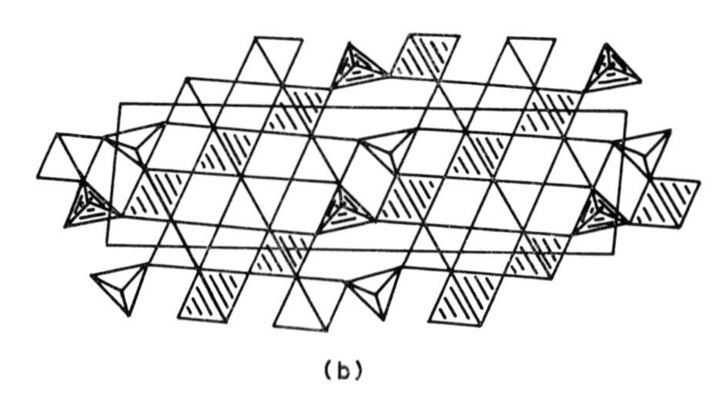

(b)

FIG. 150 The structures of Mo_4O_{11}. (a) Orthorhombic modification. (b) Monoclinic modification.

shows it to be a non-stoicheiometric compound Na_xWO_3, where x can have any value between 0·3 and 1·0 (1110).

A number of non-stoicheiometric compounds of tungsten and other early transition metal oxides are also known; they are brightly coloured and of metallic appearance and are usually referred to as bronzes, although they do not all necessarily exhibit metallic conduction.

The general method of preparation is to heat mixtures of the appropriate normal tungstate with tungsten. The tungstate M_2WO_4 can be replaced by unspecified polytungstates or by mixtures of tungstate and WO_3, and the tungsten can be replaced by other reducing agents such as WO_2, hydrogen, or electrolytic reduction. The resulting bronzes are extremely chemically inert, and they are normally obtained pure from the reaction mixtures by alternately washing with alkali and concentrated boiling acids, the latter including hydrofluoric, hydrochloric, perchloric, nitric, and aqua regia.

Na_xWO_3 is soluble in molten Na_2WO_4 and single crystals can be conveniently prepared from solution by slow cooling (327), which produces the almost stoicheiometric, bright yellow, $NaWO_3$. The sodium content can then be conveniently lowered by heating the crystal embedded in some solid with a capacity to absorb sodium, for example iodine or WO_3, without destroying the single crystal. Alternatively the cation concentration can be increased by heating in the presence of metal vapour.

Using these techniques the composition and structure of tungsten bronzes has been systematically studied, and a number of structures found. They are all based on a rigid tungsten–oxygen framework formed from WO_6 octahedra sharing all corners, so that the W:O ratio is 1:3, and the variable composition is exclusively connected with the alkali metal content. They are normally all written as $M^I_xWO_3$, but it should be noted that the number of cation sites, which is equal to the theoretical maximum value of x, depends upon the particular structure.

(b) *Structures of alkali metal tungsten bronzes.* Bronzes containing relatively large amounts of lithium or sodium crystallize in fairly large crystals of the perovskite structure (Fig. 151). The tungsten–oxygen framework is the same as that in WO_3, although the latter is somewhat distorted. For the sodium tungsten bronze Na_xWO_3, the upper limit of x is close to unity, but as the sodium content diminishes the lattice contracts until x is about 0·3, when the crystals disintegrate with the formation of a new phase with a tungsten–oxygen skeleton different to that in $NaWO_3$ and WO_3. Bronzes with low values of x and having a degenerated perovskite structure intermediate between the perovskite structure and WO_3 will be described later.

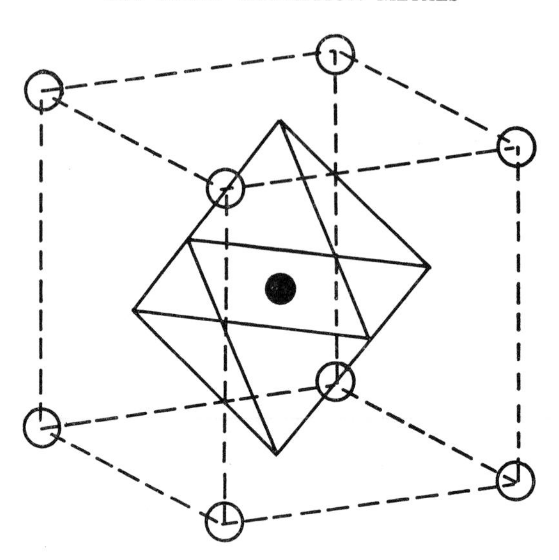

FIG. 151 The perovskite structure of Na_xWO_3. The sodium atoms (open circles) are situated in the spaces formed by the corner sharing of octahedrally coordinated tungsten atoms (black circles).

The unit cell dimensions of the cubic perovskite phase is a linear function of its composition (350, 2354):

$$a(\text{Å}) = 3{\cdot}785 + 0{\cdot}082x$$

These sodium bronzes show a remarkable series of vivid colours which depend upon composition as shown:

x = 1·0	bright yellow
x = 0·9	yellow
x = 0·8	orange
x = 0·7	red
x = 0·6–0·5	violet
x = 0·5–0·4	dark blue

The diffuse reflectance spectra show a single broad absorption peak, the wavelength of which is given by (350):

$$\lambda_{max}(\text{Å}) = 7488 - 3387x$$

Lithium forms an isomorphous series of dark blue to blue-violet tungsten bronzes (1634, 2238). The range of composition is lower than for the corresponding sodium bronze, being from x = 0·3 to x = 0·56. The unit cell is smaller than the sodium bronzes, but in this case the size of the unit cell decreases with increasing x and is given by (2354):

$$a(\text{Å}) = 3 \cdot 785 - 0 \cdot 134x$$

The cubic K_xWO_3 is found only at high pressures (236, 327).

The red-violet potassium bronze K_xWO_3 ($x = 0 \cdot 4 - 0 \cdot 6$) has the rather intricate tetragonal structure shown in Fig. 152 (1622). The ideal

FIG. 152 The structure of the tetragonal tungsten bronzes. The potassium atoms (circles) are situated in the spaces formed by WO_6 octahedra sharing all corners.

formula is $K_6W_{10}O_{30}$ or $K_{0 \cdot 6}WO_3$. The structure shows two different potassium sites. One third are of a smaller type (centre and corners of the unit cell shown), and although they appear similar to the holes in the perovskite structure, they are significantly larger due to the distortion of the octahedra. It is presumed that in the potassium deficient structures, these holes are preferentially unoccupied. Both the a and c cell dimensions are increased in approximately the same ratio as x is increased.

A tetragonal sodium bronze with structure related to the tetragonal potassium bronze is found at low sodium content (1621), although it may appear strange that the lattice accommodates less of the smaller sodium atom. However the lattice is of lower symmetry, being a superlattice of the tetragonal potassium bronze due to a puckering of the tungsten sheets. The cell dimensions are smaller than for the potassium analogue, and do not vary so much with composition. The tetragonal bronze is a brighter blue than the perovskite bronze, and occurs as needles rather than cubes.

Sodium and potassium also form a complete series of tetragonal bronzes $(Na, K)_xWO_3$, where the cations are interchangeable (327).

Rubidium and caesium have only been found to form the dark blue hexagonal tungsten bronze (Fig. 153), in which the alkali metal sites are found between rings formed from six WO_6 octahedra, in contrast to five (and four) of the potassium bronze and four of the sodium and lithium cubic bronzes. The stoicheiometry range has not been determined, the only compounds which have been isolated being $Rb_{0\cdot27}WO_3$ and $Cs_{0\cdot32}WO_3$ (1626, 1632). The theoretical formula corresponding to the occupation of all alkali metal sites is $M^I_{0\cdot33}WO_3$. Potassium also forms this bronze when x falls below 0·33. Mixed sodium-potassium (327) and lithium-potassium

FIG. 153 The structure of the hexagonal tungsten bronzes. The alkali metal atoms (circles) are situated in the spaces formed by WO_6 octahedra sharing all corners.

(140) bronzes with this structure can also be formed. In the latter case the lithium ions appear to occupy the smaller "triangular" holes shown in Fig. 153. Under conditions of hydrothermal synthesis lithium also forms the hexagonal $Li_{0\cdot3}WO_3$, where the lithium could again occupy the smaller holes with the larger holes empty, or possibly the larger holes are occupied by hydrated lithium ions (995).

Sodium tungsten bronze at low sodium content ($x \sim 0\cdot1$) forms a degenerated perovskite structure which is intermediate between the perovskite phase and WO_3 itself (1625). The colour changes from blue through green to yellow as the sodium content is decreased.

(c) *Properties of alkali metal tungsten bronzes.* The perovskite Na_xWO_3 has had its properties studied in some detail, although there remains some controversy about the properties themselves, and also their interpretation.

Although the details of the electronic structure are not simple, it is clear that there are Na^+ ions in the WO_3 lattice, and that the electrons arising from the ionization of the sodium atoms are accommodated in a conduction band.

The compounds have small temperature independent susceptibilities (1057, 1483, 2242) the magnitudes of which vary with stoicheiometry (for example, for $x = 0 \cdot 49$ and $0 \cdot 85$, $10^6 \chi_g = 0 \cdot 007$ and $0 \cdot 053$ cgsu respectively). This is too low to indicate the presence of tungsten(V) or tungsten(IV), but is in satisfactory agreement with calculations based on a metallic model.

The behaviour of the electrical conductivity is complex. Studies on single crystals of Na_xWO_3 (351, 773, 975, 1236, 1610) show that within the approximate limits of 120°K to 800°K and $x = 0 \cdot 4$ to $0 \cdot 9$, the conductivity decreases linearly with increasing temperature, which is typical for metallic conduction where the resistance is due to the thermal scattering of the atoms. The quantitative figures however vary considerably between different workers. The lowest value for the resistance was obtained (773) for crystals which were prepared by electrolysis and subsequently annealed, and which were carefully selected for their homogeneity of conduction throughout the crystal. The resistance found was only about three times that of sodium itself.

There has been some disagreement about the variation of conductance with composition. The earlier work (351, 975) indicated that the conductance passed through a maximum at approximately $x = 0 \cdot 75$. This was not due to a significant contribution from unionized sodium atoms which would contribute scattering centres without contributing conducting electrons, as measurements of the Hall coefficient showed that each sodium atom contributed one free electron throughout the entire stoicheiometry range (975). It was suggested that this maximum may be associated with the ordering of the sodium atoms observed at this composition by neutron diffraction measurements (95). However the most precise work (773), with the carefully selected crystals referred to above, showed a continuous increase in conductivity with increase in x (although even this evidence is not universally accepted (1608)). It was suggested that the fall-off at high x previously observed was due to the inclusion of sodium tungstates within the crystal, for which there was some microscopic evidence.

The electrical conductivity of cubic Li_xWO_3 also shows metallic type conduction (2150), the values for the conductance and electron mobility being similar to those for Na_xWO_3.

The n.m.r. spectra of sodium (1316) and tungsten (1755) in Na_xWO_3, and of lithium (1316) in Li_xWO_3 show negligible chemical shifts relative to the alkali metal chlorides and WO_3. This would appear to eliminate the possibility of band formation from Na(3*s*) or W(6*s*) orbitals, but the results are compatible with band formation from Na(3*p*), W(5*d*) or W(5*d*)–O(2*p*) orbitals.

Tetragonal $K_{0\cdot5}WO_3$ has a significantly higher magnetic susceptibility than the perovskite phases, and is intermediate between that expected for a completely delocalized electron model, and the model where the electron is localized on the tungsten atom (1483, 2139). However high metallic-type electrical conductivity is observed (2139).

Hexagonal Rb_xWO_3 and K_xWO_3 have low magnetic susceptibilities and high metallic-type electrical conductivities consistent with delocalization of electrons in a conduction band (2139).

A polycrystalline sample of the degenerated perovskite $Na_{0\cdot025}WO_3$ has a low conductivity, and the increase in conductance with temperature confirms non-metallic properties (1610).

FIG. 154 Mo_6O_{22} (a) and $Mo_{10}O_{36}$ (b) groups found in $K_{0\cdot26}MoO_3$ and $K_{0\cdot28}MoO_3$.

(d) *Other bronzes.* A number of other elements are capable of forming A_xWO_3 which are also termed bronzes in view of their non-stoicheiometry and chemical inertness. However not all have high electrical conductivity or are isomorphous with one of the alkali metal tungsten bronzes. In the above formula, A may represent Mg, Ca, Sr, Ba, rare earths, actinides, Cu, Ag, Cd, Ga, In, Tl, Sn or Pb (273, 373, 539, 540, 541, 995, 1316, 1821, 2138, 2313, 2354).

Alkali metal molybdenum bronzes analogous to the tungsten compounds can only be formed under high pressure. The cubic $Na_{0\cdot9}MoO_3$, $K_{0\cdot9}MoO_3$ and tetragonal $K_{0\cdot5}MoO_3$ are metallic, whereas the hexagonal $Rb_{0\cdot3}MoO_3$ shows only semiconductor properties (236).

The electrolytic reduction (2408) of molybdenum trioxide-alkali metal molybdate melts forms $Na_{0\cdot9}Mo_6O_{17}$ with a distorted perovskite structure (2214), the red semiconductor $K_{0\cdot26}MoO_3$, and the blue metallic $K_{0\cdot28}MoO_3$. The structures of the last two compounds show discrete Mo_6O_{22} groups formed from six octahedra sharing edges (2215), and $Mo_{10}O_{36}$ groups formed from ten octahedra sharing edges (1043), respectively (Fig. 154). These blocks then share corners with each other.

B. Hexahalides

1. *Fluorides*

The reaction between chromium and fluorine under the severe conditions of 400° C and 350 atm of fluorine produces the lemon-yellow CrF_6 (1012). At lower pressures only the light red pentafluoride is formed. Under normal pressures, chromium hexafluoride dissociates above −80° C to the pentafluoride and fluorine.

Molybdenum and tungsten hexafluorides are formed by the direct action of fluorine on the metals (1806, 1963). Other fluorinating agents which have been used include ClF_3 (1789), BrF_3 (595) and SF_4 (1816), and the trioxides have been used in place of the metals. The tungsten compound has also been formed from $W(CO)_6$ and ReF_6 (1136). A particularly convenient synthesis of tungsten hexafluoride is the electrical explosion of tungsten filaments in sulphur hexafluoride, since the latter is unreactive and easy to handle under normal conditions (1310).

Molybdenum and tungsten hexafluorides are colourless solids, liquids or gases at about room temperatures, the tungsten compound being more volatile than the molybdenum one (141, 402, 1961):

$$MoF_6\text{: M.P.} = 17°\text{ C} \quad \text{B.P.} = 34°\text{ C}$$
$$WF_6\text{: M.P.} = 2°\text{ C} \quad \text{B.P.} = 17°\text{ C}$$

The vapours are monomeric (1806). Electron diffraction and Raman and infrared spectra of the gases indicate octahedral structures as expected (386, 483, 984, 2098, 2262).

2. *Chlorides and bromides*

Tungsten hexachloride can be prepared from the elements at 500–600° C, or by the chlorination of WO_3 with excess CCl_4 in a bomb at about 400° C (at lower temperatures $WOCl_4$ is formed) (784, 1433). Tungsten hexachloride is a blue-black substance, M.P. = 283° C (1794), which is very unstable to air.

Molybdenum hexachloride has proven to be much more difficult to prepare and characterize. Even the pentachloride decomposes significantly at room temperature to the tetrachloride and chlorine (page 309). The low temperature chlorination of molybdenum hexacarbonyl with liquid chlorine at −78° C forms only $MoCl_2(CO)_4$, in contrast to the reaction with tungsten hexacarbonyl which yields the hexachloride (532).

Nevertheless it has been claimed that the black $MoCl_6$ can be obtained by careful fractional sublimation of the mixture of oxochlorides and lower chlorides, obtained by either direct chlorination of the metal or by chlorinating molybdic acid with refluxing thionyl chloride. The product is isomorphous with WCl_6 (1679).

The dark blue tungsten hexabromide can be formed directly from the elements, by the action of bromine on tungsten carbonyl at 0° C (1159), or by the action of boron tribromide on tungsten hexachloride at room temperature (694). The compound decomposes above 200° C in vacuo to the pentabromide and bromine.

Tungsten hexafluoride reacts with titanium tetrachloride at 5° C to give WCl_5F as a pale yellow liquid, M.P. = −34° C (519). It slowly disportionates at 25° C to WF_6 and uncharacterized chlorofluorides. Reaction between WF_6 and $TiCl_4$ at room temperature forms WCl_6, but with BCl_3 or WCl_6 the mixed WCl_3F_3 is formed (1809). This compound is thermally stable and can be sublimed under vacuum with little decomposition.

The analogous reactions with MoF_6 result in reduction with the formation of $Mo_2Cl_3F_6$ (page 323).

C. Reactions of the Hexahalides

No simple adducts of the hexahalides have been firmly established. The only formally tungsten(VI) compound of this type is the hexahydrido $WH_6(PhMe_2P)$ formed by the reaction of $[WCl_4(PhMe_2P)]$ and borohydride in methanol (1725). The compounds described as $WF_6,3Me_3N$ and WF_6,3py should be treated with caution due to the fluorinating and oxidising power of WF_6 (487).

Molybdenum hexafluoride is hydrolysed instantly by air or by imperfectly dried glassware, and reacts readily with organic matter such as tap grease. In spite of these difficulties some chemical reactions have been investigated (1807), for example:

$$2\,MoF_6 + PF_3 \rightarrow 2\,MoF_5 + PF_5$$

Tungsten hexafluoride appears more stable.

The reaction of MoF_6 or WF_6 with dry NaF forms $NaMF_7$ at 60°–80° C, and Na_2MF_8 at 150° C. The physical state of the sodium fluoride is apparently critical if the reaction is to proceed at a noticeable rate (595, 1335, 1336). The product obtained is dependent upon conditions, since the heptafluoro anions readily disproportionate to the octafluoroanions and the hexafluorides (1336):

$$2\,NaMF_7 \leftrightarrows MF_6 + Na_2MF_8$$

Similar heptafluoro and octafluoro complexes have been obtained using KF, RbF, CsF, NH_4F, NOF or NO_2F, either by themselves or by using iodine pentafluoride as an ionising solvent (487, 595, 987, 988, 1133, 1135). All compounds are white, readily hydrolysed in moist air, but stable to dry air.

The only evidence for complex haloanions, other than the fluorides, appears to be from the photometric or potentiometric titration of chloride ion against tungsten hexachloride dissolved in phosphorus oxychloride. An endpoint is obtained corresponding to $[WCl_7(POCl_3)_x]^-$ (104). Reduction of MoF_6 occurs with alkali metal and alkaline earth chlorides (1810).

Tungsten hexafluoride reacts with dimethylsulphite at —30° C to form $W(OMe)F_5$, which slowly decomposes at room temperature to WOF_4 and MeF (1795).

In many respects the reactions of tungsten hexachloride resemble those of molybdenum pentachloride. For example acetonitrile causes reduction to $WCl_4(MeCN)_2$, while oxygen abstraction as well as reduction occurs in tetrahydrofuran, the brown solution turning green and depositing the blue solid $WOCl_3(THF)_2$ (903). Many such products are described in the appropriate sections dealing with tungsten(V) and tungsten(IV).

The hexavalent state however is maintained for tungsten in a number of solvolysis reactions (for example $WCl_2(OPh)_4$ and $WCl_2(NH_2)_4$) as well as in the fluoro complexes described above.

Tungsten chloroalkoxides are formed by the addition of aliphatic alcohols to benzene solutions of tungsten hexachloride. Complete replacement of chloride is attained by the addition of anhydrous ammonia, forming the oxo compounds $WO(OR)_4$, where R is Me, Et, Pr, Bu or Bz (962). Similar compounds are obtained from $WOCl_4$ (956). The reaction of WCl_6 or $WOCl_4$ with phenol however forms the hexaphenoxide $W(OPh)_6$ (950, 956, 1723, 1887). The intermediates $WCl_2(OPh)_4$ and $WCl(OPh)_5$ have also been isolated. The hexaphenoxide is monomeric in benzene and surprisingly stable. It does not hydrolyse or exchange ligands under aqueous alkaline conditions, although it does so under acid conditions. The aromatic rings may be brominated, nitrated, and the nitro group reduced, without otherwise effecting the complex (1723). Steric hindrance by the six aromatic groups preventing direct attack on the metal atom is presumably responsible for this inertness.

A number of similar products with aliphatic and aromatic carboxylic acids of the general formula $W(RCOO)_6$ have also been prepared (1888).

Molybdenum oxotetrachloride is however reduced to molybdenum(V) with phenol (1505).

The action of ammonia on WCl_6 progressively forms $WCl_5(NH_2)$, $2NH_3$,

$WCl_4(NH_2)_2,2NH_3$, $WCl_3(NH_2)_3$ and $WCl_2(NH_2)_4$ (925). A number of organic amines have also been studied with both WCl_6 and $WOCl_4$ (333, 1886, 1889, 1890, 1891).

The reaction of tungsten hexachloride with lithium dimethylamide forms the orange $[W(NMe_2)_6]$ (284). The structure is octahedral, with planar $\genfrac{}{}{0pt}{}{C}{C}\!\!>\!N\!-\!W\!-\!N\!<\!\!\genfrac{}{}{0pt}{}{C}{C}$ groups.

Tungsten hexachloride reacts with thiocyanate in appropriate solvents to form $W(NCS)_6,2Me_2CO$, $W(NCS)_6,2MeCO.Et$ and $W(NCS)_6,2$ Dioxan (952). The formation of these adducts is in marked contrast to the behaviour of the hexachloride itself.

D. Oxohalides and Adducts

1. *Oxohalides*

(a) *Chromium*. The fluorination of chromium in a glass apparatus leads to the formation of $CrOF_4$, the oxygen atom coming from the fluorination of the glass (728). This appears to be the only known oxotetrahalide, and chromium(VI) compounds containing ligands other than oxygen or fluorine can apparently only be formed if there are two or more oxygen atoms also bound to the chromium.

The fluorination of CrO_3 or $K_2Cr_2O_7$ with, for example, HF (779), CoF_3 (887), SF_4 (1469), or SeF_4 (156), yields the red-violet CrO_2F_2. The compound melts at 31·6° C, but reaches a vapour pressure of one atm at 29·6° C; the vapour pressure at the triple point is 885 mm Hg (779). It is immediately hydrolysed by water, and vigorously oxidizes organic compounds.

Chromyl chloride, CrO_2Cl_2, is conveniently prepared by distillation of an alkali metal dichromate (or a mixture of an alkali metal chloride and chromium trioxide), in concentrated sulphuric acid. The chromium–oxygen stretching frequencies (980 and 995 cm^{-1}) are lower than for CrO_2F_2 (1005 and 1015 cm^{-1}) (152, 1682, 2204, 2334).

The mixed CrO_2FCl, and CrO_2Br_2, are also known (887).

(b) *Molybdenum and tungsten*. The known oxohalides of molybdenum(VI) and tungsten(VI) are shown below:

$MoOF_4$	MoO_2F_2	WOF_4	WO_2F_2		
$MoOCl_4$	MoO_2Cl_2	$WOCl_4$	WO_2Cl_2	$WOCl_3Br$	$WOCl_2Br_2$
—	MoO_2Br_2	$WOBr_4$	WO_2Br_2		
—	—	—	WO_2I_2		

These are now discussed in the order fluorides, chlorides, bromides, and iodide.

The oxofluorides $MoOF_4$, MoO_2F_2, WOF_4 and WO_2F_2 were first prepared as white solids by the reaction of the corresponding oxochloro compounds with anhydrous hydrogen fluoride (1962, 1964). The $MoOF_4$ and WOF_4 have also been prepared by the action of oxygen–fluorine mixtures on the metals (403), by the partial hydrolysis of the hexafluorides (1751), and from the reaction between WF_6 and SO_2 (487). The reaction of MoO_3 or WO_3 with SeF_4 forms compounds such as MoO_2F_2, SeF_4 and WOF_4, SeF_4 which can be sublimed unchanged (156), in contrast to CrO_3 which forms CrO_2F_2. The melting points and boiling points are given below (403, 1962, 1964):

$MoOF_4$	M.P. = 97° C	B.P. = 180° C
WOF_4	M.P. = 105° C	B.P. = 186° C

The structures of WOF_4 and two different modifications of $MoOF_4$ are composed of octahedrally coordinated metal atoms, each linked to two neighbouring octahedra through the sharing of two *cis* corners, but the way the octahedra link together is quite different in the three cases.

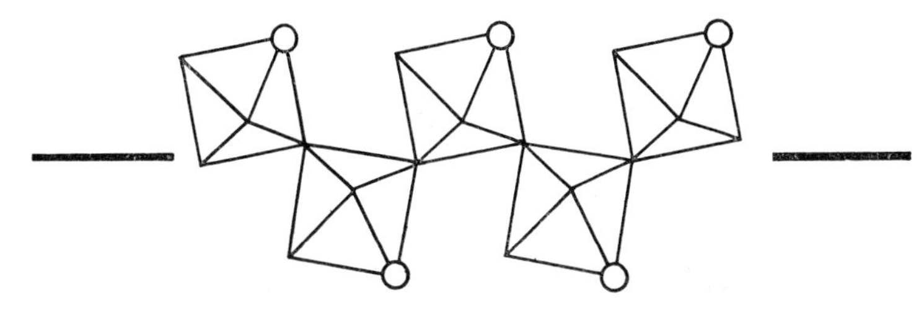

FIG. 155 The structure of $MoOF_4$ (I).

The structure of the first form of $MoOF_4$ consists of infinite strings of octahedra joined by bridging fluorine atoms (Fig. 155) (736, 742). The molybdenum–nonbridging fluorine bond length of 1·82 Å is the same as in MoF_6, while the molybdenum–oxygen bond length is 1·64 Å. The molybdenum-bridging fluorine bond *trans* to the oxygen atom is considerably longer than the one *cis* to the oxygen atom (2·29 and 1·94 Å respectively). The total effect can be described by considering the metal atom drawn 0·3 Å towards the oxygen atom inside a relatively undistorted octahedron of light atoms. The Mo–F–Mo angle is 151°. This structure is similar to that of VF_5, CrF_5, TcF_5, ReF_5 and $ReOF_4$.

A second form of $MoOF_4$ is isostructural with $TcOF_4$, and consists of a trimeric arrangement formed by the linking of MOF_5 octahedra through *cis*-fluorine atoms as before (Fig. 156) (734).

An X-ray structural determination on WOF_4 shows a tetrameric structure similar to that found for NbF_5, TaF_5 and MoF_5 (Fig. 95, page

157) (733, 736). This structure however does not distinguish whether the atoms bridging the metal atoms are oxygen or fluorine (177), although the latter is favoured because of the similarity of the infrared spectrum with that of $MoOF_4$.

All four oxychlorides $MoOCl_4$, MoO_2Cl_2, $WOCl_4$ and WO_2Cl_2 have been described. MoO_2Cl_2 and $WOCl_4$ are well known and were prepared early in the last century by the partial chlorination of the oxides, or by heating the metals with a mixture of chlorine and oxygen. However these and similar methods often lead to mixtures which must be separated by fractional sublimation, during which the desired compound may decompose, for example WO_2Cl_2 decomposes to WO_3 and $WOCl_4$. Particular methods have therefore been devised for these preparations.

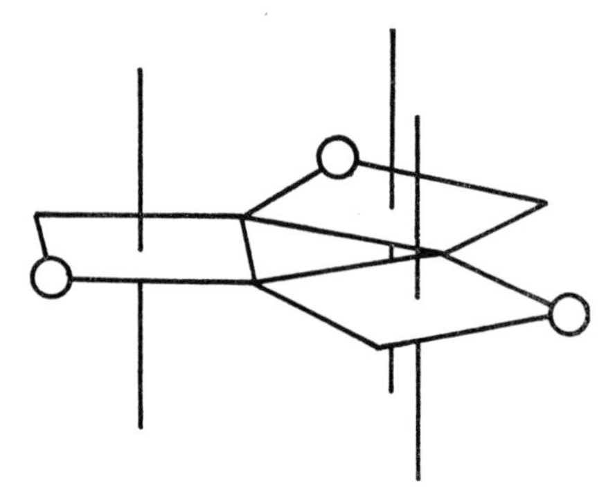

FIG. 156 The structure of $MoOF_4$ (II).

The partial chlorination of the oxide to form a particular oxychloride may be conveniently controlled either by the use of lower temperatures than those required for complete chlorination, or by using stoicheiometric quantities of chlorinating agent. For example:

(i) The reaction of tungsten trioxide with carbon tetrachloride at 300°–320° C under pressure, in the appropriate stoicheiometric ratios, leads to the following reactions:

$$WO_3 + CCl_4 \rightarrow WO_2Cl_2 + COCl_2$$
$$WO_3 + 2\,CCl_4 \rightarrow WOCl_4 + 2\,COCl_2$$
$$WO_3 + 3\,CCl_4 \rightarrow WCl_6 + 3\,COCl_2$$

At 250° C only $WOCl_4$ is formed, even with excess CCl_4 (783).

(ii) When CCl_4 vapour entrained in oxygen is passed over MoO_3 at 550° C, a quantitative yield of MoO_2Cl_2 is obtained (2432). The corresponding WO_2Cl_2 has been similarly obtained by passing CCl_4

entrained in oxygen over WO_3 at 370° C (2432), or CCl_4 entrained in HCl over WO_3 at 600° C (2073).

(iii) Alternatively higher boiling point chlorinated hydrocarbons may be used as the chlorinating agent. For example refluxing excess octachlorocyclopentene (B.P. = 285° C) with WO_3 forms only $WOCl_4$ (144).

(iv) Pure $WOCl_4$ is also obtained in almost quantitative yields by the reaction of WO_3 (1157) or H_2WO_4 (1679) with thionyl chloride:

$$WO_3 + 2\,SOCl_2 \rightarrow WOCl_4 + 2\,SO_2$$

Similarly the reaction of MoO_3 with refluxing thionyl chloride forms $MoOCl_4$ (536).

(v) The desired oxychloride can also sometimes be prepared by heating stoicheiometric ratios of the oxide and chloride, for example:

$$2\,WCl_6 + WO_3 \rightarrow 3\,WOCl_4$$
$$WCl_6 + 2\,WO_3 \rightarrow 3\,WO_2Cl_2$$

(vi) The reaction of $MoCl_5$ as a slurry in carbon tetrachloride with Cl_2O, which is a strong oxidizing agent in addition to chlorinating agent, forms $MoCl_4$:

$$2\,MoCl_5 + 2\,Cl_2O \rightarrow 2\,MoOCl_4 + 3\,Cl_2$$

The Cl_2O is prepared by the action of Cl_2 on HgO, the Cl_2O with excess Cl_2 is passed into the molybdenum chloride slurry at −30° C, which is slowly allowed to warm to room temperature. The reaction is convenient as it proceeds readily at room temperature, is free from side reactions, and gives no product other than $MoOCl_4$ and chlorine.

(vii) Tungsten hexachloride reacts with liquid sulphur dioxide at room temperature in a sealed tube to form $WOCl_4$ (903). Similar oxygen abstraction reactions from sulphur dioxide with molybdenum pentachloride and tungsten pentabromide are noted elsewhere.

The green $MoOCl_4$ melts at 101°–103° C, rapidly turns blue in air or water, but readily dissolves in anhydrous carbon tetrachloride (solubility = 7% at 25° C), chloroform (13%) and dichloromethane (19%) (1505). Like $MoCl_5$ which readily forms $MoCl_4$ in aromatic solvents (page 309), $MoOCl_4$ reacts with benzene to form $MoOCl_3$, hydrogen chloride and poly-*p*-phenylene, and with chlorobenzene to give $MoOCl_3$, hydrogen chloride and dichlorobenzene (1505):

$$2\,MoOCl_4 + C_6H_5Cl \rightarrow 2\,MoOCl_3 + HCl + C_6H_4Cl_2$$

It is most interesting to note that $MoOCl_4$ is monomeric in carbon tetrachloride and in the gas phase, and the Mo=O stretching frequency at 1009 cm^{-1} in carbon tetrachloride and at 1015 cm^{-1} in the gas phase is retained at 997 cm^{-1} in the solid state (1259, 1505, 2334).

The orange $WOCl_4$ (M.P. = 209° C) on the other hand shows only broad tungsten–oxygen absorption bands in the infrared spectrum below 900 cm^{-1}, characteristic of W–O–W bonding rather than W=O bonding (10, 903). Bands at 1020–1030 cm^{-1} for the gaseous state and for solution) in thionyl chloride and carbon disulphide indicate W=O bonding (2334s. The polymeric nature of the solid state has been confirmed by an X-ray structural determination (1188), which shows octahedrally coordinated

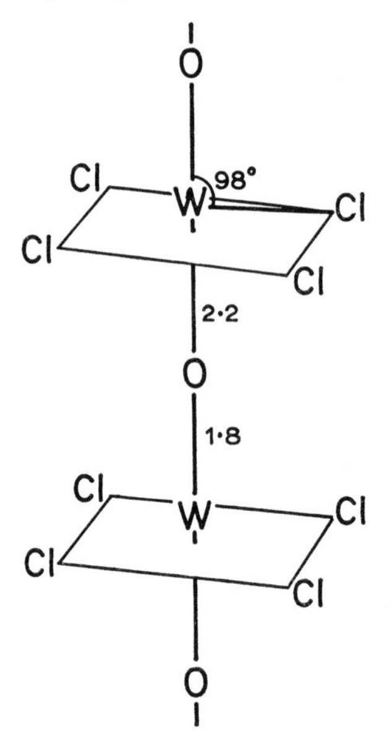

FIG. 157 The structure of $WOCl_4$.

metal atoms sharing oxygen atoms to form infinite strings (Fig. 157). It is thus intermediate between WCl_6 (octahedral coordination, no corners bridging) and WO_3 (octahedral coordination, all corners bridging). The octahedron in $WOCl_4$ is distorted in such a way as to approach a square pyramid, the tungsten-oxygen distances being alternately 1·8 and 2·2 Å, and the O–W–Cl bond angle being 98° (Fig. 157). The oxygen and chlorine atoms are arranged in cubic close packed layers all of which have the composition OCl_4.

The mass spectra of MoO_2Cl_2 and WO_2Cl_2 show significant amounts of dimer formation, dimers contributiong about 40% of the ion current in the case of MoO_2Cl_2, and about 1% in the case of WO_2Cl_2 (153). These mass spectra further show that the dimers are oxygen bridged rather than halogen bridged, since species such as $M_2O_2^+$ are present, but not $M_2Cl_2^+$. The observation of two metal-oxygen stretching frequencies and two metal-halogen stretching frequencies in the infrared spectra of gaseous MO_2X_2 (where M is Cr, Mo or W and X is Cl or Br) has however been simply interpreted in terms of tetrahedral monomers (153, 1259, 2334).

In the solid state the four MO_2X_2 compounds, where M is Mo or W and X is Cl or Br, are isomorphous (96, 153). The structure (96, 1291) of WO_2Cl_2 shows octahedral coordination about each metal atom, each octahedron being linked to four adjacent octahedra through four coplanar oxygen atoms to form an infinite sheet structure. The bridging oxygen atoms are again not equidistant between the tungsten atoms, but each has short (1·63 or 1·70 Å) and long (2·22 or 2·34 Å) oxygen–tungsten bonds. The W–O–W angle is 160°. There is a progressive increase in tungsten–chlorine bond lengths along the series WCl_6 (2·24 Å), $WOCl_4$ (2·29 Å) and WO_2Cl_2 (2·31 Å).

The dioxodichlorides are not very soluble in organic solvents. The molecular weight in boiling chloroform indicates the persistence of polymers (153). The most notable feature in the infrared spectra is the appearance of bands which are assigned to M=O vibrations. For example MoO_2Cl_2 in diethyl ether shows bands at 336 and 342 cm^{-1} (Mo–Cl), 763 and 821 cm^{-1} (Mo–O–Mo), and 926 and 963 cm^{-1} (Mo=O), compared with solid MoO_2Cl_2 in which the last bands are missing (10, 153).

Like MoO_2Cl_2 and $WOCl_4$, MoO_2Br_2 and $WOBr_4$ were prepared before 1900 by the bromination of the appropriate oxide, or by heating the metal or oxide in oxygen/ bromine mixtures (535). The purple-black $WOBr_4$ is also formed from WBr_5 and SO_2 in a sealed tube (903). The red WO_2Br_2 has been prepared by heating WO_3 with the stoicheiometric amount of CBr_4 in a sealed tube at 200° C; excess CBr_4 forms $WOBr_4$ together with a small amount of WBr_6 (1882).

$WOBr_4$ is iosomorphous with $WOCl_4$ (see above), with O–W–Br angles of 98·3° (1188). The dioxodibromides are isomorphous with the dioxodichlorides (see above).

The mixed oxochlorobromides $WOCl_3Br$ and $WOCl_2Br_2$ have been prepared by the reaction of bromine (plus some adventitious source of oxygen) on $K_3W_2Cl_9$, and by the halogen exchange between phosphorus pentabromide and $WOCl_4$, respectively (270, 1358). Both compounds can be sublimed unchanged. Table 26 shows that the *a*-axis of the unit cell expands as chlorine atoms are progressively replaced by bromine atoms, whereas the *c*-axis, which is parallel to the infinite –W–O–W–O– strings, remains approximately constant. The tungsten-oxygen stretching frequency simultaneously decreases indicating double bond character.

TABLE 26

Cell dimensions and W–O–W stretching frequencies of tungsten oxotetrahalides

	a(Å)	*c*(Å)	$\nu_{W-O-W}(cm^{-1})$
$WOCl_4$	8·48	3·99	878
$WOCl_3Br$	8·52	3·98	860
$WOCl_2Br_2$	8·78	3·94	852
$WOBr_4$	8·96	3·93	830

The dark green WO_2I_2 can be formed by the reaction of tungsten and tungsten trioxide (mole ratio 1 : 2) with excess iodine under an iodine pressure of several atm by using a 300°/800° C temperature gradient (2280) or by using a 216°/300°/720° C three-temperature furnace (2279). The compound slowly decomposes in air, is insoluble in both aqueous and organic solvents, and appears to be structurally related to WO_2Cl_2 (96, 2280). The vapour pressure of WO_2I_2 (2279), and its equilibrium with iodine and solid WO_2 (653), over the range 500°–900° C have been studied in some detail due to their importance in the tungsten-iodine lamp:

T(°C)	$K_p = \frac{p_{WO_2I_2}}{p_{I_2}}$
500	0·001
600	0·005
700	0·024
800	0·076
900	0·16

Tungsten-iodine lamps have been available for about ten years, and offer a number of advantages over the conventional incandescent tungsten lamp (1903). A small amount of iodine ($\sim$ 0·25 μmol cm^{-3}) in addition to the inert gas is incorporated in the bulb, which is also much smaller (volume ~1–2 cc) and constructed from silica so that the wall temperature is much higher ($\sim$ 600° C). Tungsten evaporating from the filament is thus not deposited on cold walls decreasing the brightness, but is maintained in the gaseous phase as WO_2I_2. The small size also results in greater mechanical strength allowing the lamp to be operated at higher gas pressures. This combination of small size, high pressures and temperatures results in a high gas viscosity, so that the gas in the bulb is in quasi-stationary layers which also decrease the rate of evaporation. The filament is operated at higher tempeatures ($\sim 3000°C$) than in the conventional incandescent lamp, with a consequent increase in brightness and luminous efficiency, with particular enhancement towards the blue end of the spectrum.

2. *Oxohalide adducts*

Molybdenum oxotetrachloride is readily reduced to $MoOCl_3$ on refluxing with benzene or other aromatic solvents, and $MoOCl_3$ adducts are similarly readily formed on addition of ligands to $MoOCl_4$ in carbon tetrachloride. With $WOCl_4$ however, whether or not reduction occurs is dependent upon both ligand and conditions. The molybdenum(V) and tungsten(V) compounds are discussed elsewhere. Complexes of MoO_2Cl_2 and WO_2Cl_2 are much less prone to reduction. That is the ease of reduction is as expected:

$$MCl_6 > MOCl_4 > MO_2Cl_2$$

and

$$Mo > W$$

The only adduct known for $MoOCl_4$ appears to be $MoOCl_4(POCl_3)$ (1017).

With oxygen donors, $WOCl_4$ forms $WOCl_4$(ligand) with diethyl ether, tetrahydrofuran, tetrahydropyran or acetone; $WOCl_4$ and $WOBr_4$ form $WOX_4,\frac{1}{2}$Dioxan, where the infrared evidence suggests that both oxygen atoms of the ligand are coordinated to metal atoms (903, 956). Alkyl cyanides form $WOCl_4(RCN)$ (where R is Me, Et, Bu, Ph or Bz) and $WOBr_4(MeCN)$. These compounds as well as the ether adducts show strong infrared bands at 995–1010 cm^{-1} attributable to W=O vibrations which is in contrast to the oxyhalides themselves which show broad bands at about 800 cm^{-1} due to W–O–W vibrations. The simple octahedral monomeric structure of these complexes is confirmed by molecular weight and conductance studies on $WOCl_4(MeCN)$ (903, 956). Pyridine and dipyridyl initially form $WOCl_4(py)_2$ and $WOCl_4(dipy)$, although reduction occurs with excess pyridine, dipyridyl, triphenylphosphine and dimethylsulphide (903, 956). *o*-Phenylenebisdimethylarsine forms the seven coordinate, pentagonal bipyramidal, $WOCl_4(diars)$ (1359).

A number of adducts of the type $MoO_2Cl_2(ligand)_2$ have been prepared using the liquid ligand, or by using an appropriate solvent such as dichloromethane. The ligands used include the oxygen donors tetrahydrofuran, dimethylsulphoxide, dimethylformamide, dimethylacetamide, triphenylphosphine oxide, triphenylarsine oxide, pyridine–N–oxide and various esters, aldehydes and ketones (1218, 1336, 1356, 1505), and the nitrogen donors acetonitrile, benzonitrile and pyridine (217, 1505). The triphenylarsine oxide compound has also been prepared from $MoOCl_3$ in dichloromethane (1356) or from molybdenum(III) compounds in the solid state (427). The compounds are monomeric indicating a simple octahedral structure, which has been confirmed by the structure of the dimethylformamide complex (891). The oxygen atoms are mutually *cis* with unusually short molybdenum-oxygen distances of 1·68 Å. The organic ligands are also mutually *cis*, while the two chlorine atoms are *trans* to each other.

The corresponding WO_2Cl_2(ligand)$_2$ have also been prepared from WO_2Cl_2 (with MeCN, Ph_3PO, $\frac{1}{2}$ dipy, $\frac{1}{2}$ *o*-phen) (330), by oxygen abstraction using acetone or dimethylformamide solutions (with DMF, Me_2SO, C_4H_8SO, $(Me_2N)_3PO$) (330), by chlorine oxidation of the corresponding carbonyl complex (with Ph_3PO, $\frac{1}{2}$ $Ph_2P(O).CH_2.CH_2.P(O)Ph_2$) (1531), and from a solution prepared by dissolving freshly precipitated WO_3 in ethanol saturated with hydrogen chloride (with Ph_3PO) (1638). In all cases the infrared spectra show that the two oxygen atoms occupy *cis*-positions.

E. Anionic Oxohalocomplexes

No purely halo complex anions of the type $(M^{VI}X_{6+x})^{x-}$ appear to be known.

1. *Oxofluoro complexes*

The action of concentrated hydrofluoric acid on $K_2Cr_2O_7$ or $Cs_2Cr_2O_7$ forms the red salts $KCrO_3F$ and $CsCrO_3F$ respectively. These salts have the $CaWO_4$ structure, that is they contain the tetrahedral CrO_3F^- ion (1368).

The chemistry of the oxofluoro complexes of molybdenum and tungsten is fairly complex, and there is an almost bewildering variety of species which have been found, a selection of which is briefly given.

Phase studies of the binary systems molybdenum trioxide-alkali metal fluoride show the formation of $A^I_3MoO_3F_3$ (2059); the corresponding $A^I_3WO_3F_3$ are similarly observed (2058).

Complex oxyfluoro species are more often obtained from aqueous hydrofluoric acid. For example the acid $H_2MoO_3F_2,H_2O$ is obtained from 17–35% hydrofluoric acid at 0° C, and $H_2MoO_2F_4,1\frac{1}{2}H_2O$ from 35–62% hydrofluoric acid at 0° C (1789). The ternary system MoO_3–KF–H_2O shows in addition to $K_3MoO_3F_3$, $K_2MoO_3F_2,H_2O$ and $K_2MoO_2F_4,H_2O$, the more complex $K_3Mo_4O_{13}F,3H_2O$ and $K_{10}Mo_6O_{11}F_{24}$ (392, 2060); $K_2WO_3F_2$, H_2O and $K_2WO_2F_4,H_2O$ are similarly observed. The anion which is isolated from aqueous solution is dependent not only on metal, fluoride and hydrogen ion concentration, but also upon the cation. In addition to difluoro, trifluoro and tetrafluoro complexes such as those mentioned above, other oxofluorides which have been obtained under the appropriate conditions include the pentafluoro $AMoOF_5$ (where A is Na, Rb or Cs) (1135) and $AWOF_5$ (where A is K or Cs) (156, 1133), monofluoro Na_3MoO_4F (2059), $RbMoO_3F,\frac{1}{2}H_2O$ (1812) and Na_3WO_4F (2058), and $Cs_{11}W_5O_{15}F_{11},5H_2O$ (1815).

The crystal structure of $K_2[MoO_2F_4]$ unexpectedly shows a *trans* arrangement of the two oxygen atoms, with one molybdenum-oxygen bond considerably shorter than the other (1·67 and 1·99 Å respectively) (1870). The structure of "$(NH_4)[Cu(H_2O)_6][WO_2F_5]$" shows that the tungsten atom is again only six coordinate, the formula being more correctly written $[Cu(H_2O)_6][WO_2F_4], NH_4F$ (636).

Raman spectra of aqueous hydrofluoric acid solutions of molybdenum and tungsten have been interpreted as showing a *cis*-arrangement of oxy groups about the metal, the species apparently being $[MO_2F_2(H_2O)_2]$ or $[MO_2F_3(H_2O)]^-$ (1067). However the infrared spectrum of solid salts of $[MoO_3F_3]^{3-}$ and $[WO_3F_3]^{3-}$ have been interpreted as indicating both *cis*- (393, 1069) and *trans*- (649) arrangements of oxygen atoms.

2. *Oxochloro and oxobromo complexes*

Acidification of chromate solutions with concentrated hydrochloric acid forms, in addition to $HCrO_4^-$, H_2CrO_4 and $Cr_2O_7^{2-}$ (page 288), the chloroanion CrO_3Cl^-.

$$CrO_3Cl^- + H_2O \leftrightarrows HCrO_4^- + Cl^- + H^+$$
$$K = 0{\cdot}1 \qquad \text{(ref. 1118, 1574, 2286)}$$

The structure of $KCrO_3Cl$ shows the existence of discrete CrO_3Cl^- anions, with a distorted tetrahedral structure (1171). Salts containing the CrO_3Br^- anion can also be readily obtained.

The addition of excess hydrochloric acid to aqueous polymolybdate solutions leads to $MoO_2Cl_2(H_2O)_2$ above about $2M$ HCl, then $[MoO_2Cl_3(H_2O)]^-$ at higher acid concentrations, and finally the yellow *cis*-$[MoO_2Cl_4]^{2-}$ above about $12M$ HCl (1067, 1167, 1774, 2443). In hydrobromic acid, molybdenum(VI) is slowly reduced to molybdenum(V) (38).

Addition of the appropriate cation to solutions prepared by dissolving tungsten hexachloride or freshly precipitated WO_3 in concentrated hydrochloric acid or ethanol saturated with hydrogen chloride, precipitates *cis*-$A_2[WO_2Cl_4]$, where A is NH_4, Rb, Cs or Ph_3PH (1638, 1863).

The scarlet octahedral monooxo ion $[WOCl_5]^-$ is obtained by addition of chloride to $WOCl_4$ in chloroform. The light brown $(Et_4N)[WOBr_5]$ was similarly prepared (902).

F. Complexes with other Anionic Ligands

Very few complexes have been obtained from nonaqueous solutions. Examples include the nitrato compounds $MO_2(NO_3)_2$ obtained by the action of N_2O_5 on $KCrO_3Cl$, MoO_2Cl_2, WCl_4 or WCl_6 (2055), and $[MoO_2Br_2(dipy)]$ prepared from $[Mo(CO)_4(dipy)]$ and bromine in ethanol (1237). The structure of this compound shows a very distorted octahedral stereochemistry with a large O–Mo–O angle of 103°, a very small N–Mo–N angle of 67°, while the *trans* bromine atoms are pushed away from the oxygen atoms such that the Br–Mo–Br angle is 160° (833). The bond lengths differ widely: Mo–O = 1·64 and 1·83 Å, Mo–Br = 2·46 and 2·78 Å, Mo–N = 2·26 and 2·45 Å.

Solvolysis of $WOCl_4$ to form $WO(OR)_4$ has been referred to above ($MoOCl_4$ is reduced under the same conditions), while $WOCl_4$ and thiocyanate in dioxane or acetone form $WOCl(NCS)_3(diox)$, $WO(NCS)_4(diox)_2$ and $WO(NCS)_4(Me_2CO)_2$ respectively (952).

Two types of complex, which can be obtained from aqueous solution and which have been studied in some detail, are those of molybdenum with multidentate ligands, particularly oxygen donors, and peroxo complexes of chromium, molybdenum and tungsten. These are now summarized in turn.

1. *Molybdenum complexes*

A large number of complexes using bidentate and tridentate ligands can be obtained from aqueous solutions of molybdenum(VI), such as molybdate, acidified molybdate, freshly precipitated MoO_3, or other sources such as MoO_2Cl_2. In some cases the products have no obvious stoicheiometric composition and appear to be polymeric with bridging oxygen atoms. However a considerable number of simple octahedral monomeric complexes have also been characterized. All have two or three oxygen atoms which are mutually *cis* as expected to maximize $O(p_\pi) \rightarrow Mo(d_\pi)$ bonding. The molybdenum-oxygen bond lengths vary from about 1·7 to about 2·3 Å, and the corresponding stretching frequencies from about 940 to about 840 cm^{-1} respectively (1693).

The structure of the 8-hydroxyquinoline complex $[MoO_2(oxine)_2]$ (98), used for the standard gravimetric determination of molybdenum, shows that the free oxygen atoms are *cis* (Mo–O = 1·71 Å), the nitrogen atoms of the two chelates are also *cis* (Mo–N = 2·32 Å), but the oxygen atoms of the two chelates are *trans* (Mo–O = 1·98 Å).

Other unicharged bidentate ligands which form *cis*-[MoO_2(ligand)$_2$] complexes include β-diketones (1505, 1705, 1710), dialkyldithiocarbamates (1704, 1705), and cysteine esters ROOC.CH(NH_2).CH_2.S (1678).

Oxalate forms two different complexes containing one bidentate oxalato group per molybdenum atom. The structure of $K_2[Mo_2O_5(C_2O_4)_2(H_2O)_2]$ is shown in Fig. 158 (572). The bridging oxygen atom lies on a crystallo-

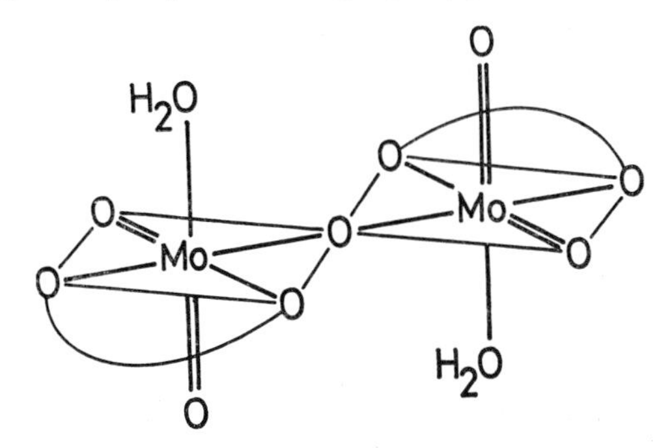

FIG. 158 The structure of $[\{MoO_2(C_2O_4)(H_2O)\}_2O]^{2-}$.

graphic centre of symmetry, and the Mo–O–Mo bond is therefore strictly linear indicating considerable $O(p_\pi) \rightarrow Mo(d_\pi)$ bonding. The Mo–O and Mo–O– bond lengths are 1·69 and 1·88 Å respectively. The structure of $Na(NH_4)[MoO_3(C_2O_4)],2H_2O$ shows an octahedral structure with infinite, bent, –O–Mo–O–Mo– bonding (97). The Mo=O and two unequal Mo–O– bond lengths are 1·83, 1·87 and 2·23 Å respectively.

The structure of the MoO_3(dien) complex obtained with the tridentate diethylenetriamine shows three mutually *cis* oxygen atoms with unusually long Mo–O bonds of 2·32 Å (compared with normal values of 1·7–1·9 Å). This is also reflected in unusually low molybdenum-oxygen stretching frequencies of 839 cm^{-1}. This weak molybdenum-oxygen bonding may be due to the extensive hydrogen bonding between the oxygen atoms in one molecule and the amine groups of adjacent molecules. The octahedron is again considerably distorted, with O–Mo–O and N–Mo–N angles of approximately 106° and 75° respectively (559).

A similar arrangement is found for the EDTA complex $Na_4[O_3Mo(OOC)_2N.CH_2.CH_2.N(COO)_2MoO_3],8H_2O$, where each end of the EDTA ion acts as a tridentate to different MoO_3 units. In this case the molybdenum-oxygen bonds lengths are 1·74 Å (1842).

There have been many other solution and preparative studies on the reaction of other oxygen-donor chelates with molybdate and tungstate, but in most cases only the ligand : metal stoicheiometry has been established (121, 363, 365).

2. *Peroxo complexes*

Peroxo compounds of chromium(VI), molybdenum(VI) and tungsten (VI) have been studied in considerable detail, and show many examples of seven coordination.

The addition of hydrogen peroxide to acid dichromate solutions leads to the blue $Cr^{VI}(O_2)_2O$. Although unstable in water (where it probably exists as the hydrated complex $[Cr(O_2)_2O(H_2O)]$), it can be extracted into organic solvents such as ethers, alcohols, ketones or nitriles, where it is considerably more stable. Solids derived from these solutions are unstable, for example the dimethyl ether complex $[Cr(O_2)_2O(Me_2O)]$ explodes above $-30°$ C. However addition of organic bases such as pyridine, dipyridyl, *o*-phenanthroline, quinoline or aniline form the stable complexes $[Cr(O_2)_2O(ligand)]$ whose structures are discussed below. However addition of ammonia or ethylenediamine leads to the formation of chromium (IV) peroxo complexes (page 330), while if the hydrogen peroxide is added to aqueous alkaline chromate solutions, chromium(V) peroxo complexes are formed (page 318) (2231).

The structures of $[Cr^{VI}(O_2)_2O(py)]$ (1855, 2227) and $[Cr^{VI}(O_2)_2O(o\text{-phen})]$ (2230) are shown in Fig. 159. $[Cr(O_2)_2O(py)]$ has a

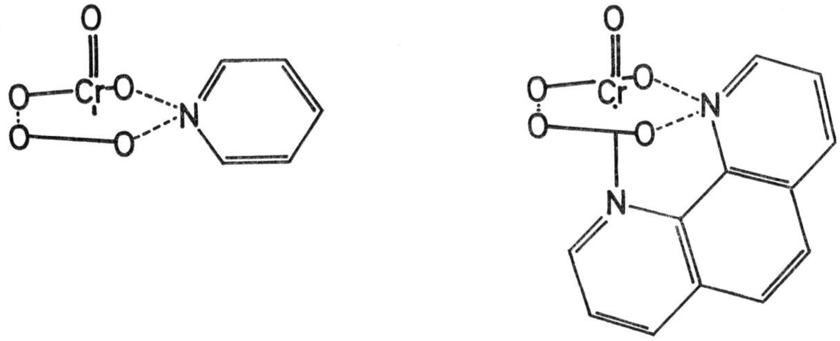

FIG. 159 The structure of $[Cr(O_2)_2O(py)]$ and $[Cr(O_2)_2O(o\text{-phen})]$.

pentagonal pyramidal structure, with two bidentate peroxo groups and the nitrogen atom in the plane, and the metal atom shifted 0·51 Å out of the plane towards the single apical oxygen atom (Cr–O = 1·58 Å). In the analogous *o*-phenanthroline (and dipyridyl (2233)) complex, the stereochemistry is pentagonal bipyramidal, with the second nitrogen atom loosely bonded in the second apical position (Cr–N = 2·26 Å, compared with 2·11 Å for the in-plane nitrogen atom). This also draws in the chromium atom so that it is only 0·27 Å out of the pentagonal plane.

Similar peroxo compounds may be formed with anionic ligands. For example addition of hydrogen peroxide to acid solutions of the yellow

$(Ph_4As)[Cr^{VI}O_3Cl]$ forms the blue $(Ph_4As)[Cr^{VI}(O_2)_2OCl]$ (2308) which presumably has a similar structure to the pyridine complex.

Hydrogen peroxide and acid molybdate or tungstate solutions form salts of $[Mo_2O_{11}(H_2O)_2]^{2-}$. The structures of $K_2[Mo_2O_{11}(H_2O)_2]$, $2H_2O$ (2232), $(pyH)_2[Mo_2O_{11}(H_2O)_2]$ (1696) and $K_2[W_2O_{11}(H_2O)_2]$, $2H_2O$ (762) have been determined. Two pentagonal bipyramids similar to those observed in the peroxo chromium compounds are joined with a bridging oxygen atom in such a way that the two planes are approximately at right angles to one another (Fig. 160). The metal atom is again displaced out of

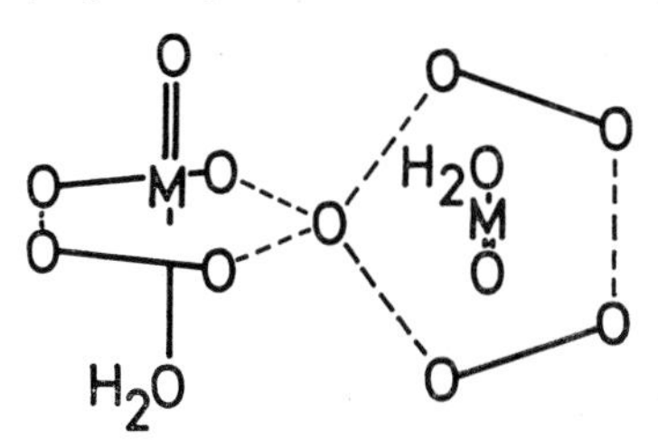

FIG. 160 The structure of $[M_2O_{11}(H_2O)_2]^{2-}$ (where M is Mo or W).

the pentagonal plane (by 0·35 Å) towards the double-bonded oxygen atom, which also decreases the M–O–M angle from 180° to about 140°.

The monoperoxo complex $K_2[Mo(O_2)OF_4]$, H_2O also has a pentagonal bipyramidal structure, with the peroxo group and three of the fluorine atoms in the pentagonal plane (1044). The infrared and ^{19}F-n.m.r. spectrum of the tungsten analogue also shows that it has this structure (791, 1061).

Other molybdenum and tungsten diperoxo and monoperoxo complexes which can be obtained from aqueous peroxide solutions, and which presumably have related structures, include $[M(O_2)_2OF_2]^{2-}$, $[M(O_2)_2O(C_2O_4)]^{2-}$, $[M(O_2)OCl_4]^{2-}$ and $[M(O_2)O_2(C_2O_4)(H_2O)]^{2-}$ (791, 1061, 1066, 1068).

The triperoxo complexes $[M_2(O_2)_4(HO_2)_2O_2]^{2-}$ (where M is Mo or W) were formulated as having hydrogen peroxide of crystallization on the basis of infrared measurements (1068), but the structure (1696) shows the presence of bridging HO_2^- groups:

```
                OH
                |
                O
               / \
[O(O2)2Mo         Mo(O2)2O]2-.
               \ /
                O
                |
                OH
```

G. Polychromates, Polymolybdates and Polytungstates

1. *Polychromates*

Acidification of chromate, CrO_4^{2-}, forms $HCrO_4^-$ in equilibrium with $Cr_2O_7^{2-}$, the monomer predominating only below 0·01 *M*. The relevant equilibrium constants at 25° C are:

$$HCrO_4^- \leftrightarrows H^+ + CrO_4^{2-} \quad K_a = 3{\cdot}2 \times 10^{-7} \quad \text{(ref. 1555)}$$
$$2\,HCrO_4^- \leftrightarrows Cr_2O_7^{2-} + H_2O \quad K = 34 \quad \text{(ref. 1554)}$$

Both $HCrO_4^-$ and $Cr_2O_7^{2-}$ are further protonated under more acid conditions:

$$HCr_2O_7^- \leftrightarrows Cr_2O_7^{2-} + H^+ \quad K_a = 0{\cdot}85 \quad \text{(ref. 2287)}$$
$$H_2CrO_4 \leftrightarrows HCrO_4^- + H^+ \quad K_a = 4{\cdot}1 \quad \text{(ref. 1118, 2286)}$$

It is not possible to obtain pure H_2CrO_4 in aqueous solution, as complexes are formed with, for example, hydrochloric acid (forming CrO_3Cl^-) and sulphuric acid (forming $CrSO_7^{2-}$), and in concentrated perchloric acid there is rapid reduction to chromium (III) by water.

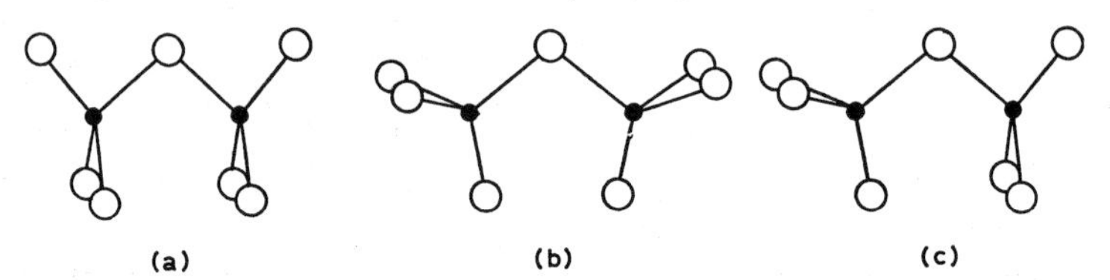

FIG. 161 The three different conformations for $Cr_2O_7^{2-}$.

The rate constant for the formation of $Cr_2O_7^{2-}$ from $HCrO_4^-$ is $1{\cdot}8\ M^{-1}\ sec^{-1}$ (1677, 2256). The rate constants for the base hydrolysis of dichromate to form chromate are $5 \times 10^{-4}\ M^{-1}\ sec^{-1}$ and $4 \times 10^2\ M^{-1}\ sec^{-1}$ for hydrolysis by water (1706, 1859, 2256) and hydroxide ion (1538, 1706) respectively. The relatively slow rates compared with other base-catalysed ionic reactions are due to the very large negative entropies of activation of —44·2 and —30·2 e.u. respectively.

Three different conformations can be readily envisaged for the dichromate ion, in which two tetrahedrally coordinated chromium atoms are linked by a bent Cr–O–Cr bridge. One has four oxygen atoms eclipsed (Fig. 161a), one has two oxygen atoms elipsed (Fig. 161b), and one has the staggered structure (Fig. 161c). The anions in $K_2Cr_2O_7$, $Rb_2Cr_2O_7$ and $(NH_4)_2Cr_2O_7$ are of the first type, but in $SrCr_2O_7$ and $PbCr_2O_7$ there are two crystallographically independent anions, one of the first type and one of the second type. (More precisely, the anions in $K_2Cr_2O_7$ are not exactly eclipsed, but are twisted away from this position by 5° and 10° respectively for two crystallographically independent anions). It is presumed that different crystallographic packing is responsible for these different isomers. The conformation in solution is not known.

Data for these different salts are collected in Table 27, which indicate that there may be significant chromium-oxygen-chromium (d_π–p_π–d_π) bonding across the dimer, since it is observed that as the Cr–O–Cr unit becomes more linear, the chromium-bridging oxygen bond becomes shorter. These changes also appear to be reflected in changes in the infrared spectra, which have Cr–O–Cr modes at about 770 and about 560 cm^{-1}. The nonbridging oxygen–chromium bond lengths may be compared with those of 1·66 Å in $(NH_4)_2CrO_4$ (982).

TABLE 27

Structural parameters of $Cr_2O_7^{2-}$

	Cr–O–Cr (angle)	$Cr–O_{bridge}$	$Cr–O_{non\text{-}bridge}$	Reference
$Rb_2Cr_2O_7$	116°	1·82 Å	1·61 Å	2386
$Ag_2Cr_2O_7$	121°	1·78 Å	1·62 Å	1156
$K_2Cr_2O_7$	126° (average)	1·79 Å (average)	1·63 Å (average)	315
$SrCr_2O_7$	136° (average)	1·73 Å (average)	1·60 Å (average)	2386

Solid trichromates such as $K_2Cr_3O_{10}$ may be obtained from highly acid concentrated solutions, the structure of which is shown in Fig. 162.

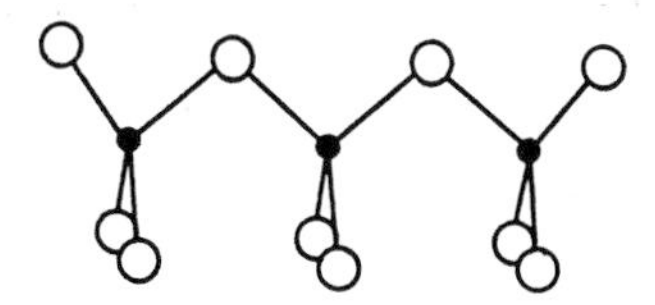

FIG. 162 A schematic representation of $Cr_3O_{10}^{2-}$.

The Cr–O–Cr stretching frequencies are decreased to 517 and 489 cm^{-1}, while additional Cr=O stretching frequencies at 893 and 818 cm^{-1} are presumably due to the central chromium atom. Solution studies using the latter band show the presence of $Cr_3O_{10}^{2-}$ provided the dichromate ion concentration is greater than 3 *M*, and provided the pH is less than zero (1008).

2. *Isopolymolybdates*

In alkaline solution molybdenum (VI) forms the molybdate ion MoO_4^{2-}, which is tetrahedral in solution and in those salts whose structure is known. The molybdenum–oxygen bond length is 1·76 Å (7, 981).

It is firmly established that $Mo_7O_{24}^{6-}$ is the first polymer of any importance formed on acidification of MoO_4^{2-} (102, 1994):

$$8\,H^+ + 7\,MoO_4^{2-} \leftrightarrows Mo_7O_{24}^{6-} + 4\,H_2O$$
$$\log K = 57{\cdot}74 \qquad \text{(refs. 1992, 1993)}$$

Further protonation to $HMo_7O_{24}^{5-}$, $H_2Mo_7O_{24}^{4-}$ and $H_3Mo_7O_{24}^{3-}$ is also significant (1992, 1993, 1994), but the formation of $Mo_8O_{26}^{4-}$ should also be noted (see later).

Ultracentrifuge measurements show that the average degree of condensation is about six (102) or seven (1009).

FIG. 163 The structure of $Mo_7O_{24}^{6-}$.

Unpolymerized protonated forms are important only below $10^{-4}M$ (1992, 1993):

$$MoO_4^{2-} + H^+ \rightleftarrows HMoO_4^- \quad \log K = 3{\cdot}9$$
$$HMoO_4^- + H^+ \leftrightarrows H_2MoO_4 \quad \log K = 3{\cdot}6$$

Salts such as $(NH_4)_6(Mo_7O_{24}),4H_2O$ or $K_4Ca(Mo_7O_{24}),7H_2O$ can be obtained from these solutions (1935). The structure (795, 890, 1548, 2131) of the former, which is isomorphous with the potassium and rubidium analogues, consists of seven MoO_6 octahedra sharing edges (Fig. 163). The central molybdenum atom is displaced from its ideal position towards the outside of the molecule, so that the Mo–Mo–Mo angle for the central three octahedra is reduced from 180° to about 160°. Although there is fair agreement concerning the position of the molybdenum atoms, there is not good agreement about the position of the oxygen atoms, for example the average nonbridging oxygen–molybdenum bond length has been quoted as 1·94 (2131) and 1·73 Å (795). The analogy to $Ti_7O_4(OEt)_{20}$ (2342) has been noted previously (page 101).

A study of the e.s.r. spectrum of X-ray irradiated $(NH_4)_6(Mo_7O_{24}),4H_2O$ was interpreted as showing that only one of the molybdenum atoms gave rise to hyperfine splitting with the spin of the generated unpaired electron (398). Reduction of isopolymolybdates in aqueous solution progressively forms blue, red, and yellow-brown solutions containing mixtures of molybdenum(V) and molybdenum(VI) (1822).

Further acidication of these molybdate solutions forms $Mo_8O_{26}^{4-}$ (102, 1009):

$$12\,H^+ + 8\,MoO_4^{2-} \rightleftharpoons Mo_8O_{26}^{4-} + 6\,H_2O$$

Salts such as $(NH_4)_4(Mo_8O_{26}),5H_2O$ can again be obtained from aqueous solution. The Raman spectrum of solid $(NH_4)_6(Mo_7O_{24}),4H_2O$ shows a very strong molybdenum-terminal oxygen stretching frequency at 934 cm^{-1}, whereas solid $(NH_4)_4(Mo_8O_{26}),5H_2O$ has this band at the significantly higher energy of 963 cm^{-1}. (Additional weaker absorptions due to bridging oxygen atoms of course occur at lower energies in both cases). In solution these bands are only slightly shifted to 940 and 961 cm^{-1} respectively. The aqueous solutions can therefore be directly analysed for MoO_4^{2-}, $Mo_7O_{24}^{6-}$ and $Mo_8O_{26}^{4-}$ (Fig. 164) (102). On the other hand

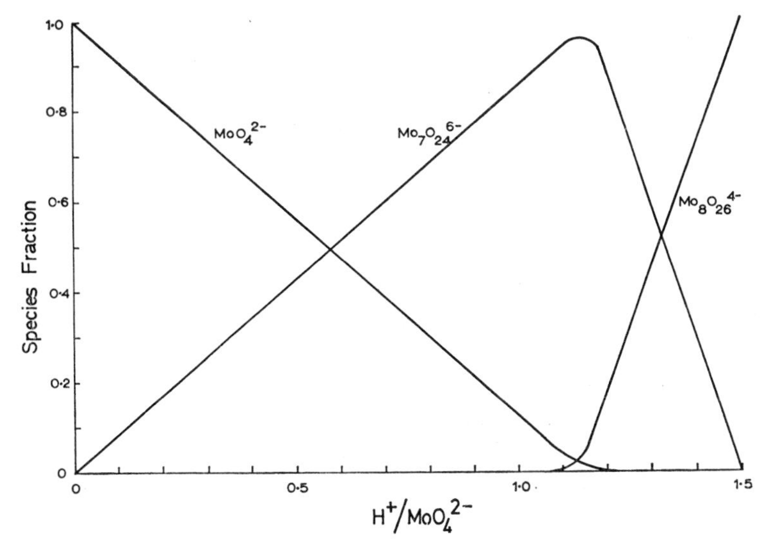

FIG. 164 Range of existence of MoO_4^{2-}, $Mo_7O_{24}^{6-}$, and $Mo_8O_{26}^{4-}$.

the existence of this octamolybdate ion could not be confirmed using precise potentiometric techniques, and it was suggested that this ion was only formed on heating (1994).

The structure of the octamolybdate ion in $(NH_4)_4(Mo_8O_{26}),5H_2O$ shows eight MoO_6 octahedra sharing edges (Fig. 165) (1549). It has been pointed out (1354) that both $Mo_7O_{24}{}^{6-}$ and $Mo_8O_{26}{}^{4-}$ can be considered to be fragments of the hypothetical $Mo_{10}O_{28}{}^{4+}$ which would be isostructural with $V_{10}O_{28}{}^{6-}$ (page 185), but this is not found due to the unfavourable high positive charge on the metal and on the ion (page 57).

FIG. 165 The structure of $Mo_8O_{26}{}^{4-}$.

The polymerization process continues as progressively more acid is added to these molybdate solutions, but these higher polymers have not yet been characterized. For example after the addition of 1·8 moles of acid to each mole of molybdate, molecular weight and potentiometric measurements show that the average anion contains approximately 19 molybdenum atoms (1272, 1994), corresponding to approximately $Mo_{19}O_{59}{}^{4-}$.

Hydrated molybdenum oxide precipitates when excess acid has been added to molybdate. It is soluble in excess acid, the solubility depending upon the particular acid used (1994, 2322).

3. *Isopolytungstates*

In a similar manner to chromium(VI) and molybdenum(VI), tungsten (VI) exists in alkaline solution as the tetraoxo ion $WO_4{}^{2-}$. The tetrahedral structure persists in many salts, the tungsten–oxygen bond length being about 1·78 Å (1339, 2268, 2436).

The structure of hydrated lithium tungstate, $Li_2WO_4, 4/7H_2O$ however shows the presence of both WO_4^{2-} anions and discrete $W_4O_{16}^{8-}$ anions. The latter consist of four WO_6 octahedra sharing edges, with the four metal atoms at the corners of a distorted tetrahedron (Fig. 166) (1239, 1272). The formula can then be written $Li_{14}(WO_4)_3(W_4O_{16})$, $4H_2O$.

The WO_4^{2-} ion polymerizes in acid solution, and these isopolytungstates have been reviewed elsewhere (1351).

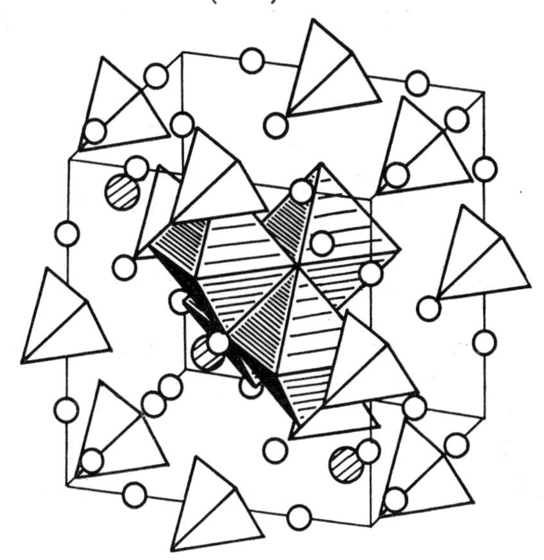

FIG. 166 The structure of Li_2WO_4, 4/7 H_2O (reference 1272). Hatched circles: H_2O. Small circles: Li^+. Tetrahedra: WO_4^{2-}. Group of four octahedra: $W_4O_{16}^{8-}$.

Below about pH6, the WO_4^{2-} condenses to form the so-called "paratungstate A" anion, $HW_6O_{21}^{5-}$:

$$7\,H^+ + 6\,WO_4^{2-} \leftrightarrows HW_6O_{21}^{5-} + 3\,H_2O$$

There is little doubt that the paratungstate A is hexameric (101, 1006, 1010, 1272, 1279, 2178, 2420), but its structure is not known and no solid crystalline derivatives appear to have been isolated.

This six-fold condensed ion slowly forms the twelve-fold condensed paratungstate B or paratungstate Z, $H_2W_{12}O_{42}^{10-}$:

$$2\,HW_6O_{21}^{5-} \rightleftarrows H_2W_{12}O_{42}^{10-}$$

These aged paratungstate solutions are less reactive to, for example, hydroxide ion or hydrogen peroxide, than freshly prepared paratungstate A solutions. Salts such as $Na_{10}H_2W_{12}O_{42}$, $27H_2O$ crystallize from these solutions over a period of several days.

A direct analysis of these solutions for WO_4^{2-}, $HW_6O_{21}^{5-}$ and $H_2W_{12}O_{42}^{10-}$ can be made using paper chromatography (707, 708) or by polarography. The latter depends on the different polarographic behaviour of 12-tungstosilicic acid and 11-tungstosilicic acid, and the rapid formation of the 12-acid from the 11-acid in the presence of WO_4^{2-} or $HW_6O_{21}^{5-}$, but not from the non-labile $H_2W_{12}O_{42}^{10-}$ (2177, 2181). However the most precise study of these equilibria is from potentiometric measurements (101, 1010, 1351, 1991):

$$K = \frac{[HW_6O_{21}^{5-}]}{[H^+]^7[WO_4^{2-}]^6} = 10^{54.0}$$

$$K = \frac{[H_2W_{12}O_{42}^{10-}]}{[H^+]^{14}[WO_4^{2-}]^{12}} = 10^{110.0}$$

$$\left(\text{or } K = \frac{[H_2W_{12}O_{42}^{10-}]}{[HW_6O_{21}^{5-}]^2} = 10^{2.1}\right)$$

The interpretation is complicated by the overlap of metatungstate into this "paratungstate region", as indicated in Fig. 167.

The rates of interconversion of paratungstate A, $HW_6O_{21}^{5-}$, and paratungstate Z, $H_2W_{12}O_{42}^{10-}$, have been measured using polarographic (2177, 2181) and paper chromatographic (707, 708) techniques. Equilibrium is

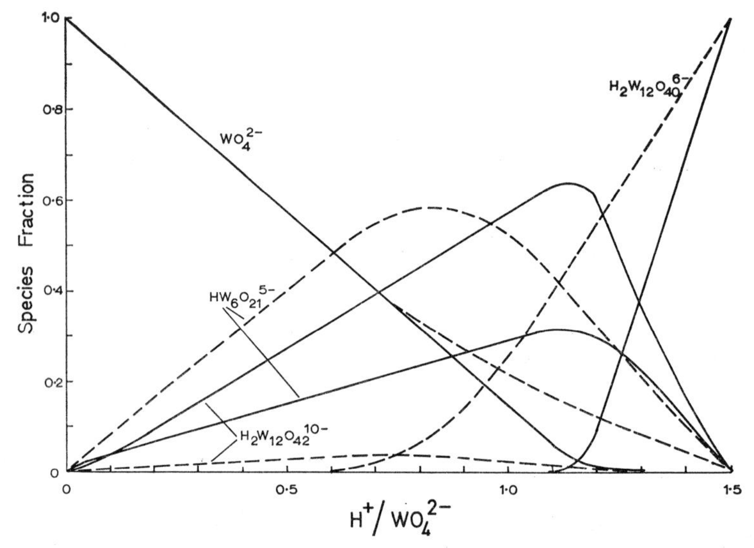

FIG. 167 Fraction of total tungsten(VI) as WO_4^{2-}, $HW_6O_{21}^{5-}$, $H_2W_{12}O_{42}^{10-}$ and $H_2W_{12}O_{40}^{6-}$ in 3 *M* LiCl for 0·2 *M* tungsten(VI) (solid lines), and 0·002 *M* tungsten(VI) (broken lines).

only established after 20–30 days. The hydrolysis of paratungstate Z at 100° C follows first order kinetics and is a two stage process; the first is complete in 5–10 minutes (activation energy of 12 kcal mol^{-1}) and the second in 10–15 hours (activation energy of 23 kcal mol^{-1}) (1327).

The structure of the anion in sodium and ammonium paratungstate Z is shown in Fig. 168 (1550, 1564, 2364). The twelve WO_6 octahedra share edges with each other forming two different types of tritungstate group, which then share corners with each other to form the complete anion.

FIG. 168 The structure of the paratungstate Z anion, $H_2W_{12}O_{42}^{10-}$.

A study of the ternary system Na_2WO_4–WO_3–H_2O at 25° C (2154) shows the existence of the ditungstate $Na_2W_2O_7, 5H_2O$ in addition to $Na_2WO_4, 2H_2O$ and $Na_{10}H_2W_{12}O_{42}, 27H_2O$. The pentahydrate $Na_2W_2O_7, 5H_2O$ readily loses water at room temperature to form the monohydrate $Na_2W_2O_7, H_2O$, which is not completely dehydrated to $Na_2W_2O_7$ until 330° C. The crystal structure of $Na_2W_2O_7$ shows it to consist of zig-zag chains of WO_6 octahedra with an equal number of WO_4 tetrahedra linked between adjacent octahedra of a chain (Fig. 169) (1547). A plausible structure for the monohydrated $W_2O_7.H_2O^{2-}$ ion appears to be a WO_6 octahedron and a WO_4 tetrahedron sharing a common edge.

There have been numerous reports of intermediates between these WO_4^{2-}, $W_2O_7.xH_2O^{2-}$, $HW_6O_{21}^{5-}$ and $H_2W_{12}O_{42}^{10-}$ ions, but they have yet to be convincingly demonstrated.

Further addition of acid to paratungstate solutions results in the formation of the pseudo metatungstate anion $(HW_6O_{20}^{3-})_n$, which is slowly transformed into the metatungstate anion $[H_2W_{12}O_{40}]^{6-}$.

$$9n\,H^+ + 6n\,WO_4^{2-} \rightleftarrows (HW_6O_{20}^{3-})_n + 4n\,H_2O$$

$$2\,(HW_6O_{20}^{3-})_n \leftrightarrows n\,H_2W_{12}O_{40}^{6-}$$

The molecular complexity of metatungstate solutions corresponds to a twelve-fold condensed ion, which is in agreement with the known structure (see below). However there has been some debate on the complexity of the pseudo metatungstate anion whose structure is not known, and formulae containing from 6 to 24 tungsten atoms have been proposed (600, 1007, 1009, 1010, 1272, 1279). The equilibria between paratungstates and metatungstates have been studied by polarographic (1994, 2177) and potentiometric (101, 1010) techniques, as well as qualitative tests such as the reaction with strong acid.

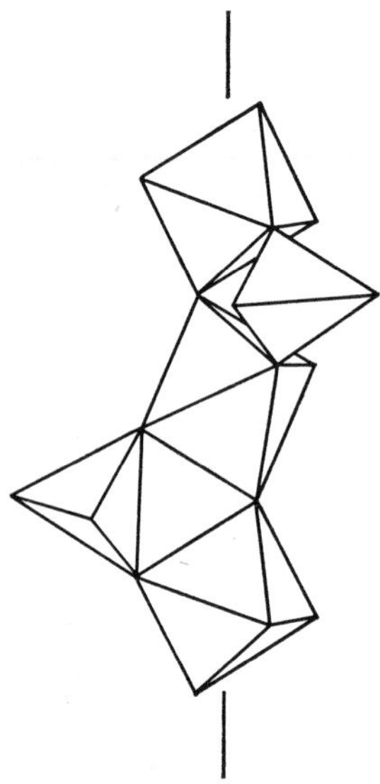

FIG. 169 The linking of WO_4 tetrahedra and WO_6 octahedra in $Na_2W_2O_7$.

$$18\,H^+ + 12\,WO_4^{2-} \rightleftharpoons H_2W_{12}O_{40}^{6-} + 8\,H_2O$$

$$\log K = 132{\cdot}5 \qquad \text{(ref. 101)}$$

$$\log K = 142 \qquad \text{(ref. 1010)}$$

The tungsten–oxygen skeleton in the metatungstate anion $[H_2W_{12}O_{40}]^{6-}$ is the same as that in heteropolytungstates such as $[CoW_{12}O_{40}]^{5-}$, $[SiW_{12}O_{40}]^{4-}$ and $[PW_{12}O_{40}]^{3-}$ (1252, 1342, 2418). The structures of these heteropolytungstates consists of twelve WO_6 octahedra arranged around the tetrahedrally coordinated heteroatom. The octahedra are divided into four tritungstate groups by the edge sharing of octahedra. The tetrahedrally coordinated atom is linked to these four tritungstate groups, which also share corners with each other (Fig. 170). In the most hydrated form, as in $H_3PW_{12}O_{40}$, $29H_2O$, the water molecules are grouped in spherical aggregates of slightly larger size than the heteropolyanions, and keep the heteropolyanions separated (279, 1557), The pentahydrated acids $H_6[H_2W_{12}O_{40}], 5H_2O$, $H_5[BW_{12}O_{40}], 5H_2O$, $H_4[SiW_{12}O_{40}], 5H_2O$ and $H_3[PW_{12}O_{40}], 5H_2O$ are isomorphous (2144).

The Raman spectrum of $[H_2W_{12}O_{40}]^{6-}$ is the same in aqueous solution as in the solid state, showing no change of structure occurs on dissolution.

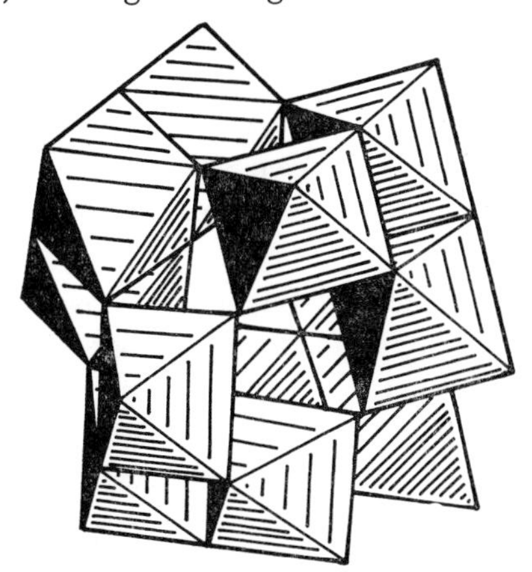

FIG. 170 The structure of the metatungstate anion, $H_2W_{12}O_{40}^{6-}$.

Dehydration experiments on the solids, titrations against alkali, and ^1H–n.m.r. studies in solution, show that two protons are retained in the metatungstate ion $[H_2W_{12}O_{40}]^{6-}$ under all conditions (1879). These are presumably embedded in the heteroatom site at the centre of the metatungstate cage.

4. *Heteropolymolybdates and heteropolytungstates*

(a) *Introduction.* Acidification of molybdate or tungstate solutions containing other ions leads to the formation of polymolybdates or polytungstates in which these extra ions are incorporated into the polyanion structure. The classical examples are those heteropolyanions obtained with phosphate or silicate. For example:

$$PO_4^{3-} + 12\ MoO_4^{2-} + 27\ H^+ \rightarrow H_3[PMo_{12}O_{40}] + 12\ H_2O$$

$$SiO_4^{4-} + 12\ WO_4^{2-} + 28\ H^+ \rightarrow H_4[SiW_{12}O_{40}] + 12\ H_2O$$

At first sight this subject appears to be very complex and rather bewildering, due to the following features:

(i) Over 30 elements can function as this heteroatom. The list includes beryllium, boron, aluminium, gallium, silicon, germanium, tin, phosphorus, arsenic, antimony, bismuth, selenium, tellurium, iodine, all the first row transition elements from titanium to zinc, second and third row transition elements as different as niobium and platinum, cerium, thorium, and other *f*-block elements. Many of these elements can exist in different and often rare oxidation states within the heteropolyanion.

(ii) The stereochemistry of this heteroatom, for the known structural types, can be tetrahedral, octahedral, or icosahedral.

(iii) The stoicheiometric ratio between heteroatom and molybdenum or tungsten can be 1:12 (as above), 1:11, 1:9, 1:$8\frac{1}{2}$, or 1:6. A number of other stoicheiometries have also been proposed, but whether they are real or due to mixtures and/or imprecise analyses has not been proven.

(iv) The structures of the molybdenum–oxygen and tungsten–oxygen skeletons in the heteropolyanions are not necessarily the same as in the isopolyanions.

(v) The heteroatom is normally in the centre of the molybdenum–oxygen or tungsten–oxygen skeleton, which allows the continued polymerization of the molybdate or tungstate before encountering the energy barrier due to molybdenum(VI)–molybdenum(VI) or tungsten(VI)–tungsten(VI) Coulombic repulsions. These complexes may be described as true heteropolyanions and are discussed in more detail below. However other types of association between the free isopolyanion and heteroatoms can occur (1272). The association between alkali metal ions and isopolyvanadates has been referred to on page 185. Similarly the behaviour of the polyniobate ion as a tridentate has been referred to on page 189. Another type of association is found in the "1:3" anion $(NH_4)_6[TeMo_6O_{24}][Te(OH)_6], 7H_2O$, where the $[Te(OH)_6]$ molecules are strongly hydrogen-bonded to the $[TeMo_6O_{24}]^{6-}$ anions (795).

The structurally characterized heteropolyanions can be divided into three groups, depending upon whether the heteroatom is tetrahedrally coordinated, octahedrally coordinated, or icosahedrally coordinated.

(b) *Heteropoly anions with tetrahedrally coordinated heteroatoms.* The most commonly studied heteropolyanions with heteroatom: molybdenum or tungsten ratios of 1:12, are of this type. The best known examples are $[PMo_{12}O_{40}]^{3-}$, $[PW_{12}O_{40}]^{3-}$, $[SiMo_{12}O_{40}]^{4-}$, $[SiW_{12}O_{40}]^{4-}$ and $[BW_{12}O_{40}]^{5-}$.

The molybdenum compounds are yellow, whereas the tungsten analogues are colourless.

The free acids, and many of their salts, are very soluble in water, which combined with their high molecular weights produce very dense solutions. The alkali metal salts however are relatively insoluble, for example potassium, rubidium and caesium can be precipitated from aqueous solution as the phosphotungstates, or rubidium and caesium as the silicotungstates. The insolubility of ammonium phosphomolybdate is also used in the analytical determination of phosphate.

These heteropolyacids also show a very unusual affinity for other oxygen–donor ligands, and show particularly remarkable behaviour with diethyl ether, which is used in the preparation of these complexes. When an aqueous solution of these heteropolyacids is shaken with diethyl ether, three phases separate; the least dense is an ethereal layer containing some dissolved heteropolyacid, which is then followed by an aqueous layer which contains very little of the heteropolyacid, but in addition there is also a very dense oil which is the heteropolyacid–ether complex. The quaternary system phosphomolybdic acid-sulphuric acid–water–ether has been studied in some detail (2179). The extractive behaviour of heteropolyacids using oxygen or nitrogen containing molecules has also been reviewed (30).

The heteropolymolybdates decompose above acid concentrations of about one molar, but the heteropolytungstates are more stable to acid conditions. Salts of the former anions are similarly more easily decomposed by heat than those of the latter anions. For different heteroatoms the order of stability is, for example, $[PW_{12}O_{40}]^{3-} > [SiW_{12}O_{40}]^{4-} > [BW_{12}O_{40}]^{5-}$.

The 1:12 heteropolyanions are composed of twelve octahedrally coordinated molybdenum or tungsten atoms grouped around the tetrahedrally coordinated heteroatom, as has been discussed on page 297 (Fig. 170) when dealing with the structure of metatungstate. Ionic weight (1366, 1664), size (132, 1484), and X-ray scattering (1521) measurements show that this structure is retained in solution.

The behaviour of heteropolyanions upon addition of base in aqueous solution has been studied for many years, but it is only recently that these reactions have been understood. The problem is that more moles of base per mole of heteropolyacid are consumed than there are protons in the free acid. The resultant high charge on the anion and the stoicheiometry of the salts obtained was also difficult to explain. It is now believed that these observations are due to the formation of the 1:11 anions (2181):

$$6\,H_3[PW_{12}O_{40}] + 45\,OH^- \rightarrow 6\,[PW_{11}O_{39}]^{7-} + HW_6O_{20}{}^{3-} + 13\,H_2O$$

Fig. 171 shows the growth of the 1:11 phosphomolybdate at the expense

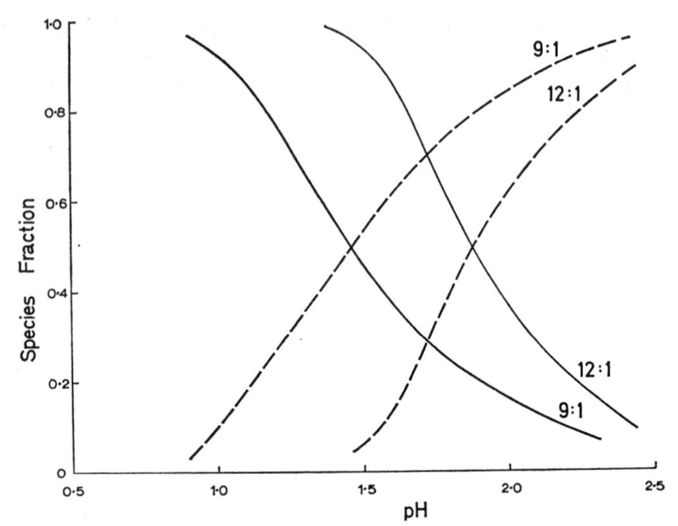

FIG. 171 Proportion of $PMo_{11}O_{39}^{7-}$ (broken lines) and $PMo_{12}O_{40}^{3-}$ (full lines) as a function of pH for molybdate : phosphate stoicheiometric ratios of 12:1 and 9:1.

of the 1:12 phosphomolybdate as the pH is increased. Fig. 172 summarizes the results for a number of heteropolyacids.

By the appropriate control of pH, mixed heteropolyanions can be synthesized. For example if a solution of $[SiW_{12}O_{40}]^{4-}$ is adjusted to pH 7, and cobalt(II) added, the anion $[SiCoW_{11}O_{40}]^{8-}$ is obtained, where one of the tungsten(VI) ions has been replaced by a cobalt(II) ion (128).

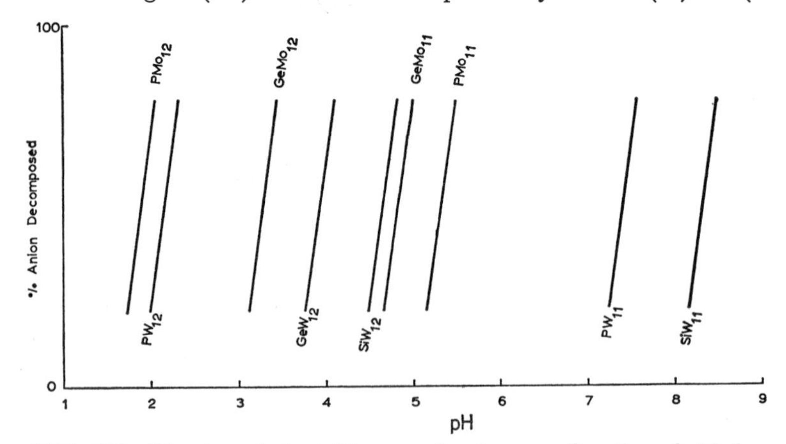

FIG. 172 The degradation of heteropolyanions as a function of pH, for molybdenum and tungsten concentrations of about $0 \cdot 1$ *M*. SiW_{11} signifies $[SiW_{11}O_{39}]^{8-}$, etc.

The similar replacement of one tungsten atom in $[BW_{12}O_{40}]^{5-}$, $[SiW_{12}O_{40}]^{4-}$, $[GeW_{12}O_{40}]^{4-}$, $[PW_{12}O_{40}]^{3-}$, $[AsW_{12}O_{40}]^{3-}$ and $[H_2W_{12}O_{40}]^{6-}$ by a transition metal ion (Mn^{II}, Fe^{II}, Co^{II}, Ni^{II}, Cu^{II}, Zn^{II}, Cr^{III}, Mn^{III}, Fe^{III}, Co^{III}, (Ga^{III}), Ti^{IV}, V^{IV}, V^{V}, Mo^{VI}) appears to be a general reaction, and occurs under conditions where the 12-heteropolytungstates are partly degraded (128, 1642, 1864, 1927, 2205, 2289, 2292). Replacement of additional atoms can be achieved with vanadium(V) (page 188).

When solutions of the 1:12 heteropolyanions $[PMo_{12}O_{40}]^{3-}$,

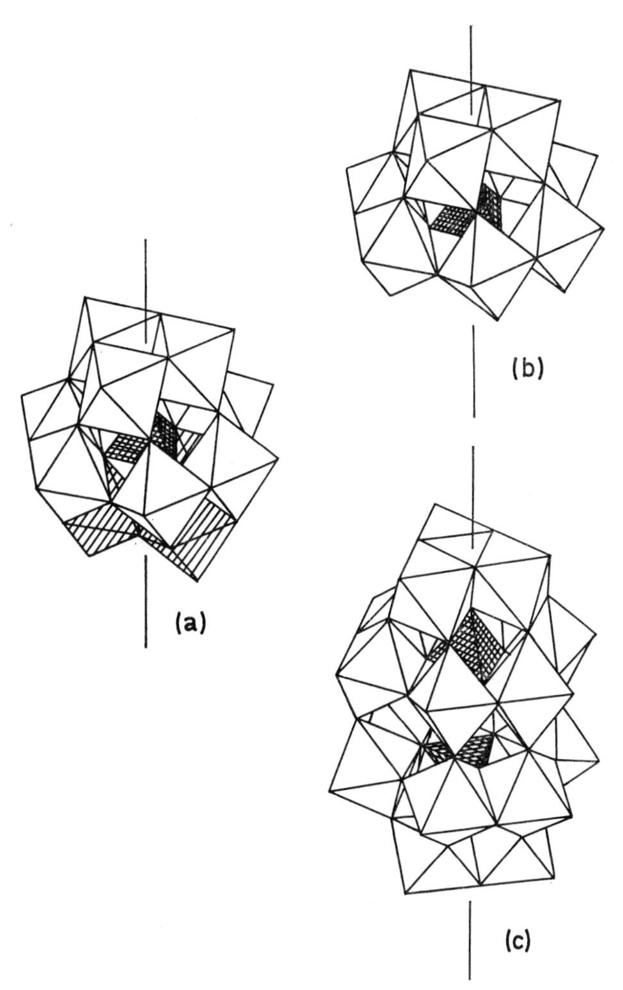

FIG. 173 (a) The 12-tungstophosphate anion $[PW_{12}O_{40}]^{3-}$. (b) The complete "half unit". (c) The 2:18 tungstophosphate anion.

$[PW_{12}O_{40}]^{3-}$, $[AsMo_{12}O_{40}]^{3-}$ and $[AsW_{12}O_{40}]^{3-}$ are aged for some weeks, or gently heated, the 2:18 heteropolyanions are formed (2187). For example:

$$3\,[PW_{12}O_{40}]^{3-} + PO_4^{\,3-} \rightarrow 2\,[P_2W_{18}O_{62}]^{6-}$$

The structure (632) (Fig. 173c) is closely related to that of $[PW_{12}O_{40}]^{3-}$. The 2:18 anion consists of two identical half-units related by a plane of symmetry perpendicular to the three-fold rotation axis. Each half unit is equivalent to the residue obtained by removing three corner-sharing WO_6 octahedra (shaded in Fig. 173a) from the 1:12 anion.

These 2:18 heteropolyanions can form 2:17 heteropolyanions by the suitable control of pH, in the same way that 1:11 heteropolyanions are formed from 1:12 heteropolyanions (2180). Similarly it is found that these 2:17 hetereopolyanions can incorporate metal ions forming anions of the general types $[H_2P_2M^{II}W_{17}O_{62}]^{8-}$ (where M^{II} is Mn, Co, Ni or Zn) and $[H_2P_2M^{III}W_{17}O_{62}]^{7-}$ (where M^{III} is Mn or Co) (1642, 2289, 2292).

The 1:12 heteropolyanions, and also metatungstate, are relatively easily reduced producing characteristic very intense blue colours. These are used analytically for the determination of, for example, silicon, germanium, phosphorus and arsenic (1292). The reduction occurs in three stages corresponding to the total addition of one, two, and four electrons per heteropolyanion (1661, 1880, 2185, 2188, 2189). Many of these reduced anions have been isolated as salts (1113, 1114, 1115, 1116, 2290, 2291). The related 1:11, 2:18 and 2:17 heteropolyanions are similarly reduced (1877, 1878, 2182).

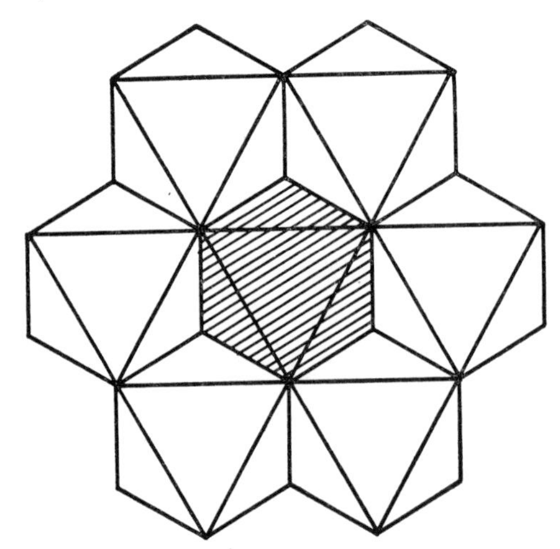

FIG. 174 The structure of $[TeMo_6O_{24}]^{6-}$. The TeO_6 octahedron is shaded.

(c) *Heteropolyanions with octahedrally coordinated heteroatoms.* Acidication of mixtures of molybdenum(VI) and tellurium(VI) or many trivalent metals form the 1:6 heteropolyanions $[Te^{VI}Mo_6O_{24}]^{6-}$ and $[M^{III}Mo_6O_{24}]^{9-}$ (where M is Al, Ga, Cr, Fe, Co or Rh) respectively (129, 1936, 2410). The structure (Fig. 174) shows an octahedrally coordinated tellurium atom surrounded by a planar annulus of MoO_6 octahedra sharing edges (792, 795, 1860).

Manganese(IV) and nickel(IV) form the 1:9 heteropolymolybdates $[MnMo_9O_{32}]^{6-}$ and $[NiMo_9O_{32}]^{6-}$ respectively (131, 2352). The structure

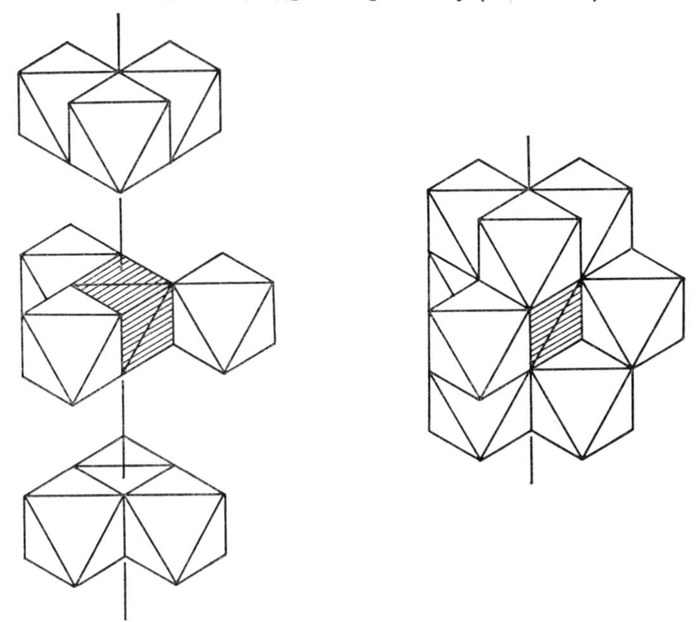

FIG. 175 The structure of the $[MnMo_9O_{32}]^{6-}$ anion. Left: an "exploded" view of the ion. The MnO_6 octahedron is shaded.

is shown in Fig. 175, together with a structure which has been "exploded" along the three-fold axis.

It has been pointed out (page 58) (1354) that the structures of both these 1:6 and 1:9 types are those in which the Coulombic repulsion between the metal atoms is minimized.

Polymerization can continue further if two octahedrally coordinated heteroatoms are incorporated into the structure, as in $[H_4Co_2^{III}Mo_{10}O_{38}]^{6-}$ (Fig. 40, page 60) (798).

(d) *Heteropolyanions wtih icosahedrally coordinated heteroatoms.* Cerium(IV), zirconium(IV) and thorium(IV) form 1:12 heteropolymolybdates of the type $[CeMo_{12}O_{42}]^{8-}$ (130). The structure (Fig. 176) shows that the cerium atom is surrounded by six Mo_2O_9 units which share corners with each other; each Mo_2O_9 unit is constructed of two MoO_6 octahedra sharing a common face (658).

FIG. 176 The structure of $[CeMo_{12}O_{42}]^{8-}$ showing linking of the MoO_6 octahedra.

H. Organometallic Complexes

As would be expected, organometallic derivatives of this oxidation state are rare, and are restricted to oxo derivatives such as $(C_5H_5)MoO_2Cl$, $(C_5H_5)MoO_2Br$, $[(C_5H_5)MoO_2]_2O$, $(C_5H_5)WO(OMe)Cl_2$, $(C_5H_5)WO(OMe)_2Cl$ and $(C_5H_5)_2WOCl_2$ (51, 581, 582).

2. OXIDATION STATE V

A. Pentahalides

1. *Pentafluorides*

Chromium pentafluoride has been formed by the direct fluorination of the metal, or of some chromium salts, preferably under high pressure (489, 728, 1012, 2340). It is a bright red volatile compound, melting at 30° C. The structure is the same as that found for VF_5, TcF_5 and ReF_5, and is closely related to that of $MoOF_4$, consisting of chains of corner sharing octahedra with the two bridging fluorine atoms in a mutually *cis* position (Fig. 94, page 155) (728, 736).

Molybdenum pentafluoride was first formed by the action of fluorine on molybdenum hexacarbonyl at −75 °C, and heating the resultant olive green solid to 170° C forming the pentafluoride as a yellow sublimate and leaving the tetrafluoride as a light green residue (1848). It has also been formed directly from the elements at 400° C (731), and by reduction of the hexafluoride with molybdenum metal at 60° C (1681). Possibly the most satisfactory preparation is the reduction of molybdenum hexafluoride with phosphorus trifluoride (1807):

$$2\,MoF_6 + PF_3 \rightarrow 2\,MoF_5 + PF_5$$

The yellow solid melts at 67° C (403, 741), and disproportionates to MoF_4 and MoF_6 above about 150° C.

Molybdenum pentafluoride is isostructural with NbF_5 and TaF_5 (Fig. 95 page 157). Octahedrally coordinated metal atoms share two mutually *cis* fluorine atoms with two neighbouring metal atoms as noted for CrF_5 above, but the octahedra are linked into tetrameric molecules rather than infinite strings (741). The Mo–F–Mo bonds are almost linear, which has been interpreted as being due to significant fluorine–metal π bonding (418). The room temperature effective magnetic moment is 1·0 B.M.

Molybdenum(V) also forms the mixed halides $MoCl_2F_3$ (1438, 1813), $MoBrF_4$ (1681) and probably $MoCl_4F$ (1813). A mixed chloride–fluoride with a formal oxidation state of +4·5, namely $Mo_2Cl_3F_6$, is mentioned further under molybdenum(IV) (page 323).

Tungsten pentafluoride has been formed by heating tungsten wires to 500°–700° C in a WF_6 atmosphere, and collecting the yellow product on the walls at −50° C to −60° C (2074). It is very unstable to air, and disproportionates to the tetrafluoride and hexafluoride at 50°–70° C. It is isomorphous with molybdenum pentafluoride (730).

2. *Pentachlorides and bromides*

Molybdenum pentachloride is readily prepared by direct chlorination of the metal at 300°–400° C (530, 1853, 2337). The compound is black when pure, but often appears green due to a surface coating of $MoOCl_4$, which can be removed by washing with carbon tetrachloride (solubility of the pentachloride is 1–2% at 25° C), or by vacuum sublimation at 80°–90° C. The brown colour also often observed is due to the tetrachloride, formed by the ready disproportionation of the pentachloride to the tetrachloride and chlorine at relatively low temperatures.

Molybdenum pentachloride has also been formed by the action of carbon tetrachloride on molybdenum trioxide in sealed tubes at 350°–400° C (1433):

$$2\,MoO_3 + 6\,CCl_4 \rightarrow 2\,MoCl_5 + 6\,COCl_2 + Cl_2$$

The product again appears to contain molybdenum oxychlorides and tetrachloride. For example in a typical experiment at 400° C, the product analysed as $MoCl_{4 \cdot 93}$ although the pressure of chlorine must have been of the order of ten atmospheres.

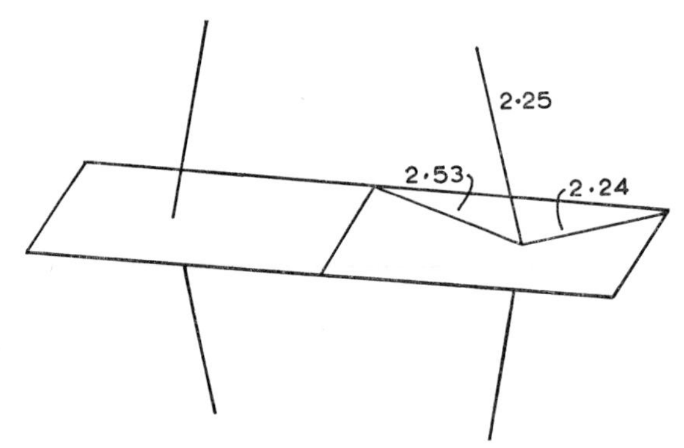

FIG. 177 The structure of $MoCl_5$.

The structure of molybdenum pentachloride is shown in Fig. 177 (1987), and may be compared with that of niobium pentachloride (Fig. 96, page 159).

The magnetic susceptibility obeys the Curie–Weiss law with $\theta \sim 0°$ C and $\mu_{eff} = 1 \cdot 67$ B.M. (530). The lower values around $1 \cdot 5$ B.M. obtained earlier were due to contamination by $MoOCl_4$ and/or $MoCl_4$ (1424, 1430).

The very high entropy of vaporization of $35 \cdot 6$ e.u. suggests that the dimeric structure of the solid is broken and the molecule is monomeric in the vapour. This has been confirmed by the mass spectrum (2029). The Raman spectrum is consistent with a trigonal bipyramidal structure (176).

Chlorine is slowly evolved from solutions of molybdenum pentachloride in carbon tetrachloride, leaving residues which appear to be mixtures of lower chlorides and oxychlorides (1853). In spite of the complex equilibria which appear to be involved, and the sensitivity of the system to water, the ultra violet, visible and infrared spectra of these solutions, and of $MoCl_5$ vapour, have been interpreted as being due to a monomeric trigonal bipyramidal molecule, with Jahn–Teller distortions both in the ground and electronically excited states (109, 110, 451).

Tungsten hexachloride can be reduced to the dark green pentachloride with hydrogen at 380°–410° C (35, 229, 332), or with red phosphorus (1800) or stannous chloride (1459). Alternatively the hexachloride can be slowly disproportionated to the pentachloride and chlorine at its boiling point (531). Another convenient synthesis is from the hexachloride and tetrachloroethylene, with the formation of hexachloroethane (369).

The compound is isomorphous with molybdenum pentachloride (269). The magnetic moment at room temperature is about 1·1 B.M., but is dependent upon the preparation. It increases slightly at lower temperatures (214, 1424), although the reverse behaviour has also been noted. (531)

Tungsten pentabromide can be conveniently prepared directly from the elements above about 700° C (331, 332, 1424). Alternatively the tungsten hexabromide formed by the bromination of tungsten hexacarbonyl below 0° C can be disproportionated above 250° C to the pentabromide and bromine.

B. Reactions of the Pentahalides

The reactions of molybdenum pentachloride are complex, and have been extensively studied. Less work has been carried out on the other pentahalides.

In addition to the formation of anionic halo and oxohalo complexes described in the following sections, the reactions of the pentahalides may be conveniently classified under the following headings:

1. *Adduct formation.* The formation of $MX_5(\text{ligand})_n$ is relatively rare for molybdenum pentachloride, but not for tungsten pentachloride. Molybdenum pentafluoride forms both 1:1 and 1:2 adducts with ligands such as ammonia, pyridine, acetonitrile, dimethyl ether, and dimethyl sulphide (1681).

2. *Oxygen abstraction.* Oxygen containing molecules readily form oxomolybdenum complexes such as $MoOCl_3(\text{ligand})_n$ and eventually $MoO_2Cl_2(\text{ligand})_n$, the properties of which are given more fully in the appropriate sections.

3. *Reduction and/or disproportionation.* Potential reducing agents readily form molybdenum(IV) and tungsten(IV) complexes, the properties of which are also given more fully in the appropriate section.

4. *Alkoxides and amides.* Molecules containing OH or NH groups readily eliminate hydrogen halide leaving in many instances complicated mixtures of products containing the pentavalent metal.

5. *Other reactions.*

These reactions will now be considered individually in more detail. In many instances a single reagent produces a mixture of these products. For example the reaction of molybdenum pentachloride with ethers, even when scrupulously dry and free from dissolved oxygen, usually forms a mixture of $MoCl_5$(Ether), molybdenum(IV) compounds, and oxomolybdenum(V) compounds (1356).

1. *Adduct formation*

This is certainly one of the reactions occurring with many ligands, but pure adducts have been characterized only in a few cases. The best example with an oxygen donor is the highly air-sensitive, dark violet $MoCl_5$(dioxan), for which the infrared spectrum and room temperature effective magnetic moment of 1·36 B.M. are consistent with octahedral molybdenum(V) (1356). However as with other oxygen-containing ligands, the molybdenum atom slowly abstracts the coordinated oxygen atom to form $MoOCl_3,1\frac{1}{2}$dioxan and 2,2′-dichlorodiethyl ether (see below).

The compound "$MoCl_5,Ph_3AsO$" (1218) is not a simple adduct, as oxygen abstraction has again occurred forming $(Ph_3AsCl)[MoOCl_4]$ (1356).

Another example may be the dark green $MoCl_5, POCl_3$ prepared by the action of phosphorus pentachloride on molybdenum trioxide, reported in 1879 (1873). This compound may be an oxomolybdenum compound, or it may be similar to the octahedral $NbCl_5, POCl_3$ (page 163). It is relevant that $NbCl_5$ does not abstract oxygen atoms from $POCl_3$, although it does from a number of other ligands.

Nitrogen–donor ligands can produce molybdenum(IV) complexes with molybdenum pentachloride, but there are examples where reduction does not occur. 2,4,6-Trimethylpyridine forms $[MoCl_4(Me_3C_5H_2N)_2]Cl$, for which the room temperature effective magnetic moment is 1·41 B.M. (372). Reaction with trimethylamine for several days forms the black $[MoCl_5(Me_3N)]$, while reaction in a sealed tube for six months forms the red $MoCl_5,2Me_3N$; the high magnetic moments of 1·80 and 1·85 B.M. respectively indicate some reduction has occurred. Further reduction occurs if trimethylamine vapour is passed into molybdenum pentachloride dissolved in carbon tetrachloride (746). Triethylamine similarly forms the dark brown $[MoCl_5(Et_3N)]$ (746).

The tungsten pentahalides are a little more resistant to reduction. Complexes $[WX_4(ligand)_2]X$ (where X is Cl or Br, and the ligand is 2,4,6-trimethylpyridine, benzonitrile, pyridine, $\frac{1}{2}$(dipyridyl), $\frac{1}{2}$(*o*-phenanthroline) or $\frac{1}{2}(Ph_2P.CH_2.CH_2.PPh_2)$] have been obtained, with magnetic moments in the range 0·63 –1·45 B.M. (268, 372). Diethyl ether forms $WBr_5(Et_2O)$ (959).

2. *Oxygen abstraction*

Molybdenum pentachloride reacts with oxygen–containing molecules, with abstraction of the oxygen atom and the formation of green oxomolybdenum complexes (1356).

In dioxan the compound $MoCl_5$(dioxan) is initially obtained, but after leaving in dioxan at room temperature for some weeks, $MoOCl_3,1\frac{1}{2}$ dioxan and 2,2′-dichlorodiethylether are obtained. Similarly tetrahydrofuran forms $MoOCl_3(THF)_2$ and 1,4-dichlorobutane (1356). Similar but less well defined reactions occur with a large number of aliphatic and aromatic ethers (1356, 2337).

Oxygen abstraction occurs more rapidly with molecules such as triphenylarsine oxide (forming $(Ph_3AsCl)[MoOCl_4]$), triphenylphosphine oxide, dimethylsulphoxide, dimethylformamide, acetone, and sulphur dioxide, although in most cases the organic product has not been characterized (34, 743, 1218, 1356). Perfectly dry molecular oxygen however does not react with molybdenum pentachloride in carbon tetrachloride.

A solution of molybdenum pentachloride in acetone, ethanol, or similar solvent, is therefore a convenient route for the synthesis of a number of green oxomolybdenum(V) complexes (see later).

Oxygen abstraction reactions with tungsten pentahalides have not been so extensively studied.

3. *Reduction and/or disproportionation*

Molybdenum(IV) compounds are often obtained from reactions involving molybdenum pentachloride, and may be qualitatively distinguished from molybdenum(V) products, since on hydrolysis in water the former give red solutions while the latter give green solutions (1356).

Molybdenum tetrachloride has been prepared by reduction of the pentachloride with benzene, the tetrachloride precipitating according to the reaction:

$$2\,MoCl_5 + C_6H_6 \rightarrow 2\,MoCl_4 + C_6H_5Cl + HCl$$

Similarly chlorobenzene produced dichlorobenzenes (1463, 1504). Although this is a convenient method for the preparation of relatively large quantities of molybdenum tetrachloride, the product is contaminated with polyphenylene carbonaceous materials. Hydrocarbon solvents also slowly produce lower valence molybdenum compounds (1356, 1853, 2106). There is usually some free chlorine also evolved in these reactions (1355), and it appears that the first step of the reaction is the loss of a chlorine atom. A particularly suitable reagent to remove this chlorine atom is tetrachloroethylene (see $MoCl_4$, page 324).

Under some conditions the reverse reaction, the formation of pentachloride from the tetrachloride and chlorine, occurs in solution. For example if chlorine is passed through a slurry of MoO_2 in hexachlorobutadiene, mixtures of $MoCl_4$ and $MoCl_5$ are obtained (43, 100, 1136, 1357).

Molybdenum pentachloride and alkyl cyanides produce the molybdenum(IV) compounds $MoCl_4(RCN)_2$, where R is Me, Et or Pr (35, 1356). Similarly pentamethylene oxide forms $MoCl_4[(CH_2)_5O]_2$ in addition to $MoOCl_3[(CH_2)_5O]_2$ (820). These compounds have effective magnetic moments in the range 2·37–2·52 B.M. Reduction also occurs with pyridine (2337).

Reduction and/or disproportionation to tungsten(IV) also occurs with acetonitrile, pyridine, dipyridyl, trimethylamine, triethylamine, triphenylphosphine and tetrahydrofuran (35, 268, 332, 1584).

4. *Alkoxides and amides*

Compounds containing –OH or –NH groups react vigorously with molybdenum pentachloride with the evolution of hydrogen halide. This reaction always occurs unless solvents are thoroughly dry and free from alcohols. The products are significantly different to those obtained, for example, from niobium pentachloride (pages 174 and 178).

Molybdenum pentachloride with methanol forms $MoCl_3(OMe)_2$, $MoCl_3(OMe)_2(MeOH)$, $MoCl_3(OMe)_2(MeOH)_3$, $MoOCl_3(MeOH)_2$, $MoCl_2(OMe)_3$, $MoCl(OMe)_4$, or $MoO(OMe)_3$, $\frac{1}{2}$MeOH depending upon conditions (301, 954, 960). In the presence of organic base or cation, anionic complexes such as $[MoCl_4(OMe)_2]^-$ are formed. The e.s.r. spectra of these compounds indicate that the alkoxide groups are *trans*; magnetic moments lie within the range 1·71–1·88 B.M. at room temperature (301, 960, 1595).

Tungsten pentachloride (or tungsten hexachloride) behaves in a similar manner (958, 1046, 1418, 1924, 1925). Methanol produces the green $WCl_3(OMe)_2(MeOH)_3$, and in the presence of bases diamagnetic pentavalent compounds of the type $W_2Cl_2(OMe)_8$, $W_2Cl_4(OMe)_6$ and $W_2Cl_6(OMe)_4$ are obtained. This last compound reacts with acetonitrile or tetrahydrofuran to form the apparently octahedral $[WCl_3(OMe)_2(\text{ligand})]$. The dimeric compounds also form complexes with molecular nitrogen, which can then be reduced with sodium naphthalide to ammonia. In the presence of base, salts containing anions of the type $[WCl_5(OMe)]^-$, $[WCl_4(OMe)_2]^-$ and $[WOCl_4]^-$ can be obtained. In one instance the seven coordinate anion $[WCl_6(OEt)]^{2-}$ was obtained.

Tungsten pentabromide forms $WBr_3(O\phi)_2$, $WBr_2(O\phi)_3$ and $WBr_2(O\phi)_3(\phi OH)$ with a number of substituted phenols in carbon tetrachloride, chloroform, carbon disulphide or diethyl ether (959).

Elimination of hydrogen halide similarly occurs with carboxylic acids forming, for example, $MoCl_3(OAc)_2$ (1502).

The reaction of molybdenum pentachloride with N–H groups is more complex due to the simultaneous formation of the amine hydrochloride.

Dimethylamine forms $[MoCl_3(Me_2N)_2(Me_2NH)]$, whereas diethylamine and dipropylamine form $[MoCl_3(R_2N)_2]_2$ (304, 746). The tungsten pentahalides form the diamagnetic $[WX_2(Me_2N)_2(Me_2NH)]_2$ and $[WX_3(Et_2N)(Et_2NH)_2]$, although reduction as well as solvolysis is observed (332). Similar behaviour is observed with primary amines (332, 746).

The extraction of molybdenum pentachloride with liquid ammonia forms the orange, paramagnetic $MoCl_3(NH_2)_2(NH_3)$, which on heating successively forms $MoCl_3(NH_2)_2$, $MoCl(NH_2)_2$, and finally Mo_2N. Under other conditions $MoCl_2(NH)(NH_2)$ and $MoCl(NH)(NH_2)_2$ can be formed (747).

5. *Other reactions*

At 0° C, molybdenum pentachloride is oxidized by chlorine azide as shown:

$$MoCl_5 + ClN_3 \rightarrow MoCl_4(N_3)_2 + Cl_2$$

At 20° C, this decomposes to $MoNCl_3$, nitrogen and chlorine. Tungsten hexachloride similarly forms $WNCl_3$ (650).

At room temperature in acetone, methyl ethyl ketone or dioxan, alkali metal thiocyanates are reported to form $MoCl(NCS)_4(solvent)_2$, while on warming complete replacement of the chloride occurs forming $Mo(NCS)_5(solvent)_2$ (952).

C. Oxohalides and Adducts

1. *Oxohalides*

The oxytrihalides $CrOF_3$, $CrOCl_3$, $MoOCl_3$, $MoOBr_3$ $WOCl_3$ and $WOBr_3$, and also the mixed $WOCl_{1\frac{1}{2}}Br_{1\frac{1}{2}}$ are known. In addition dioxodihalides MoO_2Cl and WO_2I have been described.

The reaction of BrF_3, BrF_5 or ClF_3 on CrO_3 at room temperature formed $CrOF_3,0\cdot25BrF_3$, $CrOF_3,0\cdot25BrF_5$ and $CrOF_3,0\cdot30ClF_3$ respectively; attempts to obtain $CrOF_3$ free from halogen fluoride were unsuccessful. These red compounds had room temperature effective magnetic moments within the range 1·76–2·02 B.M. (490).

The reaction of CrO_3 or CrO_2Cl_2 with thionyl chloride, sulphuryl chloride (1467) or boron trichloride (1300) forms the dark red $CrOCl_3$. The room temperature effective magnetic moment is 1·80 B.M. The compound is monomeric in nitrobenzene, but disproportionates above 0° C into CrO_2Cl_2 and chromium(III).

There are two different structural isomers of $MoOCl_3$, one containing terminal Mo=O units and the other bridging Mo–O–Mo units. The former is prepared by the reduction of $MoOCl_4$ with aluminium at

130° C followed by vacuum sublimation from the $MoOCl_2$ also formed (1680), by the thermal decomposition of $MoOCl_4$ (530), or by the action of liquid sulphur dioxide (743) or antinomy trioxide (610) on $MoCl_5$. The structure (Fig. 178) shows an octahedral arrangement of four bridging chlorine atoms, one terminal chlorine atom, and one terminal oxygen atom, about the molybdenum atom. The metal atom is displaced away from the centre of the octahedron towards the oxygen atom (Mo–O = 1·60 Å), so that the molybdenum–chlorine bond *trans* to the oxygen atom is longer than the other molybdenum-bridging chlorine bonds

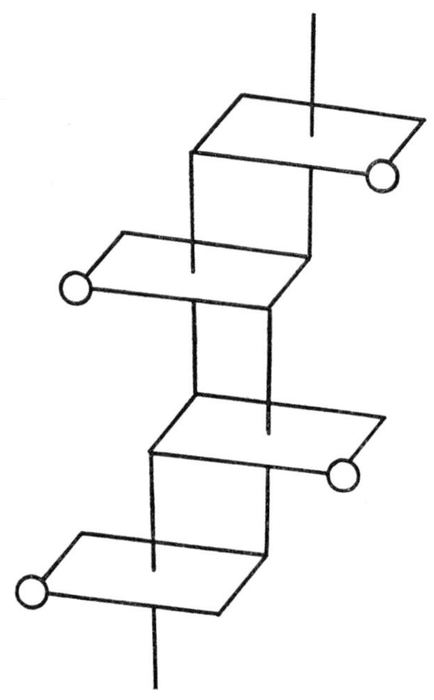

FIG. 178 The structure of $MoOCl_3$ (isomer I). (Compare with Fig. 56).

(2·81 Å compared with 2·37–2·48 Å respectively), or the molybdenum-terminal chlorine bond (2·28 Å). The two pairs of bridging chlorine atoms link the octahedra into infinite strings parallel to the *c*-axis of the monoclinic unit cell. This form has an effective magnetic moment of 1·62–1·69 B.M. at room temperature (535, 608, 743).

The second form of $MoOCl_3$ is prepared by the reduction of $MoOCl_4$ with HI in liquid sulphur dioxide at −23° C. It is isomorphous with the tetragonal $NbOCl_3$ (Fig. 98, page 167), and thus contains infinite -Mo–O–Mo–O- strings (1680).

The oxobromide $MoOBr_3$ has also been reported, and is isomorphous with the tetragonal form of $MoOCl_3$, that is the structure contains bridging oxygen atoms (530, 608, 711).

Tungsten oxotetrachloride is reduced to the oxotrichloride with aluminium in a sealed tube at 100°–140° C (608, 901, 1680), by benzene or tetrachloroethylene under reflux for several days (369, 1680), or by hydrogen iodide in liquid sulphur dioxide at —23° C (1680). It is also prepared by the action of antimony trioxide on the pentachloride (610). In addition $WOCl_{1\frac{1}{2}}Br_{1\frac{1}{2}}$ (1358) and $WOBr_3$ (535, 608, 610, 2281) are formed by similar methods. These three tungsten oxotrihalides are isomorphous with the tetragonal $MoOCl_3$ and $NbOCl_3$.

The structurally uncharacterized MoO_2Cl is formed by the reduction of MoO_2Cl_2 with aluminium at 200° C (1680), or by the thermal decomposition of $MoCl_3(OEt)_2$ *in vacuo* (535).

The only pentavalent oxoiodide known, WO_2I, is prepared by the transport reaction between tungsten, tungsten trioxide, and iodine (2015, 2279).

2. *Oxohalide adducts*

A very large number of adducts of the type $MoOCl_3$(ligand) or $MoOCl_3(\text{ligand})_2$ have been prepared by a wide variety of methods. These include the direct addition to $MoOCl_3$ (820), from $MoCl_5$ in moist solvents (1691), from $MoCl_5$ using oxygen abstraction reactions from either the ligand (820) or solvents such as acetone (1218), by the reduction of $MoOCl_4$ with the ligand used (1505), or by the oxidation of zerovalent substituted carbonyls (1531). The monoadducts are formed with R_2S (where R is Me, Et or Pr), Ph_2CO and $POCl_3$, whereas the *bis*-adducts are formed with $(CH_2)_4O$, $(CH_2)_5O$, MeOH, Ph_3PO, Ph_3AsO, Me_2SO, $(CH_2)_4S$, $(CH_2)_5S$, C_5H_5N, and Ph_3P (745, 820, 960, 1017, 1218, 1436, 1505). The mixed $MoOCl_3(Ph_3AsO)(Me_2CO)$ and $MoOCl_3(Ph_3PO)$(dioxan) are also known (1218, 1356). Bidentates, or potential bidentates, which similarly form $MoOCl_3$(bidentate) include $MeO.CH_2.CH_2.OMe$, dioxan, thioxan, $Ph_2P(O).CH_2.CH_2.P(O)Ph_2$, $(Me_2N)_2P(O).O.P(O)(NMe_2)_2$, acacH, dipyridyl, *o*-phenanthroline, and *o*-$C_6H_4(AsMe_2)_2$ (745, 820, 1298, 1359, 1436, 1505, 1531, 1691).

All these compounds have magnetic moments close to the spin-only value (1·64–1·75 B.M.) and follow the Curie law. A strong band is shown at 950–1000 cm^{-1} in the infrared spectrum due to the terminal oxygen atom. All compounds are green with absorptions at about 14 000 and 22 000 cm^{-1} assigned to the $^2B_2 \rightarrow {}^2E(I)$ transition involving the π-orbitals of the oxygen atom, and to the ligand field transition $^2B_2 \rightarrow {}^2B_1$ respectively These spectra confirm that the tetragonal distortion due to the oxygen atom is the most important feature of the molecular orbital scheme.

Similar complexes of the type $MoOBr_3(ligand)_2$ and $MoOBr_3$(bidentate) have also been described (427, 609, 1436, 1531).

Slightly different behaviour is observed for adducts of tungsten oxotrichloride (609, 903). In addition to $WOCl_3(ligand)_2$ (with $(CH_2)_4O$, dioxan, MeCN and C_5H_5N) and $WOCl_3$(bidentate) (with dipyridyl and dithian), which have tungsten–oxygen stretching frequencies at about 1000 cm^{-1} and effective magnetic moments of about 1·5 B.M., the formation of dark blue $WOCl_3(py)$ and $WOCl_3(Me_3N)$ also occurs. These have much lower effective magnetic moments of 0·79 and 0·58 B.M. respectively, and although the infrared spectra were not recorded they were formulated as having infinite –W–O–W–O– chains.

D. Anionic Halo and Oxohalo Complexes

1. *Fluoro and oxofluoro complexes*

Complex fluoro anions of chromium(V) do not appear to be known. The action of BrF_3 on $K_2Cr_2O_7$ forms $K(CrOF_4)$ (2112), for which the room temperature effective magnetic moment is 1·76 B.M. (1803).

Salts of MoF_6^- and WF_6^- have been prepared by the reduction of the hexafluorides with liquid sulphur dioxide in the presence of an alkali metal fluoride (1132, 1134, 1220), or with nitric oxide forming the nitrosyl salts (988). They have also been prepared by the action of an alkali metal fluoride (or iodide) on the hexacarbonyls in liquid IF_5 (1133, 1135). Salts such as K_3MoF_8 and K_3WF_8 may also be obtained (1133); these differences in stoicheiometry, and also whether tungsten(VI) or tungsten(V) complex fluorides are obtained, depend upon the particular reaction conditions.

The hexafluorometallate anions are octahedral, although the details of the ionic lattice depend upon the cation present. $Li(MF_6)$ and $Na(MF_6)$, where M is Mo, W, V, Nb or Ta, have the sodium chloride or a closely related structure, whereas the K^+, Rb^+, Cs^+, Ag^+ and Tl^+ salts of these anions have the caesium chloride or a closely related structure (1132, 1347). In $NaMoF_6$ the Mo–F distance is 1·74 Å (739). K_3MoF_8 has a cubic structure, although the structural details are not known (1135).

The room temperature magnetic moments are in the range 1·24–1·75 B.M. for the molybdenum salts, and 0·5–0·6 B.M. for the tungsten salts. In both cases the Curie–Weiss law is followed with very large values of θ, which is dependent upon cation, indicating some sort of antiferromagnetic interaction (1132, 1133, 1134, 1135).

The only apparent complex oxofluoride of molybdenum(V) or tungsten (V) to have been prepared is K_2MoOF_5, which was formed by fusing $KMoF_6$ and KHF_2, and extracting with a water/acetone mixture (1132).

2. *Chloro, bromo, oxochloro and oxobromo complexes*

(a) *Chromium.* Complex chloro anions of chromium(V) are again not known.

Reduction of CrO_3 in glacial acetic acid with dry hydrogen chloride at 0° C, followed by the requisite amount of the appropriate chloride ACl (where A is Me_4N, pyH, Cs, etc.), precipitates the orange-yellow $A[CrOCl_4]$ or the red to brown $A_2[CrOCl_5]$. The room temperature effective magnetic moments of these compounds are in the range 1·80–1·93 B.M., and they follow the Curie–Weiss law (91, 352, 1467). The very narrow e.s.r. line, the strong infrared band at 925–950 cm^{-1}, and the structure of $Cs_2(CrOCl_5)$, all show the presence of a Cr=O bond (352, 1447, 1448).

(b) *Molybdenum and tungsten.* The reaction of Et_4NCl with $MoCl_5$ in dichloromethane in sealed tubes at 65°–75° C forms the black, crystalline, $(Et_4N)(MoCl_6)$. If the Et_4NCl was even slightly impure, the yellow $(Et_4N)_2(Mo^{IV}Cl_6)$ was obtained (336). The room temperature magnetic moment has been given as 1·31 (336) and 1·55 (214) B.M.

Dark green salts such as $(Et_4N)(WCl_6)$ have similarly been obtained from, for example, Et_4NCl and WCl_6 or WCl_5 in sulphuryl chloride or chloroform, or by the reaction of WCl_6 with an alkali metal iodide at 80°–130° C, with the evolution of iodine (9, 114, 336, 661). Heating these salts at 280°–300° C *in vacuo* causes disproportionation to $M^I_2W^{IV}Cl_6$ and $W^{VI}Cl_6$; the reaction is reversible on grinding at room temperature (661). The room temperature effective magnetic moments of these complexes are in the range 0·66–1·17 B.M.

The corresponding WBr_6^- is also known (214, 331, 336).

The energy of the first low energy transition in the visible spectrum of these d^1 octahedral anions, corresponding to 10 Dq, is given below (214, 366):

	$^2T_{2g} \rightarrow {}^2E_g$		$^2T_{2g} \rightarrow {}^2E_g$
MoF_6^-	24 000 cm^{-1}	WF_6^-	32 400 cm^{-1}
$MoCl_6^-$	21 700	WCl_6^-	23 300
		WBr_6^-	18 900

Molybdenum(V) in concentrated hydrochloric acid (>10 *M*) exists as the green $(MoOCl_5)^{2-}$, from which the green salts can be readily obtained (34, 470, 743, 1286, 1417, 1800, 2153). The molybdenum(V) solution can be readily obtained from $MoCl_5$ or $MoOCl_3$, or by electrolytic or tin reduction of molybdenum(VI); in the latter case the initial products appear to be an equimolar mixture of molybdenum(V) and molybdenum

(III), which is slowly followed by complete reduction to molybdenum(III) (214).

The caesium salt has the K_2PtCl_6 structure, showing octahedral co-ordination about the metal atom (352). These salts have magnetic moments close to the spin-only value ($\mu_{eff} = 1 \cdot 65–1 \cdot 74$ B.M.) and follow the Curie–Weiss law with low values of θ, indicating that the orbital angular momentum is effectively quenched by the large distortion from octahedral stereochemistry caused by the oxygen atom (214, 352, 979, 1131).

The visible spectrum has been interpreted (1047) in terms of a grossly distorted octahedron (C_{4v} symmetry). If the z-axis is defined as being along the molybdenum–oxygen bond, then there is substantial π bonding from the oxygen p_x and p_y orbitals to the metal d_{xz} and d_{yz} orbitals, so that the molybdenum–oxygen bond approaches a triple bond. The unpaired electron on the molybdenum(V) atom then remains in the d_{xy} orbital (assuming no metal-to-chlorine π bonding). Bands in the visible spectrum at about 14 000 and 23 000 cm^{-1} are assigned as originating from the d_{xy} orbital, that is, to ligand field bands. Bands at about 27 000, 32 000 and 42 000 cm^{-1} were originally assigned to transitions from the oxygen–molybdenum π bonding orbitals into orbitals located mainly on the metal atom, but were later reassigned as being halogen-to-metal charge transfer bands (34, 1222).

At lower hydrochloric acid concentration, 2–7 M, the green molybdenum(V) solutions turn brown due to a change in the relative intensities of the bands in the visible spectrum, and detailed spectral studies show the existence of at least two species in equilibrium with $MoOCl_5^{2-}$ (1283, 2153). At 6 M hydrogen chloride, the optical density is proportional to the square of the molybdenum concentration, due to the existence of dimeric species (1117, 2153).

As the acid concentration is decreased, the magnetic susceptibility (979, 1975) and the intensity of the e.s.r. signal (1131) decrease sharply. Taking the intensity of the e.s.r. signal as being directly proportional to the $MoOCl_5^{2-}$ concentration, the results as a function of acid concentration are compared with the magnetic susceptibility results in Table 28. It has

TABLE 28

E.S.R. and magnetic susceptibility results on molybdenum(V) in hydrochloric acid

[HCl] M	12	10	8·4	7·1	6·0	5·2	4·0	2·0
% $MoOCl_5^{2-}$ from e.s.r.	100	100	85	74	46	16	~1	0
% Total susceptibility	100	100	99	95	75	41	9	0

been concluded (1131) that a paramagnetic species other than $(MoOCl_5)^{2-}$ is present between 4 and 10 *M* hydrochloric acid, which in conjunction with the spectral results indicates the formation of paramagnetic and diamagnetic dimers.

The only spectral band which is markedly effected by dilution is the band assigned to the $^2B_2 \rightarrow {}^2B_1$ transition (d_{xy} non-bonding to $d_{x^2-y^2}$ antibonding) which is localized in the plane perpendicular to the Mo=O axis. It is therefore reasonable to formulate the dimer(s) as containing bridging group(s), presumably oxo or hydroxo groups, in the *xy* plane. In agreement with this suggestion, it is generally found (see later) that for compounds which have more than one oxygen atom bonded to a molybdenum atom, this *cis*-configuration is preferred to the *trans*-configuration because of the greater opportunities for oxygen–molybdenum π bonding to different molybdenum *d* orbitals.

The anhydrous green complexes $A^I[MoOCl_4]$ have been prepared by the addition of the appropriate alkali metal, or substituted ammonium, chloride to molybdenum pentachloride in liquid sulphur dioxide. The room temperature effective magnetic moments are in the range 1·61–1·74 B.M. (34, 1356). These compounds appear to hydrate in the presence of water vapour forming the *trans*-$[MoOCl_4(H_2O)]^-$; the molybdenum–oxygen stretching frequency simultaneously falls from 1012 to 981 cm^{-1} (for the tetraphenylarsonium salts) (2000). Other ligands form similar anionic complexes, namely $[MoOCl_4(MeOH)]^-$ (954, 960), $[MoOCl_4(MeCN)]^-$ (1220), and $[MoOCl_3(\beta\text{-diketonate})]^-$ (1856). The compound previously formulated as $MoCl_5(Ph_3AsO)$ (1218) has been shown to be $(Ph_3AsCl)[MoOCl_4]$ (1356).

The oxygen-to-metal π bonding becomes more important as the negative charge on the complex is increased, as is indicated by the molybdenum–oxygen stretching frequencies of 1020 cm^{-1} for $MoOCl_3$, 1000–1010 cm^{-1} for $MoOCl_4^-$ salts, and 960–970 cm^{-1} for $MoOCl_5^{2-}$ (1693).

Molybdenum(V) in hydrobromic acid is significantly different to molybdenum(V) in hydrochloric acid. At high acid concentrations (8–10 *M*), e.s.r. results indicate that there is an equilibrium mixture of $(MoOBr_5)^{2-}$ and $(MoOBr_4)^-$, the former predominating. At lower acidity the concentration of the oxopentabromo species falls, and that of the oxotetrabromo species increases, reaching a maximum at about 7 *M* (684). Spectrophotometric work however indicated only the presence of $[MoOBr_4]^-$ (38).

Brown or green salts of the $MoOBr_5^{2-}$, $MoOBr_4^-$ and $[MoOBr_4(H_2O)]^-$ anions can be obtained depending upon conditions (75, 1355, 1947, 2000, 2357). The last two may also be conveniently obtained from molybdenum hexacarbonyl, bromine, and the appropriate bromide in chloroform (1355).

These salts have room temperature effective magnetic moments in the range 1·48–1·76 B.M. and follow the Curie–Weiss law with low values of θ (34, 979, 1355). The structure of $(Ph_4As)[MoOBr_4(H_2O)]$ (1999) shows that the water molecule *trans* to the oxygen atom is only weakly held (Mo–OH_2 = 2·39 Å; M=O = 1·78 Å) and is lost at 40° C *in vacuo* to form $(Ph_4As)[MoOBr_4]$. This is accompanied by an increase in the molybdenum–oxygen stretching frequency from 981 to 1007 cm^{-1}.

In an analogous manner to the hydrochloric acid solutions, these paramagnetic monomers disappear below about 6 *M* hydrobromic acid (684, 979, 1296).

Salts of the oxohalotungstate(V) anions $(WOCl_5)^{2-}$, $(WOCl_4)^-$, $(WOBr_5)^{2-}$ and $(WOBr_4)^-$ are obtained in a similar manner to the corresponding molybdenum complexes (523, 2321). The salts are yellow or brown, have magnetic moments in the range 1·35–1·56 B.M. which are lower than for the molybdenum complexes as expected, and follow the Curie–Weiss law with low values of θ (34, 214, 352). Below about 7 *M* hydrochloric acid the tungsten(V) species become diamagnetic indicating polymerization (1295), as was noted above for molybdenum(V). The infrared and far infrared spectra of these oxo complexes has been studied in some detail (1974).

E. Complexes with other Anionic Ligands

In this section attention is concentrated on the eight coordinate peroxochromate, the molybdenum and tungsten octacyanides, and the considerable number of dimeric molybdenum and tungsten complexes containing one or two bridging oxygen atoms. All of these complexes are obtained from aqueous solution.

The addition of hydrogen peroxide to alkaline chromate solutions lead to the formation of tetraperoxochromate(V), $[Cr(O_2)_4]^{3-}$. The same compounds can be prepared by solid state reactions (2070). The structure of $K_3[Cr(O_2)_4]$ shows that each peroxo group functions as a bidentate, and spans the m edges of a dodecahedron (Fig. 7, page 16, Table 5 page 15) (2225, 2234).

Oxidation of $[Mo^{IV}(CN)_8]^{4-}$ and $[W^{IV}(CN)_8]^{4-}$ with strong oxidizing agents such as permanganate yields $[Mo^{V}(CN)_8]^{3-}$ and $[W^{V}(CN)_8]^{3-}$. The crystal structures have not been determined, but the observation that $g_{//} > g_{\perp}$ in the e.s.r. spectrum of frozen solutions is not consistent with a perfect dodecahedral structure. It is however consistent with square antiprismatic or distorted dodecahedral structures (1150, 1599, 1841), or with a rapid intramolecular conversion between square antiprismatic and dodecahedral structures (247, 1202) (see also page 21).

The thiocyanate complexes are markedly different to the cyanide complexes. The reduction of molybdenum(VI) or tungsten(VI) in the presence of thiocyanate forms deep red colours which are apparently complex equilibrium mixtures. Salts obtained from these solutions contain either the monomeric anions $[MoO(NCS)_5]^{2-}$ and $[WO(NCS)_5]^{2-}$ with normal magnetic moments of about 1·6 B.M., or dinuclear ions with much lower magnetic moments of 0–1 B.M. These can probably be written $[(SCN)_4(O)M{-}O{-}M(O)(NCS)_4]^{4-}$ or

$[(SCN)_3(O)M\langle^{O}_{O}\rangle M(O)(NCS)_3]^{4-}$ (263, 951, 1693, 1695).

There are a large number of oxo complexes which have been obtained from aqueous solutions of molybdenum(V). In all cases the molybdenum atom appears to be octahedrally coordinated and to be attached to one non-bridging oxygen atom. The complexes can be conveniently classified into three groups:

(i) Monomeric paramagnetic $MoOX_5^{2-}$, the best examples of which are when X is Cl, Br or NCS. Other compounds also mentioned above include $[MoOCl_4(ligand)]^-$ and $[MoOCl_3(\beta\text{-diketonate})]^-$.

(ii) Dimeric diamagnetic $[X_3Mo(=O)(\mu\text{-}O)_2Mo(=O)X_3]^{x-}$, where there is considerable direct molybdenum–molybdenum bonding in addition to the two bridging oxo groups.

(iii) Dimeric $[X_4Mo(=O){-}O{-}Mo(=O)X_4]^{x-}$, in which the Mo–O–Mo units are close to linear, and which again have low magnetic moments.

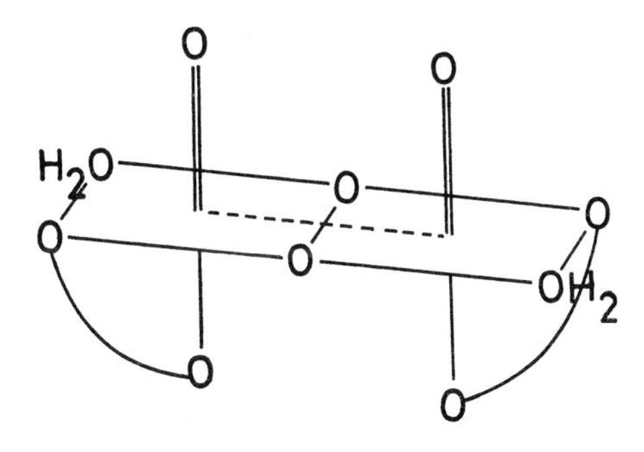

FIG. 179 The structure of $[Mo_2O_4(C_2O_4)_2(H_2O)_2]^{2-}$.

The best characterized example of a compound with two bridging oxygen atoms is the oxalate complex $[Mo_2^VO_4(C_2O_4)_2(H_2O)_2]^{2-}$ (1641), the structure of which is shown in Fig. 179. Two octahedrally coordinated molybdenum atoms are linked by two oxygen bridges; the short molybdenum–molybdenum distance of 2·54 Å suggests that the low effective magnetic moment of ~0·4 B.M. is due to direct molybdenum–molybdenum bonding. The two non-bridging oxygen atoms attached to the two molybdenum atoms are on the same side of the molecule, which possesses a two-fold rotation axis. The two water molecules can be replaced by pyridine to form $[Mo_2^VO_4(C_2O_4)_2(py)_2]^{2-}$, which presumably has an analogous structure (1692).

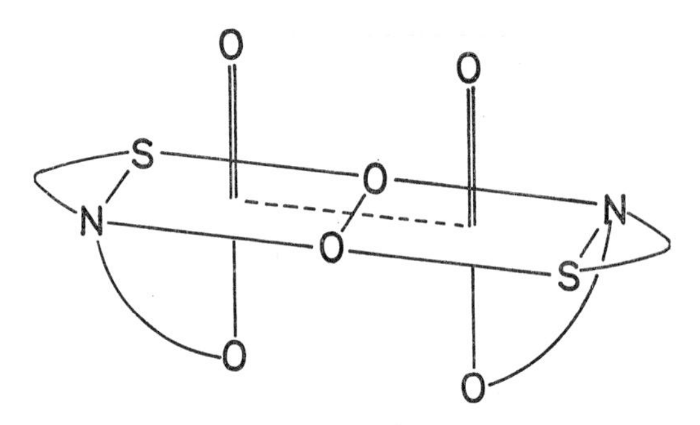

FIG. 180 The structure of $[Mo_2O_4(cysteine)_2]^{2-}$.

Cysteine forms a related complex with oxomolybdenum(V) which has been studied in some detail as the enzyme xanthine oxidase is thought to contain a related structure (1678). The structure of $[Mo_2O_4(cysteine)_2]^{2-}$ is shown schematically in Fig. 180 (1428). The short molybdenum–molybdenum distance of 2·57 Å is again consistent with the observed diamagnetism.

Similar complexes include those with ethylenediaminetetraacetic acid (Fig. 181) (1151, 1692) dipyridyl (1691, 2402), and dipyridyl-N,N′-dioxide (1616).

Other bidentate ligands form dimeric complexes containing two bidentate ligands per molybdenum atom, leaving only one bridging oxygen atom. For example the ethyl xanthate complex $[(EtO.CS_2)_2(O)Mo–O–Mo(O)(S_2C.OEt)_2]$ has a linear Mo–O–Mo structure (238), which arises from a three-centre molecular orbital formed from the p_π electrons on the oxygen atom and the d_π electrons on the molybdenum atom forming filled bonding and non-bonding, but empty antibonding, molecular orbitals.

Other complexes containing Mo:O ratios of 2 :3 and which presumably have similar structures include the oxine (1692), acetylacetonate (1503), dithiocarbamate (1641, 1704) and dithiophosphate analogues of the xanthate complex, cationic complexes such as $[Mo_2O_3(urea)_8]Cl_4$ (2201), mixed ligand complexes such as $[Mo_2O_3(C_2O_4)_2(o\text{-phen})_2]$, and anionic complexes such as $[Mo_2O_3(C_2O_4)_2(oxine)_2]^{2-}$ (1692).

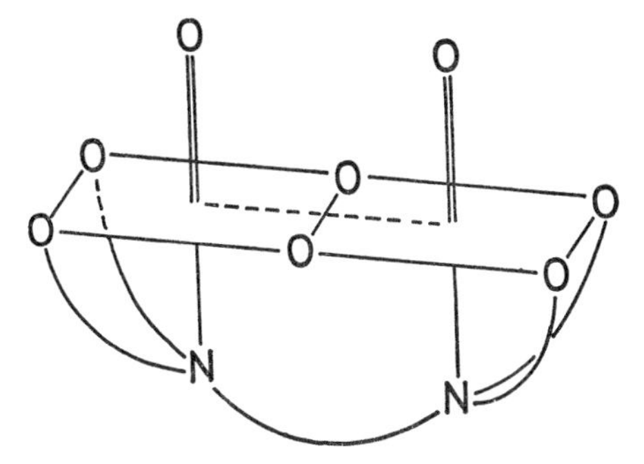

FIG. 181 The structure of $[Mo_2O_4(EDTA)]^{2-}$.

F. Organometallic Complexes

Oxidation of lower valent monocyclopentadienyl molybdenum compounds such as $[(C_5H_5)Mo(CO)_3]_2$ leads to a number of pentavalent compounds in addition to the hexavalent compounds mentioned on page 304. These include paramagnetic monomers such as $(C_5H_5)MoOBr_2$ and $(C_5H_5)MoX_4$ (where X is Cl or Br) (682), and diamagnetic dimers such as $(C_5H_5)_2Mo_2O_4$ (581) and $(C_5H_5)_2Mo_2O_2S_2$ (2300). The structure of the latter shows two bridging sulphur atoms (2218). The molybdenum–molybdenum bond length of 2·89 Å is longer than in the oxygen bridged oxalato (2·54 Å) and cysteinato (2·57 Å) complexes referred to above.

3. OXIDATION STATE IV

A. Oxides

Chromium dioxide has the undistorted rutile structure, which is described on page 61. The structures of MoO_2 and WO_2 are related, but are distorted so that the metal–metal distances along the *c*-axis are not equidistant, but are alternately short and long:

CrO_2	Cr–Cr = 2·92 Å (2387)
MoO_2	Mo–Mo = 2·511 and 3·112 Å (316)
WO_2	W–W = 2·49 and 3·08 Å (1629)

The relationship to other dioxides containing metal–metal bonds is discussed more fully in Chapter 3 (page 190). The metal–metal bonding is again weaker for mixed dioxides.

Chromium dioxide is ferromagnetic below 121° C (1089), indicating that the electrons are ordered *away* from the adjacent metal atoms along the c-axis, which is also indicated by the negative temperature coefficient of this cell length (2159).

Both MoO_2 and WO_2 show only a small temperature independent paramagnetism ($10^6\chi'_M = 40$ and 60 cgsu respectively) consistent with metal–metal bonding. The susceptibility increases and becomes temperature dependent when finely dispersed on alumina, for example for MoO_2,50 Al_2O_3, $10^6\chi_M = 830$, 1250 and 3500 cgsu at 295°, 195° and 78° K respectively (1951).

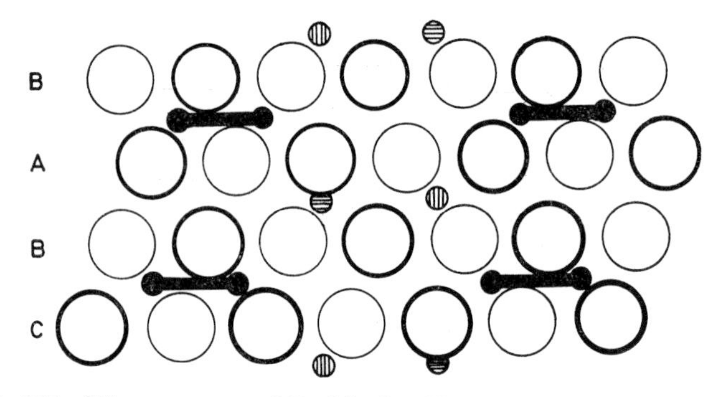

FIG. 182 The structure of $Zn_2Mo_3O_8$. Zinc atoms in tetrahedral sites (horizontal hatching) and in octahedral sites (vertical hatching) are situated between the oxygen-molybdenum-oxygen layers shown in Fig. 19 (page 28).

All three dioxides exhibit metallic conductivity (460, 1476, 1934) which has been interpreted as arising from a partially filled conduction band formed from the overlap of the empty d_{xy} orbital on the metal atom (taking the rutile c-axis as the atomic z-axis) and filled p_π-orbitals on the oxygen atoms.

Triangular clusters of molybdenum(IV) atoms also connected by metal–metal bonding are observed in $A_2^{II}Mo_3^{IV}O_8$ (where A is Mg, Mn, Fe, Co, Ni, Zn or Cd) (81, 1591, 1592).

The structure shows double hexagonal close packing of oxygen atoms (that is, abcb sequence of layers) with the molybdenum atoms occupying every alternate layer of octahedral holes, and the bivalent cations distributed between both octahedral and tetrahedral interstices. The molybdenum atoms within each layer are arranged as discrete triangles (Fig.

182), and considerable distortion of the lattice occurs due to metal–metal bonding within these triangles. For $Zn_2Mo_3O_8$:

$$\begin{aligned} \text{Mo–Mo (within triangle)} &= 2{\cdot}524\ \text{Å} \\ \text{Mo–Mo (between triangles)} &= 3{\cdot}235\ \text{Å} \\ \text{Mo–Mo (inter layer)} &= 5{\cdot}775\ \text{Å} \end{aligned}$$

These compounds are black, electrical insulators, and virtually diamagnetic.

A similar structure involving Mo_3 groups has been found for $LiScMo_3O_8$, $LiYMo_3O_8$ and $LiGaMo_3O_8$, except that in these cases the *c*-axes are halved indicating that the oxygen atoms are arranged in simple hexagonal close packing. The cations are ordered so that the lithium atoms are in tetrahedral holes and the trivalent metal atoms in octahedral holes (635, 679).

Attempts to prepare tungsten and rhenium analogues of these layer oxides have been unsuccessful.

The bonding proposed (554) uses molecular orbitals formed by a combination of atomic orbitals from the three molybdenum atoms. Two electrons are accommodated in a strongly bonding A_1 molecular orbital, and four in weakly bonding E molecular orbitals.

Related structures containing triangular clusters of metal atoms between close packed anion layers are described on pages 28 and 223.

B. Tetrahalides

1. *Tetrafluorides*

Chromium tetrafluoride was first prepared by the fluorination of the trifluoride or trichloride (2340). It is also readily prepared from the elements at 350° C (489). The amorphous green solid hydrolyses in air, but is insoluble in organic solvents. It can be sublimed at 150° C to give a blue vapour. The magnetic susceptibility follows the Curie–Weiss law, $\mu_{eff} = 3{\cdot}02$ B.M., $\theta = 70°$.

The fluorination of molybdenum carbonyl at −75° C forms the olive green Mo_2F_9, which on heating decomposes to the volatile MoF_5 and the light green involatile MoF_4 (1848). It has also been obtained from MoS_2 and SF_4 at 350° C (1816). The room temperature effective magnetic moment is 2·3 B.M. (1135).

The mixed halide $Mo_2Cl_3F_6$ may be related to the above Mo_2F_9. It is most conveniently prepared from $MoCl_5$ and HF, or from MoF_6 with a transition metal chloride (for example $MoCl_5$ or $TiCl_4$), a non-metal chloride (for example BCl_3, CCl_4, $SiCl_4$, PCl_3 or $AsCl_3$), or an alkali metal or alkaline earth chloride (1808, 1809, 1810). A polymeric ionic formula,

$(Mo_3^{IV}Cl_9)(Mo^VF_6)_3$, was suggested. No reaction was noted with the less reactive WF_6 and the above chlorides.

The reduction of tungsten hexafluoride with, for example, benzene or phosphorus trifluoride, forms the tetrafluoride (1900). It is also prepared by the disproportionation of the pentafluoride at 50°–70° C (2074). The involatile compound is stable up to high temperatures, but is again readily hydrolysed by water.

2. *Tetrachlorides, tetrabromides, and tetraiodides*

Chromium tetrachloride, tetrabromide and tetraiodide do not appear to be known in the solid state. The equilibria between solid and gaseous chromium trichloride, chlorine, and gaseous chromium tetrachloride over the temperature range 400°–1000° C have been studied, and the thermodynamic constants for the gaseous tetrachloride evaluated (1801, 1819).

The α-form of $MoCl_4$, as well as $MoBr_4$, WCl_4 and WBr_4 are isomorphous with the tetrachlorides and tetrabromides of niobium and tantalum, and contain pairs of metal atoms between close packed layers of chloride ions (Fig. 17, page 26). A second isomer of molybdenum tetrachloride, β-$MoCl_4$, can be prepared at higher temperatures, and does not contain molybdenum–moybdenum bonds.

α-$MoCl_4$ is most conveniently prepared by the prolonged refluxing of tetrachloroethylene with molybdenum pentachloride (369, 1357). Less pure material is obtained by refluxing with benzene (1504), or indeed any organic molecule capable of reacting with a chlorine atom, but in these cases the molybdenum tetrachloride is often contaminated with polymerized carbonaceous material. The black compound has a significant magnetic susceptibility of 0·85 B.M. at room temperature, in contrast to the other isostructural tetrahalides where the electrons are completely paired off through metal–metal bonding. Moreover the variation of the susceptibility with temperature cannot be interpreted as due to interactions within pairs of molybdenum atoms, indicating that magnetic exchange is spread along the string of metal atoms (1357).

β-$MoCl_4$ is prepared at higher temperatures by the reduction of the pentachloride with molybdenum metal in a 450°–375° C temperature gradient, which maintains a sufficiently high partial pressure of $MoCl_5$ to prevent disproportionation of the tetrachloride to the pentachloride and trichloride (2029). This form of molybdenum tetrachloride has a much higher magnetic moment (2·31 and 2·54 B.M. for two different preparations) indicating the absence of significant molybdenum–molybdenum bonding. The structure is related to that of α-$TiCl_3$ or $AlCl_3$ (page 125), but only $\frac{3}{4}$ of the metal sites are occupied. The structure appears to contain localized domains in which half the metal sites in the trihalide structure are vacant in every alternate layer, so that any one domain can be thought of as sheets of composition $(Mo_2Cl_6)^{2+}$ alternating with sheets of composition $(MoCl_6)^{2-}$. These domains are then statistically disordered throughout the crystal.

Both forms of molybdenum tetrachloride are unstable to air and water. They are completely soluble in aqueous hydrochloric acid, in contrast to molybdenum trichloride with which they have been confused.

The black molybdenum tetrabromide is the highest known molybdenum bromide, and can be obtained from the direct combination of the elements. An early measurement (1424) of the magnetic susceptibility gave similar values to α-$MoCl_4$ ($\mu_{eff} = 1 \cdot 3$ and $1 \cdot 0$ B.M. at 293 and 90° K respectively).

Tungsten tetrachloride was first prepared by the chlorination of WO_2 with carbon tetrachloride in sealed tubes at 280° C (1684). It is conveniently prepared as pure black crystals by the reduction of the hexachloride with aluminium in a 475°–225° C temperature gradient (1584). The tetrabromide is similarly prepared from the pentabromide and aluminium in a 475–240° C temperature gradient (1583, 1584). Both these tungsten tetrahalides are diamagnetic.

The mixed halide WCl_3Br has been claimed to have been formed by the bromination of $K_3W_2Cl_9$ (2426), but the product has been shown to be $W^{IV}OCl_3Br$ (270).

There have been occasional references to molybdenum and tungsten tetraiodides, but it must be concluded that their existence has not been confirmed (451).

C. Adducts of the Tetrahalides

Molybdenum tetrafluoride forms the *diamagnetic* adducts MoF_4(ligand) (with Me_3N, $PhMe_2N$ and C_5H_5N) and $MoF_4(\text{ligand})_2$ (with $Me_2N.CHO$ and Me_2SO) (1729).

The reduction of molybdenum pentachloride with many ligands to form adducts of the type $MoCl_4(\text{ligand})_2$ has been referred to previously (page 309). Compounds of the same stoicheiometry have also been prepared directly from molybdenum(IV), or by the oxidation of $Mo(CO)_4(\text{ligand})_2$ with halogen.

Using $MoCl_4$ or $MoCl_4(PrCN)_2$, the adducts $MoCl_4(\text{ligand})_2$ have been formed with the oxygen donors $(CH_2)_4O$, $(CH_2)_5O$ and Ph_3PO, the sulphur donor thioxane, the nitrogen donors MeCN, EtCN, PhCN, C_5H_5N, γ-picoline, pyrazine and dimethylpyrazine, the phosphorus donor Ph_3P, and the arsenic donor Ph_3As. Reduction of molybdenum(IV) with excess γ-picoline was noted after prolonged reaction, but no compound was isolated. Bidentates which form $MoCl_4$(bidentate) include dipyridyl, *o*-phenanthroline, tetramethylethylenediamine and *o*-phenylenebisdimethylarsine (37, 371, 1356, 1359, 1504). These compounds range in colour from yellow through red to brown. The room temperature effective magnetic moments are mostly in the range 2·3–2·5 B.M., which is lower than the free ion value due to spin-orbit coupling, but higher than that of the more symmetrical $MoCl_6^{2-}$.

The only adduct of molybdenum tetrachloride which appears different is the so-called $MoCl_4,4Ph_3AsO$ (1219). This compound was white, diamagnetic, and apparently stable to boiling water, and was formulated as the only example of a neutral eight coordinate molybdenum(IV) complex. It has been suggested that this compound is in fact a mixture of $MoO_2Cl_2(Ph_3AsO)_2$ and/or $MoOCl_4(Ph_3AsO)$ with $Ph_3As(OH)Cl$ resulting from "$MoCl_4$" which was largely a mixture of $MoCl_5$ and MoO_2Cl_2 (or $MoOCl_4$) (43). More precise work under strictly anhydrous conditions has shown that this product was a mixture of $(Ph_3AsCl)(MoOCl_4)$, $MoO_2Cl_2(Ph_3AsO)_2$ and Ph_3AsO (1359).

The analogous reactions with tungsten are more complex and have not yet been so comprehensively studied. Acetonitrile reduces WCl_6 and WCl_5 to $WCl_4(MeCN)_2$ (35, 246). Pyridine reacts with WCl_6, WCl_4, $WCl_4(MeCN)_2$, WCl_6^{2-}, WBr_5, WBr_4 or WBr_6^{2-} to form $WX_4(py)_2$, but reduction to tungsten(III) also occurs readily (246, 1348, 1584). Similarly $Ph_2P.CH_2.CH_2.PPh_2$ and *o*-$(C_6H_4)(AsMe_2)_2$ form $[WCl_4(\text{bidentate})]$ or more complex products depending upon conditions (267, 1359). These tungsten(IV) compounds have room temperature effective magnetic moments in the range 1·1–1·9 B.M.

Halogen oxidation of $Mo(CO)_4$(dipy), $Mo(CO)_4[C_6H_4(AsMe_2)_2]$, $W(CO)_4$(dipy) and $W(CO)_4(PhMe_2P)_2$ forms MoX_4(dipy) (where X is Cl or Br), $MoBr_4[C_6H_4(AsMe_2)_2]$, WX_4(dipy) (where X is Cl or Br), and $WCl_4(PhMe_2P)_2$ respectively (1237, 1725, 1788). In addition intermediate carbonyl halides of molybdenum(II) and tungsten(II) can be isolated which are dealt with in the appropriate section, while pentavalent and hexavalent oxo complexes are obtained in the presence of oxygen.

Reaction of $[WCl_4(PhMe_2P)_2]$ with borohydride forms the hexahydride $WH_6(PhMe_2P)_2$ (1725).

D. Oxohalides

Molybdenum oxodichloride, $MoOCl_2$, can be prepared by heating MoO_3 with two or three moles of $MoCl_3$ at 300°–350° C or by the reduction of $MoOCl_4$ with aluminium, followed by the removal of the $MoOCl_3$ impurities by sublimation to leave the yellow brown involatile product (610, 1680, 2036).

The corresponding $WOCl_2$ is prepared by the thermal disproportionation of $WOCl_3$ to involatile $WOCl_2$ and volatile $WOCl_4$ at 400° C, or by the reduction of $WOCl_4$ with aluminium or stannous chloride (610, 772, 1680).

Both these oxychlorides are isomorphous with $NbOCl_2$ and $TaOCl_2$ (Fig. 116, page 205), that is there are infinite –M–O–M–O– strings with metal–metal bonding between the strings. As expected, the compounds exhibit only a small temperature independent paramagnetism.

The analogous $WOBr_2$ was obtained using a temperature gradient of 580°–450° C, according to the stoicheiometry:

$$2\,W + WO_3 + 3\,Br_2 \rightarrow 3\,WOBr_2$$

Using a 550°–470° C temperature gradient, the mixed tungsten(IV)–tungsten(V) oxobromide $W_2O_3Br_3$ was obtained (2282).

E. Anionic Halo and Oxohalo Complexes

1. *Fluoro complexes*

The reaction of chromium tetrafluoride with an equimolar quantity of alkali metal chloride in bromine trifluoride forms $ACrF_5$ (where A is K, Rb or Cs) (489). The structures of these interesting compounds are not known, but the anion is presumably polymeric.

All other tetravalent Group VI complex halides are of the type $A^I_2MX_6$. The chromium complexes may be obtained as above, but using two moles of the alkali metal salt (257, 489, 1243).

These chromium compounds have high magnetic moments (3·1–3·6 B.M.).

The reduction of molybdenum hexafluoride with an alkali metal iodide in liquid sulphur dioxide forms $A^IMo^VF_6$ (see page 314), but excess sodium iodide forms the dark brown Na_2MoF_6 (738). In contrast to the six coordinate anion in Na_2CrF_6, the larger molybdenum atom forms an eight coordinate structure (364).

2. *Chloro, bromo and oxochloro complexes*

The reaction of molybdenum pentachloride with the halides of the lighter alkali metals in liquid sulphur dioxide forms dark green oxomolybdenum(V) compounds of the type $A^I(Mo^VOCl_4)$, but in the case of rubidium and caesium chlorides the tetravalent $Rb_2(MoCl_6)$ and $Cs_2(MoCl_6)$ are obtained (34). The potassium, rubidium, caesium and thallium salts have similarly been obtained from liquid iodine monochloride (740). The caesium salt has also been prepared from the solid state reaction between molybdenum tetrachloride and caesium chloride (1249).

These salts are isomorphous with the corresponding hexachloroplatinates, confirming octahedral coordination of the molybdenum atom (740).

Salts with organic cations can be formed with acetonitrile or dichloromethane as solvents, although in these cases solvated products were obtained (1220). For example with tetramethylammonium chloride in acetonitrile, the yellow $(Me_4N)_2(MoCl_6)$, 0·5 MeCN was obtained. This compound was a 2 :1 electrolyte in nitrobenzene, but the infrared spectrum appears typical for coordinated acetonitrile. Similarly $(Me_4N)_2(MoCl_6)$, 1/3 CH_2Cl_2 was obtained from dichloromethane.

Reaction of rubidium or caesium bromides with molybdenum tribromide (740) or molybdenum hexacarbonyl (1817) in iodine monobromide forms the analogous $A^I_2(Mo^{IV}Br_6)$. They are again isomorphous with the corresponding hexabromoplatinates.

The room temperature effective magnetic moments of these hexahalomolybdate salts lie in the range 2·1–2·6 B.M.

Salts containing the WCl_6^{2-} anion are prepared in a similar manner to the molybdenum analogues. For example reaction of tungsten hexachloride with the appropriate chloride in a sealed tube at 300° C leaves the involatile $A^I_2(W^{IV}Cl_6)$ (red for A = K, Rb or Cs, green for A = Tl or $\frac{1}{2}$ Ba) (1348). Alternatively iodine monochloride can be used as a supporting solvent. Similar reactions with tungsten hexabromide form the green A_2WBr_6 (where A is K, Rb or Cs) (1348).

Under certain conditions secondary and tertiary amines reduce WCl_6 to WCl_6^{2-} (333). For example secondary amines at −34° C are reported to form the buff coloured $(Me_2NH_2)_2(WCl_6)$, $(Me_2NH_2)_2(WCl_6)$, Me_2NH and $(Et_2NH_2)_2(WCl_6)$. However at room temperature simple adducts are formed which are slowly reduced to the amido compounds $WCl_3(Me_2N)(Me_2NH)_2$ and $WCl_3(Et_2N)(Et_2NH)_2$. Tertiary amines again initially form simple addition compounds followed by reduction to $(Me_3NH)_2WCl_6$, $(Me_3NH)_2WCl_6$,Me_3N and $(Et_3NH)WCl_6$.

The room temperature effective magnetic moments of the tungsten compounds (1·34–1·72 B.M.) are lower than for the molybdenum complexes as expected.

The reduction of tungsten(VI) in concentrated hydrochloric acid, electrolytically or with tin, leads to the formation of the deep violet $K_4[W_2OCl_{10}]$ (1451). This was earlier formulated as "$K_2[W(OH)Cl_5]$". The solutions however are unstable, and disproportionation to tungsten(III) and tungsten(V) subsequently occurs. The compound is not isomorphous with $K_4[Re_2OCl_{10}]$ in which the Re–O–Re bond is linear due to oxygen(p_π)-to-rhenium(d_π) bonding. Similarly the W–O–W stretching frequency at 652 cm^{-1}, and the magnetic moment of 2·40 B.M. at 294° K dropping to 1·72 B.M. at 77° K, are consistent only with a bent

$$W\overset{O}{\diagup\diagdown}W$$

bridge. The main unexpected feature in the visible spectrum is an intense band at 19 100 cm^{-1} (extinction coefficient of 10^4 per tungsten atom), which could be due to electron transfer between metal atoms in two different oxidation states. The compound has therefore been formulated as $K_4[Cl_5W^{III}.O.W^{V}Cl_5]$, which is also consistent with the above disproportionation reaction.

F. Complexes with other Anionic Ligands

1. *Introduction*

There are a number of complexes of chromium(IV), molybdenum(IV) and tungsten(IV) which can be obtained by oxidation or reduction from the more common lower and higher oxidation states respectively.

As was also noted for the pentavalent state, the best characterized compounds obtained from aqueous solution, which do not contain co-ordinated oxo or hydroxo groups, again include seven coordinate para-magnetic chromium peroxo complexes, and eight coordinate diamagnetic cyano complexes of molybdenum and tungsten, $[M(CN)_8]^{4-}$. The latter slowly hydrolyse to the diamagnetic $[MO_2(CN)_4]^{4-}$. The diamagnetic and hence presumably eight coordinate $[W(oxine)_4]$ is formed by the sealed tube reaction between $(NH_4)_3W_2Cl_9$ and excess 8-hydroxyquinoline (86).

Other paramagnetic chromium complexes of the types $Cr(OR)_4$ and $Cr(NR_2)_4$ can be obtained under strictly anhydrous conditions.

Of particular interest are the diamagnetic dithiolate or dithiolene complexes which are discussed in this section as a matter of convenience, although the most appropriate formal oxidation state is a matter of opinion.

All the above compounds are discussed in more detail in the following paragraphs.

One of the few octahedral complexes is $[Mo(NCS)_6]^{2-}$, also obtained from aqueous solution, the pyridinium salt of which has a magnetic moment of 2·45 B.M. (1695).

A number of compounds containing oxo groups are obtained from aqueous or ethanolic solutions. A structural determination of "$MoO_2(acac)(EtOH)$" reveals a trimeric molecule with the molybdenum atoms in the average oxidation state of 3·67, and which should be re-formulated $[Mo_3O_4(acac)_3(EtOH)_3]$. The three octahedrally coordinated molybdenum atoms share octahedral edges, and are in the form of an equilateral triangle with short molybdenum–molybdenum distances of 2·47 Å. These metal–metal bonds account for the observed diamagnetism (1183, 2002). It is possible that other complexes, such as the trimeric oxalato complex formulated as $[Mo_3^{IV}O_4(C_2O_4)_3(H_2O)_x]^{2-}$, are similar (2202, 2370).

Monomeric oxomolybdenum compounds include the diamagnetic MoO(phthalonitrile), obtained from MoO_2Cl_2 and phthalonitrile in dimethylformamide (1194). This compound has a molybdenum–oxygen stretching vibration at 972 cm^{-1}, and the strength of this bond is also suggested by the unchanged visible spectrum in the presence of potential ligands such as pyridine. Bidentate sulphur donors similarly form monomeric diamagnetic complexes such as the dithiocarbamato $[MoO(S_2CNEt_2)_2]$ and the dithiophosphato $[MoO(S_2P(OEt)_2)_2]$ (1321, 1546).

2. *Alkoxides*

Chromium tetra(tertiary butoxides) such as $Cr(OBu^t)_4$ have been prepared by the action of $Bu_2^tO_2$ on $Cr(C_6H_6)_2$ (1112), by the action of Bu^tOH on $Cr(NR_2)_4$, and by the oxidation of $Cr(OBu^t)_3$ with, for example, oxygen, bromine, or $Bu_2^tO_2$ (159). Probably the most convenient route is the oxidation of chromium(III) with the stoicheiometric amount of cuprous chloride (1468):

$$CrCl_3(THF)_3 + 4\,NaOBu^t + CuCl \rightarrow Cr(OBu^t)_4 + 4\,NaCl + Cu + 3\,THF$$

The deep blue, sublimable solid is monomeric in organic solvents, but very unstable to air and water. The effective magnetic moment is 2·88 B.M. at room temperature, confirming the d^2 configuration.

3. *Amido complexes*

Chromium(IV) forms the tetrakisdiethylamide $Cr(NEt)_4$ which is closely related to the chromium(IV) alkoxides above. The volatile, viscous green liquid is formed from chromium trichloride and lithium diethylamide under anhydrous conditions (159).

4. *Peroxo complexes*

In alkaline solution the chromium(VI) peroxo complex $Cr(O_2)_2O$ is reduced to the chromium(V) peroxo complex $[Cr(O_2)_4]^{3-}$, which on warming in aqueous ammonia is further reduced to the chromium(IV) $[Cr(O_2)_2(NH_3)_3]$. The ammonia molecules are readily replaced by neutral or anionic ligands (790, 2231).

The structures of $[Cr(O_2)_2(NH_3)_3]$ (1609, 2227), $[Cr(O_2)_2(H_2O)(en)]$ (2229), and $K_3[Cr(O_2)_2(CN)_3]$ (2228) reveal pentagonal bipyramidal structures in which two peroxo groups occupy four of the five planar sites.

The magnetic moments of 2·8–2·9 B.M. confirm chromium(IV) (226, 2229).

5. *Cyano complexes*

The octacyanomolybdate anion $[Mo(CN)_8]^{4-}$, is readily prepared by the addition of cyanide ion to aqueous solutions of molybdenum(III) (in the presence of oxygen), or molybdenum(V). A well-tried method is from a thiocyanato–molybdenum(V) complex(es) to form the dihydrate $K_4[Mo(CN)_8]$, $2H_2O$ (968).

The molybdenum atom in this salt has a dodecahedral stereochemistry (1202). However there is no significant difference in the molybdenum–carbon bond lengths to the "A" corners and to the "B" corners of the dodecahedron ($2{\cdot}16_5$ and $2{\cdot}16_2$ Å respectively). The angles θ_A and θ_B, the angles the molybdenum–carbon bonds make to the four-fold inversion axis, are 36·0° and 72·9° respectively. A more detailed examination of this structure is included in Chapter 1.

The structure of this anion in aqueous solution has received considerable attention. The three most likely possibilities (see Chapter 1) are:

(a) dodecahedral stereochemistry,

(b) square antiprismatic stereochemistry, or

(c) "fluxional" stereochemistry.

An examination of the visible spectrum does not provide any information as the expected *d–d* bands at 20 000–30 000 cm^{-1} are obscured by ligand(π) to metal(d) transitions (1015, 1021, 1022, 1449, 1861). Infrared and Raman spectra of aqueous solutions are consistent with a square antiprismatic structure, but do not prove it (1145, 1371, 1839, 1841). The ^{13}C–n.m.r. spectrum gives a single sharp band indicating that within the limits of resolution the eight cyanide groups are equivalent, or alternatively there is a rapid (on the n.m.r.-time scale) intramolecular conversion between chemically different cyanide groups (1730).

Photolysis of aqueous solutions of the yellow $[Mo(CN)_8]^{4-}$ and $[W(CN)_8]^{4-}$ produces blue and violet solutions respectively, with the loss of four cyanide groups per metal atom. This photolysis reaction has been extensively studied, although there is considerable disagreement between different workers (15, 138, 1282, 2207, 2333). The salts isolated from these solutions are dependent upon pH, and include green $K_2[Mo(OH)_2(CN)_4]$, blue $K_3[MoO(OH)(CN)_4]$, red $K_4[MoO_2(CN)_4]{,}6H_2O$, purple $K_3[WO(OH)(CN)_4]$ and yellow-brown $K_4[WO_2(CN)_4]{,}6H_2O$. All these compounds are diamagnetic. These compounds were earlier formulated as, for example, the eight coordinate $K_3[Mo(OH)_3(CN)_4(H_2O)]$ and

$K_4[Mo(OH)_4(CN)_4],4H_2O$. The same compounds can be formed directly from aqueous solution (1560, 1561, 1876).

The structures of $K_4[MoO_2(CN)_4],6H_2O$ (1560, 1561) and $NaK_3[MoO_2(CN)_4],6H_2O$ (633) show that the oxo groups are *trans* to each other, with molybdenum–oxygen distances of 1·83 Å. The molybdenum–carbon bond length of 2·20 Å is significantly longer than in $K_4[Mo(CN)_8],2H_2O$ (2·16 Å), in spite of the reduced steric crowding in going from an eight coordinate to a six coordinate complex. This increase is accompanied by an increase in the carbon–nitrogen bond length (1·17 and 1·15 Å respectively) and a decrease in its stretching frequency (2070 and 2120 cm^{-1} respectively).

In the presence of hydrazine, salts such as $Cd_2[Mo(CN)_8],2N_2H_4,4H_2O$ are obtained, the structure of which shows that neither the water or hydrazine molecules are bonded to the metal atom (472). Similarly the formation of the adduct $K_4[Mo(CN)_8]$, $7·3BF_3$ does not appear to alter the environment around the metal atom (2132).

A particularly interesting series of isonitrile complexes are made by the reaction of $Ag_4[W(CN)_8]$ with RX (where R is Et, Pr, Bu or Bz):

$$Ag_4[W(CN)_8] + 4\,RX \rightarrow [W(CN)_4(CNR)_4] + 4\,AgX$$

The yellow to brown complexes are diamagnetic (1508).

6. *Dithiolate complexes*

Like a number of other transition metals, chromium, molybdenum and tungsten form a series of very interesting tris-dithiolate or tris-dithiolene complexes of the general type $[M(S_2C_2R_2)_3]$, where R is H, CH_3, CF_3, Ph or CN (631, 1593, 2072, 2211).

In these complexes the metal is in the +VI oxidation state if the ligand is considered to be the dithiolate anion

R S−
 \ /
  C
  ||
  C
 / \
R S−

, which is unrealistic in view of the formation of similar complexes with vanadium (765, 2344), or in the zero oxidation state if the ligand is considered to be the neutral dithioketone

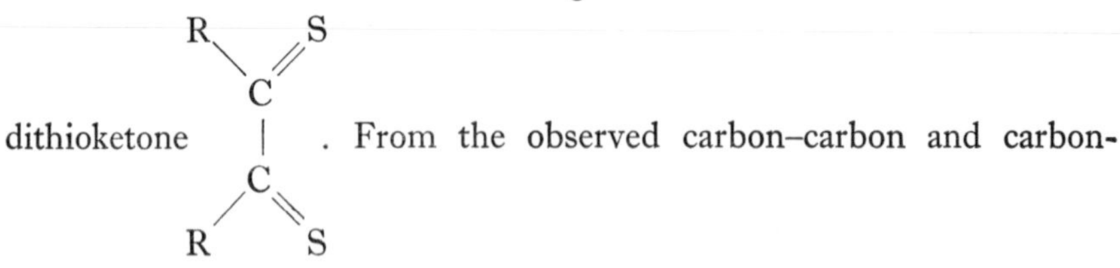

. From the observed carbon–carbon and carbon-sulphur bond lengths in the coordinated ligand, it appears that the true formulation should be intermediate between these two extremes. A proposed molecular orbital scheme (2072) is consistent with a formal oxidation state of +IV for the metal atom, and the valence bond resonance structures shown in Fig. 183. These calculations show that there is considerable

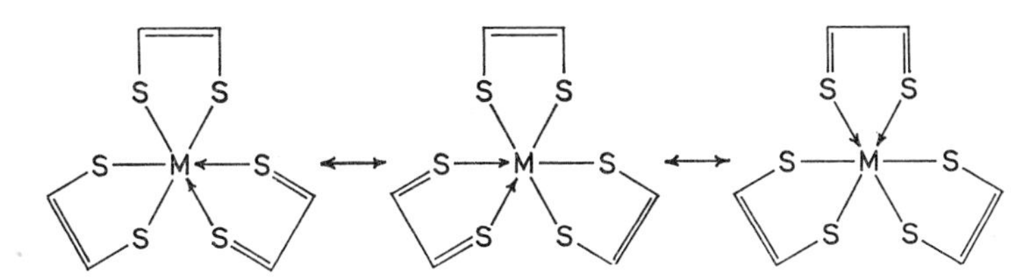

FIG. 183 Valence bond resonance structures for $[M(S_2C_2R_2)_3]$ (where M is Cr, Mo or W).

electron delocalization into molecular orbitals containing approximately 40% metal character, 20% sulphur character, and 40% ligand π bond character, and any simple bonding scheme is to some extent a matter of opinion. This picture will undoubtedly change as magnetic, e.s.r., and spectral results accumulate.

Compounds of this type are readily reduced to $[M(S_2C_2R_2)_3]^{2-}$ and $[M(S_2C_2R_2)_3]^{3-}$. Reduction can be considered as either lowering the formal oxidation state of the metal, or as increasing the contribution of the dithiolate form of the ligand relative to the dithioketone form of the ligand.

Similar compounds are formed with the benzene dithiolene (o-$C_6H_4S_2$), toluene dithiolene, and tetrafluorodithiolene ligands (415, 2211, 2212).

Another interesting aspect of these compounds is their stereochemistry, which for $V(S_2C_2Ph_2)_3$ (763), $Mo(S_2C_2H_2)_3$ (2170), and $Re(S_2C_2Ph_2)_3$

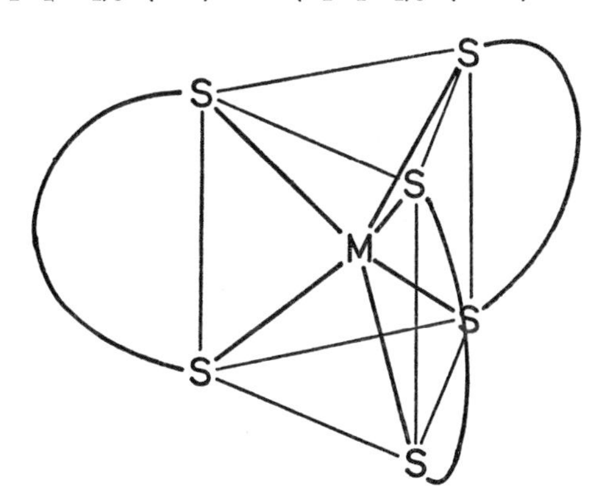

FIG. 184 The structure of $[M(S_2C_2R_2)_3]$.

(764), is trigonal prismatic (Fig. 184). As far as can be ascertained from X-ray powder photographs, the neutral molecules $[M(S_2C_2R_2)_3]$ appear to be generally isostructural for M = V, Cr, Mo, W and Re. The reasons for this unusual stereochemistry rather than the more common octahedral stereochemistry are not yet clear; interligand interactions through the sulphur atoms may play an important role.

The reduced anion in $(Me_4N)_2[V(S_2C_2(CN)_2)_3]$ however is intermediate between a trigonal prism and an octahedron, which may be due to increasing interligand repulsions due to the increase in charge (2210). The salts $(Ph_4P)_2[M(S_2C_2(CN)_2)_3]$ are isomorphous for M = V, Cr, Mo, W, but not for M = Mn or Fe which are more likely to be octahedral. Similarly all three $(Ph_4P)_3[M(S_2C_2(CN)_2)_3]$ are isomorphous for M = Cr, Fe and Co, and also probably octahedral (1593).

G. Organometallic Complexes

The tetravalent organometallic compounds of molybdenum and tungsten can be divided into two classes, depending upon whether the molecules contain one or two cyclopentadienyl groups.

The former group consists only of the halo carbonyls $[(C_5H_5)MoX_3(CO)_2]$ (where X is Cl, Br or I) and the maleonitriledithiolates $[(C_5H_5)M^{IV}(S_2C_2(CN)_2)_2]$ (where M is Mo or W). The carbonyl halides are formed by oxidation of $[(C_5H_5)Mo(CO)_3]_2$ or $[(C_5H_5)MoX(CO)_3]$ with excess halogen (1050, 1119). The maleonitrile dithiolates are similarly obtained from $[(C_5H_5)M^{II}I(CO)_3]$ (1570). All compounds are diamagnetic.

Reaction of molybdenum pentachloride or tungsten hexachloride with sodium cyclopentadienyl forms $[(C_5H_5)_2MoCl_2]$ and $[(C_5H_5)_2WCl_2]$ (574).

A relatively large number of biscyclopentadienyl derivatives have been studied. The reaction of $[(C_5H_5)_2MoCl_2]$ and $[(C_5H_5)_2WCl_2]$ with thiols RSH (where R is Me or Et) or carboxylic acids RCOOH (where R is H, Ph or CF_3) forms the diamagnetic red $[(C_5H_5)_2M(SR)_2]$ and green $[(C_5H_5)_2M(RCOO)_2]$ respectively (1051, 1139). Bidentate ligands similarly form $[(C_5H_5)_2MB]^+$ (where BH is acacH, $H_2N.CH_2.CH_2.SH$, or *o*-$C_6H_4(NH_2)(SH)$) and $[(C_5H_5)_2MB]$ (where BH_2 is $HS.CH_2.CH_2.SH$, $HS.CH_2.CH_2.CH_2.SH$, $(NC)_2C_2(SH)_2$, *o*-$C_6H_4(SH)_2$, *o*-$C_6H_4(OH)_2$, *o*-$C_{14}H_8(OH)_2$, *o*-$C_6H_4(NH_2)(OH)$, or *o*-$C_6H_4(NH_2)(SH)$) (1031, 1051, 1139). These formulations have been confirmed by an X-ray structural determination of $[(C_5H_5)_2Mo(S_2.C_6H_3.CH_3)]$ (1428).

Reaction of $(C_5H_5)_2MCl_2$ (where M is Mo or W) with sodium borohydride forms $(C_5H_5)_2MH_2$ (858, 1052), the structure of which has been determined (5, 990).

Both $(C_5H_5)_2MH_2$ and $(C_5H_5)_2M(SR)_2$ can behave as electron donors to a wide variety of acceptors, to form complexes of the type

$$(C_5H_5)_2M\langle{}^{H}_{H}\rangle AX_3$$ (where A is a Group III element and X is F or Cl),

$$(C_5H_5)_2M\langle{}^{H}_{H}\rangle M'(CO)_5$$ (where M′ is Cr, Mo or W),

$$(C_5H_5)_2M\langle{}^{(SMe)}_{(SMe)}\rangle M'X_2$$

and $(C_5H_5)_2M\langle{}^{(SMe)}_{(SMe)}\rangle M'\langle{}^{(SMe)}_{(SMe)}\rangle M(C_5H_5)_2$ (where M′ is Ni, Pd or Pt and X is Cl, Br or SCN), and $(C_5H_5)_2M\langle{}^{(SPh)}_{(SPh)}\rangle M'(CO)_3$ (where M′ is Cr, Mo or W) (654, 660, 1052, 1309).

4. OXIDATION STATE III

A. Trihalides

1. *Trifluorides*

The green chromium trifluoride can be prepared by passing anhydrous hydrogen fluoride over heated chromium oxide or, preferably, chromium trichloride, at about 500° C (2245). The structure consists of CrF_6 octahedra sharing corners, and it is isomorphous with VF_3 (1267, 1429).

The reduction of molybdenum pentafluoride with molybdenum forms the trifluoride. The colour has been reported to range from light green through tan, grey and dark red to black. However all preparations apparently have the rhombahedral VF_3 and CrF_3 structure (1510).

It has been shown that an air stable "trifluoride" with the ReO_3 structure (927, 1099) is a higher valence oxofluoride (2026, 2164).

2. *Trichlorides, bromides and iodides*

Anhydrous chromium trichloride can be prepared from chromium and chlorine, from chromium sesquioxide, carbon and chlorine, from chromium oxide and carbon tetrachloride, or from the dehydration of the hexahydrate with refluxing thionyl chloride. It can be purified by sublimation in a stream of chlorine which prevents decomposition to the dichloride. The structure (1717) is based on cubic close packing of chloride ions, with chromium atoms in $\frac{2}{3}$ of the octahedral holes in every second layer. At about $-50°$ C the anion packing changes to hexagonal close packing (see Fig. 74, page 125). The magnetic moment is 3·90 B.M. (2206).

Chromium tribromide and chromium triiodide are best prepared from the elements. The structure of $CrBr_3$ at room temperature is the same as the low temperature form of the trichloride, but changes at about 150° C to one based on cubic close packing of halide ions (1717).

Molybdenum trichloride is fairly well known, being first prepared in 1857 by the reduction of the liquid pentachloride with hydrogen at about 200° C (250). This remains the normal method of preparation (578). Stannous chloride has also been used as a reductant (1643). On further heating ($\sim$ 500° C) the trichloride disproportionates to the dichloride and tetrachloride. Molybdenum trichloride has been obtained as black crystals by the transport reaction (2029, 2067):

$$MoCl_{3(s)} + MoCl_{5(g)} \rightleftharpoons 2\ MoCl_{4(g)}$$

The compound is stable in dry air at ordinary temperatures, and insoluble in organic solvents. It is slowly hydrolysed in alkaline solution to the hydroxide, which is readily oxidised by air. Although it is insoluble in hydrochloric acid, the chloroanions $[Mo^{III}Cl_6]^{3-}$ and $[Mo^{III}Cl_5(H_2O)]^{2-}$ can be obtained by reduction of the higher oxidation states.

The effective magnetic moment of 0·49 B.M. at 295 °K falling to 0·25 B.M. at 90° K indicates considerable electron pairing between the molybdenum atoms. This is confirmed by the crystal structure (2029), which shows that like $AlCl_3$ and $CrCl_3$, $\frac{2}{3}$ of the octahedral holes are occupied between every second layer of a cubic close packed chloride lattice. However the structure is strongly distorted as shown in Fig. 18 (page 27). Instead of each metal atom being surrounded by three equidistant metal atoms, each molybdenum atom has one neighbour at only 2·76 Å, and the other two at 3·71 Å, so that pairs of molybdenum atoms lie parallel to the *b*-axis of the monoclinic unit cell. This bonding reduces the *b*-cell dimension, so that the b/a axial ratio is only 1·60, compared with $3^{\frac{1}{2}} = 1·73$ if the lattice had been undistorted by molybdenum–molybdenum bonding.

It is also possible to obtain β-$MoCl_3$, which is based on hexagonal close packing of chloride ions, but is otherwise similar to the first form, retaining pairs of molybdenum atoms. The exact stoicheimoetry appears to correspond to $MoCl_{3\cdot08\pm0\cdot01}$ (2029).

Molybdenum tribromide and molybdenum triiodide are prepared directly from the elements at about 400° and 300° C respectively (1424, 1525). They have also been prepared from molybdenum pentachloride and boron tribromide or anhydrous hydrogen iodide respectively (1416). Molybdenum triiodide is also produced by the action of iodine on molybdenum carbonyl (673).

The effective magnetic moment of molybdenum triiodide prepared from the elements is about 1·4 B.M. at room temperature, but decreases markedly as the temperature is lowered (1525). The iodide prepared from the carbonyl is much more finely divided and has a significantly higher susceptibility, the actual values depending upon the solvent used in the preparation; the susceptibility is also dependent upon field strength.

The structures of molybdenum tribromide and triiodide are related to that of β-$TiCl_3$ and the zirconium and hafnium trihalides (Fig. 75, page 126). Infinite polymeric strings parallel to the c-axis of the hexagonal unit cell are formed by each MoX_6 octahedron sharing opposite faces with the two adjacent octahedra. The c/a axial ratios of the unit cell are 2·75 and 2·69 respectively (compare with page 127) (106, 1845). The structure may be the same as for the ruthenium trihalides, where $c/a \sim 2\cdot7$, and the metal–metal distances along the c-axis are alternately short and long, for example 2·73 and 3·12 Å in $RuBr_3$ (343, 344).

"Tungsten trichloride", $[(W_6Cl_{12})Cl_6]$ has the same structure as the $[(Nb_6Cl_{12})Cl_6]^{4-}$ ion (Fig. 129, page 238). Six tungsten atoms form an octahedron (W–W = 2·92 Å), twelve chlorine atoms lie above the twelve octahedral edges, and an additional chlorine atom is bonded to each of the tungsten atoms. It is formed by the oxidation of the "dichloride", $(W_6Cl_8)Cl_4$, with liquid chlorine at 100° C (2143). The volatilization of tungsten tetrachloride or tungsten trichloride in a mass spectrometer has lead to the observation of ions formed from W_2Cl_6 and W_3Cl_9 (1926).

Oxidation of tungsten dibromide, $(W_6Br_8)Br_4$, with bromine forms a compound of composition WBr_3. However this "tribromide" contains $(W_6Br_8)^{6+}$ groups attached to polybromide Br_4^{2-} groups; it is discussed with the other molybdenum and tungsten clusters on page 351 (1583, 2034, 2142).

Tungsten triiodide has been prepared from iodine and tungsten hexacarbonyl, by analogy with the preparation of molybdenum triiodide (673).

B. Oxohalides

The only trivalent Group VI oxohalides appear to be CrOCl (2038) and CrOBr (625). The oxochloride is prepared from Cr_2O_3 and $CrCl_3$, and can be purified by transport in the gas phase with $CrCl_3$. This chemically inert green-violet compound is isomorphous with, for example, TiOCl, VOCl and FeOCl. The magnetic properties are normal ($\mu_{eff} = 3{\cdot}9$ B.M., Curie–Weiss law obeyed), in contrast to the antiferromagnetic TiOCl and VOCl.

C. Anionic Halo Complexes

1. *Fluoro complexes*

Salts of the CrF_6^{3-} anion appear to be best prepared under anhydrous conditions, for example by fusing alkali metal fluorides or bifluorides with $CrF_3,3H_2O$ or salts of $[CrF_5(H_2O)]^{2-}$ (1432, 1849, 2245). Under aqueous conditions salts of $[CrF_5(H_2O)]^{2-}$ are obtained.

K_3MoF_6 and K_3WF_6 are likewise obtained from K_3MoCl_6, MoO_3 or WO_3 with KF or KHF_2 at high temperatures (27, 1850).

2. *Chloro, bromo and iodo complexes*

(a) *Chromium.* Salts of hexachlorochromate(III) are similarly prepared under anhydrous conditions, either in a melt or in non-aqueous solvents.

In addition to salts of the type $A_3Cr^{III}Cl_6$ (where A is Li, Na or pyH) (931), salts of the type $A_3[Cr_2^{III}Cl_9]$ (where A is K, Rb, Cs, Bu_4N or $EtNH_2$) (9, 612, 1983) can be obtained, depending upon the stoicheiometry of the reactants and the size of the cation. The spectral and magnetic properties of salts of the mononuclear anion are consistent with octahedral chromium(III). The structure (2378) of $Cs_3Cr_2Cl_9$ (Fig. 16, page 25) shows that the anion is composed of two $CrCl_6$ octahedra sharing a common face. The chromium atoms are displaced from the centres of their octahedra away from each other, the chromium–chromium distance of 3·12 Å being much too long to allow direct chromium–chromium bonding. The spectral and magnetic properties are again consistent with octahedral chromium(III) (612, 724, 931, 1983).

A third chloroanion may be $CrCl_4^-$. The reaction of PCl_5 with CrO_2Cl_2 at 140° C forms blue $CrPCl_8$ (1600):

$$3\,PCl_5 + CrO_2Cl_2 \rightarrow PCrCl_8 + 2\,POCl_3 + 1\tfrac{1}{2}\,Cl_2$$

The room temperature effective magnetic moment of 3·6 B.M. appears slightly low for octahedral chromium(III) ($\mu_{eff} \sim 3{\cdot}8 - 3{\cdot}9$ B.M.), and it may be formulated as $(PCl_4)^+(CrCl_4)^-$. The Mossbauer and infrared spectrum of the analogous $FePCl_9$ have also been interpreted as indicating $(PCl_4)^+(FeCl_4)^-$ (342). The $CrCl_4^-$ ion can apparently also be obtained from anhydrous acetonitrile (1097).

An aquo ion $[CrCl_5(H_2O)]^{2-}$ is again obtained from aqueous solution.

Complex bromides and iodides of chromium appear similar to the complex chlorides, but they have not been extensively studied (1983).

(b) *Molybdenum.* Electrolytic reduction of an approximately 10% solution of MoO_3 in 5–11 *M* hydrochloric acid results in colour changes through green and brown to red molybdenum(III) (1260, 1571). This solution is unstable to air. The products obtained from solution depend upon the acid concentration. Thus the red-brown hydrate $MoCl_3,H_2O$ can be obtained on evaporation, and addition of alkali metal or quaternary-ammonium chlorides precipitate pink or red salts $A^I_3MoCl_6$ if the acid concentration is above 7 *M*, but orange or red salts $A^I_2[MoCl_5(H_2O)]$ if the acid concentration is between 4 *M* and 7 *M*. At high molybdenum concentrations and from concentrated acids, the red $A^I_3Mo_2Cl_9$ are obtained (378, 431, 469, 1087, 1088, 1446, 1529, 2105). Unusual double salts formulated as $(NH_4)_2[MoCl_5(H_2O)],2NH_4Cl$ and $K_2[MoCl_5(H_2O)],5KCl,5H_2O$ have also been described (1260).

Considerable decomposition may occur if the reduction is carried out in hydrobromic acid, but salts of $[MoBr_6]^{3-}$, $[MoBr_5(H_2O)]^{2-}$ and $[Mo_2Br_9]^{3-}$ have been obtained (1146, 2201, 2335). Electrolytic reduction in hydroiodic acid has not lead to molybdenum(III) complexes (2208).

The magnetic properties ($\mu_{eff} \sim 3 \cdot 8$ B.M., Curie law obeyed) (785, 846, 1260, 1424) and spectral properties (bands at 9000, 15 000, 19 000 and 24 000 cm^{-1} assigned to transitions from the $^4A_{2g}$ level to the ^2E_g $^2T_{1g}$, $^2T_{2g}$, $^4T_{2g}$ and $^4T_{1g}(F)$ levels) (1446) of compounds containing the $[MoCl_6]^{3-}$ and $[MoCl_5(H_2O)]^{2-}$ anions are in accord with that expected for a d^3 ion in an octahedral ligand field. The structure of K_3MoCl_6 confirms simple octahedral coordination (48).

The dinuclear $Mo_2Cl_9{}^{3-}$ and $Mo_2Br_9{}^{3-}$ show more complex behaviour. There is a superficial resemblance between the visible spectra of these dinuclear anions with those of the hexahalomolybdates(III), but the bands are 10–20 times more intense in the former cases possibly due to the absence of a centre of symmetry, and the bands are shifted to higher energies. The effective magnetic moments are very low indicating considerable magnetic interactions between the two molybdenum atoms. However diamagnetism is not completely attained, and moreover the magnetic susceptibilities increase as the size of either the halide or cation increase. For example (1059, 1984):

	μ_{eff}(B.M.)	
	300° K	90° K
$Cs_3Mo_2Cl_9$	0·42	0·05
$(Et_4N)_3Mo_2Cl_9$	0·96	0·15
$Cs_3Mo_2Br_9$	0·52	0·03
$(Et_4N)_3Mo_2Br_9$	1·09	0·14

An exchange integral has been evaluated from this variation with temperature, and it is found that there is a linear correlation between the *c*-unit cell dimension, which is parallel to the molybdenum–molybdenum vector, and this exchange integral. The molybdenum–molybdenum distances in $Cs_3Mo_2Cl_9$ and $Cs_3Mo_2Br_9$ are 2·67 and 2·78 Å respectively.

Another dinuclear molybdenum(III) anion is $Mo_2Cl_8^{2-}$, formed from α-$MoCl_4$ in methanol. The compounds isolated were $(Ph_4As)_2Mo_2Cl_8$, $2H_2O$, $(Et_4N)_2Mo_2Cl_8$, 2MeOH and $(pyH)_2Mo_2Br_8$, MeOH (40).

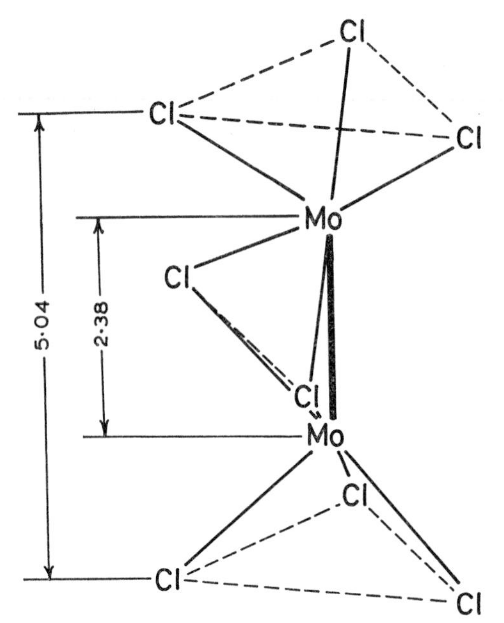

FIG. 185 The structure of $Rb_3Mo_2Cl_8$.

The mixed molybdenum(III)–molybdenum(II) anion, $Mo_2Cl_8^{3-}$, is formed by the action of concentrated hydrochloric acid on molybdenum(II) acetate, and can be precipitated as the rubidium, caesium or triphenylphosphonium salts (40, 41, 203). The corresponding bromides can also be prepared (203). The structure of $Rb_3Mo_2Cl_8$ is shown in Fig. 185 (203). The structure is closely related to the above $Mo_2Cl_9^{3-}$, but one of the bridging chlorine atoms has been removed. The molybdenum–molybdenum distance of 2·38 Å is considerably shorter than in $Mo_2Cl_9^{3-}$ (2·67 Å) or even $W_2Cl_9^{3-}$ (2·41 Å). This is presumably due to the removal of the chlorine atom, the decreased molybdenum–molybdenum repulsion because of the lower charge, the lower oxidation state, and the bonding role of the additional electron. The structure of the molybdenum(II) $Mo_2Cl_8^{4-}$ (Mo–Mo = 2·14 Å) is not closely related (page 357).

Below 4 *M* hydrochloric acid, hydrolysis of molybdenum(III) occurs with the formation of green solutions from which $MoOCl,4H_2O$ of unknown structure can be isolated. It appears to exist in both a brown and a green form (1088, 2336). The corresponding $MoOBr,4H_2O$ (2335) and $MoOF,3\frac{1}{2}H_2O$ (2388) can similarly be obtained. Below 1*M* hydrochloric acid, the dark brown hydroxide is precipitated (1088).

(c) *Tungsten.* The reduction of tungsten(VI) in concentrated hydrochloric acid with tin (1982) or electrolytically (1314) forms the green enneachloroditungstate ion $W_2Cl_9^{3-}$ which may be precipitated with an appropriate cation. The yields appear highly variable. The corresponding brown bromo complexes can be similarly prepared from hydrobromic acid solutions (2423).

The instability of the mononuclear $[WCl_6]^{3-}$ and $[WCl_5(H_2O)]^{2-}$, which are not known, is in marked contrast to molybdenum and chromium, and illustrates the increasing tendency to form the metal–metal bonded $M_2Cl_9^{3-}$.

TABLE 29

Hexagonal cell dimensions of compounds of stoicheiometry $A_3^IM_2^{III}Cl_9$

		a(Å)	c(Å)	$\frac{c}{a}$	Reference
Type 1	$Cs_3Ti_2Cl_9$	7·32	17·97	2·45	1249, 2378
	$Cs_3V_2Cl_9$	7·24	17·94	2·48	1249, 2378
	$Cs_3Cr_2Cl_9$	7·22	17·93	2·48	1249, 2378
	$Cs_3Ru_2Cl_9$	7·22	17·58	2·43	1249
	$Cs_3Rh_2Cl_9$	7·19	17·94	2·50	1249
	$Cs_3Nb_2Cl_9$	7·36	17·45	2·37	1249
	$Cs_3Mo_2Cl_9$	7·35	17·53	2·38	1249
	$Cs_3W_2Cl_9$	7·41	17·08	2·32	349
	$K_3Mo_2Cl_9$	7·12	16·66	2·34	1249
	$K_3W_2Cl_9$	7·17	16·26	2·27	2347
	$(NH_4)_3W_2Cl_9$			2·26	349
	$Rb_3W_2Cl_9$			2·34	349
	$Tl_3W_2Cl_9$			2·29	349
Type 2	$Cs_3Tl_2Cl_9$	$\frac{12·82}{\sqrt{3}} = 7·40$	18·27	2·47	1201, 1883
Type 3	$Cs_3Fe_2Cl_9$	7·28	2(8·90) = 17·80	2·44	2417
	$Cs_3As_2Cl_9$	7·37	2(8·91) = 17·82	2·42	1200
	$Cs_3Sb_2Cl_9$	7·61	2(9·32) = 18·64	2·45	2417

Compounds containing a trinuclear anion such as $K_5W_3Cl_{14}$ have also been claimed (1509), but these have been shown to be mixtures of $K_3W_2Cl_9$ and $K_4[W_2^{IV}OCl_{10}]$ (1450).

The slight paramagnetism sometimes observed for $K_3W_2Cl_9$ is due to impurities, the pure compound being diamagnetic (1355). The structure of $K_3W_2Cl_9$ (349, 2347) shows the existence of discrete $W_2Cl_9^{3-}$ groups composed of pairs of octahedra sharing a common face. The presence of tungsten–tungsten bonding is shown by the displacement of the metal atoms from the centres of their octahedra towards one another, so that the tungsten–tungsten bond length is 2·42 Å (Fig. 16, page 25). The ammonium, rubidium, caesium and thallium salts are isomorphous (349).

A number of similar double halides of stoicheiometry $A_3^IM_2^{III}Cl_9$ are known (Table 29). For purposes of comparison with these structures, the structure of $K_3W_2Cl_9$ will be considered to be derived from the close packing of sheets of composition K_xCl_{3x}. One quarter of the resultant octahedral holes are bounded by six chlorine atoms, and if $\frac{2}{3}$ of these are occupied by tungsten atoms, the composition $K_3W_2Cl_9$ is obtained. The stacking of the close packed layers is ABCAB so that the two WCl_6 octahedra share a common face.

A second type of structure is found in $Cs_3Tl_2Cl_9$, where there is simple hexagonal close packing of the Cs_xCl_{3x} layers. The length of the *c*-axis corresponds to the layers ABABAB, while the *a*-axis must be divided by $3^{\frac{1}{2}}$ to enable a direct comparison with the structures of the first type. The structure can again be considered to contain $Tl_2Cl_9^{3-}$ ions formed from two octahedra sharing a common face.

A third type of structure involving cubic close packing is found in $Cs_3As_2Cl_9$. For purposes of comparison the *c*-axis has been doubled in Table 29. In this case the structure contains discrete $AsCl_6$ units which do not share faces with similar units.

The ratio *c*/*a* of the unit cell dimensions (Table 29) reflects the extent of metal–metal bonding along the *c*-axis, and falls into one of two groups:

(i) $\frac{c}{a} > 2\cdot4$ (average value 2·46). The absence of metal–metal bonding is also reflected in the magnetic properties of these compounds. Normal behaviour for octahedrally coordinated trivalent ions is observed for $Cs_3V_2Cl_9$, $Cs_3Cr_2Cl_9$ and $Cs_3Fe_2Cl_9$, while $Cs_3Ti_2Cl_9$ ($\mu_{eff} = 1\cdot2$–$1\cdot4$ B.M.) shows only slight magnetic exchange (1004, 1249, 1983).

(ii) $\frac{c}{a} < 2\cdot4$ (average value 2·32). The compounds containing the metal–metal bonded anions $Nb_2Cl_9^{3-}$ (1249), $Mo_2Cl_9^{3-}$ and $W_2Cl_9^{3-}$ are diamagnetic (or only slightly paramagnetic).

The e.s.r. spectra of salts of $Cr_2Cl_9^{3-}$, $Mo_2Cl_9^{3-}$ and $W_2Cl_9^{3-}$ again reflect the increasing metal–metal bonding along this series (349).

The $W_2Cl_9^{3-}$ ion is quantitatively oxidised in solution by chlorine, bromine or iodine to $W_2Cl_9^{2-}$, in which the average formal oxidation state is $+3 \cdot 5$ (1985). The violet $(Bu_4N)_2W_2Cl_9$ is a 2:1 electrolyte in acetonitrile, and has an effective magnetic moment of 1·87 B.M. confirming the presence of one unpaired electron, but the structural details are not known.

There is a brief mention in the literature that heating $Cs_3W_2Cl_9$ *in vacuo* forms $Cs_3W_2Cl_8$ in which the tungsten is in the average formal oxidation state of $+2 \cdot 5$ (1984). The molybdenum analogue has been described in more detail on page 340.

D. Other Inorganic Complexes

1. *Chromium*

The trivalent state is the most common one for chromium, and the chemistry is very extensive. The substitution reactions are relatively slow, and complexes such as *cis*-$[CrCl_2(en)_2]$ and $[Cr(o\text{-phen})_3]^{3+}$ may be resolved into their optical isomers (1513). Similarly an almost complete range of hydrates and ammines can be isolated, for example $[Cr(H_2O)_x(NH_3)_{6-x}]^{3+}$ (where x is 0, 1, 2, 3, 4 or 6) and $[CrCl_x(NH_3)_{6-x}]^{(3-x)+}$ (where x is 0, 1, 2, 3, 4 or 6). Crystal structures of these complexes confirm octahedral coordination (623, 1716). Similarly pseudo-halide ions form, for example, $[Cr(CN)_6]^{3-}$, $[Cr(SCN)_6]^{3-}$, $[Cr(SeCN)_6]^{3-}$ and $[Cr(N_3)_6]^{3-}$ (1097).

Oxygen-18 isotope dilution methods show that 6·0 oxygen atoms remain attached to the metal atom in aqueous solution (1241, 1242, 1874). However an average of 6·8 water molecules per chromium atom were found using an ^{17}O–n.m.r. technique, suggesting the existence of a more rapidly exchanging second solvation shell around $[Cr(H_2O)_6]^{3+}$ (25).

The chemistry of chromium(III) is therefore more akin to that of cobalt(III) than that of a typical early transition metal. A comprehensive review is therefore far beyond the scope of this book, and only some typical reactions will be quoted. The aqueous chemistry has been reviewed elsewhere (715).

Anhydrous chromium trichloride forms $CrCl_3(ligand)_3$ with monodentate ligands such as Ph_3PO, Me_2SO, $H.CONMe_2$, $MeCO.NMe_2$ $(CH_2)_4O$, MeCN and C_5H_5N (379, 562, 906, 942, 1147, 1363, 1364, 1937, 2258). There is no doubt that these adducts are octahedral, although no structures have been reported. However trimethylamine forms the five coordinate $CrCl_3(Me_3N)_2$ with a distorted trigonal bipyramidal stereochemistry. The three chlorine atoms form the triangular plane, with *trans* axial amine groups, as in $TiCl_3(Me_3N)_2$ and $VCl_3(Me_3N)_2$.

Trialkyl phosphines form both the monomeric $[CrCl_3(R_3P)_3]$ and the dimeric $[CrCl_3(R_3P)_2]_2$ (1263, 1264). Tridentate arsines $MeAs(o\text{-}C_6H_4.AsMe_2)_2$ and $MeC(CH_2.CH_2.AsMe_2)_3$ (501, 542), and the tridentate phosphine $P(o\text{-}C_6H_4.PPh_2)_3$ (1231) form the expected $[CrCl_3(\text{terdentate})]$.

Displacement of halide also occurs. For example if only amines are considered, compounds of the types $[CrCl(\text{monodentate})_5]Cl_2$ (1933), $[CrCl_2(\text{bidentate})_2]Cl$ and $[Cr(\text{bidentate})_3]Cl_3$ (500, 1556, 2050, 2367), and $[CrCl_2(\text{tetradentate})]Cl$ (272, 1425) can be obtained. Many of these complexes with polydentate ligands can also be obtained under aqueous conditions.

Some alkoxide and amide complexes have been prepared under strictly anhydrous conditions. On the basis of spectral and magnetic properties, $Cr(OMe)_3$ and $Cr(OEt)_3$ were assigned polymeric octahedral structures (14, 362). The bulky tertiary butoxide group however forms the four coordinate $Li[Cr(OBu^t)_4]$, whereas the diisopropylamido ligand forms the monomeric three coordinate $[Cr(NPr_2)_3]$ (45). The existence of this unusual triangular stereochemistry is confirmed by the observation that $[Cr\{N(SiMe_2)_2\}_3]$ is isomorphous with the iron analogue of known structure (297).

Hexaaquochromium(III) dimerises when the solutions are boiled for about one hour, forming

$$[(H_2O)_4Cr\overset{OH}{\underset{OH}{\langle\rangle}}Cr(H_2O)_4]^{4+}$$

(1506). Oxygen-18 exchange reactions confirm that only $10{\cdot}0 \pm 0{\cdot}1$ oxygen atoms do not exchange with the solvent water, in contrast to eleven which would be expected if the oxo-bridged $[(H_2O)_5Cr\text{–}O\text{–}Cr(H_2O)_5]^{4+}$ was formed (1436). Polymerisation continues on boiling until $HCrO_2$ is precipitated after some months (681).

Chromium(III) appears relatively unusual in also being able to form dinuclear complexes with only a single hydroxo bridge. In aqueous solution ammonia forms the hydroxo bridged $[(NH_3)_5Cr\text{–}OH\text{–}Cr(NH_3)_5]^{5+}$, which can lose a proton to form the oxo bridged $[(NH_3)_5Cr\text{–}O\text{–}Cr(NH_3)_5]^{4+}$.

Boiling the former solution for 24 hours results in the loss of one ammonia molecule forming $[(NH_3)_5Cr\text{–}OH\text{–}Cr(H_2O)(NH_3)_4]^{5+}$ which can lose a proton from the water molecule forming $[(NH_3)_5Cr\text{–}OH\text{–}Cr(OH)(NH_3)_4]^{4+}$. The three hydroxo bridged complexes have normal visible spectra and magnetic moments for octahedral chromium(III). The infrared and visible spectrum, and the magnetic moment, for the oxo-bridged complex ($\mu_{eff} = 1{\cdot}3$ B.M. at room temperature compared with 3·5 B.M. for the hydroxo bridged complexes) indicate considerable magnetic exchange or π bonding through a linear, or almost linear, Cr–O–Cr three-centre bond (724, 2042, 2398).

Complexes which contain a double hydroxo bridge include

$$[(H_2O)_4Cr\begin{matrix} \diagup OH \diagdown \\ \diagdown OH \diagup \end{matrix}Cr(H_2O)_3(NH_3)]^{4+} \quad (1189)$$

and $[(\text{bidentate})_2Cr\begin{matrix} \diagup OH \diagdown \\ \diagdown OH \diagup \end{matrix}Cr(\text{bidentate})_2]^{4+}$, where the bidentate is dipyridyl, *o*-phenanthroline, glycine or phenylalanine (724, 1660).

Chromium(III) also forms tetranuclear hydroxo bridged complexes. The structure of the ethylenediamine compound $[Cr_4(OH)_6(en)_6]^{6+}$ is shown in Fig. 186 (889). The structure of the ammine

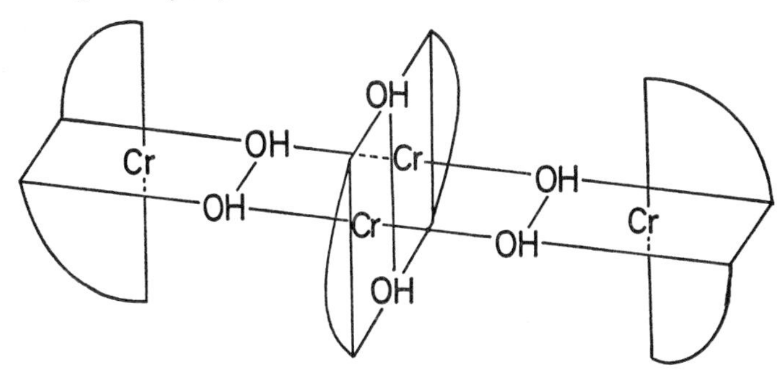

FIG. 186 The structure of $[Cr_4(OH)_6(en)_6]^{6+}$. The four octahedrally coordinated chromium atoms are held together by six hydroxo bridges.

$[Cr_4(OH)_6(NH_3)_{12}]^{6+}$ is completely analogous (139). The magnetic exchange between the octahedrally coordinated chromium atoms has been studied in some detail, but is too complex to be used to assign structures (151, 889, 2373).

The only well characterized oxo complex containing more than two chromium atoms appears to be $[OCr_3(OAc)_6(L)_3]^+$, where L represents H_2O, Me_2SO, $MeCONMe_2$, MeOH, EtOH, NH_3 or C_5H_5N. The structure (849) shows that the oxygen atom is surrounded by a planar equilateral triangle of chromium atoms. Two bridging acetate groups span each edge of the OCr_3 triangle, one above and one below the OCr_3 plane. Octahedral coordination about each metal atom is completed by the additional ligand placed *trans* to the oxygen atom. The spectral (696) and magnetic (716) properties of these complexes have been studied in considerable detail, the low magnetic moments indicating considerable exchange between the metal atoms.

2. *Molybdenum*

Compounds of molybdenum(III) are much less common than those of chromium(III), whereas those of tungsten(III) are exceedingly rare. Examples of molybdenum(III) complexes which appear analogous to those of chromium(III) include $[Mo(NCS)_6]^{3-}$, $[Mo(acac)_3]$ (137), and a number of adducts described below.

The direct action of monodentate ligands on the appropriate molybdenum trihalide forms simple six coordinate adducts such as $MoX_3(py)_3$ (where X is Cl, Br or I) (748, 1690, 1946), $MoBr_3(\gamma\text{-picoline})_3$ (37), and $MoX_3(RCN)_3$ (where X is Cl or Br, and R is Me, Et, Pr, Ph or Bz) (37, 2172). The acetonitrile adduct has been used to prepare $MoBr_3(THF)_3$, $MoBr_3(pyrazine)_3$, $MoBr_3(Ph_3P)_2(MeCN)$, $MoBr_3(Ph_3As)_2(MeCN)$, and $MoBr_3(dipy)(MeCN)$ (37). Reduction to molybdenum(II) occurs with MoI_3 and $o\text{-}C_6H_4(AsMe_2)_2$ (1528). Compounds of the same $MoX_3(ligand)_3$ stoicheiometry can be obtained from $[MoX_5(H_2O)]^{2-}$ in aqueous or ethanolic solution (where X is Cl and ligand is urea, dimethylformamide or thiourea; and where X is Br and ligand is urea or dimethylformamide) (789, 1072, 1312, 1446, 1546, 2201). These compounds have room temperature effective magnetic moments in the range 3·5–3·87 B.M.

Displacement of halide to form $[MoBr_2(urea)_4]Br$ and $[Mo(urea)_6]Br_3$ occurs under more vigorous conditions.

However in addition it is possible to obtain the interesting $Mo_2Cl_6(ligand)_3$ with thiourea and substituted thioureas, which have abnormally low magnetic moments of 0·3–0·6 B.M. Other thio ligands which form polymeric materials include monothio-β-diketonates, $Mo_2Cl_2L_4(H_2O)_3$, with low effective magnetic moments in the range 0·3–1·1 B.M. (1546).

The compounds obtained from aqueous or ethanolic solution which were formulated as $[Mo(bidentate)_3]X_3$ (where the bidentate is dipyridyl, *o*-phenanthroline or 2-aminomethylpyridine) (2209, 2251) do not exist. They are mixtures of, for example $(dipyH)[Mo^{III}Cl_4(dipy)]$, $(NH_4)[Mo^{III}Cl_4(dipy)]$, and $[Mo_2^{V}O_3Cl_4(dipy)_2]$ (698, 1657, 2046). Cationic complexes $[MoCl_2(dipy)]Cl$ and $[MoCl_2(dipy)_2][MoCl_4(dipy)]$ can also be prepared (428, 698).

Molybdenum trichloride and tribromide with liquid ammonia form $MoCl_2(NH_2)(NH_3)_2$ and $MoBr_2(NH_2)(NH_3)_3$ respectively. Similarly primary and secondary amines form compounds such as $MoBr(MeNH)_2(MeNH_2)$, $MoBr_2(MeNH)$ and $MoBr_2(NH.CH_2.CH_2.NH_2)(NH_2.CH_2.CH_2.NH_2)$. These have effective magnetic moments in the range 0·6–0·9 B.M. (744, 748, 749).

Thiocyanate forms salts containing the $[Mo(NCS)_6]^{3-}$ ion, in which the ligand is bonded to the molybdenum atom through the nitrogen atom (1427, 1527). Two thiocyanate groups may be replaced by neutral ligands forming, for example, $[Mo(NCS)_4(py)_2]^-$, $[Mo(NCS)_4(Et_3P)_2]^-$ and $[Mo(NCS)_4(Ph_2P.CH_2.CH_2.PPh_2)]^-$ (1261, 1946).

Cyanide in aqueous solutions forms $K_4Mo(CN)_7,2H_2O$ (2422). The low magnetic moment of 1·75 B.M. indicates a seven or possibly eight coordinate anion, $[Mo(CN)_7]^{4-}$ or $[Mo(CN)_7(H_2O)]^{4-}$ (1527). The report of $K_2Mo(CN)_5$ should be regarded with doubt (1636).

3. *Tungsten*

The reduction of WCl_4 or $WCl_4(MeCN)_2$ with pyridine forms paramagnetic $WCl_3(py)_2$ ($\mu_{eff} = 2·11$ B.M.) (246). A similar compound WCl_3(diphosph) was obtained with $Ph_2P.CH_2.CH_2.PPh_2$ ($\mu_{eff} = 1·29$ B.M.) (267).

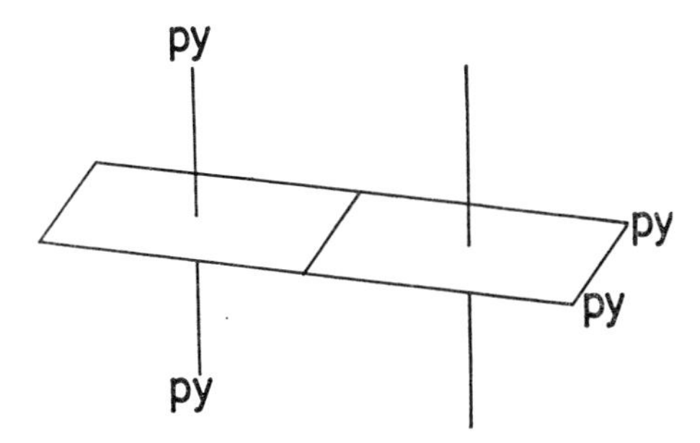

FIG. 187 The structure of $W_2Cl_6(py)_4$.

The reaction of $K_3W_2Cl_9$ with pyridine forms $W_2Cl_6(py)_4$, (and not $W_2Cl_6(py)_3$ as reported earlier (1313)) (1982). In contrast to the product with the same composition above, this isomer is diamagnetic. The structure (Fig. 187) shows that the two octahedrally coordinated metal atoms share a common edge in contrast to the face sharing $K_3W_2Cl_9$, but retain a strong tungsten–tungsten bond (1269). Similar compounds were obtained with 4-picoline and 4-isopropylpyridine.

Apparently similar compounds, $W_2Cl_4(OR)_2(ROH)_4$ are obtained from $W_2Cl_9^{3-}$ and ROH (where R is Me, Et or Pr). These diamagnetic compounds are dimeric in chloroform, which is confirmed by the mass spectrum (491).

The only other well characterized compound appears to be the low spin seven coordinate $[WBr_2(CO)_3(o\text{-}C_6H_4(AsMe_2)_2)]Br$ (1526).

The reduction of solid $K_4[W^{IV}(CN)_8]$ with hydrogen at 390° C, followed by aerial oxidation, is reported to yield $K_3[W^{III}(CN)_6]$ (1297). The magnetic moment of 1·76 B.M. is surprising when compared with 1·50 B.M. for the molybdenum analogue.

E. Organometallic Complexes

Whereas the reaction of phenylmagnesium bromide with chromium trichloride in diethyl ether yields black pyrophoric solids containing chromium(I) and chromium(O) (page 373), the same reaction in tetrahydrofuran yields $Cr(Ph)_3(THF)_3$ (2439). This is one of a series of compounds containing from one (for example $CrCl_2(Ph)(THF)_3$ (1164, 2174)) to six (for example $Li_3Cr(Ph)_6$, $2\frac{1}{2}Et_2O$ (1165)) chromium–phenyl σ bonds. This work has been reviewed (2440). The structure of $CrCl_2(p\text{-}Me.C_6H_4)(THF)_3$ shows that the two chlorine atoms are *trans* to each other in the octahedral structure. The chromium–oxygen bond length (2·21 Å) *trans* to the chromium–carbon bond is considerably longer than the other two chromium–oxygen bonds (2·04 Å) in the same plane (622).

Cyclopentadienyl chromium(III) compounds include $(C_5H_5)CrX_2(Ligand)$ (875) (where X is Cl, Br or I and the ligand is THF, py or Ph_3P), $(C_5H_5)CrBr(acac)$ (2277), and $(C_5H_5)_2CrI$ (859).

The diamagnetic $WPh_3,3LiPh,3Et_2O$ claimed (953) to have been formed from tungsten pentabromide and phenyllithium in diethylether has been shown (1990) to be the tungsten(IV) compound $WPh_4,2LiPh,3Et_2O$.

5. OXIDATION STATES II AND I

A. Chromium Dihalides

The chromium dihalides may be prepared by the reduction of the anhydrous trihalides with hydrogen at 300°–500° C (228, 781, 1162, 2045), by the action of the anhydrous hydrogen halide on the metal above the melting point of the dihalide (228, 882, 1454, 1480, 2045) (M.P. $CrCl_2$ = 820° C (228, 2130), M.P. $CrBr_2$ = 840° C (320)), or in the case of the bromide and iodide, by the direct reaction of the elements at about 700° C (2293). These compounds are very deliquescent, and when moist are easily oxidized to chromium(III).

As a first approximation the structures of all four dihalides can be considered to be based on close packing of halide ions with the chromium atoms in the octahedral interstices. However in all cases the octahedra are considerably distorted, with four planar near neighbours and two more distant *trans* halide ions:

	CrF_2 (1267)	$CrCl_2$ (2294)	$CrBr_2$ (2293)	CrI_2 (2295)
4 Cr–X (Å)	1·98 + 3(2·01)	2·39	2·54	2·74
2 Cr–X (Å)	2·43	2·93	3·00	3·24
Ratio	0·82	0·82	0·85	0·85

Such a distortion for spin-free chromium(II) (d^4) is expected from the Jahn–Teller effect.

Study of the CrF_2–CrF_3 system has shown the existence of the intermediate Cr_2F_5 (2245). Strings formed by the edge sharing of octahedra lie parallel to the *c*-axis of the monoclinic unit cell, with corner sharing along the *b*- and *c*-axes (Fig. 188) (2213). The edge shared octahedra

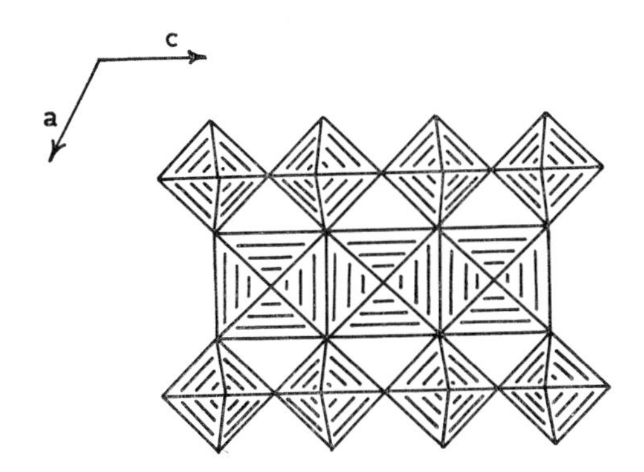

FIG. 188 The structure of Cr_2F_5.

are relatively undistorted (Cr–F = 1·88–1·90 Å) and are therefore thought to contain chromium(III) atoms, whereas the corner shared octahedra are tetragonally elongated (Cr–4(F) = 1·96–2·01 Å, Cr–2 *trans*(F) = 2·57 Å) and are therefore thought to contain chromium(II) atoms. The magnetic coupling between the chromium atoms has been studied (1820).

B. Inorganic Chromium(II) Complexes

Chromium(II) complexes can readily be made either from aqueous solutions (prepared by reduction of chromium(III) with amalgamated zinc), or by the direct reaction of the dihalide under anhydrous conditions.

By far the most common stereochemistry is octahedral, the high spin d^4

configuration leading to a tetragonal elongation by the Jahn–Teller effect. Complexes which have been characterized include hydrates (M_2SO_4, $CrSO_4,6H_2O$ where M is NH_4, Rb or Cs (719), $CrSO_4,5H_2O$, $CrCl_2,4H_2O$, $CrBr_2,6H_2O$, and $Cr(ClO_4)_2,6H_2O$ (722, 803, 1579)), complex alkali metal and ammonium fluorides ($ACrF_3$ and A_2CrF_4 with distorted perovskite and K_2NiF_4 structures respectively (659, 723, 737, 2089, 2095, 2296)), polymeric adducts of the type $CrCl_2(Ligand)_2$ (where the ligand is Ph_3PO, Me_2SO, $(CH_2)_4O$, MeCN or C_5H_5N (1210, 1211, 1363 1579, 1998)), and complexes with multidentate ligands such as $[CrCl_2(N_2H_4)_2]$, $[CrCl_2(en)_2]$, $[Cr(en)_3]^{2+}$, and $[Cr(dien)_2]^{2+}$ (717, 718, 721, 1337).

These octahedral complexes have room temperature effective magnetic moments of about 4·9 B.M., and follow the Curie–Weiss law with $\theta = 0$–30°. The effect of the tetragonal elongation is to split the low lying ^5E_g orbitals into $^5B_{1g}$ and $^5A_{1g}$ components, and the upper $^5T_{2g}$ into $^5B_{2g}$ and ^5E_g components. The absorption spectrum shows a band or shoulder at about 10 000 cm^{-1} which is assigned to this $^5B_{1g} \rightarrow {}^5A_{1g}$ transition between the split levels of the ^5E_g state. The other band at about 15 000 cm^{-1} is assigned to $^5B_{1g} \rightarrow {}^5E_g$.

Ligands such as cyanide, phthalocyanine, dipyridyl and *o*-phenanthroline form complexes with magnetic moments in the range 2·9–3·5 B.M., and have been studied in some detail (123, 225, 720, 1163, 1181, 1182, 1184, 1520, 1578, 2270).

It may be noted that whereas the chloro and bromo complexes $CrX_2(Ph_3PO)_2$ and $CrX_2(MeCN)_2$ are apparently octahedral and polymerized through halogen bridges, the iodo complexes $CrI_2(Ph_3PO)_2$ (1998) and $CrI_2(MeCN)$ (1212) have different properties and have been assigned tetrahedral structures.

The "tripod" ligand $N(CH_2.CH_2.NMe_2)_3$, which is capable of forcing trigonal bipyramidal five coordination on a number of metal ions, also forms the five coordinate [CrBr(Ligand)]Br ($\mu_{eff} = 4 \cdot 85$ B.M.) (479).

The dimeric acetato complexes $[Cr_2(OAc)_4(Ligand)_2]$ (where the ligand is water, piperidine, α-picoline, etc.) have been referred to on page 31 (Fig. 23), and contain pairs of chromium atoms linked by four bridging acetate groups. The chromium–chromium distance in the aquo complex of 2·64 Å (1786) is the same as that in $[Cu_2(OAc)_4(H_2O)_2]$, but the lower magnetic moment ($\mu_{eff} < 1$ B.M.) indicates a stronger interaction between the metal atoms (1180). However this interaction is clearly not as strong as in the diamagnetic $Mo_2(OAc)_4$. Analogous compounds with other carboxylic acids have also been reported (1179, 1180).

Finally it may be noted that whereas the double sulphate Cs_2SO_4, $CrSO_4$, $6H_2O$ has a normal magnetic moment (4·9 B.M.), the dihydrate Cs_2SO_4, $CrSO_4$, $2H_2O$ has an effective magnetic moment of only 0·9 B.M. It was suggested that the structure was related to that of $Cr_2(OAc)_4(H_2O)_2$, but that the bridging acetate groups had been replaced by bridging sulphate groups (719).

C. Molybdenum and Tungsten Clusters

1. *Halides*

Molybdenum and tungsten difluorides are not known. Unsuccessful preparative attempts include the reaction of tungsten and tungsten dibromide with hydrogen fluoride (775).

All six dihalides, $MoCl_2$, $MoBr_2$, MoI_2, WCl_2, WBr_2 and WI_2 were prepared before 1900 by the thermal decomposition and/or reduction of the higher halides. Typical conditions are given below (451, 2029):

Reaction	Temperature °C
$Mo + MoCl_3 \rightarrow MoCl_2$	550–750 gradient
$Mo + Br_2 \rightarrow MoBr_2$	600
$MoBr_3 \rightarrow MoBr_2$	600
$MoI_3 \rightarrow MoI_2$	500
$WCl_4 \rightarrow WCl_2$	500
$W + Br_2 \rightarrow WBr_2$	560–760 gradient
$W + 2{\cdot}00\ Br_2 \rightarrow WBr_2$	40–600 gradient
$W(CO)_6 + I_2 \rightarrow WI_2$	500
$W + I_2 \rightarrow WI_2$	300–450 gradient

All six dihalides are isomorphous. The structure of $MoCl_2$ shows that it is based on $(Mo_6Cl_8)^{4+}$ groups, which are structurally the same as in Fig. 127 (page 236). The six molybdenum atoms form an octahedron, with molybdenum–molybdenum distances of 2·61 Å (2029). A chlorine atom lies above each face of the octahedron bonded to three molybdenum atoms, with molybdenum-chlorine bond lengths of 2·47 Å. Each molybdenum atom is also bonded to an additional chlorine atom in a "centrifugal" position (Fig. 189). Four of these six centrifugal chlorine atoms are shared with adjacent $(Mo_6Cl_8)^{4+}$ clusters (Mo–Cl = 2·50 Å) forming an infinite sheet structure (Fig. 190). The remaining two *trans*-centrifugal chlorine atoms are not shared (Mo–Cl = 2·38 Å). The formula may therefore be written $(Mo_6Cl_8)Cl_{4/2}Cl_2$. The structure is much more open than those of the higher molybdenum chlorides, which are based on the close packing of chloride ions.

Oxidation of tungsten dibromide, $(W_6Br_8)Br_4$ with bromine forms

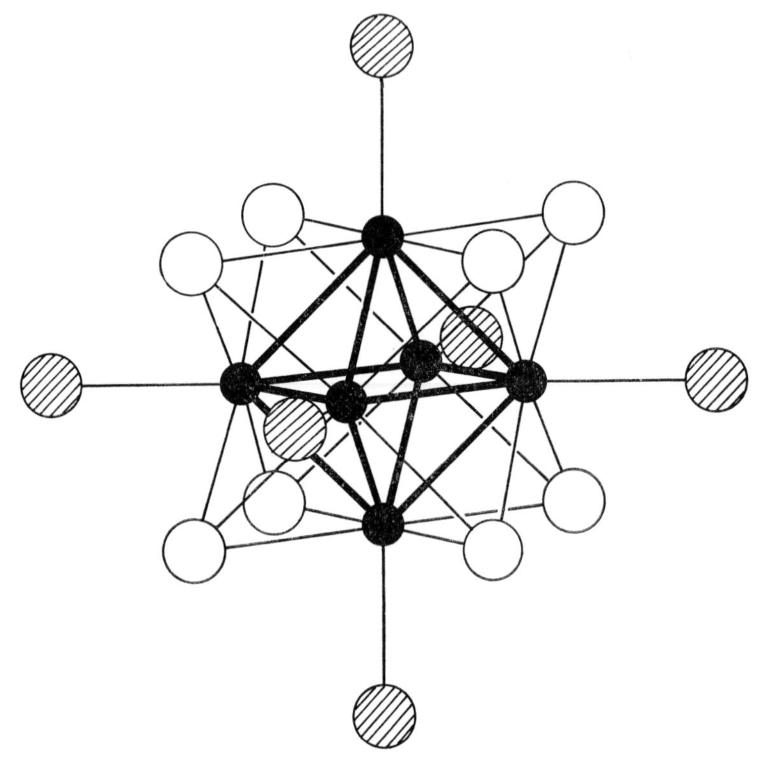

FIG. 189 The $[(Mo_6Cl_8)Cl_6]^{2-}$ ion. (Compare with Fig. 127, page 236).

W_6Br_{12}, W_6Br_{16} or W_6Br_{18} (two isomers) depending upon conditions (2028, 2033, 2034, 2141, 2142).

$$\left.\begin{array}{l}(W_6Br_8)Br_4 + Br_2 \xrightarrow{<80^\circ} \alpha\text{-}W_6Br_{18} \\ \qquad\qquad\qquad \xrightarrow{100^\circ} \beta\text{-}W_6Br_{18} \\ \qquad\qquad\qquad \xrightarrow{150^\circ} W_6Br_{16}\end{array}\right\} \xrightarrow[\text{vac.}]{\text{heat}} W_6Br_{14} \xrightarrow[\text{vac.}]{350^\circ} W_6Br_{12}$$

The structure of W_6Br_{16} shows an undistorted (W_6Br_8) cluster as before ($W\text{–}W = 2{\cdot}64$ Å, $W\text{–}Br_{bridging} = 2{\cdot}58$ Å), but now the six centrifugal sites are occupied by four non-bridging bromine atoms ($W\text{–}Br = 2{\cdot}56$ Å) and two bridging Br_4^{2-} ions ($W\text{–}Br = 2{\cdot}60$ Å) (Fig. 191). The structures of W_6Br_{12} and W_6Br_{16} can then be represented as $(W_6Br_8)^{4+}Br_{4/2}Br_2$ and $(W_6Br_8)^{6+}(Br_4^{2-})_{2/2}Br_4$ respectively. It was further suggested that

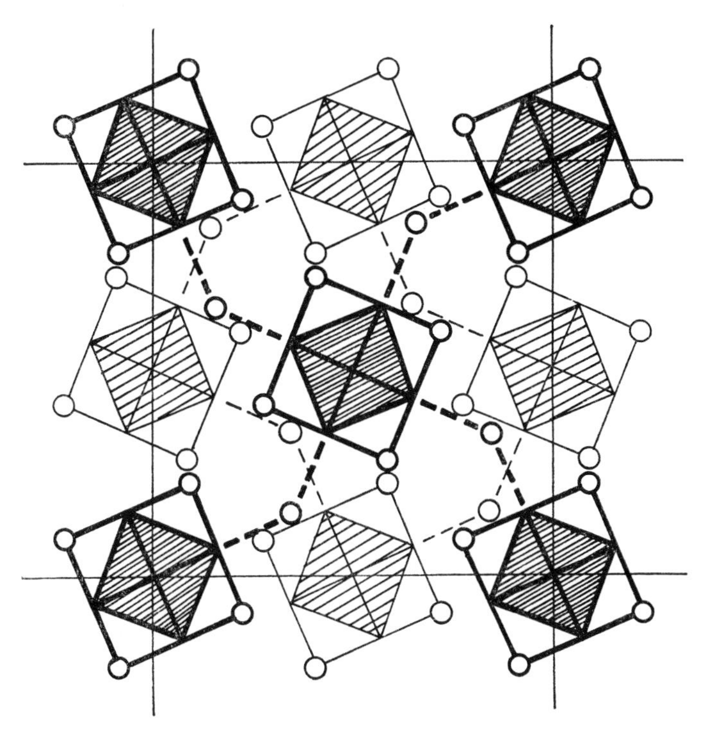

FIG. 190 The structure of $(Mo_6Cl_8)Cl_4$. The cluster-centrifugal chlorine bonds are shown by broken lines.

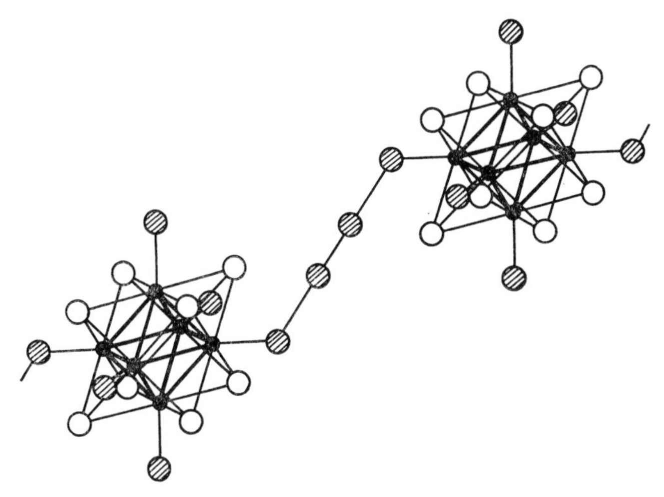

FIG. 191 The linking of two $(W_6Br_8)Br_4$ groups by a Br_4^{2-} group in W_6Br_{16}.

W_6Br_{14} and W_6Br_{18} could be formulated as $(W_6Br_8)^{6+}Br_6$ and $(W_6Br_8)^{6+}(Br_4^{2-})_{4/2}Br_2$ respectively.

Whereas W_6Br_{12} is diamagnetic, W_6Br_{14}, W_6Br_{16} and the two isomers of W_6Br_{18} show a temperature dependent paramagnetism, although the absolute values of the corrected susceptibilities are subject to relatively large uncertainties due to the high molecular weights. For example the effective magnetic moment of W_6Br_{16} at 295°, 195° and 90° K is 1·7, 1·3 and 0·4 B.M. respectively.

Oxidation of $(W_6Br_8)Br_4$ with chlorine yields $W_6Br_8Cl_8$ with the same structure as W_6Br_{16}.

Oxidation of $(W_6Cl_8)Cl_4$ with chlorine however yields the tungsten(III) cluster $(W_6Cl_{12})Cl_6$ (page 337).

2. *Complexes*

Molybdenum and tungsten dichlorides readily dissolve in hot hydrochloric acid, from which the yellow crystalline chloroacids $H_2[(Mo_6Cl_8)Cl_6],8H_2O$ and $H_2[(W_6Cl_8)Cl_6],8H_2O$ precipitate on cooling, where now all six centrifugal sites are occupied by non-bridging chlorine atoms. The salts $A_2[(Mo_6Cl_8)Cl_6]$ (where A is NH_4, Et_4N, Ph_3PH, Ph_3AsOH, $(Ph_3AsO)_2H$ or $(Ph_3PO)_2H$) and $(Et_4N)_2[(W_6Cl_8)Cl_6]$ are also readily obtained (576, 2121, 2122, 2123, 2124).

Ethanolic solutions of $[(Mo_6Cl_8)Cl_6]^{2-}$ are stable, and 6 of the 14 chlorine atoms can be precipitated with silver nitrate. Repeated recrystallization from aqueous hydrobromic or hydroiodic acid yields the corresponding acids $H_2[(Mo_6Cl_8)Br_6],8H_2O$ and $H_2[(Mo_6Cl_8)I_6],8H_2O$ (2121), or their salts such as $(Bu_4N)_2[(Mo_6Cl_8)Br_6]$ and $(Bu_4N)_2[(Mo_6Cl_8)I_6]$ (576). The chloroacid is more stable than the bromoacid, the equilibrium constant being approximately 0·8 for the following reaction (2122):

$$[(Mo_6Cl_8)Cl_6]^{2-} + 6\,Br^- \rightleftarrows [(Mo_6Cl_8)Br_6]^{2-} + 6\,Cl^-$$

The salts $K_2[(Mo_6Cl_8)(NCS)_6], 6H_2O$ and $Na_2[(Mo_6Cl_8)(OR)_6]$ (where R is Me, Et or Ph) have also been reported (1753, 1754, 2120).

These cluster compounds are stable in acid solutions, but in water are hydrolysed precipitating amorphous products, although crystalline $[(Mo_6Cl_8)Cl_4(H_2O)_2],6H_2O$ can be obtained under the appropriate conditions. The alkaline solutions are unstable and are oxidized to higher valence states with the evolution of hydrogen. The structure of $[(Mo_6Br_8)Br_4(H_2O)_2]$ (prepared by the action of aqueous hydrobromic acid on molybdenum at 700° C and 3000 atm) shows that the water molecules occupy two *trans*-centrifugal sites. The metal–metal distances belonging to these two *trans*-molybdenum atoms appear slightly shorter than those between pairs of molybdenum atoms with centrifugal bromine ligands ($2 \cdot 63_0$ and $2 \cdot 64_0$ Å respectively) (1086).

Molybdenum dichloride readily forms complexes of the type $[(Mo_6Cl_8)Cl_4(ligand)_2]$ (where the ligand is H_2O, Ph_3AsO, Ph_3PO, Me_2SO, $HCONMe_2$, C_5H_5NO, EtOH, thioxan, $(CH_2)_5S$, NH_3, Me_3N, Et_3N, C_5H_5N and other nitrogen heterocycles, alkyl nitriles, or Ph_3P) (426, 557, 744, 836, 1542, 1545, 2121, 2122, 2332).

An examination of a series of complexes with 4-substituted pyridine-1-oxides (845) of the type $[(Mo_6Cl_8)Cl_4(4\text{–}Z\text{–}C_5H_4NO)_2]$, shows that the molybdenum-oxygen stretching frequency decreases as the electron density on the donor oxygen atom decreases, that is along the series Z = Me_2N, MeO, Me, H, Cl. For substituted pyridine-1-oxide complexes of the later transition metals such as chromium(III), iron(III), cobalt(II), nickel(II), copper(II) and zinc(II), the metal-oxygen stretching frequency increases as the electron withdrawing power of the 4-substituent increases. Only titanium and zirconium tetrafluoride complexes show the same behaviour as $(Mo_6Cl_8)^{4+}$, which is consistent with a low degree of cluster-ligand π bonding.

Substitution of additional centrifugal chloride ions can occur to form, for example, $[(Mo_6Cl_8)Cl_3(Ph_3P)_3]^+$, $[(Mo_6Cl_8)Cl_2(Ph_3P)_2(py)_2]^{2+}$, and $[(Mo_6Cl_8)(Me_2SO)_6]^{4+}$ (557, 835, 836).

Multidentate ligands have also been studied, but the nature of the products is not clear (836, 2332).

When the acids $H_2[(Mo_6Cl_8)X_6]$, or those complexes with volatile ligands, are heated *in vacuo*, the anhydrous dihalide $(Mo_6Cl_8)X_4$ is reformed.

The acids $H_2[(Mo_6Br_8)X_6]$ and $H_2[(Mo_6I_8)X_6]$ (where X is Cl, Br or I) are rather ill-defined, but a number of salts have been characterized (576, 2125).

The far infrared and Raman spectra of these compounds show bands due to both metal–metal and metal-halogen vibrations (503, 576, 1143, 1666).

The $(Mo_6Cl_8)^{4+}$ core is broken up under fairly vigorous conditions (2120, 2124, 2126, 2127). Alkaline hydrolysis and subsequent acidification yielded a compound which was formulated as $(Et_4N)_2[\{Mo_6Cl_7(OH)\}Cl_6]$. When $(Mo_6Cl_8)Cl_4$ and NaOMe are refluxed in methanol to form $Na_2[(Mo_6Cl_8)(OMe)_6]$, and the reaction is continued by evaporation to dryness and heating to 150° C, the pyrophoric $Na_2[\{Mo_6(OMe)_8\}(OMe)_6]$ can be extracted with methanol/diethyl ether (1753). The phenoxide compounds $Na_2[\{Mo_6(OMe)_8\}(OPh)_6]$ and $Na_2[\{Mo_6(OMe)_4(OPh)_4\}(OPh)_6]$ were also obtained (1754).

In aqueous hydrobromic acid, the oxidized clusters W_6Br_{14}, W_6Br_{16} and W_6Br_{18} are reduced to $[(W_6Br_8)Br_4(H_2O)_2]$.

3. *Bonding*

The treatment of the bonding in the diamagnetic $(M_6X_8)^{4+}$ clusters closely parallels the treatment of the bonding in the $(M_6X_{12})^{2+}$ clusters (page 247).

The axes are chosen as before. The z-axis on each metal atom is directed centrifugally away from the centre of the octahedron with the x-and y-axes directed above the edges of the octahedron. The halogen atoms are situated between the x- and y-axes, in contrast to the $(Nb_6X_{12})^{2+}$ clusters. Of the 36 valence electrons on the six molybdenum atoms, 8 are used for bonding the halogen atoms and four to provide the net charge, leaving 24 electrons available for metal–metal bonding.

On the valence bond approach each of the twelve edges of the octahedron is occupied by an electron pair bond. The disposition of electron pairs about the metal atom is similar to that in the $(Nb_6Cl_{12})^{2+}$ clusters (d^5p^3 or $d_{x^2-y^2}\, d_{xz}\, d_{yz}\, d_{xy}\, sp^3$ hybridization) (1369).

On the molecular orbital approach the 24 electrons occupy molecular orbitals formed from the individual atomic d_{z^2}, $d_{x^2-y^2}$, d_{xz} and d_{yz} orbitals. Although there is some disagreement concerning the precise ordering of the molecular orbitals (563, 607, 1086), there is agreement about the nature of the 12 bonding orbitals which accommodate the available electrons leading to the observed diamagnetism. The d_{z^2} orbitals, which are directed towards the centre of the octahedral cluster, form a bonding A_{1g} orbital, and antibonding T_{1u} and E_g orbitals. The $d_{x^2-y^2}$ orbitals, which are directed above the octahedral edges, form bonding E_g and T_{2u} orbitals, and an antibonding A_{2g} orbital. The d_{xz} and d_{yz} orbitals, which are directed towards the middle of the octahedral faces form bonding T_{1g} and T_{1u} orbitals, and antibonding T_{2u} and T_{2g} orbitals.

D. Other Molybdenum(II) and Tungsten(II) Inorganic Complexes

The chemistry of divalent molybdenum(II) and tungsten(II), apart from that of the $(M_6X_8)^{4+}$ clusters, is not extensive.

The only simple monomeric octahedral complexes appear to be $[MoX_2(diars)_2]$ (where X is Cl, Br or I), and $[WI_2(diars)_2]$, where diars is o-$C_6H_4(AsMe_2)_2$ (673, 1528). They are prepared from aqueous solutions of molybdenum(III), or by the action of the diarsine on the triiodides. They are isomorphous with the iron, technetium and rhenium analogues, and have room temperature effective magnetic moments in the range 2·7–2·9 B.M.

Like many other dipyridyl complexes, $[Mo^{III}Cl_2(dipy)_2]^+$ is readily electrolytically reduced to $[Mo^{II}Cl_2(dipy)_2]$ and $[Mo^{I}Cl_2(dipy)_2]^-$ (698).

The diamagnetic compound obtained by the reduction of molybdenum (VI) with hydroxylamine in the presence of cyanide and formulated as the eight coordinate $K_4[Mo^{II}(OH)_2(CN)_5(NO)]$ (1063), has been shown to be $K_4[Mo^0(CN)_5(NO)]$ (1923).

The action of acetic acid on molybdenum hexacarbonyl forms the dimeric acetate $Mo_2(OAc)_4$ (2216), the structure of which (1512) contains a pair of molybdenum atoms tightly bonded together (Mo–Mo = 2·11 Å) bridged by four acetate groups (Fig. 23, page 31). All four electrons on each molybdenum atom are used to form a molybdenum–molybdenum quadruple bond ($\sigma + 2\pi + \delta$). A band at 34 000 cm^{-1} is assigned to the $\delta \rightarrow \delta^*$ transition (697). This band moves to 29 000 and 24 000 cm^{-1} for $Mo_2(C_3F_7COO)_4$ and $Mo_2(PhCOO)_4$ respectively due to weaker molybdenum–molybdenum bonding. The last two compounds will also form pyridine adducts, which is also a consequence of the weaker metal–metal bonding (see page 31 for a further discussion).

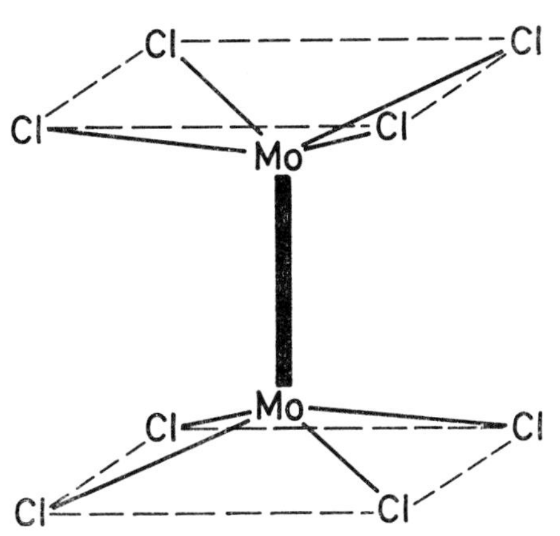

FIG. 192 The structure of $Mo_2Cl_8^{4-}$.

Molybdenum acetate reacts with hydrogen chloride to form $MoCl_2$ (42, 2216), and with hydrochloric acid to form $[Mo_2^{II}Cl_8]^{4-}$ or $[Mo^{II}Mo^{III}Cl_8]^{3-}$ depending upon conditions. The latter is more closely related to $[Mo_2^{III}Cl_9]^{3-}$ than to $[Mo_2^{II}Cl_8]^{2-}$ or $[Mo_2^{II}(OAc)_4]$, and is accordingly described on page 340.

The brown dichloride is chemically different to $(Mo_6Cl_8)Cl_4$, for example it is insoluble in hydrochloric acid, is readily oxidized, and has an effective magnetic moment of 0·49 B.M. The "somewhat faint and diffuse X-ray powder diffraction patterns" bear a "distinct resemblance" to those of $CdCl_2$.

The structure of the red $K_4[Mo_2Cl_8],2H_2O$ (323) is closely related to that of $Mo_2(OAc)_4$ (Fig. 192). The retention of the molybdenum–molybdenum δ-bond is confirmed by the eclipsed orientation of the two $MoCl_4$ units. The molybdenum–molybdenum bond length is 2·14 Å. The molybdenum atoms are displaced from their planes of chlorine atoms towards each other, so that the Cl–Mo–Mo angle is increased to 105°. The water molecules are not bonded to the molybdenum atoms.

E. Organometallic Complexes

1. *Introduction*

Organometallic compounds of the Group VI elements in the formal oxidation states of II and I may be very broadly classified under two headings:

(a) *Carbonyl halides and substituted carbonyl halides.* There are a large number of seven coordinate complexes of the type $[MX_2(CO)_3(ligand)_2]$, particularly with nitrogen, phosphorus or arsenic donor ligands. These are discussed more fully in the next section. Also included under this heading are compounds of the type $(ligand)_2(CO)_3M\overset{Cl}{\frown}SnX_3$ (Cl bridging M and SnX_3, with M–Sn bond), and arene substituted carbonyl halides $[(arene)MX(CO)_3]^+$.

(b) *Cyclopentadienyl compounds.* The five electron donor cyclopentadienyl group forms $[(C_5H_5)M(CO)_4]^+$, $[(C_5H_5)MX(CO)_3]$, and $[C_5H_5)M(CO)_3]^-$, all with 18 electrons filling the valence shell. The $(C_5H_5)M(CO)_3$ moiety, containing 17 valence shell electrons, forms a large number of compounds of the type $(C_5H_5)M(CO)_3X$, where X can be hydride, halide, a σ-bonded alkyl or aryl, or another metal atom such as $SnCl_3$, HgCl, $Au(PPh_3)$, $Fe(C_5H_5)(CO)_2$, $Mn(CO)_5$, and so on. The eighteen electron configuration can also be attained through the sharing of two electrons with the formation of $[(C_5H_5)M(CO)_3]_2$ with a metal–metal bond. It is therefore convenient to include the metal–metal bonded compounds in the formal oxidation state of +I with compounds without such bonds in the formal oxidation state of +II. This general group is also discussed in more detail in a separate section.

The 18-electron configuration can also be achieved with fewer electron donating ligands, either by the sharing of more electrons between the metal atoms as in $(CO)_2(Me_5C_5)Mo{\equiv}Mo(Me_5C_5)(CO)_2$ and related compounds (see later), or by the formation of bridged compounds such as $[(CO)_4Mo(\mu\text{-}PMe_2)_2Mo(CO)_4]$ (two bridging PMe_2 groups, with Mo–Mo bond) and a large number of closely related compounds containing strong metal–metal bonds (317, 464, 465, 1153,

$$1637), [(CO)_2(C_5H_5)Mo \underset{PR_2}{\overset{PR_2}{\rightleftarrows}} Mo(C_5H_5)(CO)_2] \text{ (see later), and}$$

$$[(CO)_2(Ph_4C_4)Mo \underset{Br}{\overset{Br}{\overline{\rightleftarrows}}} Mo(Ph_4C_4)(CO)_2] \text{ (1663).}$$

Other ways of achieving the 18-electron configuration are with the three electron donor π-allyl group, as in compounds of the type $(C_3H_5)MoCl(CO)_2(dipy)$, prepared by oxidation of $Mo(CO)_4(dipy)$ with allyl chloride (compare with the substituted carbonyl halides above) (1041, 1238).

The nitrosyl group also behaves as a three electron donor, as in

$$(NO)(C_5H_5)Cr \underset{SMe}{\overset{SMe}{\overline{\rightleftarrows}}} Cr(C_5H_5)(NO) \text{ (1899) and}$$

$$(NO)(C_5H_5)Mo \underset{I}{\overset{I}{\overline{\rightleftarrows}}} Mo(C_5H_5)(NO) \text{ (1399). A more unusual compound}$$

is $Mo(C_5H_5)_3(NO)$ (567), prepared from $[(C_5H_5)MoI_2(NO)]_2$ (1399) and thallium cyclopentadienyl. The ^1H–n.m.r. spectrum shows only a single signal for all 15 protons, but on cooling from 0° to −50° C, one cyclopentadienyl group becomes distinguishable from the other two, with further complexities at lower temperatures. The structure shows one clearly defined σ-bonded cyclopentadienyl group (Mo–C = 2·29 Å, C–C = 1·45 Å, C=C = 1·35 Å) with the other two cyclopentadienyl groups having a less clear mode of attachment (Mo–C = 2·32–2·68 Å, C–C = 1·34–1·43 Å) (414).

π-Cycloheptatrienyl complexes include $(C_7H_7)Cr(C_5H_5)$, $(CO)_2(C_7H_7)Mo\text{–}X$ (where X isCl, Br, I, $\sigma\text{–}C_6F_5$ or $Mn(CO)_5$), and $[(C_7H_7)Mo(CO)_3]^+$ (478, 485, 627, 1404). In contrast to $(C_7H_7)Cr(C_5H_5)$, molybdenum forms the dicarbonyl $(C_7H_7)Mo(C_5H_5)(CO)_2$, in which the cycloheptatrienyl group can only be behaving as a three electron donor, retaining two uncoordinated double bonds.

A number of unstable, paramagnetic chromium(I) or diamagnetic chromium(II) organometallic compounds do not achieve the 18-electron configuration. The diamagnetic bis-cyclopentadienyl chromium(II), which is isomorphous with ferrocene $Fe(C_5H_5)_2$, has been prepared from chromium(III) and sodium cyclopentadienyl (854, 2390), or from chromium hexacarbonyl and cyclopentadiene (2388). (Under similar conditions molybdenum and tungsten gave only the cyclopentadienyl carbonyls $[(C_5H_5)M(CO)_3]_2$).

Paramagnetic chromium(I) compounds include $Cr(C_5H_5)(C_6H_6)$ (861), $CrI(CO)_5$ (199), and its derivatives $[Cr(CO)_2(NH_3)_4]^+$ and $[Cr(en)_3]^+$ (191). A number of chromium(I) compounds have also been prepared by the reaction of a slurry of chromium trichloride in diethyl ether with phenyl magnesium bromide to yield salts of $[Cr(C_6H_6)_2]^+$, $[Cr(C_6H_6)(C_6H_5.C_6H_5)]^+$, and $[Cr(C_6H_5.C_6H_5)_2]^+$. This is in contrast to the same reaction in tetrahydrofuran which forms $CrPh_3(THF)_3$.

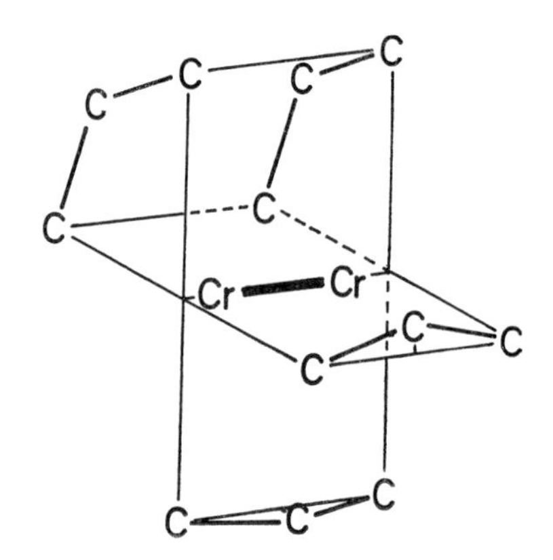

FIG. 193 The structure of $Cr_2(C_3H_5)_4$.

A particularly notable electron deficient compound is $Cr_2(C_3H_5)_4$ prepared from chromium trichloride and allylmagnesium chloride (84). The structure (Fig. 193) contains a remarkably short chromium–chromium bond of 1·97 Å which is the shortest metal–metal bond known for any organometallic compound. The chromium–chromium bond is bridged by two of the four allyl groups.

Molybdenum(I) and tungsten(I) compounds are considerably more unstable with respect to oxidation, and include $Mo(C_5H_5)(C_6H_6)$, $[Mo(C_6H_6)_2]^+$ and $[W(C_6H_6)_2]^+$ analogous to the above chromium compounds (863, 872). In addition the oxidation of $M(CO)_2(Ph_2P.CH_2.CH_2.PPh)_2$ (where M is Mo or W) with iodine yields the monomeric, paramagnetic $[M(CO)_2(Ph_2P.CH_2.CH_2.PPh)_2]^+$; the absence of a metal–metal bonded seven coordinate structure, and the trans-disposition of the carbonyl groups, were attributed to steric crowding by the bulky ligands (1530). The dimeric π-allyl complexes

$$(CO)_2(C_3H_5)Mo\underset{X}{\overset{X}{\langle - X - \rangle}}Mo(C_3H_5)(CO)_2$$ (where X is Cl, Br, OMe, OEt, SPh or OH) are obtained by the action of allylhalide on $[MoX(CO)_5]^-$ (1744, 1745).

2. *Carbonyl halides and substituted carbonyl halides*

The reaction of molybdenum or tungsten hexacarbonyls with chlorine or bromine at −78° C yields the dihalocarbonyls $MoCl_2(CO)_4$ (532), $MoBr_2(CO)_4$ (533), $WCl_2(CO)_4$ (78) and $WBr_2(CO)_4$ (78), although the last three compounds were not isolated in a pure state, being contaminated with higher valence binary halides and unreacted hexacarbonyls. These compounds are diamagnetic, which appears inconsistent with a d^4 octahedral stereochemistry, and the failure to achieve the 18-electron configurations; the infrared spectra show the absence of bridging carbonyl groups. Similarly the reaction of $Mo(CO)_6$ and $W(CO)_6$ with iodine at room temperature under ultra violet irradiation forms $MoI_2(CO)_4$ and $WI_2(CO)_4$ respectively (527).

Molybdenum and tungsten hexacarbonyls similarly react with chlorine azide to form the polymeric $[MCl(N_3)(CO)_2]_\infty$ (1491).

Chromium hexacarbonyl appears unreactive towards halogens.

There are a large number of dihalo substituted molybdenum and tungsten carbonyls of the general type $[MX_2(CO)_3(\text{monodentate})_2]$ (where the ligand is Ph_3P, Ph_3As, Ph_3Sb or C_5H_5N) (78, 532, 533, 534, 2223), or $[MX_2(CO)_3(\text{bidentate})]$ (where the ligand is dipyridyl, *o*-phenanthroline, $Ph_2P.CH_2.PPh_2$, $Ph_2P.CH_2.CH_2.PPh_2$, *o*-$C_6H_4(PPh_2)_2$, $MeC(CH_2.PPh_2)_3$ behaving as a bidentate, $Ph_2As.CH_2.CH_2.AsPh_2$, *o*-$C_6H_4(AsMe_2)_2$, or o-$C_6H_4(SMe)(PPh_2)$) (77, 198, 528, 534, 1533, 1788, 2304).

These compounds are all diamagnetic and attain the inert gas electron configuration, although none have been structurally characterized. They have been prepared by the halogenation of substituted carbonyls, by the displacement of carbon monoxide from the carbonyl halide with the appropriate ligand, and by the displacement of the arene from the arene substituted carbonyl halide.

Tridentate ligands displace either one of the halide ions to form $[MX(CO)_3(\text{tridentate})]X$, or an additional carbonyl group forming $[MX_2(CO)_2(\text{tridentate})]$ depending upon experimental conditions, metal, halogen, and tridentate ($PhP(o\text{-}C_6H_4.PPh_2)_2$, $MeAs(o\text{-}C_6H_4.AsMe_2)_2$, $MeAs(CH_2.CH_2.CH_2.AsMe_2)_2$, $MeC(CH_2.AsMe_2)_3$, $P(o\text{-}C_6H_4.SMe)_3$ behaving as a tridentate, $PhP(o\text{-}C_6H_4.SMe)_2$, or $PhAs(o\text{-}C_6H_4.SMe)_2$)

(542, 1230, 1805, 2223, 2304). Similar compounds are formed containing one bidentate and one monodentate ligand (77, 79).

Compounds containing one quadridentate ligand ($P(o\text{-}C_6H_4.PPh_2)_3$) (1230) or two bidentate ligands (dipyridyl, *o*-phenanthroline, $Ph_2P.CH_2.PPh_2$, $Ph_2P.CH_2.CH_2.PPh_2$, or $o\text{-}C_6H_4(AsMe_2)_2$) (77, 198, 525, 526, 1526, 1533, 1788, 2223) form $[MX_2(CO)(\text{tetradentate})]$, $[MX(CO)_2(\text{tetradentate})]X$ or $[MX(CO)_2(\text{bidentate})_2]X$.

Of particular interest are the carbon monoxide carriers $MX_2(CO)_3(Ph_3P)_2$ where M is Mo or W and X is Cl, Br or I. These lose one mole of carbon monoxide in boiling solvents or on simply bubbling nitrogen through the solution, and change from yellow to dark blue. The products readily reabsorb one mole of carbon monoxide at atmospheric pressure (78, 527, 529, 532):

$$MX_2(CO)_3(Ph_3P)_2 \leftrightarrows MX_2(CO)_2(Ph_3P)_2 + CO$$

It appears very likely that many of the above compounds are capable of behaving in this manner (80).

Oxidation of the substituted carbonyl with mercuric chloride or stannic chloride in place of halogen forms $W(HgCl)_2(CO)_3(dipy)$ (972) and $MoCl(SnCl_3)(CO)_3(dipy)$ (1481). In all about 30 compounds of the latter type were prepared with tungsten in place of molybdenum, and/or $RSnCl_2$ or R_2SnCl in place of $SnCl_3$, and/or Ge in place of Sn, and/or *o*-phenanthroline, dithiahexane or $Ph_2P.CH_2.CH_2.PPh_2$ in place of dipyridyl. The structures of $MoCl(SnMeCl_2)(CO)_3(dipy)$ (768, 769) and $WCl(SnMeCl_2)(CO)_3(MeS.CH_2.CH_2.SMe)$ (770) unexpectedly contain chlorine atoms bridging the metal–tin bonds:

$$(\text{bidentate})(CO)_3M\overset{Cl}{\frown}SnCl_2Me$$

The molybdenum and tungsten atoms are seven coordinate with a capped octahedral stereochemistry (page 8), the tin atom occupying the unique ligand site. The molybdenum–tin and tungsten–tin bond lengths of 2·75 and 2·76 Å respectively are quite short.

Seven coordinate anionic complexes such as $[MoX_3(CO)_4]^-$ and $[MoX_3(CO)_3(Ph_3P)]^-$ are also readily obtained (275, 973, 1532, 2304).

Halogen oxidation of arene carbonyls also produce chromium(II), molybdenum(II) and tungsten(II) complexes. The most fully characterized examples are those with hexamethylbenzene, $[(C_6Me_6)MX(CO)_3]^+$ (2176, 2223). The structure of $[(C_6Me_6)WI(CO)_3]^+$ is indicated in Fig. 194, the large iodine atom being staggered with respect to the carbon atoms of the benzene ring (2175).

3. *Cyclopentadienyl carbonyl compounds*

(a) *Dimeric compounds, hydrides, halides, and ions*, The direct reaction between chromium, molybdenum or tungsten hexacarbonyls with cyclopentadiene yields $[(C_5H_5)M^I(CO)_3]_2$. Other preparative methods include the action of carbon monoxide on the biscyclopentadienyls, and the dehydrogenation of the hydrides $(C_5H_5)M(CO)_3H$ (856, 857, 2388).

The structure of $[(CO)_3(C_5H_5)Mo–Mo(C_5H_5)(CO)_3]$ shows that the two halves of the molecule are joined by a fairly long molybdenum–molybdenum bond of 3·22 Å, there being no bridging carbonyl groups (2399). The mass spectrum of $[(C_5H_5)Mo(CO)_3]_2$ shows a number of ions of the type $[(C_5H_5)Mo(CO)_x]_2$, indicating that the dimeric structure is fairly stable. The difference in appearance potentials between

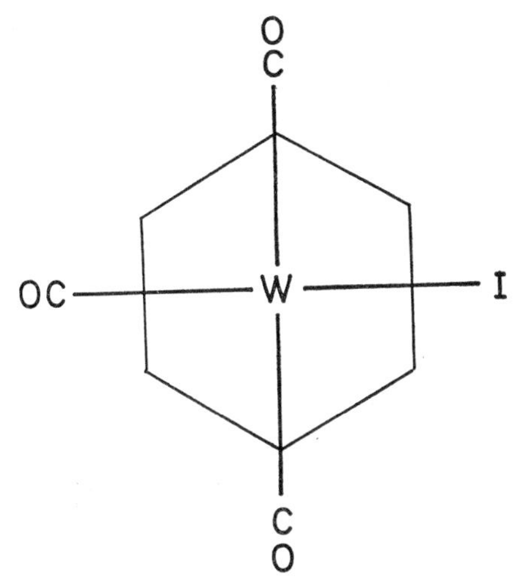

FIG. 194 The structure of $[(C_6Me_6)WI(CO)_3]^+$.

$[(C_5H_5)Mo(CO)_3]_2^+$ and $[(C_5H_5)Mo(CO)_3]^+$, which is a measure of the molybdenum–molybdenum bond energy, is 2·2 eV. The mass spectrum of the chromium analogue however shows that the ions of highest mass correspond to $[(C_5H_5)Cr(CO)_3]^+$ and $[(C_5H_5)Cr(CO)_3H]^+$, showing a less stable chromium–chromium bond (1397, 2077). Sublimation of $[(C_5H_5)Cr(CO)_3]_2$ at 80° K produced paramagnetic $(C_5H_5)Cr(CO)_3$ (1343).

One of the carbonyl groups on one or both molybdenum atoms may be replaced by phosphines (Ph_3P, $(C_6H_{11})_3P$, $(PhO)_3P$ or $(BuO)_3P$). Stronger electron donating ligands (Ph_3P, Et_3P, Bu_3P) rupture the metal–metal bond causing disproportionation to $[(C_5H_5)Mo^{II}(CO)_2(R_3P)_2]^+[(C_5H_5)Mo^O(CO)_3]^-$ (146, 1121). The diphosphine *trans*-$Ph_2P.CH{=}CH.PPh_2$ displaces three carbonyl groups forming

$(C_5H_5)_2Mo_2(CO)_3$(diphosph), suggesting donation of electron pairs from the double bond as well as both phosphorus atoms (1411). Sulphur donor ligands form compounds of the type $[(C_5H_5)Mo(SR)_2]_2$ whose structures are not known, although the molybdenum–molybdenum bond is presumably retained (1395, 2080, 2300).

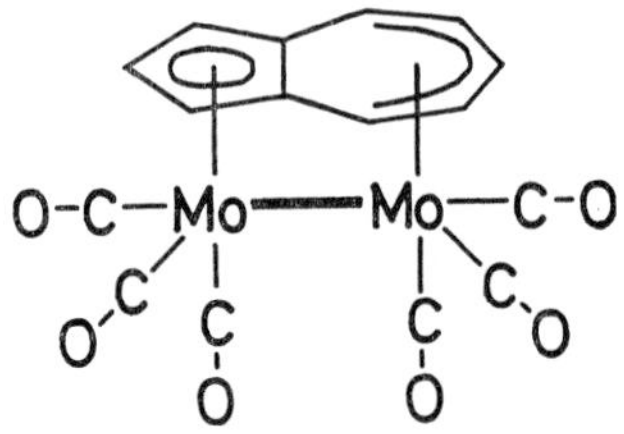

FIG. 195 The structure of $(C_{10}H_8)Mo_2(CO)_6$.

Similar compounds of the type $[(C_5H_4R)M(CO)_3]_2$ are formed with some alkyl-substituted cyclopentadienes (1, 1920). Azulene, and substituted azulenes, form $(C_{10}H_8)Mo_2(CO)_6$ where one molybdenum atom is bonded to the cyclic pentadienyl ring, and the other to a non-cyclic pentadienyl group which is part of the seven membered ring (Fig. 195) (475, 1606). The molybdenum–molybdenum bond lengths, 3·24–3·27 Å, are slightly greater than in $[(C_5H_5)Mo(CO)_3]_2$ (3·22 Å). The molybdenum–molybdenum bond length in the dihydroheptalene analogue, $(C_{12}H_{12})Mo_2(CO)_6$ (Fig. 196) (1541), which contains two such non-cyclic pentadienyl groups is only 3·19 Å, and it appears that these changes are probably associated with the steric requirements of the ligands.

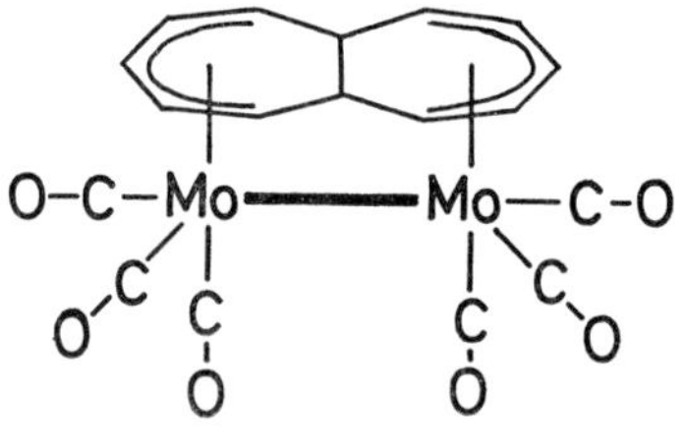

FIG. 196 The structure of $(C_{12}H_{12})Mo_2(CO)_6$.

Different behaviour is observed with the cyclopentadiene analogues pentamethylcyclopentadiene (1407) and 8,9-dihydroindene (1398). These displace an additional carbonyl group to form $[(C_5Me_5)Mo(CO)_2]_2$ and $[(C_9H_9)Mo(CO)_2]_2$ respectively. There appear to be no bridging carbonyl groups, and in order for the inert gas configuration to be retained, a molybdenum–molybdenum triple bond must be postulated (Fig. 197).

The reaction of chromium, molybdenum or tungsten hexacarbonyls with lithium cyclopentadienyl yields $[(C_5H_5)M(CO)_3]^-$, which on acidification yields the hydrides $(C_5H_5)M(CO)_3H$ (857). A more convenient synthesis is the direct reaction of $Mo(CO)_3(MeCN)_3$ or $W(CO)_3(MeCN)_3$ with cyclopentadiene (1365). Phosphine substituted compounds of the type $(C_5H_5)Mo(CO)_2(Ph_3P)H$ are also known (1669).

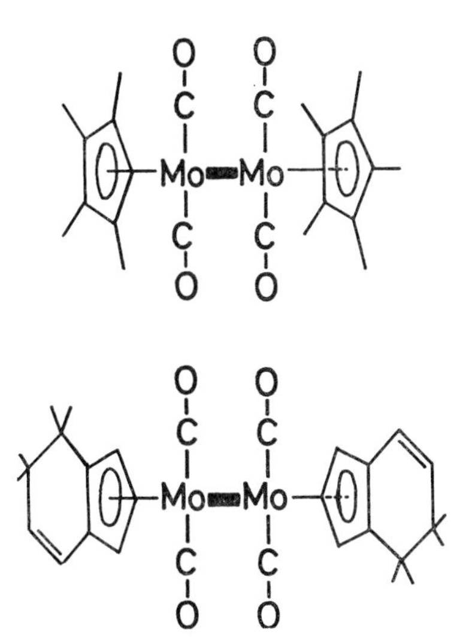

FIG. 197 Suggested structures of $[(C_5Me_5)Mo(CO)_2]_2$ and $[(C_9H_9)Mo(CO)_2]_2$.

The corresponding halides and pseudo halides can be obtained, for example by (1365, 2166):

$$[(C_5H_5)Mo(CO)_3]_2 + I_2 \rightarrow (C_5H_5)Mo(CO)_3I$$
$$(C_5H_5)Mo(CO)_3H + CCl_4 \rightarrow (C_5H_5)Mo(CO)_3Cl$$
$$(C_5H_5)Mo(CO)_3H + (SCN)_2 \rightarrow (C_5H_5)Mo(CO)_3(SCN)$$

In the last case both N-bonded and S-bonded thiocyanate isomers can be obtained.

The tetracarbonyl cations $[(C_5H_5)M(CO)_4]^+$ can be obtained from these monohalides using aluminium chloride at 60°–70° C under 240 atm of carbon monoxide (853).

The reactions of the halides with neutral ligands have been extensively studied (868, 1120, 1319, 1394, 1645), and can lead to the displacement of:

(i) one carbonyl group to form $[(C_5H_5)M(CO)_2(\text{ligand})X]$,

(ii) two carbonyl groups to form $[(C_5H_5)M(CO)(\text{ligand})_2X]$,

(iii) halide to form $[(C_5H_5)M(CO)_3(ligand)]^+$,

(iv) halide and one carbonyl group to give $[(C_5H_5)M(CO)_2(ligand)_2]^+$.

(b) *Alkyls, and compounds with other p block elements.* The σ-bonded molybdenum and tungsten alkyls can be obtained from a number of cyclopentadienyltricarbonyl compounds, for example:

$$(C_5H_5)Mo(CO)_3I + RMgX \rightarrow (C_5H_5)Mo(CO)_3R \quad (857)$$
$$(C_5H_5)Mo(CO)_3H + CH_2N_2 \rightarrow (C_5H_5)Mo(CO)_3(CH_3)$$
$$(C_5H_5)Mo(CO)_3H + C_2F_4 \rightarrow (C_5H_5)Mo(CO)_3(C_2F_5) \quad (2297)$$
$$[(C_5H_5)W(CO_3)]^- + RCl \rightarrow (C_5H_5)W(CO)_3R \quad (87, 1049)$$

Perfluorobenzoyl compounds $(C_5H_5)M(CO)_3(C_6F_5CO)$ are obtained from $[(C_5H_5)Mo(CO)_4]^+$ or $[(C_5H_5)W(CO)_4]^+$ and lithium perfluorophenyl (2298). Similarly the reaction of, for example, triphenylphosphine with $(C_5H_5)Mo(CO)_3R$ forms $(C_5H_5)Mo(CO)_2(Ph_3P)(RCO)$, where the carbonyl group has been inserted into the molybdenum–alkyl bond (146, 601). Subsequent reaction under more vigorous conditions forms the decarbonylated $(C_5H_5)Mo(CO)_2(Ph_3P)(R)$ (145).

The ethyl compound can also be formed by reduction of the two electron donating ethylene compound as shown; the reaction is reversible (580):

$$[(C_5H_5)Mo(CO)_3(CH_2{=}CH_2)]^+ \underset{Ph_3C^+}{\overset{NaBH_4}{\rightleftarrows}} [(C_5H_5)Mo(CO)_3(CH_2.CH_3)]$$

The former can be obtained directly from $(C_5H_5)Mo(CO)_3Cl$, C_2H_4 and $AlCl_3$ under pressure (852).

Yet another remarkable reaction is the direct reduction of a carbonyl group to a methyl group (2298):

$$[(C_5H_5)Mo(CO)_3(Ph_3P)]^+ \xrightarrow{NaBH_4} (C_5H_5)Mo(CO)_2(Ph_3P)(CH_3)$$

On the other hand $[(C_5H_5)Mo(CO)_4]^+$ forms the hydride $(C_5H_5)Mo(CO)_3H$.

The structures of $(C_5H_5)Mo(CO)_3(C_2H_5)$ (208), $(C_5H_5)Mo(CO)_2(Ph_3P)(CH_3CO)$ (477), and $(C_5H_5)Mo(CO)_3(C_3F_5)$ (476) have been determined, the molybdenum–alkyl bond being 2·3–2·4 Å long. The dimeric azulene derivative $[(C_{10}H_8)Mo(CO)_3(CH_3)]_2$ is linked through dimerization of the azulene, which is as expected (233).

Ultra violet irradiation of a number of these σ bonded compounds containing unsaturated alkyl groups, converts the one electron donor group into a π bonded three electron donor derivative, with the loss of one molecule of carbon monoxide:

$$(C_5H_5)Mo(CO)_3(\sigma\text{-}CH_2.CH{=}CH_2) \xrightarrow[-CO]{h\nu} (C_5H_5)Mo(CO)_2(\pi\text{-}C_3H_5) \quad (580, 1053)$$

$$(C_5H_5)Mo(CO)_3(\sigma\text{-}CH_2.S.CH_3) \xrightarrow[-CO]{h\nu} (C_5H_5)Mo(CO)_2(\pi\text{-}CH_2.S.CH_3) \quad (998, 1406)$$

$$(C_5H_5)Mo(CO)_3(\sigma\text{-}CH_2.C_6H_5) \xrightarrow[-CO]{h\nu} (C_5H_5)Mo(CO)_2(\pi\text{-}CH_2.C_6H_5) \quad (566, 570, 1408)$$

The fluxional character of these molybdenum π-allyl systems has been studied by low temperature n.m.r. (202, 570, 813).

Silicon, germanium and tin analogues of the σ bonded carbon compounds can similarly be prepared, and are of the general formulae $(C_5H_5)M(CO)_3M'R_3$ or $(C_5H_5)M(CO)_2(Ph_3P)SnR_3$ (where M is Cr, Mo or W, M′ is Si, Ge or Sn, and R is Me, Ph, Bz, Cl, Br, I, etc.) (11, 422, 423, 433, 1365, 1670, 1698, 1843).

Phosphorus and arsenic substituted complexes $(C_5H_5)Mo(CO)_3(ER_2)$ (where E is P or As, and R is CF_3 or C_6F_5), can be prepared from $[(C_5H_5)Mo(CO)_3]^-$ and the appropriate R_2ECl, or from $[(C_5H_5)Mo(CO)_3]_2$ and $R_2E.ER_2$. Irradiation with ultraviolet light leads to dimerization with the loss of carbon monoxide to form

$$(CO)_2(C_5H_5)Mo\begin{smallmatrix} \nearrow ER_2 \searrow \\ \nwarrow ER_2 \swarrow \end{smallmatrix}Mo(C_5H_5)(CO)_2$$

(where E is P or As, and R is CH_3, CF_3 or C_6F_5) (545, 613, 1152). An analogous structure occurs in $[(C_5H_5)Mo(CO)_2]_2(Me_2P)H$, where one of the Me_2P^- bridges is replaced by a hydride bridge forming a bent three-centre Mo–H–Mo bond (677).

(c) *Compounds with other metals.* A complex containing a molybdenum–titanium bond, $(C_5H_5)(CO)_3Mo\text{–}Ti(OPr)_3$, analogous to the carbon, silicon, germanium and tin complexes above, has also been prepared (423).

Other organometallic complexes containing metal–metal bonds include

$(CO)_3(C_5H_5)Cr\text{–}Ti(C_5H_5)_2$ (1698), $(CO)_3(C_5H_5)Cr\text{–}V(C_5H_5)_2$ (1698),

$(CO)_3(C_5H_5)Mo\text{–}W(C_5H_5)(CO)_3$ (1), $(CO)_3(C_5H_5)Mo\text{–}Fe(C_5H_5)(CO)_2$

(1413), and $(CO)_3(C_5H_5)Mo\text{–}Mn(CO)_5$ (1770). The structure of the last compound shows a molybdenum–manganese bond length of 3·08 Å, which is equal to the average of the metal–metal bond lengths in $[(C_5H_5)Mo(CO)_3]_2$ and $Mn_2(CO)_{10}$.

Finally complexes containing the $M(C_5H_5)(CO)_3$ moiety (where M is Cr, Mo or W) will form metal–metal bonds with the d^{10} systems cadmium (II), mercury(II), copper(I) and gold(I). The mercury complexes are of the general formulae $[(C_5H_5)M(CO)_3]HgX$ and $[(C_5H_5)M(CO)_3]_2Hg$ (where M is Mo or W, and X is Cl, Br, I or SCN). In one case a tungsten–mercury–cobalt bond was obtained, namely $(CO)_3(C_5H_5)W–Hg–Co(CO)_4$ (1671). Phosphine substituted products, for example $[(C_5H_5)M(CO)_2(Ph_3P)]HgX$ and $[(C_5H_5)M(CO)_2(Ph_3P)]_2Hg$ were also obtained (1669). In addition to the cadmium analogues $[(C_5H_5)(CO)_3M]CdX$ and $[(C_5H_5)M(CO)_3]_2Cd$ (where M is Mo or W, and X is Cl, Br or I), stable adducts of the type $[(C_5H_5)M(CO)_3]CdX(ligand)_2$ and $[(C_5H_5)M(CO)_3]_2Cd(ligand)_2$ can be prepared (where the ligand is NH_3, C_5H_5N, $\frac{1}{2}$dipy, $\frac{1}{2}$*o*-phen or $\frac{1}{2}$en) (1672).

Gold(I) forms complexes of the type $(CO)_3(C_5H_5)M–Au(Ph_3P)$, where M is Cr, Mo or W (518, 1122). The gold-tungsten bond length in the last compound is 2·70 Å (2384). To obtain the corresponding molybdenum–copper and tungsten–copper bonds, the necessary four coordination of the copper atom is achieved by using the triarsine $MeAs(o\text{-}C_6H_4.AsMe_2)_2$ in $(CO)_3(C_5H_5)Mo–Cu(triars)$ and $(CO)_3(C_5H_5)W–Cu(triars)$ (518).

6. OXIDATION STATES $\leqslant 0$

A. Organometallic Complexes

1. *Introduction*

The principle source of zerovalent Group VI compounds are the hexacarbonyls $M(CO)_6$, prepared by the reductive carbonylation of higher valence state compounds. The hexacarbonyls are octahedral; for chromium hexacarbonyl, Cr–C = 1·91 Å, C–O = 1·14 Å, the C–Cr–C angles vary from 89·4 to 90·8°, and the Cr–C–O angles from 178·3 to 179·6° (2381).

π-Donor ligands form compounds of the type $M(CO)_{6-2x}(diene)_x$, $M(CO)_3(arene)$ and $M(arene)_2$. The carbonyl groups and/or arene groups can be displaced by a large number of σ-donor ligands. Only the general chemistry of these reactions is included in the following sections.

Complete substitution by σ-donors is rare, but examples include the heterocyclic complexes $M(o\text{-phen})_3$, $M(terpy)_2$ and $M(dipy)_3$ (190) the diphosphine and diarsine complexes $M(bidentate)_3$ (where the bidentate is $Ph_2P.CH_2.PPh_2$, $Ph_2P.CH_2.CH_2.PPh_2$, $Me_2P.CH_2.CH_2.PMe_2$, $o\text{-}C_6H_4(PEt_2)_2$ and $Ph_2As.CH_2.CH_2.AsPh_2$) and $M(N_2)_2(Ph_2P.CH_2.CH_2.PPh_2)$ (196, 466, 1190), $M(PF_3)_6$ (1472), $K_6[Cr(CN)_6]$ (195), and $M(CNR)_6$.

Chromium and tungsten carbenes can also be formally considered to be σ-donors of the type $\begin{matrix} R \\ R' \end{matrix}\!\!>\!C \rightarrow M(CO)_5$ (where R is Me, Ph, etc., and R′ is NH_2, NR_2, OR etc.) (867, 1415). The structures of these complexes however indicate some carbon–metal multiple bond character (537, 1245, 1689).

The reaction between $M(CO)_3$(arene) and mercuric chloride forms the unusual adducts (arene)$(CO)_3M \rightarrow HgCl_2$ (where M is Cr or Mo and the arene is $C_6H_3Me_3$ or 1,3,5–C_7H_8) (726). Similar reactions between $M(CO)_4$(dipy), $M(CO)_4$(diphosph) or $M(CO)_2(diphosph)_2$ (where M is Mo or W), and mercuric chloride or stannic chloride were also noted.

Reduction of the hexacarbonyls with borohydride in liquid ammonia, or with sodium amalgam in tetrahydrofuran, forms $[M_2(CO)_{10}]^{2-}$. The earlier report of different structural isomers is incorrect (187, 727, 1154). The infrared and Raman spectra are consistent with $[(CO)_5M–M(CO)_5]^{2-}$ of D_{4d} symmetry, that is no bridging carbonyl groups, and with the four equatorial carbonyl groups on one metal atom in a staggered arrangement with respect to those on the other metal atom. Thus the structures are the same as the isoelectronic $Mn_2(CO)_{10}$ and $Re_2(CO)_{10}$, and intermediate ions such as $[(CO)_5W–Mn(CO)_5]^-$ and $[(CO)_5W–Re(CO)_5]^-$ are presumably also similar (53). Other products obtained during the reduction include $[M(CO)_5]^{2-}$, which are presumably isostructural with the isoelectronic trigonal bipyramidal $Fe(CO)_5$, and a number of polymeric species such as $[M_2(CO)_9]^{4-}$ and $[M_3(CO)_{14}]^{2-}$ (188, 197, 1333).

The hydrides $[HM(CO)_5]^-$ and $[HM_2(CO)_{10}]^-$ can be obtained on hydrolysis of $[M(CO)_5]^{2-}$ and $[M_2(CO)_{10}]^{2-}$ (52, 192, 1154). The structure of $[HCr_2(CO)_{10}]^-$ shows a Cr—Cr distance of 3·41 Å (1128), and the octahedral coordination about each metal atom is regular if the hydrogen atom is assumed to be at the centre of the molecule. Thus there is a linear, three-centre, electron pair bond, $[(CO)_5Cr–H–Cr(CO)_5]^-$. The molybdenum and tungsten compounds appear analogous. The n.m.r. spectra of mixtures of $[HCr_2(CO)_{10}]^-$, $[HMo_2(CO)_{10}]^-$ and $[HW_2(CO)_{10}]^-$ show the presence of compounds containing the mixed linkages Cr–H–Mo, Cr–H–W, and Mo–H–W (1154).

In addition to the hydride ion, other groups may be inserted into these metal–metal bonds. For example $[W_2(CO)_{10}]^{2-}$ reacts with SnI_2, GeI_2 or SO_2 to form $[(CO)_5W–SnI_2–W(CO)_5]^{2-}$, $[(CO)_5W–GeI_2–W(CO)_5]^{2-}$, and $[(CO)_5W–SO_2–W(CO)_5]^{2-}$ respectively (1960).

2. *σ-Bonded compounds*

The carbonyl groups of the Group VI hexacarbonyls are readily replaced by other ligands forming compounds of the type $[M(CO)_{6-x}$

(ligand)$_x$]. The extent of replacement depends upon the metal, the ligand, and the reaction conditions, being particularly favourable for phosphines and arsines which, like carbon monoxide, are capable of participating in back π-bonding with the metal.

This displacement of carbonyl groups results in enhanced metal–carbon π-bonding to the remaining carbonyl groups, which is reflected in shorter metal–carbon bonds (for example 1·82 Å in $Cr(CO)_3$(dien) compared with 1·91 Å in $Cr(CO)_6$ (573)), and weaker carbon–oxygen bonding which is reflected in lower carbon–oxygen stretching frequencies.

Monosubstituted $M(CO)_5$(ligand) have been prepared by reaction of equimolar quantities of the hexacarbonyl and ligand, either by heating to about 150° C in an appropriate solvent, or through ultra violet irradiation. The disubstituted $M(CO)_4$(ligand)$_2$ and $M(CO)_4$(bidentate) are obtained by similar methods (252, 390, 458, 466, 900, 911, 1172, 1644, 1787, 2221, 2239, 2375).

Trisubstituted compounds $M(CO)_3$(ligand)$_3$ are often the end products of reactions with excess monodentate ligands. They are also conveniently prepared from the arenes $M(CO)_3$(arene) (911, 1618). Compounds of the type $M(CO)_3$(bidentate)(ligand) are obtained from the corresponding $M(CO)_4$(bidentate) (190, 251, 1222, 1225, 1226). Tridentate ligands similarly form $M(CO)_3$(tridentate) (2, 466, 542, 573, 575, 1644, 1804).

The tetrasubstituted $M(CO)_2$(bidentate)$_2$ and $M(CO)_2$(bidentate)(ligand)$_2$ are obtained using higher boiling point solvents than those used for $M(CO)_4$(bidentate), for example decalin compared with carbon tetrachloride respectively (187, 466, 1225, 1226, 1787). A possibly more convenient route is to start with the dicarbonyl $MoX(C_3H_5)(CO)_2(MeCN)$ (662).

Further replacement of carbonyl groups is relatively rare and occurs only with ligands capable of extensive metal–ligand π-bonding. A number of examples have been quoted above. Phosphorus trifluoride and molybdenum hexacarbonyl form all nine possible products, namely
$Mo(CO)_5(PF_3)$, *cis*-$Mo(CO)_4(PF_3)_2$, *trans*-$Mo(CO)_4(PF_3)_2$, *cis*-$Mo(CO)_3(PF_3)_3$, *trans*-$Mo(CO)_3(PF_3)_3$, *cis*-$Mo(CO)_2(PF_3)_4$, *trans*-$Mo(CO)_2(PF_3)_4$, $Mo(CO)(PF_3)_5$, and $Mo(PF_3)_6$ (492).

The carbon–oxygen stretching frequencies in the monosubstituted compounds $M(CO)_5$(ligand) have been used to evaluate the relative importance of σ and π contributions to the metal–ligand bonding. It is assumed that changes in the metal–ligand σ-bond will effect all metal–carbon and hence carbon–oxygen bonds equally, whereas changes in the metal–ligand π-bond will effect the *trans* carbon–oxygen bond twice as much as the four *cis* carbon–oxygen bonds. This is because both the d_π orbitals which interact with the ligand will also interact with the *trans* carbon–oxygen bond, whereas only one will interact with each *cis* carbon–oxygen bond. It was found that the metal–ligand σ-bonding increases, and the π-bonding decreases, along the series: CO, $P(OMe)_3$, $P(OEt)_3$, PBu^n_3, $P(OPh)_3$, PPh_3, $AsPh_3$, $C_6H_{11}.NH_2$, C_5H_5N, $(CH_2)_5NH$, CH_3CN, Pr^i_2O (1040, 2219).

A study of the equilibria:

$$M(CO)_5(\text{ligand}) + (\text{ligand})' \leftrightarrows M(CO)_5(\text{ligand})' + \text{ligand}$$

shows an order of stability which includes $Bu^n_3P > EtC(CH_2O)_3P > (Bu^nO)_3P > Ph_3P > Ph_3As > (PhO)_3P$, which is the same order as the decreasing basicity of the ligand (74).

The infrared spectra and ^{183}W–^{31}P coupling show that $(PhO)_3P$ is a better π-acceptor than Ph_3P in these compounds, which is confirmed by the structures of $Cr(CO)_5(Ph_3P)$ and $Cr(CO)_5[(PhO)_3P]$, the chromium–phosphorus bond being shorter in the latter compound (2·42 and 2·31 Å respectively). The *trans* chromium–carbon bonds (1·84 and 1·86 Å respectively) are shorter than the *cis* chromium–carbon bonds (1·88 and 1·90 Å respectively) as expected (1875).

Halide and pseudo-halide ions react with the hexacarbonyls to form $[MX(CO)_5]^-$ (3, 869, 2407). The thiocyanate complex is apparently nitrogen bonded, in contrast to the isoelectronic $Mn(SCN)(CO)_5$ which is apparently sulphur bonded. Reaction of tungsten hexacarbonyl with azide ion forms the cyanato complex $[W(NCO)(CO)_5]^-$ (180).

Further substitution occurs with cyanide forming $[M(CN)(CO)_5]^-$, $[M(CN)_2(CO)_4]^{2-}$ and $[M(CN)_3(CO)_3]^{3-}$ (190, 193, 197). Trisubstitution also occurs with potassium acetylides in liquid ammonia, forming $[Cr(C{\equiv}CR)_3(CO)_3]^{3-}$ (where R is H, Me or Ph) (1756).

Compounds containing metal–germanium and metal–tin bonds, $[(CO)_5M\text{–}M'Cl_3]^-$ (where M is Cr, Mo or W, and M′ is Ge or Sn), are similarly prepared using $(M'Cl_3)^-$ (1959).

3. π-*Bonded compounds*

Organometallic compounds containing π-electron donor ligands co-ordinated to chromium, molybdenum and tungsten can be simply prepared from the hexacarbonyl or $M(CO)_3(MeCN)_3$ and the appropriate alkene, alkyne or arene. The reactions are reversible. A detailed treatment is again outside the scope of this book, and attention will be concentrated on some four electron and six electron donor ligands.

Dienes can form $M(CO)_4(\text{diene})$, $M(CO)_2(\text{diene})_2$ or $M(\text{diene})_3$ depending upon the reaction conditions, the metal, and the diene. Dienes which have been studied include butadiene, cyclohexa-1,3-diene, cycloocta-1, 5-diene, norbornadiene, cyclobutadiene, tetramethylcyclobutadiene, tetraphenylcyclopentadienone, benzoquinone, and methylvinylketone (47, 411, 862, 1401, 1409, 1516, 2362). The chromium cycloocta-1,5-diene complex loses carbon monoxide and hydrogen to form the xylene derivative $Cr(CO)_3(o\text{-}C_6H_4Me_2)$ (1516).

Compounds containing coordinated cyclic dienes are also among the products of reaction between alkynes and the hexacarbonyls. For example diphenylacetylene forms, in addition to $M(CO)(Ph_2C_2)_3$, a number of products containing tetraphenylcyclobutadiene Ph_4C_4, and/or tetraphenylcyclopentadienone Ph_4C_5O, both behaving as four electron donors to the metal atom. For the 18-electron configuration in the $M(CO)(Ph_2C_2)_3$ compounds to be attained, the three alkyne units must donate a total of 10 electrons. The structure is considered as a resonant hybrid of structures in which two of the alkyne units are considered to be four electron donors and the other a two electron donor.

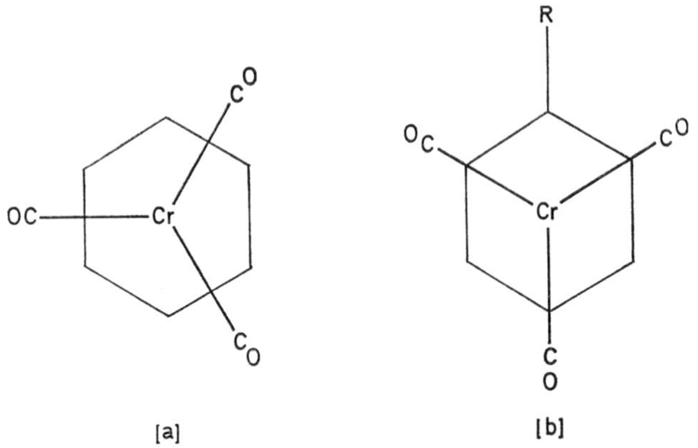

FIG. 198 (a) Staggered, and (b) eclipsed configurations for $Cr(CO)_3$(arene).

Structures have been determined of a number of compounds of the type $Cr(CO)_3$(arene), where the chromium atom is situated above the centre of a six membered aromatic ring. The benzene (120), hexamethyl benzene (118), naphthalene (1482), anthracene (1129), phenanthrene (1737) and 9,10-dihydrophenanthrene (1736) compounds have a number of features in common:

(i) The three Cr–CO bonds are staggered with respect to the carbon atoms of the bonded aromatic ring (Fig. 198a). However in the anisole (435) and *o*-toluidine (437) analogues, and in the charge transfer complex between 1,3,5-trinitrobenzene and the anisole complex (1244), the bulky alkyl groups force an eclipsed configuration (Fig. 198b). Also consistent with this eclipsed configuration is the unusually high proportion of meta-substitution which can be attributed to the steric influence of the three carbonyl groups (1268), and the restricted rotation of the arene groups containing bulky alkyl groups (1038).

(ii) There appears to be no significant differences in adjacent carbon–carbon bond lengths of the bonded aromatic ring; that is there is no localization of the three electron pairs involved in π-bonding on formation of the complex.

(iii) The carbon–carbon bond lengths are slightly greater than in the free hydrocarbons indicating a general loosening of the bonding in the arene. For example in the anthracene and phenanthrene complexes, where the chromium atom is bonded to one of the end aromatic rings, the average carbon–carbon bond length in the half of the molecule attached to the chromium atom is 0·01–0·02 Å longer than in the nonbonded half of the molecule, which is the same as in the free hydrocarbons.

(iv) The Cr–CO bond length of about 1·83 Å is shorter than in the chromium hexacarbonyl (1·91 Å), indicating greater back donation to the carbonyl groups.

Other six electron donor molecules which replace three carbonyl groups include cycloheptatriene, cyclooctatriene, 1,6-methanocyclodecapentaene, hexamethylborazole, thiophen, cyclooctatetraene and tetramethylcyclooctatetraene (116, 117, 119, 207, 710, 1090, 1409, 1605, 1901). In the cyclooctatetraene compounds only three of the four available double bonds are coordinated to the metal atom in the solid state; the n.m.r. spectra of these compounds as a function of temperature have been studied in some detail, and have been interpreted as showing oscillatory motions of the metal with respect to the ligands, followed at higher temperatures by trans-annular shifts (560, 561, 1400, 1471, 1605, 2403) Cyclooctatetraene similarly forms $M_2(C_8H_8)_3$ (where M is Cr, Mo or W) analogous to $Ti_2(C_8H_8)_3$ (page 141) (322).

Chromium dibenzene has been prepared by the reaction of PhMgBr with chromium dichloride in tetrahydrofuran or diethyl ether (2441), or with chromium trichloride in diethyl ether, followed by hydrolysis of the black pyrophoric intermediate. An intermediate chromium(III) compound, $Cr(Ph)_3(THF)_3$, is obtained from chromium trichloride in tetrahydrofuran; the tetrahydrofuran can be removed *in vacuo* or by washing with diethyl ether, and on hydrolysis again yields $Cr(C_6H_6)$ (2439). Other major products from this reaction are $Cr(C_6H_6)(C_6H_5.C_6H_5)$ and $Cr(C_6H_5.C_6H_5)_2$ indicating a biradical intermediate $Cr(\pi\text{-}C_6H_5\cdot)_2$. The additional hydrogen atom presumably originates from the water used in the hydrolysis, but hydrolysis with D_2O yields mixtures of mono-, di-, tri- and tetradeutero species, indicating a complex mechanism (2173). The same reaction in the presence of carbon monoxide yields $Cr(CO)_6$ and a proliferation of other products, mainly aromatic carbonyl compounds, again indicating a radical intermediate.

The structure of $Cr(C_6H_6)_2$ has been a subject of some controversy. An early X-analysis indicated that there was some alternation of long and short carbon–carbon bonds (1·44 and 1·37 Å) around the benzene rings, indicating some localization of the three pairs of π-electrons (1293). However subsequent studies by X-ray (558, 1247, 1372), neutron (899) and electron (1105) diffraction, vibrational spectra (1781), and heat capacity measurements (73), indicate no significant bond differences of this type.

4. *Nitrosyl compounds*

The action of nitrosyl chloride or bromide on molybdenum or tungsten hexacarbonyls forms the dinitrosyls $MX_2(NO)_2$, which were formulated as polymeric compounds with all halogen atoms acting as bridges between octahedrally coordinated metal atoms. These bridges could be cleaved with appropriate ligands to form $MX_2L_2(NO)_2$ (where L is C_5H_5N, Ph_3P, Ph_3As, $\frac{1}{2}o\text{-}C_6H_4(AsMe_2)_2$, $\frac{1}{2}$en, etc.), $[MX_4(NO)_2]^{2-}$, and $[M(S_2CNR_2)_2(NO)_2]$. In all cases the infrared spectra indicated a *cis* arrangement of nitrosyl groups (565, 823, 1305, 1306).

Similarly the action of nitrosyl chloride on biscyclopentadienyl chromium forms the dinitrosyl $(C_5H_5)CrCl(NO)_2$ (1872). This has been used to prepare the compounds $(C_5H_5)CrX(NO)_2$ (where X is SCN, CH_3, C_6H_5, etc. (1872)), $[(C_5H_5)Cr(CO)(NO)_2]^+$ (864), and $[(C_5H_5)Cr(NO)_2]_2$ (1405). The structure of $(C_5H_5)CrCl(NO)_2$ has been determined (436).

The molybdenum nitrosyl $(C_5H_5)Mo(CO)_2(NO)$ is obtained from $[(C_5H_5)Mo(CO)_3]^-$ (851, 1399, 1872). This mononitrosyl may be oxidized with loss of carbon monoxide to the dimeric $[(C_5H_5)MoI_2(NO)]_2$, which in turn forms monomeric complexes such as $(C_5H_5)MoI_2(Ph_3P)(NO)$ with a number of ligands (1399).

The structure of $[Mo(OH)(CO)(NO), Ph_3PO]_4$ shows that the four molybdenum atoms are in a tetrahedral arrangement with hydroxo groups above each of the four tetrahedral faces. The triphenylphosphine oxide molecules are hydrogen bonded to these hydroxo groups, and not to the molybdenum atoms (24).

The deep violet diamagnetic compound obtained by the reduction of molybdenum(VI) with hydroxylamine in the presence of cyanide, is apparently $K_4[Mo^0(CN)_5(NO)]$ (1923) and not $K_4[Mo(OH)_2(CN)_5(NO)]$ as formulated earlier (1063).

References

1. Abel, E. W., Singh, A. and Wilkinson, G. (1960). *J. Chem. Soc.*, 1321.
2. Abel, E. W., Bennett, M. A. and Wilkinson, G. (1959). *J. Chem. Soc.*, 2323.
3. Abel, E. W., Butler, I. S. and Reid, J. G. (1963). *J. Chem. Soc.*, 2068.
4. Abrahams, S. C. (1963). *Phys. Rev.* **130,** 2230.
5. Abrahams, S. C. and Ginsberg, A. P. (1966). *Inorg. Chem.* **5,** 500.
6. Abrahams, S. C., Ginsberg, A. P. and Knox, K. (1964). *Inorg. Chem.* **3,** 558.
7. Abrahams, S. C. and Reddy, J. M. (1965). *J. Chem. Phys.* **43,** 2533.
8. Adams, A. C. and Larsen, E. M. (1966). *Inorg. Chem.* **5,** 228, 814.
9. Adams, D. M., Chatt, J., Davidson, J. M. and Gerrat, J. (1963). *J. Chem. Soc.*, 2189.
10. Adams, D. M. and Churchill, R. G. (1968). *J. Chem. Soc. (A)*, 2310.
11. Adams, D. M., Crosby, J. N. and Kemmitt, R. D. W. (1968). *J. Chem. Soc. (A)*, 3056.
12. Adams, D. M. and Newton, D. C. (1968). *J. Chem. Soc. (A)*, 2262.
13. Adams, J. and Rogers, M. D. (1959). *Acta Cryst.* **12,** 951.
14. Adams, R. W., Bishop, E., Martin, R. L. and Winter, G. (1966). *Austral. J. Chem.* **19,** 207.
15. Adamson, A. W. and Perumareddi, J. R. (1965). *Inorg. Chem.* **4,** 247.
16. Addison, C. C., Amos, D. W., Sutton, D. and Hoyle, W. H. H. (1967). *J. Chem. Soc. (A)*, 808.
17. Addison, C. C., Boorman, P. M. and Logan, N. (1965). *J. Chem. Soc.*, 5146.
18. Addison, C. C., Garner, C. D., Simpson, W. B., Sutton, D. and Wallwork, S. C. (1964). *Proc. Chem. Soc.*, 367.
19. Addison, C. C. and Walker, A. (1963). *J. Chem. Soc.*, 1220.
20. Aebi, F. (1948). *Helv. Chim. Acta* **31,** 8.
21. Ahrland, S. and Noren, B. (1958). *Acta Chem. Scand.* **12,** 1595.
22. Ahuja, H. S., Jain, S. C. and Rivest, R. (1968). *J. Inorg. Nuclear Chem.* **30,** 2459.
23. Ainscough, J. B., Holt, R. J. and Trouse, F. W. (1957). *J. Chem. Soc.*, 1034.
24. Albano, V., Bellon, P., Ciani, G. and Manassero, M. (1969). *Chem. Comm.*, 1242.
25. Alei, M. and Jackson, J. A. (1964). *J. Chem. Phys.* **41,** 3402.
26. Alekseev, N. V. and Ronova, I. A. (1966). *J. Struct. Chem.* **7,** 91.
27. Aleonard, S. (1965). *Compt. rend.* **260,** 1977.
28. Alexander, K. M. and Fairbrother, F. (1949). *J. Chem. Soc.*, 2472.
29. Alexander, K. M. and Fairbrother, F. (1949). *J. Chem. Soc.*, S 223.
30. Alimarin, I. P., Sudakov, F. P. and Klitina, V. I. (1965). *Russ. Chem. Rev.* **34,** 574.

31. Al-Karaghouli, A. R. and Wood, J. S. (1970). *Chem. Comm.*, 135.
32. Al-Karaghouli, A. R. and Wood, J. S. (1968). *J. Amer. Chem. Soc.* **90,** 6548.
33. Allbut, M., Feenan, K. and Fowles, G. W. A. (1964). *J. Less-Common Metals* **6,** 299.
34. Allen, E. A., Brisdon, B. J., Edwards, D. A., Fowles, G. W. A. and Williams, R. G. (1963). *J. Chem. Soc.*, 4649.
35. Allen, E. A., Brisdon, B. J. and Fowles, G. W. A. (1964). *J. Chem. Soc.*, 4531.
36. Allen, E. A., Edwards, D. A. and Fowles, G. W. A. (1962). *Chem. and Ind.*, 1026.
37. Allen, E. A., Feenan, K. and Fowles, G. W. A. (1965). *J. Chem. Soc.*, 1636.
38. Allen, J. F. and Neumann, H. M. (1964). *Inorg. Chem.* **3,** 1612.
39. Allen, R. J. and Sheldon, J. C. (1965). *Austral. J. Chem.* **18,** 277.
40. Allison, G. B., Anderson, I. R., van Bronswyk, W. and Sheldon, J. C. (1969). *Austral. J. Chem.* **22,** 1097.
41. Allison, G. B., Anderson, I. R. and Sheldon, J. C. (1967). *Austral. J. Chem.* **20,** 869.
42. Allison, G. B., Anderson, I. R. and Sheldon, J. C. (1969). *Austral. J. Chem.* **22,** 1091.
43. Allison, G. B. and Sheldon, J. C. (1963). *Inorg. Chem.* **6,** 1493.
44. Allred, A. L. and Thompson, D. W. (1968). *Inorg. Chem.* **7,** 1196.
45. Alyea, E. C., Basi, J. S., Bradley, D. C. and Chisholm, M. H. (1968). *Chem. Comm.*, 495.
46. Alyea, E. C. and Bradley, D. C. (1969). *J. Chem. Soc.* (*A*), 2330.
47. Amiet, R. G., Reeves, P. C. and Pettit, R. (1967). *Chem. Comm.*, 1208.
48. Amilius, Z., van Laar, B. and Rietveld, H. M. (1969). *Acta Cryst.* **B 25,** 400.
49. Amirthalingam, V. and Muralidbaran, K. V. (1964). *J. Inorg. Nuclear Chem.* **26,** 2038.
50. Anagnostopoulos, A. and Nicholls, D. (1965). *J. Inorg. Nuclear Chem.* **27,** 339.
51. Anand, S. P., Multani, R. K. and Jain, B. D. (1969). *J. Organometallic Chem.* **17,** 423.
52. Anders, U. and Graham, W. A. G. (1965). *Chem. Comm.*, 499.
53. Anders, U. and Graham, W. A. G. (1967). *J. Amer. Chem. Soc.* **89,** 539.
54. Anderson, I. R. and Sheldon, J. C. (1968). *Inorg. Chem.* **7,** 2602.
55. Andersson, G. (1954). *Acta Chem. Scand.* **8,** 1599.
56. Andersson, G. (1956). *Acta Chem. Scand.* **10,** 623.
57. Andersson, G. and Magneli, A. (1957). *Acta Chem. Scand.* **11,** 1065.
58. Andersson, S. (1960). *Acta Chem. Scand.* **14,** 1161.
59. Andersson, S. (1964). *Acta Chem. Scand.* **18,** 2339.
60. Andersson, S. (1965). *Acta Chem. Scand.* **19,** 557.
61. Andersson, S. (1965). *Acta Chem. Scand.* **19,** 2285.
62. Andersson, S. (1967). *Z. anorg. Chem.* **351,** 106.
63. Andersson, S., Collen, B., Kuylenstierna, U. and Magneli, A. (1957). *Acta Chem. Scand.* **11,** 1641.

64. Andersson, S. and Jahnberg, L. (1964). *Arkiv Kemi* **21,** 413.
65. Andersson, S. and Magneli, A. (1956). *Naturwiss.* **43,** 495.
66. Andersson, S., Mumme, W. G. and Wadsley, A. D. (1966). *Acta Cryst.* **21,** 802.
67. Andersson, S. and Wadsley, A. D. (1960). *Nature, Lond.* **187,** 499. *Acta Chem. Scand.* **15,** 663 (1961).
68. Andersson, S. and Wadsley, A. D. (1961). *Acta Cryst.* **14,** 1245.
69. Andersson, S. and Wadsley, A. D. (1962). *Acta Cryst.* **15,** 194.
70. Andersson, S. and Wadsley, A. D. (1962). *Acta Cryst.* **15,** 201.
71. Andra, K. (1968). *J. Organometallic Chem.* **11,** 567.
72. Andresen, A. F. (1960). *Acta Chem. Scand.* **14,** 919.
73. Andrews, J. T. S., Westrum, E. F. and Bjerrum, N. (1969). *J. Organometallic Chem.* **17,** 293.
74. Angelici, R. J. and Ingemanson, C. M. (1969). *Inorg. Chem.* **8,** 83.
75. Angell, F. G., James, R. G. and Wardlaw, W. (1929). *J. Chem. Soc.*, 2578.
76. Angstadt, R. L. and Tyree, S. Y. (1962). *J. Inorg. Nuclear Chem.* **24,** 913.
77. Anker, M. W., Colton, R., Rix, C. J. and Tomkins, I. B. (1969). *Austral. J. Chem.* **22,** 1341.
78. Anker, M. W., Colton, R. and Tomkins, I. B. (1967). *Austral. J. Chem.* **20,** 9.
79. Anker, M. W., Colton, R. and Tomkins, I. B. (1968). *Austral. J. Chem.* **21,** 1143, 1159.
80. Anker, M. W., Colton, R. and Tomkins, I. B. (1968). *Rev. Pure Appl. Chem.* (*Australia*) **18,** 23.
81. Ansell, G. B. and Katz, L. (1966). *Acta Cryst.* **21,** 482.
82. Antler, M. and Laubengayer, A. W. (1955). *J. Amer. Chem. Soc.* **77,** 5250.
83. Anzenhofer, K. and de Boer, J. J. (1969). *Rec. Trav. chim.* **88,** 286.
84. Aoki, T., Furusaki, A., Tomiie, Y., Ono, K. and Tanaka, K. (1969). *Bull. Chem. Soc. Japan* **42,** 545.
85. Archambault, J. and Rivest, R. (1958). *Canad. J. Chem.* **36,** 1461; **38,** 1331 (1960).
86. Archer, R. D. and Bonds, W. D. (1967). *J. Amer. Chem. Soc.* **89,** 2236.
87. Ariyaratne, J. K. P., Bierrum, A. M., Green, M. L. H., Ishaq, M., Prout, C. K. and Swanwick, M. G. (1969). *J. Chem. Soc.* (*A*), 1309.
88. Arutyunyan, E. G., Porai-Koshits, M. A., Molodkin, A. K. and Ivanova, O. M. (1966). *J. Struct. Chem.* **7,** 760.
89. Asbrink, S., Friberg, S., Magneli, A. and Andersson, G. (1959). *Acta Chem. Scand.* **13,** 603.
90. Asbrink, S. and Magneli, A. (1957). *Acta Chem. Scand.* **11,** 1606. *Acta Cryst.* **12,** 575 (1959).
91. Asmussen, R. W. (1944). "Magnetokemiske Undersogelser over Uorganiske Kompleksforbindelser". Copenhagen.
92. Assour, J. M., Goldmacher, J. and Harrison, S. E. (1965). *J. Chem. Phys.* **43,** 159, 4542.
93. Astrom, A. (1966). *Acta Chem. Scand.* **20,** 969.

94. Atkinson, R. H., Steigman, J. and Hiskey, C. F. (1952). *Anal. Chem.* **24**, 477.
95. Atoji, M. and Rundle, R. E. (1960). *J. Chem. Phys.* **32,** 627.
96. Atovmyan, L. O., Aliev, Z. G. and Tarakanov, B. M. (1968). *J. Struct. Chem.* **9,** 985.
97. Atovmyan, L. O. and Bokii, G. B. (1963). *Zhur. strukt. Khim.* **4,** 576.
98. Atovmyan, L. O. and Sokolova, Yu. A. (1969). *Chem. Comm.*, 649.
99. Aubin, R. and Rivest, R. (1958). *Canad. J. Chem.* **36,** 915.
100. Austin, T. E. and Tyree, S. Y. (1960). *J. Inorg. Nuclear Chem.* **14,** 141.
101. Aveston, J. (1964). *Inorg. Chem.* **3,** 981.
102. Aveston, J., Anacker, E. W. and Johnson, J. S. (1964). *Inorg. Chem.* **3,** 735.
103. Aveston, J. and Johnson, J. S. (1964). *Inorg. Chem.* **3,** 1051.
104. Baaz, M., Gutmann, V. and Talant, M. Y. A. (1961). *Monatsh.* **92,** 714.
105. Babel, D. and Deigner, P. (1965). *Z. anorg. Chem.* **339,** 57.
106. Babel, D. and Rudorff, W. (1964). *Naturwiss.* **51,** 85.
107. Babko, A. K., Lukachina, V. V. and Nabivanets, B. I. (1968). *Russ. J. Inorg. Chem.* **10,** 467.
108. Bachman, H. G., Ahmed, F. R. and Barnes, W. H. (1961). *Z. Krist.* **115,** 110.
109. Bader, R. F. W. and Huang, K. P. (1965). *J. Chem. Phys.* **43,** 3760.
110. Bader, R. F. W. and Westland, A. D. (1961). *Canad. J. Chem.* **39,** 2306.
111. Baenziger, N. C. and Rundle, R. E. (1948). *Acta Cryst.* **1,** 274.
112. Bagnall, K. W. and Brown, D. (1964). *J. Chem. Soc.*, 3021.
113. Bagnall, K. W., Brown, D. and Jones, P. J. (1964). *J. Chem. Soc.*, 2396.
114. Bagnall, K. W., Brown, D. and du Preez, J. G. H. (1964). *J. Chem. Soc.*, 2603.
115. Bagnall, K. W., Brown, D. and du Preez, J. G. H. (1964). *J. Chem. Soc.*, 5523.
116. Baikie, P. E. and Mills, O. S. (1966). *Chem. Comm.*, 683.
117. Baikie, P. E., Mills, O. S., Pauson, P. L., Smith, G. H. and Valentine, J. (1965). *Chem. Comm.*, 425.
118. Bailey, M. F. and Dahl, L. F. (1965). *Inorg. Chem.* **4,** 1298.
119. Bailey, M. F. and Dahl, L. F. (1965). *Inorg. Chem.* **4,** 1306.
120. Bailey, M. F. and Dahl, L. F. (1965). *Inorg. Chem.* **4,** 1314.
121. Baillie, M. J. and Brown, D. H. (1961). *J. Chem. Soc.*, 3691.
122. Bains, M. S. and Bradley, D. C. (1962). *Canad. J. Chem.* **40,** 1350.
123. Baker, B. R. and Mehta, B. D. (1965). *Inorg. Chem.* **4,** 848.
124. Baker, K. L., Edwards, D. A., Fowles, G. W. A. and Williams, R. G. (1967). *J. Inorg. Nuclear Chem.* **29,** 1881.
125. Baker, K. and Fowles, G. W. A. (1964). *Proc. Chem. Soc.*, 362.
126. Baker, K. and Fowles, G. W. A. (1965). *J. Less-Common Metals* **8,** 47.
127. Baker, K. L. and Fowles, G. W. A. (1968). *J. Chem. Soc.* (*A*), 801.
128. Baker, L. C. W., Baker, V. C., Eriks, K., Pope, M. T., Shibata, M., Rollins, O. W., Fang, J. H. and Koh, L. L. (1966). *J. Amer. Chem. Soc.* **88,** 2329.

129. Baker, L. C. W., Foster, G., Tan, W., Scholnick, F. and McCutcheon, T. P. (1955). *J. Amer. Chem. Soc.* **77,** 2136.
130. Baker, L. C. W., Gallagher, G. A. and McCutcheon, T. P. (1953). *J. Amer. Chem. Soc.* **75,** 2493.
131. Baker, L. C. W. and Weakley, T. J. R. (1966). *J. Inorg. Nuclear Chem.* **28,** 447.
132. Baker, M. C., Lyons, P. A. and Singer, S. J. (1955). *J. Amer. Chem. Soc.* **77,** 2011; *J. Phys. Chem.* **59,** 1074.
133. Baker, W. A. and Janus, A. R. (1964). *J. Inorg. Nuclear Chem.* **26,** 2087.
134. Ballhausen, C. J., Djurinskij, B. F. and Watson, K. J. (1968). *J. Amer. Chem. Soc.* **90,** 3305.
135. Ballhausen, C. J. and de Heer, J. (1965). *J. Chem. Phys.* **43,** 4304.
136. Ballhausen, C. J. and Liehr, A. D. (1961). *Acta Chem. Scand.* **15,** 775.
137. Balthis, J. H. (1962). *J. Inorg. Nuclear Chem.* **24,** 1016.
138. Balzani, V., Manfrin, M. F. and Moggi, L. (1969). *Inorg. Chem.* **8,** 47.
139. Bang, E. (1968). *Acta Chem. Scand.* **22,** 2671.
140. Banks, E. and Goldstein, A. (1968). *Inorg. Chem.* **7,** 966.
141. Barber, E. J. and Cady, G. H. (1956). *J. Phys. Chem.* **60,** 505.
142. Barclay, G. A., Gregor, I. K., Lambert, M. J. and Wild, S. B. (1967). *Austral. J. Chem.* **20,** 1571.
143. Barclay, G. A., Sabine, T. M. and Taylor, J. C. (1965). *Acta Cryst.* **19,** 205.
144. Bardawil, A. O., Collier, F. N. and Tyree, S. Y. (1965). *J. Less-Common Metals* **9,** 20.
145. Barnett, K. W. (1969). *Inorg. Chem.* **8,** 2009.
146. Barnett, K. W. and Treichel, P. M. (1967). *Inorg. Chem.* **6,** 294.
147. Barnighausen, H. and Handa, B. K. (1964). *J. Less-Common Metals* **6,** 226.
148. Barnum, D. W. (1961). *J. Inorg. Nuclear Chem.* **21,** 221.
149. Barr, W. B. (1908). *J. Amer. Chem. Soc.* **30,** 1668.
150. Barraclough, C. G., Bradley, D. C., Lewis, J. and Thomas, I. M. (1961). *J. Chem. Soc.*, 2601.
151. Barraclough, C. G., Gray, H. B. and Dubicki, L. (1968). *Inorg. Chem.* **7,** 844.
152. Barraclough, C. G., Lewis, J. and Nyholm, R. S. (1959). *J. Chem. Soc.*, 3552.
153. Barraclough, C. G. and Stals, J. (1966). *Austral. J. Chem.* **19,** 741.
154. Barrett, P. A., Dent, C. E. and Linstead, R. P. (1936). *J. Chem. Soc.*, 1719.
155. Bartell, L. S. and Gavin, R. M. (1968). *J. Chem. Phys.* **48,** 2466.
156. Bartlett, N. and Robinson, P. L. (1961). *J. Chem. Soc.*, 3549.
157. Bartlett, P. D. and Siedel, B. (1961). *J. Amer. Chem. Soc.* **83,** 581.
158. Bartley, W. G. and Wardlaw, W. (1958). *J. Chem. Soc.*, 422.
159. Basi, J. S. and Bradley, D. C. (1963). *Proc. Chem. Soc.*, 305.
160. Basu, S. (1966). *J. Inorg. Nuclear Chem.* **28,** 2769.
161. Bateman, L. R., Blount, J. F. and Dahl, L. F. (1966). *J. Amer. Chem. Soc.* **88,** 1082.
162. Bauer, D. and Schafer, H. (1968). *J. Less-Common Metals* **14,** 476.

163. Bauer, D. and v. Schnering, H. G. (1968). *Z. anorg. Chem.* **361,** 259.
164. Bauer, D., v. Schnering, H. G. and Schafer, H. (1965). *J. Less-Common Metals* **8,** 388.
165. Baur, W. H. (1961). *Acta Cryst.* **14,** 214.
166. Bawn, C. E. H. and Gladstone, J. (1959). *Proc. Chem. Soc.*, 227.
167. Bayer, G. and Hoffmann, W. (1965). *Z. Krist.* **121,** 9.
168. Beachell, H. C. and Butler, S. A. (1965). *Inorg. Chem.* **4,** 1133.
169. Bear, I. J. (1966). *Austral. J. Chem.* **19,** 357.
170. Bear, I. J. (1969). *Austral. J. Chem.* **22,** 875.
171. Bear, I. J. and Lukaszewski, G. M. (1966). *Austral. J. Chem.* **19,** 1973.
172. Bear, I. J. and Mumme, W. G. (1968). *Chem. Comm.*, 609.
173. Bear, I. J. and Mumme, W. G. (1969). *Chem. Comm.*, 230.
174. Beattie, I. R., Gilson, T. R. and Ozin, G. A. (1968). *J. Chem. Soc.* (*A*), 2765.
175. Beattie, I. R. and Leigh, G. J. (1961). *J. Inorg. Nuclear Chem.* **23,** 55.
176. Beattie, I. R. and Ozin, G. A. (1969). *J. Chem. Soc.* (*A*), 1691.
177. Beattie, I. R. and Reynolds, D. J. (1968). *Chem. Comm.*, 1531.
178. Becconsall, J. K., Job, B. E. and O'Brien, S. (1967). *J. Chem. Soc.* (*A*), 423.
179. Beck, W., Lottes, K. and Schmidtner, K. (1965). *Angew. Chem. Internat. Edn.* **4,** 151.
180. Beck, W. and Smedal, H. S. (1966). *Angew. Chem. Internat. Edn.* **5,** 253.
181. Becke-Goehring, M. and Slawisch, A. (1966). *Z. anorg. Chem.* **346,** 295.
182. Beckett, R., Colton, R., Hoskins, B. F., Martin, R. L. and Vince, D. G. (1969). *Austral. J. Chem.* **22,** 2527.
183. Bedon, H. D., Hatfield, W. E., Horner, S. M. and Tyree, S. Y. (1965). *Inorg. Chem.* **4,** 743.
184. Bedon, H. D., Horner, S. M. and Tyree, S. Y. (1964). *Inorg. Chem.* **3,** 647.
185. Beermann, C. (1959). *Angew. Chem.* **71,** 195.
186. Beermann, C. and Bestian, H. (1959). *Angew. Chem.* **71,** 618, 627.
187. Behrens, H. and Haag, W. (1961). *Ber.* **94,** 312.
188. Behrens, H. and Haag, W. (1961). *Ber.* **94,** 320.
189. Behrens, H. and Harder, N. (1964). *Ber.* **97,** 426.
190. Behrens, H. and Harder, N. (1964). *Ber.* **97,** 433.
191. Behrens, H. and Herrmann, D. (1967). *Z. anorg. Chem.* **351,** 225.
192. Behrens, H. and Klek, W. (1957). *Z. anorg. Chem.* **292,** 151.
193. Behrens, H. and Kohler, J. (1960). *Z. anorg. Chem.* **306,** 94.
194. Behrens, H. and Lutz, K. (1968). *Z. anorg. Chem.* **356,** 225.
195. Behrens, H., Meyer, K. and Muller, A. (1965). *Z. Naturforsch.* **20b,** 74.
196. Behrens, H. and Muller, A. (1965). *Z. anorg. Chem.* **341,** 124.
197. Behrens, H. and Vogl, J. (1963). *Ber.* **96,** 2220.
198. Behrens, H. and Ziegler, W. (1969). *Z. anorg. Chem.* **365,** 269.
199. Behrens, H. and Zizlsperger, H. (1961). *Z. Naturforsch.* **16b,** 349.
200. Beineke, T. A. and Delgaudio, J. (1968). *Inorg. Chem.* **7,** 715.
201. Belyaev, I. N. and Lobas, L. M. (1968). *Russ. J. Inorg. Chem.* **13,** 601.
202. Bennett, M. A., Bramley, R. and Watt, R. (1969). *J. Amer. Chem. Soc.* **91,** 3089.

203. Bennett, M. J., Brencic, J. V. and Cotton, F. A. (1969). *Inorg. Chem.* **8,** 1060.
204. Bennett, M. J., Caulton, K. G. and Cotton, F. A. (1969). *Inorg. Chem.* **8,** 1.
205. Bennett, M. J., Cotton, F. A., Foxman, B. M. and Stokely, P. F. (1967). *J. Amer. Chem. Soc.* **89,** 2759.
206. Bennett, M. J., Cotton, F. A., Legzdins, P. and Lippard, S. J. (1968). *Inorg. Chem.* **7,** 1770.
207. Bennett, M. J., Cotton, F. A. and Takats, J. (1968). *J. Amer. Chem. Soc.* **90,** 903.
208. Bennett, M. J. and Mason, R. (1963). *Proc. Chem. Soc.*, 273.
209. Berdonosov, S. S. and Lapitskii, A. V. (1965). *Russ. J. Inorg. Chem.* **10,** 1525.
210. Berdonosov, S. S., Lapitskii, A. V. and Bakov, E. K. (1965). *Russ. J. Inorg. Chem.* **10,** 173.
211. Berdonosov, S. S., Lapitskii, A. V. and Berdonosova, D. G. (1964). *Russ. J. Inorg. Chem.* **9,** 1388.
212. Berdonosov, S. S., Lapitskii, A. V., Berdonosova, D. G. and Vlasov, L. G. (1963). *Russ. J. Inorg. Chem.* **8,** 1315.
213. van den Berg, J. M. (1966). *Acta Cryst.* **20,** 905.
214. Bergh, A. A. and Haight, G. P. (1962). *Inorg. Chem.* **1,** 688.
215. Bergman, J. G. and Cotton, F. A. (1966). *Inorg. Chem.* **5,** 1208.
216. Bergman, J. G. and Cotton, F. A. (1966). *Inorg. Chem.* **5,** 1420.
217. Bernard, J. and Camelot, M. (1966). *Compt. rend.* **263C,** 1068.
218. Bernier, J.-C. and Poix, P. (1967). *Compt. rend.* **265C,** 1164.
219. Berry, K. O., Smardzewski, R. R. and McCarley, R. E. (1969). *Inorg. Chem.* **8,** 1994.
220. Bertaut, E. F. (1956). *Bull. Soc. Franc. Mineral. Crist.* **79,** 276.
221. Berthold, H. J. and Groh, G. (1963). *Angew. Chem.* **75,** 576; *Z. anorg. Chem.* **319,** 230.
222. Berthold, H. J. and Groh, G. (1966). *Angew. Chem. Internat. Edn.* **5,** 516.
223. Bertrand, A. (1880). *Bull. Soc. chim. France* **33,** 403; **34,** 631.
224. Bertrand, J. A. and Kelley, J. A. (1966). *J. Amer. Chem. Soc.* **88,** 4746; *Inorg. Chem.* **8,** 1982 (1969).
225. Bhar, D. M. and Ray, P. (1928). *J. Indian Chem. Soc.* **5,** 497.
226. Bhatnagar, S. S., Prakash, B. and Hamid, A. (1938). *J. Chem. Soc.*, 1428.
227. Bhatti, M. A., Copley, D. B. and Shelton, R. A. J. (1969). *J. Less-Common Metals* **18,** 99.
228. Biltz, W. and Birk, E. (1924). *Z. anorg. Chem.* **134,** 134.
229. Biltz, W. and Fendius, C. (1928). *Z. anorg. Chem.* **172,** 385.
230. Biltz, W. and Klemm, W. (1926). *Z. anorg. Chem.* **152,** 267.
231. Biltz, W. and Voigt, A. (1922). *Z. anorg. Chem.* **120,** 71.
232. Bird, P. H. and Churchill, M. R. (1967). *Chem. Comm.*, 403.
233. Bird, P. H. and Churchill, M. R. (1968). *Inorg. Chem.* **7,** 349.
234. Birmingham, J. M., Fischer, A. K. and Wilkinson, G. (1955). *Naturwiss.* **42,** 96.

235. Bir'yukov, B. P., Struchkov, Yu. T., Anisimov, K. N., Kolobova, N. E. and Beschastnov, A. S. (1968). *Chem. Comm.*, 667.
236. Bither, T. A., Gillson, J. L. and Young, H. S. (1966). *Inorg. Chem.* **5,** 1559.
237. Blake, A. B. (1966). *Chem. Comm.*, 569.
238. Blake, A. B., Cotton, F. A. and Wood, J. S. (1964). *J. Amer. Chem. Soc.* **86,** 3024.
239. Blanchard, S. (1964). *J. Chem. Phys.* **61,** 747.
240. Blanchard, S. (1965). *J. chim. Phys.* **62,** 919.
241. Blankenship, F. A. and Belford, R. L. (1962). *J. Chem. Phys.*, **36,** 633.
242. Blasse, G. (1963). *Z. anorg. Chem.* **326,** 44.
243. Blasse, G. (1964). *J. Inorg. Nuclear Chem.* **26,** 1191.
244. Blasse, G. (1966). *Z. anorg. Chem.* **345,** 222.
245. Blazekova, M. and Schlafer, H. L. (1968). *Z. anorg. Chem.* **360,** 169.
246. Blight, D. G. and Kepert, D. L. (1968). *J. Chem. Soc.* (*A*), 534.
247. Blight, D. G. and Kepert, D. L. (1968). *Theor. Chim. Acta* **11,** 51.
248. Block, S. (1960). *Nature, Lond.* **186,** 540.
249. Blomberg, B., Kihlborg, L. and Magneli, A. (1953). *Arkiv Kemi* **6,** 133.
250. Blomstrand, C. W. (1857). *J. prakt. Chem.* **71,** 449.
251. Bock, H. and tom Dieck, H. (1966). *Angew. Chem. Internat. Edn.* **5,** 520.
252. Bock, H. and tom Dieck, H. (1966). *Z. anorg. Chem.* **345,** 9.
253. Bock, H., tom Dieck, H., Pyttlik, H. and Schnoller, M. S. (1968). *Z. anorg. Chem.* **357,** 54.
254. Bode, H. and Dohren, V. (1958). *Acta Cryst.* **11,** 80.
255. Bode, H. and Tuefer, G. (1956). *Acta Cryst.* **9,** 929.
256. Bode, H. and Tuefer, G. (1956). *Z. anorg. Chem.* **283,** 18.
257. Bode, H. and Voss, E. (1956). *Z. anorg. Chem.* **286,** 136.
258. de Boer, J. H. and Fast, J. D. (1930) *Z. anorg. Chem.* **187,** 177.
259. Bogdanova, N. I. and Arija, S. M. (1960). *Zhur. obshchei Khim.* **30,** 3.
260. Bohland, H. and Malitzke, P. (1965). *Z. Naturforsch* **20b,** 1126.
261. Bohland, H. and Malitzke, P. (1967). *Z. anorg. Chem.* **350,** 70.
262. Bohm, G. (1926). *Z. Krist.* **63,** 319.
263. Boland, H. and Niemann, E. (1965). *Z. anorg. Chem.* **336,** 225.
264. Boland, H., Tiede, E. and Zenker, E. (1967). *J. Less-Common Metals* **13,** 224; **14,** 397 (1968); **15,** 89 (1968).
265. Bonamico, M., Dessy, G. and Fares, V. (1969). *Chem. Comm.*, 697.
266. Bonati, F. and Cotton, F. A. (1967). *Inorg. Chem.* **6,** 1353.
267. Boorman, P. M., Greenwood, N. N. and Hildon, M. A. (1968). *J. Chem. Soc.* (*A*), 2466.
268. Boorman, P. M., Greenwood, N. N., Hildon, M. A. and Parish, R. V. (1966). *Inorg. Nuclear Chem. Letters* **2,** 377; *J. Chem. Soc.* (*A*), 2002.
269. Boorman, P. M., Greenwood, N. N., Hildon, M. A. and Whitfield, H. J. (1967). *J. Chem. Soc.* (*A*), 2017.
270. Boorman, P. M., Greenwood, N. N. and Whitfield, H. J. (1968). *J. Chem. Soc.* (*A*), 2256.
271. Boorman, P. M. and Straughan, B. P. (1966). *J. Chem. Soc.* (*A*), 1514.

272. Bosnich, B., Gillard, R. D., McKenzie, E. D. and Webb, G. A. (1966). *J. Chem. Soc. (A)*, 1331.
273. Bouchard, R. J. and Gillson, J. L. (1968). *Inorg. Chem.* **7,** 969.
274. Boucher, L. J. and Yen, T. F. (1969). *Inorg. Chem.* **8,** 689.
275. Bowden, J. A. and Colton, R. (1969). *Austral. J. Chem.* **22,** 905.
276. Bowman, A. L., Wallace, T. C., Yarnell, J. L. and Wenzel, R. G. (1966). *Acta Cryst.* **21,** 843.
277. Bowman, K. and Dori, Z. (1968). *Chem. Comm.*, 636.
278. Boyd, P. D. W., Smith, P. W. and Wedd, A. G. (1969). *Austral. J. Chem.* **22,** 653.
279. Bradley, A. J. and Illingsworth, J. W. (1936). *Proc. Roy. Soc. A* **157,** 113.
280. Bradley, D. C. (1960). "Progress in Inorganic Chemistry". Vol. 2, Interscience, New York, p. 303.
281. Bradley, D. C., Chakravarti, B. N., Chatterjee, A. K., Wardlaw, W. and Whitley, A. (1958). *J. Chem. Soc.*, 99.
282. Bradley, D. C., Chakravarti, B. N. and Wardlaw, W. (1956). *J. Chem. Soc.*, 2381.
283. Bradley, D. C., Chakravarti, B. N. and Wardlaw, W. (1956). *J. Chem. Soc.*, 4439.
284. Bradley, D. C., Chisholm, M. H., Heath, C. E. and Hursthouse, M. B. (1969). *Chem. Comm.*, 1261.
285. Bradley, D. C., Gaze, R. and Wardlaw, W. (1955). *J. Chem. Soc.*, 721, 3977; 469 (1957).
286. Bradley, D. C. and Gitlitz, M. H. (1969). *J. Chem. Soc. (A)*, 1152.
287. Bradley, D. C., Halim, F. M. A. and Wardlaw, W. (1950). *J. Chem. Soc.*, 3450.
288. Bradley, D. C., Hancock, D. C. and Wardlaw, W. (1952). *J. Chem. Soc.*, 2773.
289. Bradley, D. C. and Holloway, H. (1961). *Canad. J. Chem.* **39,** 1818.
290. Bradley, D. C. and Holloway, C. E. (1964). *Inorg. Chem.* **3,** 1163.
291. Bradley, D. C. and Holloway, C. E. (1965). *Chem. Comm.*, 284.
292. Bradley, D. C. and Holloway, C. E. (1968). *J. Chem. Soc. (A)*, 219.
293. Bradley, D. C. and Holloway, C. E. (1968). *J. Chem. Soc. (A)*, 1316.
294. Bradley, D. C. and Holloway, C. E. (1969). *J. Chem. Soc. (A)*, 282.
295. Bradley, D. C., Hursthouse, M. B. and Rendall, I. F. (1969). *Chem. Comm.*, 672.
296. Bradley, D. C., Hursthouse, M. B. and Rodesiler, P. F. (1968). *Chem. Comm.*, 1112.
297. Bradley, D. C., Hursthouse, M. B. and Rodesiler, P. F. (1969). *Chem. Comm.*, 14.
298. Bradley, D. C. and Kasenally, A. S. (1968). *Chem. Comm.*, 1430.
299. Bradley, D. C. and Mehta, M. L. (1962). *Canad. J. Chem.* **40,** 1138.
300. Bradley, D. C., Moss, R. H. and Sales, K. D. (1969). *Chem. Comm.*, 1255.
301. Bradley, D. C., Multani, R. K. and Wardlaw, W. (1958). *J. Chem. Soc.*, 4647.

302. Bradley, D. C., Sincha, R. N. P. and Wardlaw, W. (1958). *J. Chem. Soc.* 4651.
303. Bradley, D. C. and Thomas, I. M. (1958). *Chem. and Ind.*, 17.
304. Bradley, D. C. and Thomas, I. M. (1959). *Proc. Chem. Soc.*, 225.
305. Bradley, D. C. and Thomas, I. M. (1960). *J. Chem. Soc.*, 3857.
306. Bradley, D. C. and Thomas, I. M. (1962). *Canad. J. Chem.* **40,** 449.
307. Bradley, D. C. and Thomas, I. M. (1962). *Canad. J. Chem.* **40,** 1355.
308. Bradley, D. C., Wardlaw, W. and Whitley, A. (1955). *J. Chem. Soc.*, 726.
309. Bradley, D. C., Wardlaw, W. and Whitley, A. (1956). *J. Chem. Soc.*, 5.
310. Bradley, D. C., Wardlaw, W. and Whitley, A. (1956). *J. Chem. Soc.*, 1139.
311. Brand, P. and Sackmann, H. (1963). *Z. anorg. Chem.* **321,** 262.
312. Brand, P. and Schmidt, J. (1966). *Z. anorg. Chem.* **348,** 257.
313. Branden, C.-I. and Lindqvist, I. (1960). *Acta Chem. Scand.* **14,** 726.
314. Branden, C.-I. and Lindqvist, I. (1961). *Acta Chem. Scand.* **17,** 353.
315. Brandon, J. K. and Brown, I. D. (1968). *Canad. J. Chem.* **46,** 933.
316. Brandt, B. G. and Skapski, A. C. (1967). *Acta Chem. Scand.* **21,** 661.
317. Braterman, P. S. and Thompson, D. T. (1968). *J. Chem. Soc.* (*A*), 1454.
318. Brauer, G. (1941). *Z. anorg. Chem.* **248,** 1.
319. Brauer, G. (1948). *Z. anorg. Chem.* **256,** 10.
320. Brauer, G. (1954). "Handbuch der preparativ anorganische Chemie". Ferdinand Enke, Stuttgart.
321. Brauer, G. and Walz, H. (1963). *Z. anorg. Chem.* **319,** 236.
322. Breil, H. and Wilke, G. (1966). *Angew. Chem. Internat. Edn.* **5,** 898.
323. Brencic, J. V. and Cotton, F. A. (1969). *Inorg. Chem.* **8,** 7.
324. Bridgland, B. E., Fowles, G. W. A. and Walton, R. A. (1965). *J. Inorg. Nuclear Chem.* **27,** 383.
325. Bridgland, B. E. and McGregor, W. R. (1969). *J. Inorg. Nuclear Chem.* **31,** 43.
326. Bright, N. F. H. and Wurm, J. G. (1958). *Canad. J. Chem.* **36,** 615.
327. Brimm, E. O., Brantley, J. C., Lorenz, J. H. and Jellinek, M. H. (1951). *J. Amer. Chem. Soc.* **73,** 5427.
328. Brintzinger, H. (1966). *J. Amer. Chem. Soc.* **88,** 4305, 4307.
329. Brintzinger, H. H. (1967). *J. Amer. Chem. Soc.* **89,** 6871.
330. Brisdon, B. J. (1967). *Inorg. Chem.* **6,** 1791.
331. Brisdon, B. J., Edwards, D. A., Machin, D. J., Murray, K. S. and Walton, R. A. (1967). *J. Chem. Soc.*, 1825.
332. Brisdon, B. J. and Fowles, G. W. A. (1964). *J. Less-Common Metals* **7,** 102.
333. Brisdon, B. J., Fowles, G. W. A. and Osburne, B. P. (1962). *J. Chem. Soc.*. 1330.
334. Brisdon, B. J., Lester, T. E. and Walton, R. A. (1967). *Spectrochim. Acta* **23A,** 1969.
335. Brisdon, B. J., Ozin, G. A. and Walton, R. A. (1969). *J. Chem. Soc.* (*A*), 342.
336. Brisdon, B. J. and Walton, R. A. (1965). *J. Inorg. Nuclear Chem.* **27,** 1101.
337. Brisdon, B. J. and Walton, R. A. (1965). *J. Chem. Soc.*, 2274.
338. Brito, F., Ingri, N. and Sillen, L. G. (1964). *Acta Chem. Scand.* **18,** 1557.

339. Britton, D. (1963). *Canad. J. Chem.* **41,** 1632.
340. Brnicevic, N. and Djordjevic, C. (1967). *J. Less-Common Metals* **13,** 470.
341. Brnicevic, N. and Djordjevic, C. (1968). *Inorg. Chem.* **7,** 1936.
342. Brodersen, K. (1966). *Angew. Chem. Internat. Edn.* **5,** 682.
343. Brodersen, K. (1968). *Angew. Chem. Internat. Edn.* **7,** 147, 148.
344. Brodersen, K., Brietbach, H. K. and Thiele, G. (1968). *Z. anorg. Chem.* **357,** 162.
345. Brodersen, K. and Moers, F. (1965). *Naturwiss.* **52,** 205.
346. Brodersen, K., Thiele, G. and Schnering, H. G. (1965). *Z. anorg. Chem.* **337,** 120.
347. Broll, A., Simon, A., v. Schnering, H. G. and Schafer, H. (1969). *Z. anorg. Chem.* **367,** 1.
348. van Bronswyk, W. (1968). *J. Chem. Soc.* (*A*), 692.
349. Brossett, C. (1935). *Nature* **135,** 874. *Arkiv Kemi, Min., Geol.* **12A,** No. 4.
350. Brown, B. W. and Banks, E. (1954). *J. Amer. Chem. Soc.* **76,** 963.
351. Brown, B. W. and Banks, E. (1951). *Phys. Rev.* **84,** 609.
352. Brown, D. (1964). *J. Chem. Soc.*, 4944.
353. Brown, D., Easey, J. F. and du Preez, J. G. H. (1966). *J. Chem. Soc.* (*A*), 258.
354. Brown, D., Easey, J. F. and Rickard, C. E. F. (1969). *J. Chem. Soc.* (*A*), 1161.
355. Brown, D., Holah, D. G. and Rickard, C. E. F. (1969). *Chem. Comm.*, 280.
356. Brown, D., Holah, D. G. and Rickard, C. E. F. (1970). *J. Chem. Soc.* (*A*), 423.
357. Brown, D., Holah, D. G. and Rickard, C. E. F. (1970). *J. Chem. Soc.* (*A*), 786.
358. Brown, D. and Jones, P. J. (1966). *J. Chem. Soc.* (*A*), 733.
359. Brown, D. and Jones, P. J. (1967). *J. Chem. Soc.* (*A*), 247.
360. Brown, D., Petcher, T. J. and Smith, A. J. (1969). *Acta Cryst.* **B25,** 178.
361. Brown, D. and Smith, A. J. (1965). *Chem. Comm.*, 554.
362. Brown, D. A., Cunningham, D. and Glass, W. K. (1968). *J. Chem. Soc.* (*A*), 1563.
363. Brown, D. H. (1961). *J. Inorg. Nuclear Chem.* **17,** 146; *J. Chem. Soc.*, 4732.
364. Brown, D. H., Dixon, K. R., Kemmitt, R. D. W. and Sharp, D. W. A. (1965). *J. Chem. Soc.*, 1559.
365. Brown, D. H. and McCallum, J. D. (1963). *J. Inorg. Nuclear Chem.* **25,** 1483.
366. Brown, D. H., Russell, D. R. and Sharp, D. W. A. (1966). *J. Chem. Soc.* (*A*), 18.
367. Brown, G. M. and Walker, L. A. (1966). *Acta Cryst.* **20,** 220.
368. Brown, T. M. and Knox, G. F. (1967). *J. Amer. Chem. Soc.* **89,** 5296.
369. Brown, T. M. and McCann, E. L. (1968). *Inorg. Chem.* **7,** 1227.
370. Brown, T. M. and Newton, G. S. (1966). *Inorg. Chem.* **5,** 1117.
371. Brown, T. M., Pings, D. K., Lieto, L. R. and DeLong, S. J. (1966). *Inorg. Chem.* **5,** 1695.
372. Brown, T. M. and Ruble, B. (1967). *Inorg. Chem.* **6,** 1335.

373. Broyde, B. (1967). *Inorg. Chem.* **6,** 1588.
374. Brubaker, C. H. and Young, R. C. (1951). *J. Amer. Chem. Soc.* **73,** 4179.
375. Brun, L. (1966). *Acta Cryst.* **20,** 739.
376. Brunton, G. (1965). *J. Inorg. Nuclear Chem.* **27,** 1173.
377. Brunton, G. (1967). *J. Inorg. Nuclear Chem.* **29,** 1631.
378. Bucknall, W. R., Carter, S. R. and Wardlaw, W. (1927). *J. Chem. Soc.*, 512.
379. Bull, W. E., Madan, S. K. and Willis, J. E. (1963). *Inorg. Chem.* **2,** 303.
380. Burbank, R. D. (1966). *Inorg. Chem.* **5,** 1491.
381. Burbank, R. D. and Bartlett, N. (1968). *Chem. Comm.*, 645.
382. Burbank, R. D. and Bensey, F. N. (1956). U.S. At. Energy Comm. *K–1280.*
383. Burbank, R. D. and Bensey, F. N. (1957). *J. Chem. Phys.* **27,** 981.
384. Burger, H. and Neese, H.-J. (1969). *Z. anorg. Chem.* **365,** 243.
385. Burger, H., Stammreich, H. and Teixeira Sans, Th. (1966). *Monatsh.* **97,** 1276.
386. Burke, T. G., Smith, D. F. and Nielson, A. H. (1952). *J. Chem. Phys.* **20,** 447.
387. Burkov, K. A., Lilic, L. S. and Sillen, L. G. (1965). *Acta Chem. Scand.* **19,** 14.
388. Burmeister, J. L., Deardorff, E. A. and Van Dyke, C. E. (1969). *Inorg. Chem.* **8,** 170.
389. Bursill, L. A., Hyde, B. G., Terasaki, O. and Watanabe, D. (1969). *Phil. Mag.* **20,** 347.
390. Bush, M. A., Cook, V. R. and Woodward, P. (1967). *Chem. Comm.*, 630.
391. Buslaev, Yu. A. and Bochkareva, V. A. (1967). *Russ. J. Inorg. Chem.* **12,** 902.
392. Buslaev, Yu. A. and Davidovich, R. L. (1965). *Russ. J. Inorg. Chem.* **10,** 1014.
393. Buslaev, Yu. A. and Davidovich, R. L. (1968). *Russ. J. Inorg. Chem.* **13,** 656.
394. Buslaev, Yu. A., Dyer, D. S. and Ragsdale, R. O. (1967). *Inorg. Chem.* **6,** 2208.
395. Buslaev, Yu. A., Glushkova, M. A., Ershova, M. M. and Bochkareva, V. A. (1968). *Russ. J. Inorg. Chem.* **13,** 31.
396. Buslaev, Yu. A. and Gustyakova, M. P. (1963). *Izvest. Akad. Nauk S.S.S.R., Ser. khim.*, 1533.
397. Bye, J. and Haegi, W. (1953). *Compt. rend.* **236,** 381.
398. Byfleet, C. R., Herring, F. G., Lin, W. C., McDowell, C. A. and Ward, D. J. (1968). *Mol. Phys.* **15,** 239.
399. Bystrom, A. M. and Evans, H. T. (1959). *Acta Chem. Scand.* **13,** 377.
400. Bystrom, A. and Wilhelmi, K.-A. (1950). *Acta Chem. Scand.* **4,** 1131.
401. Bystrom, A., Wilhelmi, K.-A. and Brotzen, O. (1950). *Acta Chem. Scand.* **4,** 1119.
402. Cady, G. H. and Hargreaves, G. B. (1961). *J. Chem. Soc.*, 1563.
403. Cady, G. H. and Hargreaves, G. B. (1961). *J. Chem. Soc.*, 1568.

404. Calderazzo, F. (1964). *Inorg. Chem.* **3,** 810.
405. Calderazzo, F. (1964). *Inorg. Chem.* **3,** 1207.
406. Calderazzo, F. (1965). *Inorg. Chem.* **4,** 223.
407. Calderazzo, F. (1966). *Inorg. Chem.* **5,** 429.
408. Calderazzo, F. and Cini, R. (1965). *J. Chem. Soc.*, 818.
409. Calderazzo, F., Cini, R., Corradini, P., Ercoli, R. and Natta, G. (1960). *Chem. and Ind.*, 500.
410. Calderazzo, F., Cini, R. and Ercoli, R. (1960). *Chem. and Ind.*, 934.
411. Calderazzo, F. and Henzi, R. (1967). *J. Organometallic Chem.* **10,** 483.
412. Calderazzo, F., Losi, S. and Susz, P. B. (1969). *Inorg. Chim. Acta* **3,** 329.
413. Calderazzo, F., Salzmann, J. J. and Mosimann, P. (1967). *Inorg. Chim, Acta* **1,** 65.
414. Calderon, J. L., Cotton, F. A. and Legzdins, P. (1969). *J. Amer. Chem. Soc.* **91,** 2528.
415. Callaghan, A., Layton, A. J. and Nyholm, R. S. (1969). *Chem. Comm.*, 399.
416. Calvert, L. D. and Pleass, C. M. (1962). *Canad. J. Chem.* **40,** 1473.
417. Canterford, J. H. and Colton, R. (1968). "Halides of the Second and Third Row Transition Metals". Wiley, London.
418. Canterford, J. H., Colton, R. and O'Donnell, T. A. (1967). *Rev. Pure Appl. Chem.* (*Australia*), **17,** 123.
419. Canterford, J. H. and O'Donnell, T. A. (1967). *Inorg. Chem.* **6,** 541.
420. Canterford, J. H. and O'Donnell, T. A. (1968). *Austral. J. Chem.* **21,** 1421.
421. Canty, A. J., Coutts, R. S. P. and Wailes, P. C. (1968). *Austral. J. Chem.* **21,** 807.
422. Cardin, D. J., Keppie, S. A., Kingston, B. M. and Lappert, M. F. (1967). *Chem. Comm.*, 1035.
423. Cardin, D. J. and Lappert, M. F. (1966). *Chem. Comm.*, 506.
424. Carlin, R. L. and Walker, F. A. (1965). *J. Amer. Chem. Soc.* **87,** 2128.
425. Carlson, G. L. (1963). *Spectrochim. Acta* **19,** 1291.
426. Carmichael, W. M. and Edwards, D. A. (1967). *J. Inorg. Nuclear Chem.* **29,** 1535.
427. Carmichael, W. M. and Edwards, D. A. (1968). *J. Inorg. Nuclear Chem.* **30,** 2641.
428. Carmichael, W. M., Edwards, D. A. and Walton, R. A. (1966). *J. Chem. Soc.* (*A*), 97.
429. Carnell, P. J. H. and Fowles, G. W. A. (1959). *J. Chem. Soc.*, 4113.
430. Carnell, P. J. H. and Fowles, G. W. A. (1962). *J. Less-Common Metals*, **4,** 40.
431. Carobbi, G. (1928). *Gazzetta* **58,** 35.
432. Carr, P. H. and Foner, S. (1960). *J. Appl. Phys.* **31,** 344 S.
433. Carrick, A. and Glocking, F. (1968). *J. Chem. Soc.* (*A*), 913.
434. Cartan, F. and Caughlan, C. N. (1960). *J. Phys. Chem.* **64,** 1756.
435. Carter, O. L., McPhail, A. T. and Sim, G. A. (1966). *J. Chem. Soc.* (*A*), 822.
436. Carter, O. L., McPhail, A. T. and Sim, G. A. (1966). *J. Chem. Soc.* (*A*), 1095.

437. Carter, O. L., McPhail, A. T. and Sim, G. A. (1967). *J. Chem. Soc. (A)*, 228.
438. Casey, A. T. and Clark, R. J. H. (1968). *Inorg. Chem.* **7,** 1598.
439. Casey, A. T. and Clark, R. J. H. (1969). *Inorg. Chem.* **8,** 1216.
440. Caughlan, C. N., Smith, H. S., Katz, W., Hodgson, W. and Crowe, R. W. (1951). *J. Amer. Chem. Soc.* **73,** 5652.
441. Caughlan, C. N., Smith, H. M. and Watenpaugh, K. (1966). *Inorg. Chem.* **5,** 2131.
442. Cavell, R. G. and Clark, H. C. (1961). *J. Inorg. Nuclear Chem.* **17,** 257.
443. Cavell, R. G. and Clark, H. C. (1962). *J. Chem. Soc.*, 2692.
444. Cavell, R. G. and Clark, H. C. (1963). *J. Chem. Soc.*, 4261.
445. Cavell, R. G. and Clark, H. C. (1964). *Inorg. Chem.* **3,** 1798.
446. Chabrie, M. C. (1907). *Compt. rend.* **144,** 804.
447. Chadwick, B. M. and Sharpe, A. G. (1966). *Adv. Inorg. chem. Radiochem.* **8.**
448. Chaigneau, M. (1956). *Compt. rend.* **243,** 957.
449. Chaigneau, M. (1956). *Compt. rend.* **242,** 263.
450. Chaigneau, M. (1957). *Compt. rend.* **244,** 900.
451. Chaigneau, M. (1957). *Bull. Soc. chim. France*, 886.
452. Chaigneau, M. (1957). *Compt. rend.* **245,** 1805.
453. Chaigneau, M. (1958). *Compt. rend.* **247,** 300.
454. Chaigneau, M. (1959). *Compt. rend.* **248,** 3173.
455. Chakravarti, B. N. (1958). *Naturwiss.* **45,** 286.
456. Chandra, G., George, T. A. and Lappert, M. F. (1967). *Chem. Comm.*, 116.
457. Chandra, G. and Lappert, M. F. (1968). *J. Chem. Soc. (A)*, 1940.
458. Chandrasegaran, L. and Rodley, G. A. (1965). *Inorg. Chem.* **4,** 1360.
459. Chapin, W. H. (1910). *J. Amer. Chem. Soc.* **32,** 323.
460. Chapin, D. S., Kafalas, J. A. and Honig, J. M. (1965). *J. Phys. Chem.* **69,** 1402.
461. Chapman, F. W., Hummers, W. S., Tyree, S. Y. and Yolles, S. (1952). *J. Amer. Chem. Soc.* **74,** 5277.
462. Chasteen, N. D., Belford, R. L. and Paul, I. C. (1969). *Inorg. Chem.* **8,** 408.
463. Chatt, J. and Hayter, R. G. (1963). *J. Chem. Soc.*, 1343.
464. Chatt, J. and Thompson, D. T. (1964). *J. Chem. Soc.*, 2713.
465. Chatt, J. and Thornton, D. A. (1964). *J. Chem. Soc.*, 1005.
466. Chatt, J. and Watson, H. R. (1961). *J. Chem. Soc.*, 4980.
467. Chaudhari, M. A. and Stone, F. G. A. (1966). *J. Chem. Soc. (A)*, 838.
468. Chauveau, F. (1960). *Bull. Soc. chim. France*, 810.
469. Chilesotti, A. C. (1903). *Gazzetta* **33,** 349.
470. Chilesotti, A. C. (1906). *Z. Elektrochem.* **12,** 173.
471. Chizhikov, D. M. and Grin'ko, A. M. (1959). *Russ. J. Inorg. Chem.* **4,** 444.
472. Chojnacki, J., Grochowski, J., Lebioda, L., Oleksyn, B. and Stadnicka, K. (1969). *Roczniki Chem.* **43,** 273.
473. Christ, C. L., Clark, J. R. and Evans, H. T. (1954). *Acta Cryst.* **7,** 801.
474. Christ, K. and Schlafer, H. L. (1964). *Z. anorg. Chem.* **334,** 1.

475. Churchill, M. R. and Bird, P. H. (1967). *Chem. Comm.*, 746; *Inorg. Chem.* **7,** 1545 (1968).
476. Churchill, M. R. and Fennessey, J. P. (1967). *Inorg. Chem.* **6,** 1213.
477. Churchill, M. R. and Fennessey, J. P. (1968). *Inorg. Chem.* **7,** 953.
478. Churchill, M. R. and O'Brien, T. A. (1969). *J. Chem. Soc.* (*A*), 1110.
479. Ciampolini, M. (1966). *Chem. Comm.*, 47.
480. Cid-Dresdner, H. and Buerger, M. J. (1962). *Z. Krist.* **117,** 411.
481. Claassen, H. H., Garner, E. L. and Selig, H. (1968). *J. Chem. Phys.* **49,** 1803.
482. Claassen, H. H. and Selig, H. (1966). *J. Chem. Phys.* **44,** 4039.
483. Claassen, H. H., Selig, H. and Malm, J. G. (1962). *J. Chem. Phys.* **36,** 2888.
484. Clauss, K. and Beermann, C. (1959). *Angew. Chem.* **71,** 627.
485. Clark, G. R. and Palenik, G. J. (1969). *Chem. Comm.*, 667.
486. Clark, H. C. and Emeleus, H. J. (1957). *J. Chem. Soc.*, 2119.
487. Clark, H. C. and Emeleus, H. J. (1957). *J. Chem. Soc.*, 4778.
488. Clark, H. C. and Emeleus, H. J. (1958). *J. Chem. Soc.*, 190.
489. Clark, H. C. and Sadana, Y. N. (1964). *Canad. J. Chem.* **42,** 50.
490. Clark, H. C. and Sadana, Y. N. (1964). *Canad. J. Chem.* **42,** 702.
491. Clark, P. W. and Wentworth, R. A. D. (1969). *Inorg. Chem.* **8,** 1223.
492. Clark, R. J. and Hoberman, P. I. (1965). *Inorg. Chem.* **4,** 1771.
493. Clark, R. J. H. 1963). *J. Chem. Soc.*, 1377.
494. Clark, R. J. H. (1964). *J. Chem. Soc.*, 417.
495. Clark, R. J. H. (1965). *J. Chem. Soc.*, 5699.
496. Clark, R. J. H. (1968). "The Chemistry of Titanium and Vanadium". Elsevier, Amsterdam.
497. Clark, R. J. H. and Errington, W. (1966). *Inorg. Chem.* **5,** 650.
498. Clark, R. J. H. and Errington, W. (1967). *J. Chem. Soc.* (*A*), 258.
499. Clark, R. J. H., Errington, W., Lewis, J. and Nyholm, R. S. (1966). *J. Chem. Soc.* (*A*), 989.
500. Clark, R. J. H. and Greenfield, M. L. (1967). *J. Chem. Soc.* (*A*), 409.
501. Clark, R. J. H., Greenfield, M. L. and Nyholm, R. S. (1966). *J. Chem. Soc.* (*A*), 1254.
502. Clark, R. J. H., Kepert, D. L. and Nyholm, R. S. (1965). *J. Chem. Soc.*, 2877.
503. Clark, R. J. H., Kepert, D. L., Nyholm, R. S. and Rodley, G. A. (1966). *Spectrochim. Acta* **22,** 1697.
504. Clark, R. J. H., Lewis, J., Machin, D. J. and Nyholm, R. S. (1963). *J. Chem. Soc.*, 379.
505. Clark, R. J. H., Lewis, J. and Nyholm, R. S. (1962). *J. Chem. Soc.*, 2460.
506. Clark, R. J. H., Lewis, J., Nyholm, R. S., Pauling, P. and Robertson, G. B. (1961). *Nature, Lond.* 192, 222.
507. Clark, R. J. H. and Machin, D. J. (1963). *J. Chem. Soc.*, 4430.
508. Clark, R. J. H., Maresca, L. and Puddephatt, R. J. (1968). *Inorg. Chem.* **7,** 1603.
509. Clark, R. J. H. and Negrotti, R. H. U. (1968). *Chem. and Ind.*, 154.

510. Clark, R. J. H., Negrotti, R. H. U. and Nyholm, R. S. (1966). *Chem. Comm.*, 486.
511. Clark, R. J. H., Nyholm, R. S. and Scaife, D. E. (1966). *J. Chem. Soc.* (*A*), 1296.
512. Claxton, T. A. and Benson, G. C. (1966). *Canad. J. Chem.* **44,** 157.
513. Clearfield, A. (1964). *Rev. Pure Appl. Chem.* (*Australia*) **14,** 91.
514. Clearfield, A. and Vaughan, P. A. (1956). *Acta Cryst.* **9,** 555.
515. Clementi, E., Raimondi, D. L. and Reinhardt, W. P. (1967). *J. Chem. Phys.* **47,** 1300.
516. Clifford, A. F., Beachell, H. C. and Jack, W. M. (1957). *J. Inorg. Nuclear Chem.* **5,** 57.
517. Coates, J. R., Ott, J. B., Mangelson, N. F. and Jensen, R. J. (1964). *J. Phys. Chem.* **68,** 2617.
518. Coffey, E. C., Lewis, J. and Nyholm, R. S. (1964). *J. Chem. Soc.* (*A*), 1741.
519. Cohen, B., Edwards, A. J., Mercer, M. and Peacock, R. D. (1965). *Chem. Comm.*, 322.
520. Cohen, G. H. and Hoard, J. L. (1966). *J. Amer. Chem. Soc.* **88,** 3228.
521. Cohen, B. and Peacock, R. D. (1966). *J. Inorg. Nuclear Chem.* **28,** 3056.
522. Coleman, J. M. and Dahl, L. F. (1967). *J. Amer. Chem. Soc.* **89,** 542.
523. Collenberg, O. (1918). *Z. anorg. Chem.* **102,** 259. Collenberg, O. and Guthe, A. (1924). *Z. anorg. Chem.* **134,** 322.
524. Collis, R. E. (1969). *J. Chem. Soc.* (*A*), 1895.
525. Colton, R. and Howard, J. J. (1969). *Austral. J. Chem.* **22,** 2543.
526. Colton, R. and Rix, C. J. (1968). *Austral. J. Chem.* **21,** 1155.
527. Colton, R. and Rix, C. J. (1969). *Austral. J. Chem.* **22,** 305.
528. Colton, R. and Rix, C. J. (1969). *Austral. J. Chem.* **22,** 2535.
529. Colton, R., Scollary, G. R. and Tomkins, I. B. (1968). *Austral. J. Chem.* **21,** 15.
530. Colton, R. and Tomkins, I. B. (1965). *Austral. J. Chem.* **18,** 447.
531. Colton, R. and Tomkins, I. B. (1966). *Austral. J. Chem.* **19,** 759.
532. Colton, R. and Tomkins, I. B. (1966). *Austral. J. Chem.* **19,** 1143.
533. Colton, R. and Tomkins, I. B. (1966). *Austral. J. Chem.* **19,** 1519.
534. Colton, R. and Tomkins, I. B. (1967). *Austral. J. Chem.* **20,** 13.
535. Colton, R. and Tomkins, I. B. (1968). *Austral. J. Chem.* **21,** 1975.
536. Colton, R., Tomkins, I. B. and Wilson, P. W. (1964). *Austral. J. Chem.* **17,** 496.
537. Connor, J. A. and Mills, O. S. (1969). *J. Chem. Soc.* (*A*), 334.
538. Conrad, R. W. and Land, J. E. (1964). *J. Less-Common Metals* **7,** 180.
539. Conroy, L. E. and Podolsky, G. (1968). *Inorg. Chem.* **7,** 614.
540. Conroy, L. E. and Sienko, M. J. (1957). *J. Amer. Chem. Soc.* **79,** 4048.
541. Conroy, L. E. and Yokokawa, T. (1965). *Inorg. Chem.* **4,** 994.
542. Cook, C. D., Nyholm, R. S. and Tobe, M. L. (1965). *J. Chem. Soc.*, 4194.
543. Cook, C. M. (1959). *J. Amer. Chem. Soc.* **81,** 535.
544. Cook, C. M. (1959). *J. Amer. Chem. Soc.* **81,** 3828.
545. Cooke, M., Green, M. and Kirkpatrick, D. (1968). *J. Chem. Soc.* (*A*), 1507.

546. Copley, D. B., Bannerjee, A. K. and Tyree, S. Y. (1965). *Inorg. Chem.* **4,** 1480.
547. Copley, D. B., Fairbrother, F., Grundy, K. H. and Thompson, A. (1964). *J. Less-Common Metals* **6,** 407.
548. Copley, D. B., Fairbrother, F. and Thompson, A. (1964). *J. Chem. Soc.,* 315.
549. Copley, D. B., Fairbrother, F. and Thompson, A. (1965). *J. Less-Common Metals* **8,** 256.
550. Corbett, J. D. and Seabaugh, P. (1958). *J. Inorg. Nuclear Chem.* **6,** 207.
551. Corbett, M. and Hoskins, B. F. (1968). *Chem. Comm.,* 1602.
552. Corradini, P. and Allegra, G. (1959). *J. Amer. Chem. Soc.* **81,** 5510.
553. Corradini, P. and Sirigu, A. (1967). *Inorg. Chem.* **6,** 601.
554. Cotton, F. A. (1964). *Inorg. Chem.* **3,** 1217.
555. Cotton, F. A. (1965). *Inorg. Chem.* **4,** 334.
556. Cotton, F. A. and Bratton, W. K. (1965). *J. Amer. Chem. Soc.* **87,** 921.
557. Cotton, F. A. and Curtis, N. F. (1965). *Inorg. Chem.* **4,** 241.
558. Cotton, F. A., Dollase, W. A. and Wood, J. S. (1963). *J. Amer. Chem. Soc.* **85,** 1543.
559. Cotton, F. A. and Elder, R. C. (1964). *Inorg. Chem.* **3,** 397.
560. Cotton, F. A., Faller, J. W. and Musco, A. (1966). *J. Amer. Chem. Soc.* **88,** 4506.
561. Cotton, F. A., Faller, J. W and Musco, A. (1968). *J. Amer. Chem. Soc.* **90,** 1438.
562. Cotton, F. A. and Francis, R. (1960). *J. Amer. Chem. Soc.* **82,** 2986.
563. Cotton, F. A. and Haas, T. E. (1964). *Inorg. Chem.* **3,** 10.
564. Cotton, F. A. and Harris, C. B. (1965). *Inorg. Chem.* **4,** 330.
565. Cotton, F. A. and Johnson, B. F. G. (1964). *Inorg. Chem.* **3,** 1609.
566. Cotton, F. A. and LaPrade, M. D. (1968). *J. Amer. Chem. Soc.* **90,** 5418.
567. Cotton, F. A. and Legzdins, P. (1968). *J. Amer. Chem. Soc.* **90,** 6232.
568. Cotton, F. A. and Legzdins, P. (1968). *Inorg. Chem.* **7,** 1777.
569. Cotton, F. A. and Mague, J. T. (1964). *Inorg. Chem.* **3,** 1402.
570. Cotton, F. A. and Marks, T. J. (1969). *J. Amer. Chem. Soc.* **91,** 1339.
571. Cotton, F. A. and Morehouse, S. M. (1965). *Inorg. Chem.* **4,** 1377.
572. Cotton, F. A., Morehouse, S. M. and Wood, J. S. (1964). *Inorg. Chem.* **3,** 1603.
573. Cotton, F. A. and Richardson, D. C. (1966). *Inorg. Chem.* **5,** 1851.
574. Cotton, F. A. and Wilkinson, G. (1954). *Z. Naturforsch.* **9b,** 417.
575. Cotton, F. A. and Wing, R. M. (1965). *Inorg. Chem.* **4,** 314.
576. Cotton, F. A., Wing, R. M. and Zimmerman, R. A. (1967). *Inorg. Chem.* **6,** 12.
577. Cotton, F. A. and Winquist, B. H. C. (1969). *Inorg. Chem.* **8,** 1304.
578. Couch, D. E. and Brenner, A. (1959). *J. Res. Nat. Bur. Stand., Sect. A* 63, 185.
579. Courtin, P. (1968). *Bull. Soc. chim. France,* 4799.
580. Cousins, M. and Green, M. L. H. (1963). *J. Chem. Soc.,* 889.
581. Cousins, M. and Green, M. L. H. (1964). *J. Chem. Soc.,* 1567.

582. Cousins, M. and Green, M. L. H. (1969). *J. Chem. Soc.*, 16.
583. Coutts, R. S. P., Kautzner, B. and Wailes, P. C. (1969). *Austral. J. Chem.* **22,** 1137.
584. Coutts, R. S. P. and Surtees, J. R. (1966). *Austral. J. Chem.* **19,** 387.
585. Coutts, R. S. P. and Wailes, P. C. (1966). *Austral. J. Chem.* **19,** 2069.
586. Coutts, R. S. P. and Wailes, P. C. (1967). *Austral. J. Chem.* **21,** 373.
587. Coutts, R. S. P. and Wailes, P. C. (1967). *Inorg. Nuclear Chem. Letters* **3,** 1.
588. Coutts, R. S. P. and Wailes, P. C. (1968). *Austral. J. Chem.* **21,** 2199.
589. Coutts, R. S. P. and Wailes, P. C. (1968). *Chem. Comm.*, 260.
590. Coutts, R. S. P. and Wailes, P. C. (1969). *Austral. J. Chem.* **22,** 1547.
591. Cowley, A., Fairbrother, F. and Scott, N. (1958). *J. Chem. Soc.*, 3133.
592. Cowley, A. H. and White, W. D. (1969). *J. Amer. Chem. Soc.* **91,** 34.
593. Cox, B. (1954). *J. Chem. Soc.*, 3251.
594. Cox, B. (1956). *J. Chem. Soc.*, 876.
595. Cox, B., Sharp, D. W. A. and Sharpe, A. G. (1956). *J. Chem. Soc.*, 1242.
596. Cox, M., Clark, R. J. H. and Milledge, H. J. (1966). *Nature, Lond.* **212,** 1357.
597. Cox, M., Lewis, J. and Nyholm, R. S. (1964). *J. Chem. Soc.*, 6113.
598. Cox, M., Lewis, J. and Nyholm, R. S. (1965). *J. Chem. Soc.*, 2840.
599. Cozzi, D. and Vivarelli, S. (1955). *Z. anorg. Chem.* **279,** 165.
600. Craig, H. R. and Tyree, S. Y. (1965). *Inorg. Chem.* **4,** 997.
601. Craig, P. J. and Green, M. (1968). *J. Chem. Soc. (A)*, 1978.
602. Cras, J. A. (1962). *Nature, Lond.* **194,** 678.
603. Crayton, P. H. and Thompson, W. A. (1963). *J. Inorg. Nuclear Chem.* **25,** 742.
604. Creemers, H. M. J. C., Verbeek, F. and Noltes, J. G. (1968). *J. Organometallic Chem.* **15,** 125.
605. Crisp, W. P., Deutscher, R. L. and Kepert, D. L. (1970). *J. Chem. Soc. (A)*, 2199.
606. Cromer, D. T. and Herrington, K. (1955). *J. Amer. Chem. Soc.* **77,** 4708.
607. Crossman, L. D., Olsen, D. P. and Duffey, G. H. (1963). *J. Chem. Phys.* **38,** 73.
608. Crouch, P. C., Fowles, G. W. A., Frost, J. L., Marshall, P. R. and Walton, R. A. (1968). *J. Chem. Soc. (A)*, 1061.
609. Crouch, P. C., Fowles, G. W. A., Marshall, P. R. and Walton, R. A. (1968). *J. Chem. Soc. (A)*, 1634.
610. Crouch, P. C., Fowles, G. W. A., Tomkins, I. B. and Walton, R. A. (1969). *J. Chem. Soc. (A)*, 2412.
611. Crouch, P. C., Fowles, G. W. A. and Walton, R. A. (1968). *J. Chem. Soc. (A)*, 2172.
612. Crouch, P. C., Fowles, G. W. A. and Walton, R. A. (1969). *J. Chem. Soc. (A)*, 972.
613. Cullen, W. R. and Hayter, R. G. (1964). *J. Amer. Chem. Soc.* **86,** 1030.
614. Cullinane, N. M., Chard, S. J. and Leyshon, D. M. (1952). *J. Chem. Soc.*, 4106.

615. Cunningham, J. A., Sands, D. E. and Wagner, W. F. (1967). *Inorg. Chem.* **6,** 499.
616. Cunningham, J. A., Sands, D. E., Wagner, W. F. and Richardson, M. F. (1969). *Inorg. Chem.* **8,** 22.
618. Dahl, L. F., Chiang, T.-I., Seabaugh, P. W. and Larsen, E. M. (1964). *Inorg. Chem.* **3,** 1236.
619. Dahl, L. F. and Wampler, D. L. (1959). *J. Amer. Chem. Soc.* **81,** 3150; *Acta Cryst.* **15,** 903 (1962).
620. Dale, B. W., Buckley, J. M. and Pope, M. T. (1969). *J. Chem. Soc.* (*A*), 301.
621. Dale, B. W. and Pope, M. T. (1967). *Chem. Comm.*, 792.
622. Daly, J. J. and Sneedon, R. P. A. (1967). *J. Chem. Soc.* (*A*), 736.
623. Dance, I. G. and Freeman, H. C. (1965). *Inorg. Chem.* **4,** 1555.
624. Danforth, J. D. (1958). *J. Amer. Chem. Soc.* **80,** 2585.
625. Danot, M. and Rouxel, J. (1966). *Compt. rend.* **262C,** 1879.
626. Dartiguenaue, Y., Lehne, M. and Rohmer, R. (1965). *Bull. Soc. chim. France*, 62.
627. Dauben, H. J. and Honnen, H. R. (1958). *J. Amer. Chem. Soc.* **80,** 5570.
628. Davies, N., James, B. D. and Wallbridge, M. G. H. (1969). *J. Chem. Soc.* (*A*), 2601.
629. Davies, J. E. D. and Long, D. A. (1968). *J. Chem. Soc.* (*A*), 2560.
630. Davis, O. C. M. (1906). *J. Chem. Soc.*, 1575.
631. Davison, A., Edelstein, N., Holm, R. H. and Maki, A. H. (1964). *J. Amer. Chem. Soc.* **86,** 2799; *Inorg. Chem.* **4,** 55 (1965).
632. Dawson, B. (1953). *Acta Cryst.* **6,** 113.
633. Day, V. W. and Hoard, J. L. (1968). *J. Amer. Chem. Soc.* **90,** 3374.
634. Dean, P. A. W., Evans, D. F. and Phillips, R. F. (1969). *J. Chem. Soc.* (*A*), 363.
635. DeBenedittus, J. and Katz, L. (1965). *Inorg. Chem.* **4,** 1836.
636. Decian, A., Fischer, J. and Weiss, R. (1967). *Acta Cryst.* **22,** 340.
637. Dehand, J. (1965). *Bull. Soc. chim. France*, 1775.
638. Dehand, J., Guerchais, J. E. and Rohmer, R. (1966). *Bull. Soc. chim. France*, 346.
639. Dehnicke, K. (1961). *Z. anorg. Chem.* **309,** 266.
640. Dehnicke, K. (1961). *Angew. Chem.* **73,** 535; *Naturwiss.* **52,** 58 (1965).
641. Dehnicke, K. (1962). *Angew. Chem.* **74,** 495.
642. Dehnicke, K. (1964). *Berichte* **97,** 3354.
643. Dehnicke, K. (1965). *Naturwiss.* **52,** 660.
644. Dehnicke, K. (1965). *Angew. Chem. Internat. Edn.* **4,** 22.
645. Dehnicke, K. (1965). *Berichte* **98,** 290.
646. Dehnicke, K. (1965). *J. Inorg. Nuclear Chem.* **27,** 809.
647. Dehnicke, K. (1965). *Z. anorg. Chem.* **338,** 279.
648. Dehnicke, K. and Meyer, K.-U. (1964). *Z. anorg. Chem.* **331,** 121.
649. Dehnicke, K., Pausewang, G. and Rudorff, W. (1969). *Z. anorg. Chem.* **366,** 64.
650. Dehnicke, K. and Strahle, J. (1965). *Z. anorg. Chem.* **339,** 171.

651. Dehnicke, K. and Weidlein, J. (1966). *Angew. Chem. Internat. Edn.* **5,** 1041.
652. Desnoyers, J. and Rivest, R. (1965). *Canad. J. Chem.* **43,** 1879.
653. Dettingmeijer, J. H. and Meinders, B. (1968). *Z. anorg. Chem.* **357,** 1.
654. Deubzer, B. and Kaesz, H. D. (1968). *J. Amer. Chem. Soc.* **90,** 3276.
655. Deutscher, R. L. and Kepert, D. L. Unpublished data.
656. Deutscher, R. L. and Kepert, D. L. (1970). *Inorg. Chem.* **9,** 2305.
657. Deutscher, R. L. and Kepert, D. L. (1970). *Inorg. Chim. Acta.* **4,** 645.
658. Dexter, D. D. and Silverton, J. V. (1968). *J. Amer. Chem. Soc.* **90,** 3589.
659. Deyrup, A. J. (1964). *Inorg. Chem.* **3,** 1645.
660. Dias, A. R. and Green, M. L. H. (1969). *Chem. Comm.*, 962.
661. Dickenson, R. N., Feil, S. E., Collier, F. N., Horner, W. W., Horner, S. M. and Tyree, S. Y. (1964). *Inorg. Chem.* **3,** 1600.
662. tom Dieck, H. and Friedel, H. (1969). *Chem. Comm.*, 411.
663. Dietrich, H. (1963). *Acta Cryst.* **16,** 681.
664. Dietrich, H. and Dierks, H. (1966). *Angew. Chem. Internat. Edn.* **5,** 899.
665. Dijkgraaf, C. (1964). *Nature, Lond.* **201,** 1121.
666. Dijkgraaf, C. (1965). *Spectrochim. Acta* **21,** 1419.
667. Dijkgraaf, C. and Rousseau, J. P. G. (1967). *Spectrochim. Acta* **23A,** 1267.
668. Dilke, M. H. and Eley, D. D. (1949). *J. Chem. Soc.*, 2601.
669. Dilung, I. I. (1959). *Doklady Akad. Nauk S.S.S.R.* **131,** 312.
670. Dingle, R. (1969). *J. Chem. Phys.* **50,** 545.
671. Dingle, R., McCarthy, P. J., and Ballhausen, C. J. (1969). *J. Chem. Phys.* **50,** 1957.
672. Djordjevic, C. and Katovic, V. (1963). *Chem. and Ind.*, 411; *J. Inorg. Nuclear Chem.* **25,** 1099.
673. Djordjevic, C., Nyholm, R. S., Pande, C. S. and Stiddard, M. H. B. (1966). *J. Chem. Soc.* (*A*), 16.
674. Djordjevic, C. and Vuletic, N. (1968). *Inorg. Chem.* **7,** 1864.
675. Dodge, R. P., Smith, G. S., Johnson, Q. and Elson, R. E. (1967). *Acta Cryst.* **22,** 85.
676. Dodge, R. P., Templeton, D. H. and Zalkin, A. (1961). *J. Chem. Phys.* **35,** 55.
677. Doedens, R. J. and Dahl, L. F. (1965). *J. Amer. Chem. Soc.* **87,** 2576.
678. Donohue, J. (1965). *Acta Cryst.* **18,** 1018.
679. Donohue, P. C. and Katz, L. L. (1964). *Nature, Lond.* **201,** 180.
680. Doring, H. and Moliere, K. (1952). *Z. Elektrochem.* **56,** 403.
681. Douglass, R. M. (1957). *Acta Cryst.* **10,** 423.
682. Dove, M. F. A., Creighton, J. A. and Woodward, L. A. (1962). *Spectrochim. Acta* **18,** 267.
683. Downing, J. W. and Ragsdale, R. O. (1968). *Inorg. Chem.* **7,** 1675.
684. Dowsing, R. D. and Gibson, J. F. (1967). *J. Chem. Soc.* (*A*), 655.
685. Doyle, G. and Tobias, R. S. (1967). *Inorg. Chem.* **6,** 1111.
686. Doyle, G. and Tobias, R. S. (1968). *Inorg. Chem.* **7,** 2479, 2484.
687. Drago, R. S. and Whitten, K. W. (1966). *Inorg. Chem.* **5,** 677.
688. Drake, J. E. and Fowles, G. W. A. (1961). *J. Inorg. Nuclear Chem.* **18,** 136.

689. Drake, J. E. and Fowles, G. W. A. (1961). *J. Less-Common Metals* **3,** 149.
690. Drake, J. E. and Riddle, C. (1969). *Inorg. Nuclear Chem. Letters* **5,** 665.
691. Drake, J. E., Vekris, J. and Wood, J. S. (1968). *J. Chem. Soc.* (*A*), 1000.
692. Drake, J. E., Vekris, J. E. and Wood, J. S. (1969). *J. Chem. Soc.* (*A*), 345.
693. Drew, M. G. B., Fowles, G. W. A. and Lewis, D. F. (1969). *Chem. Comm.*, 876.
694. Druce, P. M., Lappert, M. F. and Riley, B. N. K. (1967). *Chem. Comm.*, 486.
695. Drummond, J. and Wood, J. S. (1970). *J. Chem. Soc.* (*A*), 226.
696. Dubicki, L. and Martin, R. L. (1969). *Austral. J. Chem.* **22,** 701.
697. Dubicki, L. and Martin, R. L. (1969). *Austral. J. Chem.* **22,** 1571.
698. DuBois, D. W., Iwamoto, R. T. and Kleinberg, J. (1969). *Inorg. Chem.* **8,** 815.
699. Duckworth, M. W. and Fowles, G. W. A. (1962). *J. Less-Common Metals* **4,** 338.
700. Duckworth, M. W., Fowles, G. W. A. and Greene, P. T. (1967). *J. Chem. Soc.* (*A*), 1592.
701. Duckworth, M. W., Fowles, G. W. A. and Hoodless, R. A. (1963). *J. Chem. Soc.*, 5665.
702. Duckworth, M. W., Fowles, G. W. A. and Williams, R. G. (1962). *Chem. and Ind.*, 1285.
703. Ducret, L. P. (1951). *Ann. Chim.* (*France*) **6,** 705.
704. Duffey, G. H. (1950). *J. Chem. Phys.* **18,** 746.
705. Duffey, G. H. (1951). *J. Chem. Phys.* **19,** 553.
706. Duffey, G. H. (1951). *J. Chem. Phys.* **19,** 963.
707. Duncan, J. F. and Kepert, D. L. (1961). *J. Chem. Soc.*, 5317.
708. Duncan, J. F. and Kepert, D. L. (1962). *J. Chem. Soc.*, 205.
709. Dunitz, J. D. and Orgel, L. E. (1953). *J. Chem. Soc.*, 2594.
710. Dunitz, J. D. and Pauling, P. (1960). *Helv. Chim. Acta* **43,** 2188.
711. Durand, C., Schaal, R. and Souchay, P. (1959). *Compt. rend.* **248,** 979.
712. Dyer, D. S. and Ragsdale, R. O. (1966). *Chem. Comm.*, 601.
713. Dyer, D. S. and Ragsdale, R. O. (1967). *Inorg. Chem.* **6,** 8.
714. Dzhabarov, F. Z. and Gorbachev, S. V. (1964). *Russ. J. Inorg. Chem.* **9,** 1297.
715. Earley, J. E. and Cannon, R. D. (1966). *Transition Metal Chem.* **1,** 33.
716. Earnshaw, A., Figgis, B. N. and Lewis, J. (1966). *J. Chem. Soc.* (*A*), 1656.
717. Earnshaw, A., Larkworthy, L. F. and Patel, K. C. (1969). *J. Chem. Soc.* (*A*), 1339.
718. Earnshaw, A., Larkworthy, L. F. and Patel, K. C. (1969). *J. Chem. Soc.* (*A*), 2276.
719. Earnshaw, A., Larkworthy, L. F., Patel, K. C. and Beech, G. (1969). *J. Chem. Soc.* (*A*), 1334.
720. Earnshaw, A., Larkworthy, L. F., Patel, K. C., Patel, K. S., Carlin, R. L. and Terazakis, E. G. (1966). *J. Chem. Soc.* (*A*), 511.
721. Earnshaw, A., Larkworthy, L. F. and Patel, K. S. (1964). *Z. anorg. Chem.* **334,** 163.

722. Earnshaw, A., Larkworthy, L. F. and Patel, K. S. (1965). *J. Chem. Soc. (A)*, 3267.
723. Earnshaw, A., Larkworthy, L. F. and Patel, K. S. (1966). *J. Chem. Soc. (A)*, 363.
724. Earnshaw, A. and Lewis, J. (1961). *J. Chem. Soc.*, 396.
725. Eberts, R. E. and Pink, F. X. (1968). *J. Inorg. Nuclear Chem.* **30,** 457.
726. Edgar, K., Johnson, B. F. G., Lewis, J. and Wild, S. B. (1968). *J. Chem. Soc. (A)*, 2851.
727. Edgell, W. F. and Pauuwe, N. (1969). *Chem. Comm.*, 284.
728. Edwards, A. J. (1963). *Proc. Chem. Soc.*, 205.
729. Edwards, A. J. (1964). *J. Chem. Soc.*, 3714.
730. Edwards, A. J. (1964). *J. Chem. Soc. (A)*, 909.
731. Edwards, A. J., Holloway, J. H. and Peacock, R. D. (1963). *Proc. Chem. Soc.*, 275.
732. Edwards, A. J. and Jones, G. R. (1968). *Chem. Comm.*, 346.
733. Edwards, A. J. and Jones, G. R. (1968). *J. Chem. Soc. (A)*, 2074.
734. Edwards, A. J., Jones, G. R. and Sills, R. J. C. (1968). *Chem. Comm.*, 1177.
735. Edwards, A. J. and Jones, G. R. (1969). *J. Chem. Soc. (A)*, 1651.
736. Edwards, A. J., Jones, G. R. and Steventon, B. R. (1967). *Chem. Comm.*, 462.
737. Edwards, A. J. and Peacock, R. D. (1959). *J. Chem. Soc.*, 4126.
738. Edwards, A. J. and Peacock, R. D. (1960). *Chem. and Ind.*, 1441.
739. Edwards, A. J. and Peacock, R. D. (1961). *J. Chem. Soc.*, 4253.
740. Edwards, A. J., Peacock, R. D. and Said, A. (1962). *J. Chem. Soc.*, 4643.
741. Edwards, A. J., Peacock, R. D. and Small, R. W. H. (1962). *J. Chem. Soc.*, 4486.
742. Edwards, A. J. and Steventon, B. R. (1968). *J. Chem. Soc. (A)*, 2503.
743. Edwards, D. A. (1963). *J. Inorg. Nuclear Chem.* **25,** 1198.
744. Edwards, D. A. (1964). *J. Less-Common Metals* **7,** 159.
745. Edwards, D. A. (1965). *J. Inorg. Nuclear Chem.* **27,** 303.
746. Edwards, D. A. and Fowles, G. W. A. (1961). *J. Chem. Soc.*, 24.
747. Edwards, D. A. and Fowles, G. W. A. (1961). *J. Less-Common Metals* **3,** 181.
748. Edwards, D. A. and Fowles, G. W. A. (1962). *J. Less-Common Metals* **4,** 512.
749. Edwards, D. A., Fowles, G. W. A. and Walton, R. A. (1965). *J. Inorg. Nuclear Chem.* **27,** 1999.
750. Ehrlich, P. and Engel, W. (1962). *Z. anorg. Chem.* **317,** 21.
751. Ehrlich, P., Gutsche, W. and Seifert, H. J. (1961). *Z. anorg. Chem.* **312,** 80.
752. Ehrlich, P., Hein, H. J. and Kuhnl, H. (1957). *Z. anorg. Chem.* **292,** 139.
753. Ehrlich, P., Kupa, G. and Blankenstein, K. (1959). *Z. anorg. Chem.* **299,** 213.
754. Ehrlich, P. and Pietzka, G. (1954). *Z. anorg. Chem.* **275,** 121.
755. Ehrlich, P., Ploger, F. and Koch, E. (1964). *Z. anorg. Chem.* **333,** 209.
756. Ehrlich, P., Ploger, F. and Pietzka, G. (1955). *Z. anorg. Chem.* **282,** 19.
757. Ehrlich, P. and Schmitt, R. (1961). *Z. anorg. Chem.*, **308,** 91.

758. Ehrlich, P. and Seifert, H. J. (1959). *Z. anorg. Chem.* **301,** 282.
759. Ehrlich, P. and Siebert, W. (1959). *Z. anorg. Chem.* **301,** 288.
760. Ehrlich, P. and Siebert, W. (1960). *Z. anorg. Chem.* **303,** 96.
761. Eick, H. A. and Kihlborg, L. (1966). *Acta Chem. Scand.* **20,** 1658.
762. Einstein, F. W. B. and Penfold, B. R. (1964). *Acta Cryst.* **17,** 1127.
763. Eisenberg, R. and Gray, H. B. (1967). *Inorg. Chem.* **6,** 1844.
764. Eisenberg, R. and Ibers, J. A. (1966). *Inorg. Chem.* **5,** 411.
765. Eisenberg, R., Stiefel, E. I., Rosenberg, R. C. and Gray, H. B. (1966). *J. Amer. Chem. Soc.* **88,** 2874.
766. Elder, M. (1969). *Inorg. Chem.* **8,** 2103.
767. Elder, M., Evans, J. G. and Graham, W. A. G. (1969). *J. Amer. Chem. Soc.* **91,** 1245.
768. Elder, M., Graham, W. A. G., Hall, D. and Kummer, R. (1968). *J. Amer. Chem. Soc.* **90,** 2189.
769. Elder, M. and Hall, D. (1969). *Inorg. Chem.* **8,** 1268.
770. Elder, M. and Hall, D. (1969). *Inorg. Chem.* **8,** 1273.
771. Elder, M. and Penfold, B. R. (1966). *Inorg. Chem.* **5,** 1197.
772. Eliseev, S. S., Glukhov, I. A. and Gaidaenko, N. V. (1969). *Russ. J. Inorg. Chem.* **14,** 328.
773. Ellerback, L. D., Shanks, H. R., Sidles, P. H. and Danielson, G. C. (1961). *J. Chem. Phys.* **35,** 298.
774. Emeléus, H. J. and Gutmann, V. (1949). *J. Chem. Soc.*, 2979.
775. Emeléus, H. J. and Gutmann, V. (1950). *J. Chem. Soc.*, 2115.
776. Emeléus, H. J. and Rao, G. S. (1958). *J. Chem. Soc.*, 4245.
777. Engebretson, G. and Rundle, R. (1963). *J. Amer. Chem. Soc.* **85,** 481.
778. Engel, G. (1935). *Z. Krist.* **90,** 341.
779. Engelbrecht, A. and Grosse, A. V. (1952). *J. Amer. Chem. Soc.* **74,** 5262.
780. English, W. D. and Sommer, L. H. (1955). *J. Amer. Chem. Soc.* **77,** 170.
781. Ephraim, F. (1934). *Helv. Chim. Acta* **17,** 291.
782. Ephraim, F. and Ammann, E. (1932). *Helv. Chim. Acta* **16,** 1273.
783. Epperson, E. R. and Frye, H. (1966). *Inorg. Nuclear Chem. Letters* **2,** 223.
784. Epperson, E. R., Horner, S. M., Knox, K. and Tyree, S. Y. (1963). *Inorg. Synth.* **7,** 163.
785. Epstein, C. and Elliott, N. (1954). *J. Chem. Phys.* **22,** 634.
786. Ercoli, R., Calderazzo, F. and Alberda, A. (1960). *J. Amer. Chem. Soc.* **82,** 2966.
787. Espenson, J. H. and Boone, D. J. (1968). *Inorg. Chem.* **7,** 636.
788. Espenson, J. H. and McCarley, R. E. (1966). *J. Amer. Chem. Soc.* **88,** 1063.
789. Evdokimov, V. B., Zelenstov, V. V., Kolli, I. D., T'ang Wen-Hsin and Spitsyn, V. I. (1962). *Doklady Akad. Nauk S.S.S.R.* **145,** 1282.
790. Evans, D. F. (1957). *J. Chem. Soc.*, 4013.
791. Evans, D. F., Griffith, W. P. and Pratt, L. (1965). *J. Chem. Soc.*, 2182.
792. Evans, H. T. (1948). *J. Amer. Chem. Soc.* **70,** 1291.
793. Evans, H. T. (1960). *Z. Krist.* **114,** 257.
794. Evans, H. T. (1966). *Inorg. Chem.* **5,** 967.

795. Evans, H. T. (1968). *J. Amer. Chem. Soc.* **90,** 3275.
796. Evans, H. T. and Block, S. (1954). *Amer. Mineralogist* **39,** 327.
797. Evans, H. T. and Block, S. (1966). *Inorg. Chem.* **5,** 1808.
798. Evans, H. T. and Showell, J. S. (1969). *J. Amer. Chem. Soc.* **91,** 6881.
799. Evans, H. T., Swallow, A. G. and Barnes, W. H. (1964). *J. Amer. Chem. Soc.* **86,** 4209.
800. Eve, D. J. and Fowles, G. W. A. (1966). *J. Chem. Soc.* (*A*), 1183.
801. Ewens, R. F. G. and Lister, M. W. (1938). *Trans. Faraday Soc.* **34,** 1358.
802. Fackler, J. P. (1962). *J. Chem. Soc.*, 1957.
803. Fackler, J. P. and Holah, D. G. (1965). *Inorg. Chem.* **4,** 954.
804. Fairbrother, F., Cowley, A. H. and Scott, N. (1959). *J. Less-Common Metals* **1,** 206.
805. Fairbrother, F. and Frith, W. C. (1951). *J. Chem. Soc.*, 3051.
806. Fairbrother, E., Frith, W. C. and Woolf, A. A. (1954). *J. Chem. Soc.*, 1031.
807. Fairbrother, F., Grundy, K. H. and Thompson, A. (1965). *J. Chem. Soc.*, 761.
808. Fairbrother, F., Grundy, K. H. and Thompson, A. (1966). *J. Less-Common Metals* **10,** 38.
809. Fairbrother, F. and Nixon, J. F. (1962). *J. Chem. Soc.*, 150.
810. Fairbrother, F., Nixon, J. F. and Prophet, H. (1965). *J. Less-Common Metals* **9,** 434.
811. Fairbrother, F., Robinson, D. and Taylor, J. B. (1958). *J. Inorg. Nuclear Chem.* **8,** 296.
812. Fairbrother, F. and Taylor, J. B. (1956). *J. Chem. Soc.*, 4946.
813. Faller, J. W. (1969). *Inorg. Chem.* **8,** 767.
814. Fast, J. D. (1938). *Z. anorg. Chem.* **239,** 145.
815. Fast, J. D. (1939). *Rec. Trav. chim.* **58,** 174.
816. Fay, R. C. and Lowry, R. N. (1967). *Inorg. Nuclear Chem. Letters* **3,** 117.
817. Fay, R. C. and Lowry, R. N. (1967). *Inorg. Chem.* **6,** 1512.
818. Feenan, K. and Fowles, G. W. A. (1964). *J. Chem. Soc.*, 2842.
819. Feenan, K. and Fowles, G. W. A. (1965). *J. Chem. Soc.*, 2449.
820. Feenan, K. and Fowles, G. W. A. (1965). *Inorg. Chem.* **4,** 310.
821. Feinleib, J. and Paul, W. (1967). *Phys. Rev.* **155,** 841.
822. Feld, R. and Cowe, P. L. (1965). "The Organic Chemistry of Titanium." Butterworth, London.
823. Feltham, R. D., Silverthorn, W. and McPherson, G. (1969). *Inorg. Chem.* **8,** 344.
824. Feltz, A. (1963). *Z. anorg. Chem.* **323,** 35.
825. Feltz, A. (1964). *Z. anorg. Chem.* **332,** 35.
826. Feltz, A. (1965). *Z. anorg. Chem.* **334,** 242.
827. Feltz, A. (1965). *Z. anorg. Chem.* **335,** 304.
828. Feltz, A. (1965). *Z. anorg. Chem.* **338,** 147.
829. Feltz, A. (1965). *Z. anorg. Chem.* **338,** 155.
830. Feltz, A. (1967). *Z. anorg. Chem.* **354,** 225.
831. Feltz, A. (1967). *Z. anorg. Chem.* **355,** 120.

832. Feltz, A. (1968). *Z. anorg. Chem.* **358,** 21.
833. Fenn, R. H. (1969). *J. Chem. Soc.* (*A*), 1764.
834. Ferguson, G., Mercer, M. and Sharp, D. W. A. (1969). *J. Chem. Soc.* (*A*), 2415.
835. Fergusson, J. E., Penfold, B. R., Elder, M. and Robinson, B. H. (1965). *J. Chem. Soc.*, 5500.
836. Fergusson, J. E., Robinson, B. H. and Wilkins, C. J. (1967). *J. Chem. Soc.* (*A*), 486.
837. Fergusson, J. E., Wilkins, C. J. and Young, J. F. (1962). *J. Chem. Soc.*, 2136.
838. Ferretti, A., Rogers, D. B. and Goodenough, J. B. (1965). *J. Phys. and Chem. Solids* **26,** 2007.
839. Field, B. O. and Hardy, C. J. (1962). *Proc. Chem. Soc.*, 76.
840. Field, B. O. and Hardy, C. J. (1963). *Proc. Chem. Soc.*, 11.
841. Field, B. O. and Hardy, C. J. (1963). *J. Chem. Soc.*, 5278.
842. Field, B. O. and Hardy, C. J. (1964). *J. Chem. Soc.*, 4428.
843. Field, B. O. and Hardy, C. J. (1964). *Quart. Rev.* **18,** 361.
844. Field, R. A. and Kepert, D. L. (1967). *J. Less-Common Metals* **13,** 378.
845. Field, R. A., Kepert, D. L. and Taylor, D. (1970). *Inorg. Chim. Acta* **4,** 113.
846. Figgis, B. N., Lewis, J. and Mabbs, F. E. (1961). *J. Chem. Soc.*, 3138.
847. Figgis, B. N., Lewis, J. and Mabbs, F. E. (1963). *J. Chem. Soc.*, 2473.
848. Figgis, B. N. and Martin, R. L. (1956). *J. Chem. Soc.*, 3837.
849. Figgis, B. N. and Robertson, G. B. (1965). *Nature, Lond.* **205,** 694.
850. Fischer, A. K. and Wilkinson, G. W. (1956). *J. Inorg. Nuclear Chem.* **2,** 149.
851. Fischer, E. O., Beckert, O., Hafner, W. and Stahl, H. O. (1955). *Z. Naturforsch.* **10b,** 598.
852. Fischer, E. O. and Fichtel, K. (1961). *Berichte* **94,** 1200.
853. Fischer, E. O., Fichtel, K. and Ofele, K. (1962). *Berichte* **95,** 249.
854. Fischer, E. O. and Hafner, W. (1953). *Z. Naturforsch.* **8b,** 444.
855. Fischer, E. O. and Hafner, W. (1954). *Z. Naturforsch.* **9b,** 503.
856. Fischer, E. O. and Hafner, W. (1955). *Z. Naturforsch.* **10b,** 140.
857. Fischer, E. O., Hafner, W. and Stahl, H. O. (1955). *Z. anorg. Chem.* **282,** 47.
858. Fischer, E. O. and Hristidu, Y. (1960). *Z. Naturforsch.* **15b,** 135.
859. Fischer, E. O. and Kogler, H. P. (1956). *Angew. Chem.* **68,** 462.
860. Fischer, E. O. and Kogler, H. P. (1957). *Berichte* **90,** 250.
861. Fischer, E. O. and Kogler, H. P. (1958). *Z. Naturforsch.* **13b,** 197.
862. Fischer, E. O., Kogler, H. P. and Kuzel, P. (1960). *Berichte* **93,** 3006.
863. Fischer, E. O. and Kohl, F. J. (1964). *Angew. Chem. Internat. Edn.* **3,** 134.
864. Fischer, E. O. and Kuzel, P. (1962). *Z. anorg. Chem.* **317,** 226.
865. Fischer, E. O. and Lochner, A. (1960). *Z. Naturforsch.* **15b,** 266.
866. Fischer, E. O., Louis, E. and Schneider, R. J. J. (1968). *Angew. Chem. Internat. Edn.* **7,** 136.

867. Fischer, E. O. and Maasbol, A. (1964). *Angew. Chem. Internat. Edn.* **3,** 580.
868. Fischer, E. O. and Moser, E. (1964). *J. Organometallic Chem.* **2,** 230; **5,** 63 (1966).
869. Fischer, E. O. and Ofele, K. (1960). *Berichte* **93,** 1156.
870. Fischer, E. O. and Palm, C. (1958). *Berichte* **91,** 1725.
871. Fischer, E. O. and Rohrscheid, F. (1966). *J. Organometallic Chem.* **6,** 53.
872. Fischer, E. O., Scherer, F. and Stahl, H. O. (1960). *Berichte* **93,** 2065.
873. Fischer, E. O. and Schneider, R. J. J. (1967). *Angew. Chem. Internat. Edn.* **6,** 569.
874. Fischer, E. O. and Treiber, A. (1961). *Berichte* **94,** 2193.
875. Fischer, E. O., Ulm, K. and Kuzel, P. (1963). *Z. anorg. Chem.* **319,** 253.
876. Fischer, E. O. and Vigoureux, S. (1958). *Berichte* **91,** 1342.
877. Fischer, E. O. and Vigoureux, S. (1958). *Berichte* **91,** 2205.
878. Fischer, J., Elchinger, R. and Weiss, R. (1967). *Chem. Comm.*, 329.
879. Fischer, J., Keib, G. and Weiss, R. (1967). *Acta Cryst.* **22,** 338.
880. Fischer, J. and Weiss, R. (1967). *Chem. Comm.*, 328.
881. Fischer, J. and Weiss, R. (1968). *Chem. Comm.*, 1137.
882. Fischer, W. and Gewehr, R. (1935). *Z. anorg. Chem.* **222,** 309.
883. Fitzwater, D. R. and Rundle, R. E. (1959). *Z. Krist.* **112,** 362.
884. Fleischer, E. and Hawkinson, S. (1967). *J. Amer. Chem. Soc.* **89,** 720.
885. Fleming, P. B., Dougherty, T. A. and McCarley, R. E. (1967). *J. Amer. Chem. Soc.* **89,** 159.
886. Fleming, P. B., Mueller, L. A. and McCarley, R. E. (1967). *Inorg. Chem.* **6,** 1.
887. Flesch, G. D. and Svec, H. J. (1958). *J. Amer. Chem. Soc.* **80,** 3189.
888. Flood, M. T., Barraclough, C. G. and Gray, H. B. (1969). *Inorg. Chem.* **8,** 1855.
889. Flood, M. T., Marsh, R. E. and Gray, H. B. (1969). *J. Amer. Chem. Soc.* **91,** 193.
890. Flood, M. T., Marsh, R. E. and Gray, H. B. (1969). *J. Amer. Chem. Soc.* **91,** 193.
891. Florian, L. R. and Corey, E. R. (1968). *Inorg. Chem.* **7,** 722.
892. Flynn, C. M. and Pope, M. T. (1970). *J. Amer. Chem. Soc.* **92,** 85.
893. Flynn, C. M. and Stucky, G. D. (1969). *Inorg. Chem.* **8,** 178.
894. Flynn, C. M. and Stucky, G. D. (1969). *Inorg. Chem.* **8,** 332.
895. Flynn, C. M. and Stucky, G. D. (1969). *Inorg. Chem.* **8,** 335.
896. Fordyce, J. S. and Baum, R. L. (1966). *J. Chem. Phys.* **44,** 1159.
897. Foex, M. and Loriers, J. (1948). *Compt. rend.* **226,** 901.
898. Foex, M. and Wucher, J. (1955). *Compt. rend.* **241,** 184.
899. Forster, E., Albrecht, G., Durselen, W. and Kurras, E. (1969). *J. Organometallic Chem.* **19,** 215.
900. Forster, A., Cundy, C. S., Green, M. and Stone, F. G. A. (1966). *Inorg. Nuclear Chem. Letters* **2,** 233.
901. Fowles, G. W. A. and Frost, J. L. (1966). *Chem. Comm.*, 252.
902. Fowles, G. W. A. and Frost, J. L. (1966). *J. Chem. Soc.* (*A*), 1631.

903. Fowles, G. W. A. and Frost, J. L. (1967). *J. Chem. Soc. (A)*, 671.
904. Fowles, G. W. A. and Greene, P. T. (1966). *Chem. Comm.*, 784.
905. Fowles, G. W. A. and Greene, P. T. (1967). *J. Chem. Soc. (A)*, 1869.
906. Fowles, G. W. A., Greene, P. T. and Lester, T. E. (1967). *J. Inorg. Nuclear Chem.* **29,** 2365.
907. Fowles, G. W. A., Greene, P. T. and Wood, J. S. (1967). *Chem. Comm.*, 971.
908. Fowles, G. W. A. and Hoodless, R. A. (1963). *J. Chem. Soc.*, 33.
909. Fowles, G. W. A., Hoodless, R. A. and Walton, R. A. (1963). *J. Chem. Soc.*, 5873.
910. Fowles, G. W. A., Hoodless, R. A. and Walton, R. A. (1965). *J. Inorg. Nuclear Chem.* **27,** 391.
911. Fowles, G. W. A. and Jenkins, D. K. (1964). *Inorg. Chem.* **3,** 257.
912. Fowles, G. W. A., Lanigan, P. G. and Nicholls, D. (1961). *Chem. and Ind.*, 1167.
913. Fowles, G. W. A. and Lanigan, P. G. (1964). *J. Less-Common Metals* **6,** 396.
914. Fowles, G. W. A. and Lester, T. E. (1967). *Chem. Comm.*, 47. Fowles, G. W. A., Lester, T. E. and Walton, R. A. (1968). *J. Chem. Soc. (A)*, 1081.
915. Fowles, G. W. A. and Lester, T. E. (1968). *J. Chem. Soc. (A)*, 1180.
916. Fowles, G. W. A., Lester, T. E. and Russ, B. J. (1968). *J. Chem. Soc. (A)*, 805.
917. Fowles, G. W. A., Lester, T. E. and Wood, J. S. (1969). *J. Inorg. Nuclear* Chem. **31,** 657.
918. Fowles, G. W. A., Lester, T. E. and Walton, R. A. (1968). *J. Chem. Soc. (A)*, 198.
919. Fowles, G. W. A., Lewis, D. F. and Walton, R. A. (1968). *J. Chem. Soc. (A)*, 1468.
920. Fowles, G. W. A. and McGregor, W. R. (1958), *J. Chem. Soc.*, 136.
921. Fowles, G. W. A. and Nicholls, D. (1958). *J. Chem. Soc.*, 1687.
922. Fowles, G. W. A. and Nicholls, D. (1959). *J. Chem. Soc.*, 990.
923. Fowles, G. W. A. and Nicholls, D. (1961). *J. Chem. Soc.*, 95.
924. Fowles, G. W. A. and Nicholls, D. (1961). *J. Inorg. Nuclear Chem.* **18,** 130.
925. Fowles, G. W. A. and Osburne, B. P. (1959). *J. Chem. Soc.*, 2275.
926. Fowles, G. W. A. and Pleass, C. M. (1957). *J. Chem. Soc.*, 1674.
927. Fowles, G. W. A. and Pleass, C. M. (1957). *J. Chem. Soc.*, 2078.
928. Fowles, G. W. A. and Pollard, F. H. (1952). *J. Chem. Soc.*, 4938.
929. Fowles, G. W. A. and Pollard, F. H. (1953). *J. Chem. Soc.*, 2588.
930. Fowles, G. W. A. and Pollard, F. H. (1953). *J. Chem. Soc.*, 4128.
931. Fowles, G. W. A. and Russ, B. J. (1967). *J. Chem. Soc. (A)*, 517.
932. Fowles, G. W. A., Russ, B. J. and Willey, G. R. (1967). *Chem. Comm.*, 646. Fowles, G. W. A. and Willey, G. R. (1968). *J. Chem. Soc. (A)*, 1435.
933. Fowles, G. W. A., Tidmarsh, D. J. and Walton, R. A. (1969). *Inorg. Chem.* **8,** 631.
934. Fowles, G. W. A., Tidmarsh, D. J. and Walton, R. A. (1969). *J. Chem. Soc. (A)*, 1546.

935. Fowles, G. W. A., Tidmarsh, D. J. and Walton, R. A. (1969). *J. Inorg. Nuclear Chem.* **31,** 2373.
936. Fowles, G. W. A. and Walton, R. A. (1963). *J. Less-Common Metals* **5,** 510.
937. Fowles, G. W. A. and Walton, R. A. (1964). *J. Chem. Soc.*, 2840.
938. Fowles, G. W. A. and Walton, R. A. (1964). *J. Chem. Soc.*, 4330.
939. Fowles, G. W. A. and Walton, R. A. (1964). *J. Chem. Soc.*, 4953.
940. Fowles, G. W. A. and Walton, R. A. (1965). *J. Inorg. Nuclear Chem.* **27,** 735.
941. Fowles, G. W. A. and Walton, R. A. (1965). *J. Less-Common Metals* **9,** 457.
942. Frazer, M. J., Gerrard, W. and Twaits, R. (1963). *J. Inorg. Nuclear Chem.* **25,** 637.
943. Frazer, M. J. and Goffer, Z. (1966). *J. Inorg. Nuclear Chem.* **28,** 2410.
944. Frazer, M. J. and Goffer, Z. (1966). *J. Chem. Soc.* (*A*), 544.
945. Frazer, M. J. and Rimmer, B. (1968). *J. Chem. Soc.* (*A*), 2273.
946. Frere, P. and Michel, A. (1961). *Compt. rend.* **252,** 740; *Ann. Chim.* (*France*) **7,** 85.
947. Frevel, L. K. and Rinn, H. W. (1956). *Acta Cryst.* **9,** 626.
948. Frost, G. H., Hart, F. A., Heath, C. and Hursthouse, M. B. *Chem. Comm.*, 1421.
949. Fuchs, R. (1965). *J. Chem. Phys.* **42,** 3781.
950. Funk, H. and Baumann, W. (1937). *Z. anorg. Chem.* **231,** 264.
951. Funk, H. and Bohland, H. (1962). *Z. anorg. Chem.* **318,** 169.
952. Funk, H. and Bohland, H. (1963). *Z. anorg. Chem.* **324,** 168.
953. Funk, H. and Hanke, W. (1959). *Angew. Chem.* **71,** 408; *Z. anorg. Chem.* **307,** 157 (1961).
954. Funk, H., Hesselbarth, M. and Schmeil, F. (1962). *Z. anorg. Chem.* **318,** 318.
955. Funk, H. and Masthoff, A. (1956). *J. prakt. Chem.* **4,** 35.
956. Funk, H. and Mohaupt, G. (1962). *Z. anorg. Chem.* **315,** 204.
957. Funk, H., Mohaupt, G. and Paul, A. (1959). *Z. anorg. Chem.* **302,** 199.
958. Funk, H. and Naumann, H. (1966). *Z. anorg. Chem.* **343,** 294.
959. Funk, H. and Scharer, H. (1960). *Z. anorg. Chem.* **306,** 203.
960. Funk, H., Schmeil, F. and Scholz, H. (1961). *Z. anorg. Chem.* **310,** 86.
961. Funk, H. and Weiss, W. (1958). *Z. anorg. Chem.* **295,** 327.
962. Funk, H., Weiss, W. and Mohaupt, G. (1960). *Z. anorg. Chem.* **304,** 238.
963. Funk, H., Weiss, W. and Roethe, K. P. (1959). *Z. anorg. Chem.* **301,** 271.
964. Funk, H., Weiss, W. and Zeising, M. (1958). *Z. anorg. Chem.* **296,** 36.
965. Furlani, C., Porta, P., Sgamellotti, A. and Tomlinson, A. A. G. (1969). *Chem. Comm.*, 1046.
966. Furlani, C. and Zinato, E. (1967). *Z. anorg. Chem.* **351,** 210.
967. Furman, S. S. and Garner, C. S. (1951). *J. Amer. Chem. Soc.* **73,** 4528.
968. Furman, N. H. and Miller, C. O. (1950). *Inorg. Synth.* **3,** 160.
969. Gado, P., Holmberg, B. and Magneli, A. (1965). *Acta Chem. Scand.* **19,** 2010.

970. Galasso, F. and Darby, W. (1962). *J. Phys. Chem.* **66,** 1318.
971. Gal'perin, E. L. and Sandler, R. A. (1962). *Soviet Phys. Cryst.* **7,** 169.
972. Ganorka, M. C. and Stiddard, M. H. B. (1965). *Chem. Comm.*, 22.
973. Ganorka, M. C. and Stiddard, M. H. B. (1965). *J. Chem. Soc.*, 3494.
974. Gardner, H. J. (1964). *Nature, Lond.* **204,** 282.
975. Gardner, W. R. and Danielson, G. C. (1954). *Phys. Rev.* **93,** 46.
976. Garner, C. D., Sutton, D. and Wallwork, S. C. (1967). *J. Chem. Soc.* (*A*), 1949.
977. Garner, C. D. and Wallwork, S. C. (1966). *J. Chem. Soc.*, (*A*), 1496.
978. Garner, C. D. and Wallwork, S. C. (1969). *Chem. Comm.*, 108.
979. Garside, J. H. (1965). *J. Chem. Soc.*, 6634.
980. Gatehouse, B. M. and Leverett, P. (1968). *Chem. Comm.*, 901.
981. Gatehouse, B. M. and Leverett, P. (1969). *J. Chem. Soc.* (*A*), 849.
982. Gatehouse, B. M. and Leverett, P. (1969). *J. Chem. Soc.* (*A*), 1857.
983. Gatehouse, B. M. and Wadsley, A. D. (1964). *Acta Cryst.* **17,** 1545.
984. Gaunt, J. (1963). *Trans. Faraday Soc.* **49,** 1122.
985. Gaunt, J. and Ainscough, J. B. (1957). *Spectrochim. Acta* **10,** 52.
986. Gavin, R. M. and Bartell, L. S. (1968). *J. Chem. Phys.* **48,** 2460.
987. Geichman, J. T., Smith, E. A. and Ogle, P. R. (1963). *Inorg. Chem.* **2,** 1012.
988. Geichman, J. R., Smith, E. A., Trond, S. S. and Ogle, P. R. (1962) *Inorg. Chem.* **1,** 661.
989. Gerloch, M., McKenzie, E. D. and Towl, A. D. C. (1968). *Nature, Lond.* **220,** 906.
990. Gerloch, M. and Mason, R. (1965). *J. Chem. Soc.*, 296.
991. Gerstein, B. C., Thomas, L. D. and Silver, D. M. (1967). *J. Chem. Phys.* **46,** 4288.
992. Giannini, U. and Zucchini, U. (1968). *Chem. Comm.*, 940.
993. Giddings, S. A. (1964). *Inorg. Chem.* **3,** 684.
994. Giddings, S. A. (1967). *Inorg. Chem.* **6,** 849.
995. Gier, T. E., Pease, D. C., Sleight, A. W. and Bither, T. A. (1968). *Inorg. Chem.* **7,** 1646.
996. Giggenbach, W. and Brubaker, C. H. (1968). *Inorg. Chem.* **7,** 129.
997. Giggenbach, W. and Brubaker, C. H. (1969). *Inorg. Chem.* **8,** 1131.
998. de Gil, E. R. and Dahl, L. F. (1969). *J. Amer. Chem. Soc.* **91,** 3751.
999. Gillespie, R. J. (1960). *Canad. J. Chem.* **38,** 818.
1000. Gillis, E. (1964). *Compt. rend.* **258,** 4765.
1001. Gimblett, F. G. R. (1963). "Inorganic Polymer Chemistry". Butterworth, London.
1002. Ginsberg, A. P. (1964). *Inorg. Chem.* **3,** 567.
1003. Ginsberg, A. P., Koubek, E. and Williams, H. J. (1966). *Inorg. Chem.* **5,** 1656.
1004. Ginsberg, A. P. and Robin, M. B. (1963). *Inorg. Chem.* **2,** 817.
1005. Glemser, O. and Baeckmann, A. (1962). *Z. anorg. Chem.* **316,** 105.
1006. Glemser, O. and Holtje, W. (1965). *Z. Naturforsch.* **20b,** 398.
1007. Glemser, O. and Holtje, W. (1965). *Z. Naturforsch.* **20b,** 492.

1008. Glemser, O., Holtje, W. and Stockburger, M. (1968). *Z. Naturforsch.* **23b,** 1137.
1009. Glemser, O. and Holznagel, W. (1960). *Angew. Chem.* **72,** 918.
1010. Glemser, O., Holznagel, W., Holtje, W. and Schwarzmann, E. (1965). *Z. Naturforsch.* **20b,** 725.
1011. Glemser, O. and Preisler, E. (1960). *Z. anorg. Chem.* **303,** 303.
1012. Glemser, O., Roesky, H. and Hellberg, K. H. (1963). *Angew. Chem. Internat. Edn.* **2,** 266.
1013. Glen, G. L., Silverton, J. V. and Hoard, J. L. (1963). *Inorg. Chem.* **2,** 250.
1014. Gliemann, G. (1962). *Theor. Chim. Acta* **1,** 14.
1015. Gliemann, G. (1966). *Theor. Chim. Acta* **4,** 332.
1016. Gloor, M. and Wieland, K. (1961). *Helv. Chim. Acta* **44,** 1098.
1017. Glukhov, I. A., Eliseev, S. S. and Igamberdyeva, L. D. (1968). *Russ. J. Inorg. Chem.* **13,** 190.
1018. Glushkova, M. A., Ershova, M. M. and Buslaev, Yu. A. (1965). *Russ. J. Inorg. Chem.* **10,** 1060.
1019. Glushkova, M. A., Ershova, M. M. and Buslaev, Yu. A. (1965). *Russ. J. Inorg. Chem.* **10,** 1290.
1020. Glushkova, M. A. and Evteeva, M. M. (1961). *Russ. J. Inorg. Chem.* **6,** 9.
1021. Golding, R. M. and Carrington, A. (1962). *Mol. Phys.* **5,** 377.
1022. Golebiewski, A. and Kowalski, H. (1968). *Theor. Chim. Acta* **12,** 293.
1023. Golibersuch, E. W. and Young, R. C. (1949). *J. Amer. Chem. Soc.* **71,** 2402.
1024. Golub, A. M. and Sych, A. M. (1964). *Russ. J. Inorg. Chem.* **9,** 593.
1025. Goodenough, J. B. (1960). *Phys. Rev.* **117,** 1442.
1026. Goodenough, J. B. (1965). *Bull. Soc. chim. France*, 1200.
1027. Goodenough, J. B. (1963). "Magnetism and the Chemical Bond". Interscience, New York.
1028. Goodgame, D. M. L., Hill, N. J., Marsham, D. F., Skapski, A. C., Smart, M. L. and Troughton, P. G. H. (1969). *Chem. Comm.*, 629.
1029. Goodman, G. (1962). *Phys. Rev. Letters* **9,** 305.
1030. Gorbunova, Yu. E., Kuznetsov, V. G. and Kovaleva, E. S. (1968). *J. Struct. Chem.* **9,** 815.
1031. Gore, E., Green, M. L. H., Harriss, M. G., Lindsell, W. E. and Shaw, H. (1969). *J. Chem. Soc. (A)*, 1981.
1032. Goroshchenko, Ya. G. and Andreeva, M. I. (1965). *Russ. J. Inorg. Chem.* **10,** 517.
1033. Goroshchenko, Ya. G. and Andreeva, M. I. (1966). *Russ. J. Inorg. Chem.* **11,** 1197.
1034. Goroshchenko, Ya. G. and Spasibenko, T. P. (1965). *Russ. J. Inorg. Chem.* **10,** 1173.
1035. Goroshchenko, Ya. G. and Spasibenko, T. P. (1967). *Russ. J. Inorg. Chem.* **12,** 156.
1036. Gorsich, R. D. (1958). *J. Amer. Chem. Soc.* **80,** 4744.
1037. Gortsema, F. P. and Didchenko, R. (1965). *Inorg. Chem.* **4,** 182.

1038. Gracey, D. E. F., Jackson, W. R., Jennings, W. B., Rennison, S. C. and Spratt, R. (1966). *Chem. Comm.*, 231.
1039. Graddon, D. P. (1962). *Nature, Lond.* **195,** 891.
1040. Graham, W. A. G. (1968). *Inorg. Chem.* **7,** 315.
1041. Graham, A. J. and Fenn, R. H. (1969). *J. Organometallic Chem.* **17,** 405.
1042. Graham, J. and Wadsley, A. D. (1961). *Acta Cryst.* **14,** 379.
1043. Graham, J. and Wadsley, A. D. (1966). *Acta Cryst.* **20,** 93.
1044. Grandjean, D. and Weiss, R. (1965). *Compt. rend.* **261,** 448.
1045. Graven, W. M. and Peterson, R. V. (1969). *J. Inorg. Nuclear Chem.* **31,** 1743.
1046. Gray, D. R. and Brubaker, C. H. (1969). *Chem. Comm.*, 1239.
1047. Gray, H. B. and Hare, C. R. (1962). *Inorg. Chem.* **1,** 363.
1048. Grdenic, D. and Matkovic, B. (1959). *Acta Cryst.* **12,** 817.
1049. Green, M. L. H., Ishaq, M. and Whiteley, R. N. (1967). *J. Chem. Soc.* (*A*), 1508.
1050. Green, M. L. H. and Lindsell, W. E. (1967). *J. Chem. Soc.* (*A*), 686.
1051. Green, M. L. H. and Lindsell, W. E. (1967). *J. Chem. Soc.* (*A*), 1455.
1052. Green, M. L. H., McCleverty, J. A. M., Pratt, L. and Wilkinson, G. (1961). *J. Chem. Soc.*, 4854.
1053. Green, M. L. H. and Stear, A. N. (1963). *J. Organometallic Chem.* **1,** 230.
1054. Greenberg, E., Natke, C. A. and Hubbard, W. N. (1965). *J. Phys. Chem.* **69,** 2089.
1055. Greene, P. T. and Orioli, P. L. (1969). *J. Chem. Soc.* (*A*), 1621.
1056. Greenwood, N. N., Parish, P. V. and, Thornton, P. (1966). *J. Chem. Soc.* (*A*), 320.
1057. Greiner, J. D., Shanks, H. R. and Wallace, D. C. (1962). *J. Chem. Phys.* **36,** 772.
1058. Grey, I. E. and Smith, P. W. (1968). *Chem. Comm.*, 1525.
1059. Grey, I. E. and Smith, P. W. (1969). *Austral. J. Chem.* **22,** 121.
1060. Griffith, J. S. (1956). *J. Inorg. Nuclear Chem.* **3,** 15.
1061. Griffith, W. P. (1964). *J. Chem. Soc.*, 5248.
1062. Griffith, W. P. (1967). *J. Chem. Soc.* (*A*), 905.
1063. Griffith, W. P., Lewis, J. and Wilkinson, G. (1959). *J. Chem.* Soc., 872.
1064. Griffith, W. P., Lewis, J. and Wilkinson, G. (1959). *J. Chem. Soc.*, 1632.
1065. Griffith, W. P. and Wickins, T. D. (1966). *J. Chem. Soc.* (*A*), 1087.
1066. Griffith, W. P. and Wickins, T. D. (1967). *J. Chem. Soc.* (*A*), 590.
1067. Griffith, W. P. and Wickins, T. D. (1967). *J. Chem. Soc.* (*A*), 675.
1068. Griffith, W. P. and Wickins, T. D. (1968). *J. Chem. Soc.* (*A*), 397.
1069. Griffith, W. P. and Wickins, T. D. (1968). *J. Chem. Soc.* (*A*), 400.
1070. Griffiths, J. E. (1968). *J. Chem. Phys.* **49,** 642.
1071. Griffiths, J. H. E., Owen, J. and Ward, I. M. (1953). *Proc. Roy. Soc.* **A219,** 526.
1072. Grigor'ev, A. I., Tam Wen-hsia, Kolli, I. D. and Spitsyn, V. I. (1964). *Russ. J. Inorg. Chem.* **9,** 1397.
1073. Grossman, G., Proskurenko, O. W. and Arija, S. M. (1960). *Z. anorg. Chem.* **305,** 121.

1074. Grubb, E. L. and Belford, R. L. (1963). *J. Chem. Phys.* **39,** 244.
1075. Gruehn, R. (1966). *J. Less-Common Metals* **11,** 119.
1076. Gruehn, R. (1967). *Naturwiss*, **54,** 645.
1077. Gruen, D. M. and McBeth, R. L. (1962). *J. Phys. Chem.* **66,** 57.
1078. Gruehn, R. and Norin, R. (1967). *Z. anorg. Chem.* **355,** 176.
1079. Gruehn, R. and Norin, R. (1969). *Z. anorg. Chem.* **367,** 209.
1080. Guenther, K. F. (1964). *Inorg. Chem.* **3,** 923.
1081. Guenther, K. F. (1964). *Inorg. Chem.* **3,** 1788.
1082. Guerchais, J. E. and Rohmer, R. (1964). *Compt. rend.* **259,** 1135.
1083. Guerchais, J. E. and Spinner, B. (1965). *Bull. Soc. chim. France*, 58.
1084. Guerchais, J. E. and Spinner, B. (1965). *Bull. Soc. chim. France*, 1122.
1085. Guerchais, J. E., Spinner, B. and Rohmer, R. (1965). *Bull. Soc. chim. France*, 55.
1086. Guggenberger, L. J. and Sleight, A. W. (1969). *Inorg. Chem.* **8,** 2041.
1087. Guibe, L. and Souchay, P. (1957). *J. Chim. phys.* **54,** 684.
1088. Guibe, L. and Souchay, P. (1957). *Compt. rend.* **244,** 780.
1089. Guillaud, C., Michel, A., Bernard, J. and Fallot, M. (1944). *Compt. rend.* **219,** 58.
1090. Gunther, H. and Grimme, W. (1966). *Angew. Chem. Internat. Edn.* **5,** 1043.
1091. Gusev, A. I. and Struchkov, Yu. T. (1969). *J. Struct. Chem.* **10,** 426.
1092. Gut, R. and Schwarzenbach, G. (1959). *Helv. Chim. Acta* **42,** 2156.
1093. Gutmann, V. (1951). *Z. anorg. Chem.* **264,** 151.
1094. Gutmann, V. (1951). *Z. anorg. Chem.* **266,** 331.
1095. Gutmann, V. (1951). *Monatsh.* **82,** 156.
1096. Gutmann, V. (1954). *Monatsh.* **85,** 1077.
1097. Gutmann, V. (1966). *Angew. Chem. Internat. Edn.* **5,** 142.
1098. Gutmann, V. and Emeleus, H. J. (1950). *J. Chem. Soc.* 1046.
1099. Gutmann, V. and Jack, K. H. (1951). *Acta Cryst.* **4,** 244.
1100. Gutmann, V. and Mairinger, F. (1958). *Monatsh.* **89,** 724.
1101. Gutmann, V., Nowotny, H. and Ofner, G. (1955). *Z. anorg. Chem.* **278,** 78.
1102. Gutmann, V. and Tannenberger, H. (1956). *Monatsh.* **87,** 421.
1103. Gutmann, V. and Tannenberger, H. (1956). *Monatsh.* **87,** 769.
1104. Gutmann, V. and Tannenberger, H. (1957). *Monatsh.* **88,** 292.
1105. Haaland, A. (1965). *Acta Chem. Scand.* **19,** 41.
1106. Haarhoff, P. C. and Pistorius, C. W. F. T. (1959). *Z. Naturforsch.* **14a,** 972.
1107. Haas, T. E. and Marram, E. P. (1965). *J. Chem. Phys.* **43,** 3985.
1108. Haas, H. and Skeline, R. K. (1966). *J. Amer. Chem. Soc.* **88,** 3219.
1109. Haendler, H. M., Bartram, S. F., Becker, R. S., Bernard, W. J. and Bukaday, S. W. (1954). *J. Amer. Chem. Soc.* **76,** 2177.
1110. Hagg, G. (1935). *Z. phys. Chem.* **29,** 192; *Nature, Lond.* **135,** 874.
1111. Hagg, G. and Sucksdorff, I. (1933). *Z. phys. Chem.* **22,** 444.
1112. Hagihara, N. and Yamazaki, H. (1959). *J. Amer. Chem. Soc.* **81,** 3160.
1113. Hahn, H. and Becker, W. (1962). *Naturwiss.* **49,** 513.

1114. Hahn, H. and Becker, W. (1963). *Naturwiss.* **50,** 402.
1115. Hahn, H. and Hahn, F. (1962). *Naturwiss.* **49,** 539.
1116. Hahn, H. and Schmidt, G. (1962). *Naturwiss.* **49,** 513.
1117. Haight, G. P. (1962). *J. Inorg. Nuclear Chem.* **24,** 663.
1118. Haight, G. P., Richardson, D. C. and Colburn, N. H. (1964). *Inorg. Chem.* **3,** 1777.
1119. Haines, R. J., Nyholm, R. S. and Stiddard, M. H. B. (1966). *J. Chem. Soc.* (*A*), 1606.
1120. Haines, R. J., Nyholm, R. S. and Stiddard, M. H. B. (1967). *J. Chem. Soc.* (*A*), 94.
1121. Haines, R. J., Nyholm, R. S. and Stiddard, M. H. B. (1968). *J. Chem. Soc.* (*A*), 43.
1122. Haines, R. J., Nyholm, R. S. and Stiddard, M. H. B. (1968). *J. Chem. Soc.* (*A*), 46.
1123. Hall, D. and Holland, R. V. (1969). *Inorg. Chim. Acta* **3,** 235.
1124. Hall, D., Rae, A. D. and Waters, T. N. (1965). *Acta Cryst.* **19,** 389.
1125. Hall, D., Rickard, C. E. F. and Waters, T. N. (1965). *Nature, Lond.* **207,** 405.
1126. Hamilton, P. M., McBeth, R., Bekebrede, W. and Sisler, H. H. (1953). *J. Amer. Chem. Soc.* **75,** 2881.
1127. Hampson, G. C. and Pauling, L. (1938). *J. Amer. Chem. Soc.* **60,** 2702.
1128. Handy, L. B., Treichel, P. M. and Dahl, L. F. (1966). *J. Amer. Chem. Soc.* **88,** 366.
1129. Hanic, F. and Mills, O. S. (1968). *J. Organometallic Chem.* **11,** 151.
1130. Hardy, C. J., Field, B. O. and Scargill, D. (1966). *J. Inorg. Nuclear Chem.* **28,** 2408.
1131. Hare, C. R., Bernal, I. and Gray, H. B. (1962). *Inorg. Chem.* **1,** 831.
1132. Hargreaves, G. B. and Peacock, R. D. (1957). *J. Chem. Soc.*, 4212.
1133. Hargreaves, G. B. and Peacock, R. D. (1958). *J. Chem. Soc.*, 2170.
1134. Hargreaves, G. B. and Peacock, R. D. (1958). *J. Chem. Soc.*, 3776.
1135. Hargreaves, G. B. and Peacock, R. D. (1958). *J. Chem. Soc.*, 4390.
1136. Hargreaves, G. B. and Peacock, R. D. (1960). *J. Chem. Soc.*, 1099.
1137. Harned, H. S. (1913). *J. Amer. Chem. Soc.* **35,** 1078.
1138. Harned, H. S., Pauling, C. and Corey, R. B. (1960). *J. Amer. Chem. Soc.* **82,** 4815.
1139. Harriss, M. G., Green, M. L. H. and Lindsell, W. E. (1969). *J. Chem. Soc.* (*A*), 1453.
1140. Harris, W. S. (1958). *U.S. At. Energy Comm.* UCRL-8381.
1141. Hart, W. 't and Meyer, G. (1965). *Rec. Trav. chim.* **84,** 1155.
1142. Hart, W. 't and Meyer, G. (1967). *Rec. Trav. chim.* **86,** 85.
1143. Hartley, D. and Ware, M. J. (1967). *Chem. Comm.*, 912.
1144. Hartmann, H., Schlafer, H. L. and Hanson, K. H. (1956). *Z. anorg. Chem.* **284,** 153; **289,** 40 (1957).
1145. Hartman, K. O. and Miller, F. A. (1968). *Spectrochim. Acta* **24A,** 669.
1146. Hartmann, H. and Schmidt, H.-J. (1957). *Z. phys. Chem.* **11,** 234.
1147. Hathaway, B. J. and Holah, D. G. (1965). *J. Chem. Soc.*, 537.

1148. Hayden, J. L. and Wentworth, R. A. D. (1968). *J. Amer. Chem. Soc.* **90,** 5291.
1149. Hayek, E., Puschmann, J. and Czaloun, A. (1954). *Monatsh.* **85,** 359.
1150. Hayes, R. G. (1966). *J. Chem. Phys.* **44,** 2210.
1151. Haynes, L. V. and Sawyer, D. T. (1967). *Inorg. Chem.* **6,** 2146.
1152. Hayter, R. G. (1963). *Inorg. Chem.* **2,** 1031.
1153. Hayter, R. G. (1964). *Inorg. Chem.* **3,** 711.
1154. Hayter, R. G. (1966). *J. Amer. Chem. Soc.* **88,** 4376.
1155. Hazell, A. C. (1963). *J. Chem. Soc.*, 5745.
1156. Hazell, R. G. (1969). *Acta Cryst.* **25,** S116.
1157. Hecht, H., Jander, G. and Schlapmann, H. (1947). *Z. anorg. Chem.* **254,** 260.
1158. Heiber, W., Peterhand, J. and Winter, E. (1961). *Berichte* **94,** 2572.
1159. Heiber, W. and Ranberg, E. (1935). *Z. anorg. Chem.* **221,** 321.
1160. Heiber, W. and Winter, E. (1964). *Berichte* **97,** 1037.
1161. Heiber, W., Winter, E. and Schubert, E. (1962). *Berichte* **95,** 3070.
1162. Hein, F. (1931). *Z. anorg. Chem.* **201,** 314.
1163. Hein, F. and Herzog, S. (1952). *Z. anorg. Chem.* **267,** 337.
1164. Hein, F. and Schmiedeknecht, K. (1967). *Z. anorg. Chem.* **352,** 138.
1165. Hein, F. and Weiss, R. (1958). *Z. anorg. Chem.* **295,** 145.
1166. Heintz, E. A. (1963). *Nature, Lond.* **197,** 690.
1167. Heitner-Wirguin, C. and Cohen, R. (1965). *J. Inorg. Nuclear Chem.* **27,** 1989.
1168. Heitner-Wirguin, C. and Selbin, J. (1968). *J. Inorg. Nuclear Chem.* **30,** 3181.
1169. Heitner-Wirguin, C. and Selbin, J. (1969). *Israel J. Chem.* **7,** 27.
1170. Helmholtz, L. (1939). *J. Amer. Chem. Soc.* **61,** 1544.
1171. Helmholtz, L. and Foster, W. R. (1950). *J. Amer. Chem. Soc.* **72,** 4971.
1172. Hendricker, D. G., McCarley, R. E., King, R. W. and Verkade, J. G. (1966). *Inorg. Chem.* **5,** 639.
1173. Hengge, E. and Zimmermann, H. (1968). *Angew. Chem. Internat. Edn.* **7,** 142.
1174. Henrici-Olive, G. and Olive, S. (1968). *Angew. Chem. Internat. Edn.* **7,** 386.
1175. Herak, R. M., Malcic, S. S. and Manojlovic, L. M. (1965). *Acta Cryst.* **18,** 520.
1176. Herman, D. F. and Nelson, W. K. (1953). *J. Amer. Chem. Soc.* **75,** 3877.
1177. Herman, D. F. and Nelson, W. K. (1953). *J. Amer. Chem. Soc.* **75,** 3882.
1178. Herzog, S. (1956). *Naturwiss.* **43,** 35, 349; *Z. anorg. Chem.* **294,** 155 (1958).
1179. Herzog, S. and Kalies, W. (1964). *Z. anorg. Chem.* **329,** 83.
1180. Herzog, S. and Kalies, W. (1967). *Z. anorg. Chem.* **351,** 237.
1181. Herzog, S. and Renner, K. (1959). *Berichte* **92,** 872.
1182. Herzog, S., Renner, K. and Schon, W. (1957). *Z. Naturforsch.* **12b,** 809.
1183. Herzog, M. and Scavnicar, S. (1967). *Croat. Chem. Acta* **39,** 137.
1184. Herzog, S. and Schon, W. (1958). *Z. anorg. Chem.* **297,** 323.
1185. Herzog, S. and Schuster, R. (1962). *Z. Naturforsch.* **17b,** 62.

1186. Herzog, S. and Taube, R. (1960). *Z. anorg. Chem.* **306,** 159.
1187. Herzog, S. and Zuhlke, H. (1966). *Z. Chem.* **6,** 382, 434.
1188. Hess, H. and Hartung, H. (1966). *Z. anorg. Chem.* **344,** 157.
1189. Hewkins, J. D. and Griffith, W. P. (1966). *J. Chem. Soc.* (*A*), 472.
1190. Hidai, M., Tominari, K., Uchida, Y. and Misona, A. (1969). *Chem. Comm.*, 1392.
1191. Hietanen, S. and Sillen, L. G. (1964). *Acta Chem. Scand.* **18,** 1015.
1192. Hietanen, S. and Sillen, L. G. (1964). *Acta Chem. Scand.* **18,** 1018.
1193. Hihara, T., Murakami, M. and Hirahara, E. (1957). *J. Phys. Soc. Japan* **12,** 743.
1194. Hill, H. A. O. and Norgett, M. M. (1966). *J. Chem. Soc.* (*A*), 1476.
1195. Hirahara, E. and Murakami, M. (1958). *J. Phys. and Chem. Solids* **7,** 281.
1196. Hirone, T., Maeda, S. and Tsuya, N. (1954). *J. Phys. Soc. Japan* **9,** 503.
1197. Ho, R. K. Y., Livingstone, S. E. and Lockyer, T. N. (1966). *Austral. J. Chem.* **19,** 1179.
1198. Hoard, J. L. (1939). *J. Amer. Chem. Soc.* **61,** 1252.
1199. Hoard, J. L., Glen, G. L. and Silverton, J. V. (1961). *J. Amer. Chem. Soc.* **83,** 4293.
1200. Hoard, J. L. and Goldstein, L. (1935). *J. Chem. Phys.* **3,** 117.
1201. Hoard, J. L. and Goldstein, L. (1935). *J. Chem. Phys.* **3,** 199.
1202. Hoard, J. L., Hamor, T. A. and Glick, M. D. (1968). *J. Amer. Chem. Soc.* **90,** 3177.
1203. Hoard, J. L., Lee, B. and Lind, M. D. (1965). *J. Amer. Chem. Soc.* **87,** 1612.
1204. Hoard, J. L. and Martin, W. (1941). *J. Amer. Chem. Soc.* **63,** 11.
1205. Hoard, J. L., Martin, W. J., Smith, M. E. and Whitney, J. F. (1954). *J. Amer. Chem. Soc.* **76,** 3820.
1206. Hoard, J. L., Silverton, E. W. and Silverton, J. V. (1968). *J. Amer. Chem. Soc.* **90,** 2300.
1207. Hoard, J. L. and Silverton, J. V. (1963). *Inorg. Chem.* **2,** 235.
1208. Hoard, J. L., Willstadter, E. and Silverton, J. V. (1965). *J. Amer. Chem. Soc.* **87,** 1610.
1209. Hoekstra, H. R. and Katz, J. J. (1949). *J. Amer. Chem. Soc.* **71,** 2488.
1210. Holah, D. G. and Fackler, J. P. (1965). *Inorg. Chem.* **4,** 1112.
1211. Holah, D. G. and Fackler, J. P. (1965). *Inorg. Chem.* **4,** 1721.
1212. Holah, D. G. and Fackler, J. P. (1966). *Inorg. Chem.* **5,** 479.
1213. Holloway, J. H. and Knowles, J. G. (1969). *J. Chem. Soc.* (*A*), 756.
1214. Holm, R. H., King, R. B. and Stone, F. G. A. (1963). *Inorg. Chem.* **2,** 219.
1215. Hon, P.-K., Belford, R. L. and Pfluger, C. E. (1965). *J. Chem. Phys.* **43,** 1323.
1216. Hon, P.-K., Belford, R. L. and Pfluger, C. E. (1965). *J. Chem. Phys.* **43,** 3111.
1217. Horner, S. M., Clark, R. J. H., Crociani, B., Copley, D. B., Horner, W. W., Collier, F. N. and Tyree, S. Y. (1968). *Inorg. Chem.* **7,** 1859.
1218. Horner, S. M. and Tyree, S. Y. (1962). *Inorg. Chem.* **1,** 122.

1219. Horner, S. M. and Tyree, S. Y. (1962). *Inorg. Chem.* **1,** 947.
1220. Horner, S. M. and Tyree, S. Y. (1963). *Inorg. Chem.* **2,** 568.
1221. Horner, S. M. and Tyree, S. Y. (1964). *Inorg. Chem.* **3,** 1173.
1222. Horner, S. M. and Tyree, S. Y. (1965). *Inorg. Nuclear Chem. Letters* **1,** 43.
1223. Horner, S. M., Tyree, S. Y. and Venezky, D. L. (1962). *Inorg. Chem.* **1,** 844.
1224. Hoschek, E. and Klemm, W. (1939). *Z. anorg. Chem.* **242,** 63.
1225. Houk, L. W. and Dobson, G. R. (1966). *J. Chem. Soc.* (*A*), 317.
1226. Houk, L. W. and Dobson, G. R. (1966). *Inorg. Chem.* **5,** 2119.
1227. Hovey, R. J. and Martell, A. E. (1960). *J. Amer. Chem. Soc.* **82,** 2697.
1228. Hovey, R. J., O'Connell, J. J. and Martell, A. E. (1959). *J. Amer. Chem. Soc.* **81,** 3189.
1229. Howarth, O. W. and Richards, R. E. (1965). *J. Chem. Soc.*, 864.
1230. Howell, I. V. and Venanzi, L. M. (1964). *Inorg. Chim. Acta* **3,** 121.
1231. Howell, I. V., Venanzi, L. M. and Goodall, D. C. (1967). *J. Chem. Soc.* (*A*), 395.
1232. Hubel, W. and Merenyi, R. G. (1964). *J. Organometallic Chem.* **2,** 213.
1233. Huber, K. and Barnok, I. (1961). *Chimia* (*Switz.*) **15,** 365.
1234. Huber, K., Jost, E., Neuenschwander, E., Studer, M. and Roth, B. (1958). *Helv. Chim. Acta* **41,** 2411.
1235. Hughes, M. A. (1964). *J. Less-Common Metals* **6,** 232.
1236. Huibregtse, E. J., Barker, D. B. and Danielson, G. C. (1951). *Phys. Rev.* **84,** 142.
1237. Hull, C. G. and Stiddard, M. H. B. (1966). *J. Chem. Soc.* (*A*), 1633.
1238. Hull, C. G. and Stiddard, M. H. B. (1967). *J. Organometallic Chem.* **9,** 519.
1239. Hullen, A. (1964). *Naturwiss.* **51,** 508.
1240. Hummers, W. S., Tyree, S. Y. and Yolles, S. (1952). *J. Amer. Chem. Soc.* **74,** 139.
1241. Hunt, J. P. and Plane, R. A. (1954). *J. Amer. Chem. Soc.* **76,** 5960.
1242. Hunt, J. P. and Taube, H. (1950). *J. Chem. Phys.* **18,** 757; **19,** 603 (1951).
1243. Huss, E. and Klemm, W. (1950). *Z. anorg. Chem.* **262,** 25.
1244. Huttner, G., Fischer, E. O., Fischer, R. D., Carter, O. L., McPhail, A. T. and Sim, G. A. (1966). *J. Organometallic Chem.* **6,** 288.
1245. Huttner, G., Schelle, S. and Mills, O. S. (1969). *Angew. Chem. Internat. Edn.* **8,** 515.
1246. Ibers, J. A. (1963). *Nature, Lond.* **197,** 686.
1247. Ibers, J. A. (1964). *J. Chem. Phys.* **40,** 3129.
1248. Ibers, J. A. and Holm, C. H. (1957). *Acta Cryst.* **10,** 139.
1249. IJdo, D. J. W. (1960). Thesis, University of Leiden.
1250. Ikeda, S., Yamamoto, A., Kurita, S., Takahashi, K. and Watanabe, T. (1966). *Inorg. Chem.* **5,** 611.
1251. Il'inskii, A. L., Aslanov, L. A., Ivanov, V. I., Khalilov, A. D. and Petrukhin, O. M. (1969). *J. Struct. Chem.* **10,** 263.
1252. Illingsworth, J. W. and Keggin, J. F. (1935). *J. Chem. Soc.*, 575.

1253. Ilse, F. E. and Hartmann, H. (1951). *Z. phys. Chem.* **197,** 239.
1254. Iman, S. A. (1965). *Nature, Lond.* **206,** 1146.
1255. Iman, S. A. and Rao, B. R. (1964). *Naturwiss.* **51,** 263.
1256. Ingram, D. J. E. and Bennett, J. E. (1954). *J. Chem. Phys.* **22,** 1136; *Discuss. Faraday Soc.* **19,** 140.
1257. Ingri, N. and Brito, F. (1959). *Acta Chem. Scand.* **13,** 1971.
1258. Intorre, B. I. and Martell, A. E. (1960). *J. Amer. Chem. Soc.* **82,** 358; **83,** 3618 (1961).
1259. Iorns, T. V. and Stafford, F. E. (1966). *J. Amer. Chem. Soc.* **88,** 4819.
1260. Irving, R. J. and Steele, M. C. (1957). *Austral. J. Chem.* **10,** 490.
1261. Issleib, K. and Biermann, B. (1966). *Z. anorg. Chem.* **347,** 39.
1262. Issleib, K. and Bohn, G. (1959). *Z. anorg. Chem.* **301,** 188.
1263. Issleib, K. and Doll, G. (1960). *Z. anorg. Chem.* **305,** 1.
1264. Issleib, K. and Frohlich, H. O. (1959). *Z. anorg. Chem.* **298,** 84.
1265. Isupov, V. A., Agranovskaya, A. I. and Bryzhina, M. F. (1963). *Kristallografiya* **8,** 108.
1266. Jack, K. H. and Gutmann, V. *Acta Cryst.* **4,** 246.
1267. Jack, K. H. and Maitland, R. (1957). *Proc. Chem. Soc.*, 232.
1268. Jackson, W. R. and Jennings, W. B. (1966). *Chem. Comm.*, 824.
1269. Jackson, R. B. and Streib, W. E. To be published. Quoted in references 1148 and 1983.
1270. Jaffray, J. and Dumas, A. (1954). *J. recherches centre natl. recherche sci. Labs. Bellevue* (*Paris*) **5,** 360.
1271. Jahnberg, L. and Andersson, S. (1967). *Acta Chem. Scand.* **21,** 615.
1272. Jahr, K. F. and Fuchs, J. (1966). *Angew. Chem. Internat. Edn.* **5,** 689.
1273. Jahr, K. F., Fuchs, J. and Preuss, F. (1963). *Berichte* **96,** 558.
1274. Jahr, F. and Preuss, F. (1965). *Berichte* **98,** 3297.
1275. Jahr, K. F. and Schoepp, L. (1959). *Z. Naturforsch.* **14b,** 467.
1276. Jahr, K. F. and Schoepp, L. (1959). *Z. Naturforsch.* **14b,** 468.
1277. Jahr, K. F., Schoepp, L. and Fuchs, J. (1959). *Z. Naturforsch.* **14b,** 469.
1278. Jahr, K. F., Schroth, H. and Fuchs, J. (1963). *Z. Naturforsch.* **18b,** 1133.
1279. Jain, D. V. S. and Dogra, S. K. (1966). *J. Chem. Soc.* (*A*), 284.
1280. Jain, S. C. and Rivest, R. (1963). *Canad. J. Chem.* **41,** 2130.
1281. Jain, S. C. and Rivest, R. (1969). *J. Inorg. Nuclear Chem.* **31,** 399.
1282. Jakob, W. and Jakob, Z. (1962). *Roczniki Chem.* **36,** 601.
1283. Jakob, W., Ogorzalek, M. and Sikorski, H. (1961). *Roczniki Chem.* **35,** 3.
1284. James, B. D., Nanda, R. K. and Wallbridge, M. G. H. (1966). *J. Chem. Soc.* (*A*), 182.
1285. James, B. D., Nanda, R. K. and Wallbridge, M. G. H. (1967). *Inorg. Chem.* **6,** 1979.
1286. James, R. G. and Wardlaw, W. (1927). *J. Chem. Soc.*, 2145.
1287. Jander, G. and Ertel, D. (1960). *J. Inorg. Nuclear Chem.* **14,** 71.
1288. Jander, G. and Ertel, D. (1960). *J. Inorg. Nuclear Chem.* **14,** 77.
1289. Jander, G. and Ertel, D. (1960). *J. Inorg. Nuclear Chem.* **14,** 85.
1290. Jander, G. and Fiedler, B. (1961). *Z. anorg. Chem.* **308,** 155.
1291. Jarchow, O., Schroder, F. and Schulz, H. (1968). *Z. anorg. Chem.* **363,** 58.

1292. Jean, M. (1955). *Anal. Chim.* **37,** 125, 163; **44,** 195, 243 (1962).
1293. Jellinek, F. (1960). *Nature, Lond.* **187,** 871; *J. Organometallic Chem.* **1,** 43.
1294. Jellinic, I., Grdenic, D. and Bezjak, A. (1964). *Acta Cryst.* **17,** 758.
1295. Jezowska-Trzebiatowska, B. and Rudolf, M. (1967). *Roczniki Chem.* **41,** 1879.
1296. Jezowska-Trzebiatowska, B. and Rudolf, M. (1968). *Roczniki Chem.* **42,** 1221.
1297. Jin Sun Yoo, Griswold, E. and Kleinberg, J. (1965). *Inorg. Chem.* **4,** 365.
1298. Joesten, M. D. (1967). *Inorg. Chem.* **6,** 1598.
1299. Johannesen, R. B. (1960). *Inorg. Synth.* **7,** 119.
1300. Johannesen, R. B. and Krauss, H.-L. (1963). *Berichte* **97,** 2094.
1301. Johannesen, R. B., Candela, G. A. and Tung Tsang (1968). *J. Chem. Phys.* **48,** 5544.
1302. Johansson, G. (1962). *Acta Chem. Scand.* **16,** 403.
1303. Johansson, G. (1963). *Arkiv Kemi* **20,** 305, 321.
1304. Johansson, G. (1968). *Acta Chem. Scand.* **22,** 389.
1305. Johnson, B. F. G. (1967). *J. Chem. Soc.* (*A*), 475.
1306. Johnson, B. F. G., Al-Obaidi, K. H. and McCleverty, J. A. (1969). *J. Chem. Soc.* (*A*), 1668.
1307. Johnson, J. S. and Kraus, K. A. (1956). *J. Amer. Chem. Soc.* **78,** 3937.
1308. Johnson, J. S., Kraus, K. A. and Holmberg, R. W. (1956). *J. Amer. Chem. Soc.* **78,** 26.
1309. Johnson, M. P. and Shriver, D. F. (1966). *J. Amer. Chem. Soc.* **88,** 301.
1310. Johnson, R. L. and Siegel, B. (1969). *J. Inorg. Nuclear Chem.* **31,** 955.
1311. Johnson, S. A., Hunt, H. R. and Neumann, H. M. (1963). *Inorg. Chem.* **2,** 960.
1312. Jonassen, H. B. and Bailin, L. J. (1963). *Inorg. Synth.* **7,** 140.
1313. Jonassen, H. B., Cantor, S. and Tarsey, A. R. (1956). *J. Amer. Chem. Soc.* **78,** 271; *Rec. Trav. chim.* **75,** 609.
1314. Jonassen, H. B., Tarsey, A. R., Cantor, S. and Felfrich, G. F. (1957). *Inorg. Synth.* **5,** 139.
1315. Jones, M. M. (1954). *J. Amer. Chem. Soc.* **76,** 5995.
1316. Jones, W. H., Gabarty, E. A. and Barnes, R. G. (1962). *J. Chem. Phys.* **36,** 494.
1317. Jørgensen, C. K. (1957). *Acta Chem. Scand.* **11,** 73.
1318. Jørgensen, C. K. (1963). "Inorganic Complexes". Academic Press, London and New York.
1319. Joshi, K. K., Pauson, P. L. and Stubbs, W. H. (1963). *J. Organometallic Chem.* **1,** 53.
1320. Jotham, R. W. and Kettle, S. F. A. (1969). *Chem. Comm.*, 258; *J. Chem. Soc.* (*A*), 2816, 2821.
1321. Jowitt, R. N. and Mitchell, P. C. H. (1966). *Chem. Comm.*, 605; *J. Chem. Soc.* (*A*), 2632 (1969).
1322. Junkins, J. H., Farrar, R. L., Barber, E. J. and Bernhardt, H. A. (1952). *J. Amer. Chem. Soc.* **74,** 3464.
1323. Juvinall, G. L. (1964). *J. Amer. Chem. Soc.* **86,** 4202.

1324. Juza, D., Giegling, D. and Schafer, H. (1969). *Z. anorg. Chem.* **366,** 121.
1325. Juza, R. and Friedrichsen, H. (1964). *Z. anorg. Chem.* **332,** 173.
1326. Juza, R. and Heners, J. (1964). *Z. anorg. Chem.* **332,** 159.
1327. Kabanov, Y. Ya and Spitsyn, V. I. (1964). *Russ. J. Inorg. Chem.* **9,** 999.
1328. Kachi, S., Takada, T. and Kosuge, K. (1963). *J. Phys. Soc. Japan* **18,** 1839.
1329. Kakos, G. A. and Winter, G. (1968). *Austral. J. Chem.* **21,** 793.
1330. Kamigaichi, T., Hihara, T., Tazaki, H. and Hirahara, E. (1956). *J. Phys. Soc. Japan* **11,** 606, 1123.
1331. Karpinskaya, N. M. and Andreev, S. N. (1968). *Russ. J. Inorg. Chem.* **13,** 25.
1332. Kasenally, A. S., Nyholm, R. S., O'Brien, R. J. and Stiddard, M. H. B. (1964). *Nature, Lond.* **204,** 871.
1333. Kaska, W. C. (1968). *J. Amer. Chem. Soc.* **90,** 6340.
1334. Katsura, T. and Hasegawa, M. (1967). *Bull. Chem. Soc. Japan* **40,** 561.
1335. Katz, S. (1964). *Inorg. Chem.* **3,** 1598.
1336. Katz, S. (1966). *Inorg. Chem.* **5,** 666.
1337. Kauffman, G. B. and Sugisaka, N. (1966). *Z. anorg. Chem.* **344,** 92.
1338. Kawakube, T., Yanagi, T. and Nomura, S. (1960). *J. Phys. Soc. Japan* **15,** 2102.
1339. Kay, M. I., Frazer, B. C. and Almodovar, I. (1964). *J. Chem. Phys.* **40,** 504.
1340. Keblys, K. A. and Dubeck, M. (1964). *Inorg. Chem.* **3,** 1646.
1341. Keesmann, I. (1966). *Z. anorg. Chem.* **346,** 30.
1342. Keggin, J. F. (1933). *Nature, Lond.* **131,** 909; **132,** 351; *Proc. Roy. Soc.* **A144,** 75 (1934).
1343. Keller, H. J. (1968). *Z. Naturforsch.* **23b,** 133.
1344. Keller, O. L. (1963). *Inorg. Chem.* **2,** 783.
1345. Keller, O. L. and Chetham-Strode, A. (1966). *Inorg. Chem.* **5,** 367.
1346. Kelmers, A. D. (1961). *J. Inorg. Nuclear Chem.* **21,** 45.
1347. Kemmitt, R. D. W., Russell, D. R. and Sharp, D. W. A. (1963). *J. Chem. Soc.*, 4408.
1348. Kennedy, C. D. and Peacock, R. D. (1963). *J. Chem. Soc.*, 3392.
1349. Kennedy, J. H. (1961). *J. Inorg. Nuclear Chem.* **20,** 53.
1350. Kenworthy, J. G., Myatt, J. and Todd, P. F. (1969). *Chem. Comm.*, 263.
1351. Kepert, D. L. (1962). *Progr. Inorg. Chem.* **4,** 199.
1352. Kepert, D. L. (1964). Ph.D. Thesis, London.
1353. Kepert, D. L. (1965). *J. Chem. Soc.*, 4736.
1354. Kepert, D. L. (1969). *Inorg. Chem.* **8,** 1556.
1355. Kepert, D. L. Unpublished.
1356. Kepert, D. L. and Mandyczewsky, R. (1968). *J. Chem. Soc.* (*A*), 530.
1357. Kepert, D. L. and Mandyczewsky, R. (1968). *Inorg. Chem.* **7,** 2091.
1358. Kepert, D. L. and Mandyczewsky, R. (1969). *J. Chem. Soc.* (*A*), 2990.
1359. Kepert, D. L. and Mandyczewsky, R. Unpublished data.
1360. Kepert, D. L. and Nyholm, R. S. (1965). *J. Chem. Soc.*, 2871.
1361. Kepert, D. L. and Vrieze, K. (1967). "Halogen Chemistry". Ed. V. Gutmann. Academic Press, London and New York.

1362. Kepert, D. L. and Vrieze, K. (197). "Comprehensive Inorganic Chemistry". Ed. Bailar, J. C., Emeléus, M. J., Nyholm, R. S. and Trotman-Dickenson, A. F. Pergamon, Oxford.
1363. Kern, R. J. (1962). *J. Inorg. Nuclear Chem.* **24,** 1105.
1364. Kern, R. J. (1963). *J. Inorg. Nuclear Chem.* **25,** 5.
1365. Keppie, S. A. and Lappert, M. A. (1969). *J. Organometallic Chem.* **19,** P5.
1366. Kerker, M., Lee, D. and Chow, A. (1958). *J. Amer. Chem. Soc.* **80,** 1539.
1367. Kestigan, M. and Ward, R. (1955). *J. Amer. Chem. Soc.* **77,** 6199.
1368. Ketelaar, J. A. A. and Wegerif, E. (1938). *Rec. Trav. chim.* **57,** 1269; **58,** 948 (1939).
1369. Kettle, S. F. A. (1965). *Theor. Chim. Acta* **3,** 211.
1370. Kettle, S. F. A. (1966). *Nature* **209,** 1021.
1371. Kettle, S. F. A. and Parish, R. V. (1965). *Spectrochim. Acta* **21,** 1087.
1372. Keulen, E. and Jellinek, F. (1966). *J. Organometallic Chem.* **5,** 490.
1373. Keys, L. K. and Mulay, L. N. (1967). *Phys. Rev.* **154,** 453.
1374. Khalilova, N. K. and Morozov, I. S. (1968). *Russ. J. Inorg. Chem.* **13,** 517.
1375. Khalilova, N. K. and Morozov, I. S. (1968). *Russ. J. Inorg. Chem.* **13,** 700.
1376. Kharitonov, Yu. Ya., Bochkarev, G. S. and Zaitsev, L. M. (1964). *Russ. J. Inorg. Chem.* **9,** 745.
1377. Kharlamova, E. N. and Gur'yanova, E. N. (1965). *Zhur. strukt. Khim.* **6,** 859.
1378. Khedekar, A. V., Lewis, J., Mabbs, F. E. and Weigold, H. (1967). *J. Chem. Soc. (A)*, 1561.
1379. Kiehl, S. J., Fox, R. L. and Hardt, H. B. (1937). *J. Amer. Chem. Soc.* **59,** 2395.
1380. Kihlborg, L. (1960). *Acta Chem. Scand.* **14,** 1612; **17,** 1485 (1963).
1381. Kihlborg, L. (1963). *Arkiv Kemi* **21,** 357.
1382. Kihlborg, L. (1963). *Arkiv Kemi* **21,** 365.
1383. Kihlborg, L. (1963). *Arkiv Kemi* **21,** 427.
1384. Kihlborg, L. (1963). *Arkiv Kemi* **21,** 443.
1385. Kihlborg, L. (1963). *Arkiv Kemi* **21,** 471.
1386. Kihlborg, L. (1967). *Acta Chem. Scand.* **21,** 2495.
1387. Kihlborg, L. (1969). *Acta Chem. Scand.* **23,** 1834.
1388. Kilbourn, B. T. and Dunitz, J. D. (1967). *Inorg. Chim. Acta* **1,** 209.
1389. Kilty, P. A. and Nicholls, D. (1965). *J. Chem. Soc.*, 4915.
1390. Kilty, P. A. and Nicholls, D. (1966). *J. Chem. Soc. (A)*, 1175.
1391. Kimball, G. E. (1940). *J. Chem. Phys.* **8,** 188.
1392. Kimura, M., Kimura, K., Aoki, M. and Shibata, S. (1956). *Bull. Chem. Soc. Japan* **29,** 95.
1393. King, R. B. (1963). *Z. Naturforsch.* **18b,** 157.
1394. King, R. B. (1963). *Inorg. Chem.* **2,** 936.
1395. King, R. B. (1963). *J. Amer. Chem. Soc.* **85,** 1587.
1396. King, R. B. (1966). *Inorg. Chem.* **5,** 2231.
1397. King, R. B. (1966). *J. Amer. Chem. Soc.* **88,** 2075.
1398. King, R. B. (1967). *Chem. Comm.*, 986.
1399. King, R. B. (1967). *Inorg. Chem.* **6,** 30.

1400. King, R. B. (1967). *J. Organometallic Chem.* **8,** 129.
1401. King, R. B. (1967). *J. Organometallic Chem.* **8,** 139.
1402. King, R. B. (1968). *Inorg. Chem.* **7,** 1044.
1403. King, R. B. (1969). *J. Amer. Chem. Soc.* **91,** 7211.
1404. King, R. B. and Bisnette, M. B. (1964). *Inorg. Chem.* **3,** 785.
1405. King, R. B. and Bisnette, M. B. (1964). *Inorg. Chem.* **3,** 791.
1406. King, R. B. and Bisnette, M. B. (1965). *Inorg. Chem.* **4,** 486.
1407. King, R. B. and Bisnette, M. B. (1967). *J. Organometallic Chem.* **8,** 287.
1408. King, R. B. and Fronzaglia, A. (1966). *J. Amer. Chem. Soc.* **88,** 709.
1409. King, R. B. and Fronzaglia, A. (1966). *Inorg. Chem.* **5,** 1837.
1410. King, R. B. and Kapoor, R. N. (1968). *J. Organometallic Chem.* **15,** 457.
1411. King, R. B. and Kapoor, R. N. (1969). *J. Organometallic Chem.* **18,** 357.
1412. King, R. B. and Stone, F. G. A. (1959). *J. Amer. Chem. Soc.* **81,** 5263.
1413. King, R. B., Treichel, P. M. and Stone, F. G. A. (1961). *Chem. and Ind.*, 747.
1414. Kitchens, J. and Bear, J. L. (1969). *J. Inorg. Nuclear Chem.* **31,** 2415.
1415. Klabunde, U. and Fischer, E. O. (1967). *J. Amer. Chem. Soc.* **89,** 7141.
1416. Klanberg, F. and Kohlschutter, H. W. (1960). *Z. Naturforsch.* **15b,** 616.
1417. Klason, P. (1901). *Berichte* **34,** 148.
1418. Klejnot, O. J. (1965). *Inorg. Chem.* **4,** 1668.
1419. Klemm, W. and Grimm, L. (1942). *Z. anorg. Chem.* **249,** 198.
1420. Klemm, W. and Grimm, L. (1942). *Z. anorg. Chem.* **249,** 209.
1421. Klemm, W. and Hoschek, E. (1936). *Z. anorg. Chem.* **226,** 359.
1422. Klemm, W. and Krose, E. (1947). *Z. anorg. Chem.* **253,** 209, 218.
1423. Klemm, W. and Schnering, H. G. (1965). *Naturwiss.* **52,** 12.
1424. Klemm, W. and Steinberg, H. (1936). *Z. anorg. Chem.* **227,** 193.
1425. Kling, O. and Schlafer, H. L. (1961). *Z. anorg. Chem.* **313,** 187.
1426. Knox, G. F. and Brown, T. M. (1969). *Inorg. Chem.* **8,** 1401.
1427. Knox, J. R. and Eriks, K. (1968). *Inorg. Chem.* **7,** 84.
1428. Knox, J. R. and Prout, C. K. (1968). *Chem. Comm.*, 1277; *Acta Cryst.* **B25,** 1857.
1429. Knox, K. (1960). *Acta Cryst.* **13,** 507.
1430. Knox, K. and Coffey, C. E. (1959). *J. Amer. Chem. Soc.* **81,** 5.
1431. Knox, K. and Ginsberg, A. P. (1964). *Inorg. Chem.* **3,** 555.
1432. Knox, K. and Mitchell, D. W. (1961). *J. Inorg. Nuclear Chem.* **21,** 253.
1433. Knox, K., Tyree, S. Y., Srivastava, R. D., Norman, V., Bassett, J. Y. and Holloway, J. H. (1957). *J. Amer. Chem. Soc.* **79,** 3358.
1434. Knyazeva, E. M. and Luk'yanov, V. F. (1965). *Russ. J. Inorg. Chem.* **10,** 599.
1435. Kogan, V. A., Osipov, O. A., Minkin, V. I. and Sokolov, V. P. (1965). *Russ. J. Inorg. Chem.* **10,** 45.
1436. Kohl, F. J., Lewis, J. and Whyman, R. (1966). *J. Chem. Soc.* (*A*), 630.
1437. Kolaczkowski, R. W. and Plane, R. A. (1964). *Inorg. Chem.* **3,** 322.
1438. Kolditz, L. and Calov, U. (1966). *Z. Chem.* **6,** 431.
1439. Kolditz, L. and Degenkolb, P. (1966). *Z. Chem.* **6,** 347.
1440. Kolditz, L. and Feltz, A. (1961). *Z. anorg. Chem.* **310,** 204.

1441. Kolditz, L. and Feltz, A. (1961). *Z. anorg. Chem.* **310,** 217.
1442. Kolditz, L. and Furcht, G. (1961). *Z. anorg. Chem.* **312,** 11.
1443. Kolditz, L., Kurschnev, C. and Calov, U. (1964). *Z. anorg. Chem.* **329,** 172.
1444. Kolditz, L., Neumann, V. and Kilch, G. (1963). *Z. anorg. Chem.* **325,** 275.
1445. Kolditz, L. and Schmidt, W. (1958). *Z. anorg. Chem.* **296,** 188.
1446. Komorita, T., Miki, S. and Yamada, S. (1965). *Bull. Chem. Soc. Japan* **38,** 123.
1447. Kon, H. (1963). *J. Inorg. Nuclear Chem.* **25,** 933.
1448. Kon, H. and Sharpless, N. E. (1965). *J. Chem. Phys.* **43,** 1081.
1449. Konig, E. (1962). *Theor. Chim. Acta* **1,** 23.
1450. Konig, E. (1963). *Inorg. Chem.* **2,** 1238.
1451. Konig, E. (1969). *Inorg. Chem.* **8,** 1278.
1452. Kopf, H. (1968). *J. Organometallic Chem.* **14,** 353.
1453. Kopf, H. and Schmidt, M. (1965). *Z. anorg. Chem.* **340,** 139; *Angew. Chem. Internat. Edn.* **4,** 953.
1454. Koppel, I. (1905). *Z. anorg. Chem.* **45,** 361.
1455. Koppel, J. and Goldman, R. (1903). *Z. anorg. Chem.* **36,** 281.
1456. Koppel, J., Goldman, R. and Kaufmann, A. (1905). *Z. anorg. Chem.* **45,** 345.
1457. Korol'kov, D. V. and Kudryashova, G. N. (1968). *Russ. J. Inorg. Chem.* **13,** 850.
1458. Korosy, F. (1939). *J. Amer. Chem. Soc.* **61,** 838.
1459. Korshunov, B. G. and Bezuevskaya, V. V. (1967). *Russ. J. Inorg. Chem.* **12,** 1736.
1460. Korshunov, B. G. and Safanov, V. V. (1961). *Russ. J. Inorg. Chem.* **6,** 385.
1461. Kosuge, K. (1967). *J. Phys. and Chem. Solids* **28,** 1613.
1462. Kosuge, K., Takada, T. and Kachi, S. (1963). *J. Phys. Soc. Japan* **18,** 318.
1463. Kovacic, P. and Lange, R. M. (1964). *J. Org. Chem.* **28,** 968; **29,** 2416; **30,** 4251 (1965).
1464. Koyama, H. and Saito, Y. (1954). *Bull. Chem. Soc. Japan* **27,** 112.
1465. Koz'min, P. A. (1964). *J. Struct. Chem.* **5,** 60.
1466. Krauss, H.-L. and Huber, W. (1961). *Berichte* **94,** 2864.
1467. Krauss, H.-L. and Munster, G. (1962). *Z. Naturforsch.* **17b,** 344; Krauss, H.-L., Leder, M. and Munster, G. (1963). *Berichte* **96,** 3008.
1468. Krauss, H.-L. and Munster, G. (1967). *Z. anorg. Chem.* **352,** 24.
1469. Krauss, H.-L. and Schwarzbach, F. (1961). *Berichte* **94,** 1205.
1470. Krebs, B. (1969). *Angew. Chem. Internat. Edn.* **8,** 146.
1471. Kreiter, C. G., Maasbol, A., Anet, F. A. L., Kaesz, H. D. and Winstein, S. (1960). *J. Amer. Chem. Soc.* **88,** 3444.
1472. Kruck, Th. (1964). *Berichte* **97,** 2018. Kruck, Th. and Prasch, A. (1964). *Z. Naturforsch.* **19b,** 669. Kruck, Th., Lang, W. and Engelmann, A. (1965). *Angew. Chem. Internat. Edn.* **4,** 148.
1473. Krylov, E. I. (1958). *Zhur. neorg. Khim.* **3,** 1487.
1474. Krylov, E. I. and Kalugina, N. N. (1959). *Russ. J. Inorg. Chem.* **4,** 1139.

1475. Krylov, E. I., Lapitskii, A. V., Pinaeva, M. M. and Berdonosov, S. S. (1966). *Russ. J. Inorg. Chem.* **11,** 541.
1476. Kubota, B. and Hirota, E. (1960). *J. Phys. Soc. Japan* **16,** 345.
1477. Kubota, M. and Schulze, S. R. (1964). *Inorg. Chem.* **3,** 853.
1478. Kudrak, D. R. and Sienko, M. J. (1967). *Inorg. Chem.* **6,** 880.
1479. Kuhn, P. J. and McCarley, R. E. (1965). *Inorg. Chem.* **4,** 1482.
1480. Kuhnl, H. and Ernst, W. (1962). *Z. anorg. Chem.* **317,** 84.
1481. Kummer, R. and Graham, W. A. G. (1968). *Inorg. Chem.* **7,** 310.
1482. Kunz, V. and Nowacki, W. (1967). *Helv. Chim. Acta* **50,** 1052.
1483. Kupka, F. and Sienko, M. J. (1950). *J. Chem. Phys.* **18,** 1296.
1484. Kureksev, T., Sargeson, A. M. and West, B. O. (1957). *J. Phys. Chem.* **61,** 1567.
1485. Kurras, E. (1959). *Naturwiss.* **46,** 171.
1486. Kurras, V. (1967). *Z. anorg. Chem.* **351,** 268.
1487. Kust, M. A., Corbett, J. D. and Friedman, R. M. (1968). *Inorg. Chem.* **7,** 2081.
1488. Lalancetto, R. A., Cefola, M., Hamilton, W. C. and La Placa, S. J. (1967). *Inorg. Chem.* **6,** 2127.
1489. La Lau, C. (1965). *Rec. Trav. chim.* **84,** 429.
1490. Lange, G. and Dehnicke, K. (1966). *Naturwiss.* **53,** 38.
1491. Lange, G. and Dehnicke, K. (1966). *Z. anorg. Chem.* **344,** 167.
1492. Langs, D. A. and Hare, C. R. (1967). *Chem. Comm.*, 890.
1493. Lappert, M. F. (1962). *J. Chem. Soc.*, 542.
1494. Lappert, M. F. and Srivastava, G. (1966). *J. Chem. Soc. (A)*, 210.
1495. Larkworthy, L. F., Murphy, J. M., Patel, K. C. and Phillips, D. J. (1968). *J. Chem. Soc. (A)*, 2936.
1496. Larkworthy, L. F., Patel, K. C. and Phillips, D. J. (1968). *Chem. Comm.*, 1667.
1497. Larsen, E. M. and Gammill, A. M. (1950). *J. Amer. Chem. Soc.* **72,** 3615.
1498. Larsen, E. M., Howatson, J., Gammill, A. M. and Wittenberg, L. W. (1952). *J. Amer. Chem. Soc.* **74,** 3489; **77,** 5850 (1955).
1499. Larsen, E. M. and Leddy, J. J. (1956). *J. Amer. Chem. Soc.* **78,** 5983.
1500. Larson, A. C. and Cromer, D. T. (1961). *Acta Cryst.* **14,** 128.
1501. Larson, A. C., Roof, R. B. and Cromer, D. T. (1964). *Acta Cryst.* **17,** 555.
1502. Larson, M. L. (1960). *J. Amer. Chem. Soc.* **82,** 1223.
1503. Larson, M. L. and Moore, F. W. (1962). *Inorg. Chem.* **1,** 856; **2,** 887 (1963).
1504. Larson, M. L. and Moore, F. W. (1964). *Inorg. Chem.* **3,** 285.
1505. Larson, M. L. and Moore, F. W. (1966). *Inorg. Chem.* **5,** 801.
1506. Laswick, J. A. and Plane, R. A. (1959). *J. Amer. Chem. Soc.* **81,** 3564.
1507. Latham, R. and Drago, R. S. (1964). *Inorg. Chem.* **3,** 291.
1508. Latka, H. (1967). *Z. anorg. Chem.* **353,** 243.
1509. Laudise, R. A. and Young, R. C. (1955). *J. Amer. Chem. Soc.* **77,** 5288; *Inorg. Synth.* **6,** 149 (1960).
1510. LaValle, D. E., Steele, R. M., Wilkinson, M. K. and Yakel, H. L. (1960). *J. Amer. Chem. Soc.* **82,** 2433.

1511. LaVilla, R. E. and Bauer, S. H. (1960). *J. Chem. Phys.* **33,** 182.
1512. Lawton, D. and Mason, R. (1965). *J. Amer. Chem. Soc.* **87,** 921.
1513. Lee, C. S., Gorton, E. M., Neumann, H. M. and Hunt, H. R. (1966). *Inorg. Chem.* **5,** 1397.
1514. Lehne, M. and Goetz, H. (1961). *Bull. Soc. chim. France*, 334.
1515. Lehovec, K. (1964). *J. Less-Common Metals* **7,** 397.
1516. Leigh, G. J. and Fischer, E. O. (1965). *J. Organometallic Chem.* **4,** 461.
1517. Lenz, W., Schlafer, H. L. and Ludi, A. (1969). *Z. anorg. Chem.* **365,** 55.
1518. Leshchenko, A. V., Panyushkin, V. T., Garnovskii, A. D. and Osipov, O. A. (1966). *Russ. J. Inorg. Chem.* **11,** 1155.
1519. Levashova, L. B. and Zolotavin, V. L. (1965). *Russ. J. Inorg. Chem.* **10,** 77.
1520. Lever, A. B. P. (1965). *J. Chem. Soc.*, 1821.
1521. Levy, H. A., Agron, P. A. and Danford, M. D. (1959). *J. Chem. Phys.* **30,** 1486.
1522. Levy, H. A., Danford, M. D. and Agron, P. A. (1959). *J. Chem. Phys.* **31,** 1458.
1523. Lewis, J., Mabbs, F. E. and Richards, A. (1967). *J. Chem. Soc.* (*A*), 1014.
1524. Lewis, J., Machin, D. J., Newnham, I. E. and Nyholm, R. S. (1962). *J. Chem. Soc.*, 2036.
1525. Lewis, J., Machin, D. J., Nyholm, R. S., Pauling, P. and Smith, P. W. (1960). *Chem. and Ind.*, 259.
1526. Lewis, J., Nyholm, R. S., Pande, C. S. and Stiddard, M. H. B. (1963). *J. Chem. Soc.*, 3600.
1527. Lewis, J., Nyholm, R. S. and Smith, P. W. (1961). *J. Chem. Soc.*, 4590.
1528. Lewis, J., Nyholm, R. S. and Smith, P. W. (1962). *J. Chem. Soc.*, 2592.
1529. Lewis, J., Nyholm, R. S. and Smith, P. W. (1969). *J. Chem. Soc.* (*A*), 57.
1530. Lewis, J. and Whyman, R. (1965). *Chem. Comm.*, 159.
1531. Lewis, J. and Whyman, R. (1965). *J. Chem. Soc.*, 6027.
1532. Lewis, J. and Whyman, R. (1967). *J. Chem. Soc.*, 77.
1533. Lewis, J. and Whyman, R. (1967). *J. Chem. Soc.*, 5486.
1534. Libby, W. F. (1967). *J. Chem. Phys.* **46,** 399.
1535. Liebe, W., Weise, E. and Klemm, W. (1961). *Z. anorg. Chem.* **311,** 280.
1536. de Liefde Meijer, H. J., Jansson, M. J. and van der Kerk, G. J. M. (1961). *Rec. Trav. chim.* **80,** 831.
1537. de Liefde Meijer and van der Kerk, G. J. M. (1965). *Rec. Trav. chim.* **84,** 1418.
1538. Lifshitz, A. and Perlmutter-Hayman, B. (1961). *J. Phys. Chem.* **65,** 2098.
1539. Lind, M. D., Hoard, J. L., Hamor, M. J. and Hamor, T. A. (1964). *Inorg. Chem.* **3,** 34.
1540. Lind, M. D., Lee, B. and Hoard, J. L. (1965). *J. Amer. Chem. Soc.* **87,** 1611.
1541. Lindley, P. F. and Mills, O. S. (1969). *J. Chem. Soc.* (*A*), 1286.
1542. Lindner, K. (1922). *Berichte* **55,** 1458.
1543. Lindner, K. and Feit, H. (1924). *Z. anorg. Chem.* **132,** 10.
1544. Lindner, K. and Feit, H. (1924). *Z. anorg. Chem.* **137,** 66.

1545. Lindner, K., Heller, E. H. and Helwig, H. (1923). *Z. anorg. Chem.* **130,** 209.
1546. Lindoy, L. F., Livingstone, S. E. and Lockyer, T. N. (1965). *Austral. J. Chem.* **18,** 1549.
1547. Lindqvist, I. (1950). *Acta Chem. Scand.* **4,** 1066.
1548. Lindqvist, I. (1950). *Arkiv Kemi* **2,** 325.
1549. Lindqvist, I. (1950). *Arkiv Kemi* **2,** 349.
1550. Lindqvist, I. (1952). *Acta Cryst.* **5,** 667.
1551. Lindqvist, I. (1953). *Arkiv Kemi* **5,** 247.
1552. Lindqvist, I. and Aronsson, B. (1955). *Arkiv Kemi* **7,** 49.
1553. Lindqvist, I. (1963). "Inorganic Adduct Molecules of Oxo-Compounds". Springer-Verlag, Berlin.
1554. Linge, H. G. and Jones, A. L. (1968). *Austral. J. Chem.* **21,** 1445.
1555. Linge, H. G. and Jones, A. L. (1968). *Austral. J. Chem.* **21,** 2189.
1556. Linhard, M. and Weigel, M. (1952). *Z. anorg. Chem.* **271,** 115.
1557. Linnett, L. W. (1961). *J. Chem. Soc.*, 3796.
1558. Lipatova, N. P. and Morozov, I. S. (1965). *Russ. J. Inorg. Chem.* **10,** 231.
1559. Lipatova, N. P. and Morozov, I. S. (1965). *Russ. J. Inorg. Chem.* **10,** 1528.
1560. Lippard, S. J., Nozaki, H. and Russ, B. J. (1967). *Chem. Comm.*, 118.
1561. Lippard, S. J. and Russ, B. J. (1967). *Inorg. Chem.* **6,** 1943.
1562. Lippard, S. J. and Russ, B. J. (1968). *Inorg. Chem.* **7,** 1686.
1563. Lippard, S. J., Schugar, H. and Walling, C. (1967). *Inorg. Chem.* **6,** 1825.
1564. Lipscomb, W. N. (1965). *Inorg. Chem.* **4,** 132.
1565. Lipscomb, W. N. and Whittaker, A. G. (1945). *J. Amer. Chem. Soc.* **67,** 2019.
1566. Lister, B. A. J. and McDonald, L. A. (1952). *J. Chem. Soc.*, 4315.
1567. Lister, R. L. and Flengas, S. N. (1965). *Canad. J. Chem.* **43,** 2947.
1568. Littke, W. and Brauer, G. (1963). *Z. anorg. Chem.* **325,** 122.
1569. Lock, C. J. L. and Wilkinson, G. (1964). *J. Chem. Soc.*, 2281.
1570. Locke, J. and McCleverty, J. A. (1966). *Inorg. Chem.* **5,** 1157.
1571. Lohmann, K. H. and Young, R. C. (1953). *Inorg. Synth.* **4,** 97.
1572. Longo, J. M. and Sleight, A. W. (1968). *Inorg. Chem.* **7,** 108.
1573. Loopstra, B. O. and Boldrini, P. (1966). *Acta Cryst.* **21,** 158. Loopstra, B. O. and Reitveld, H. M. (1969). *Acta Cryst.* **B25,** 1420.
1574. Lukkari, O. (1962). *Suomen Kem.* **B35,** 91.
1575. Lundberg, M. (1965). *Acta Chem. Scand.* **19,** 2274.
1576. Lundgren, G. (1956). *Rec. Trav. chim.* **75,** 585.
1577. Lundgren, G. (1958). *Arkiv Kemi* **13,** 59.
1578. Lutz, P. M., Long, G. J. and Baker, W. A. (1969). *Inorg. Chem.* **8,** 2529.
1579. Lux, H., Eberle, L. and Sarre, D. (1964). *Berichte* **97,** 503.
1580. Lynton, H. and Fleming, J. E. (1959). *Chem. and Ind.*, 1409.
1581. McCarley, R. E. and Boatman, J. C. (1963). *Inorg. Chem.* **2,** 547.
1582. McCarley, R. E. and Boatman, J. C. (1965). *Inorg. Chem.* **4,** 1486.
1583. McCarley, R. E. and Brown, T. M. (1962). *J. Amer. Chem. Soc.* **84,** 3216.
1584. McCarley, R. E. and Brown, T. M. (1964). *Inorg. Chem.* **3,** 1232.

1585. McCarley, R. E., Hughes, B. G., Boatman, J. C. and Torp, B. A. (1963). "Reactions of Coordinated Ligands and Homogeneous Catalysis". A.C.S., Washington.
1586. McCarley, R. E., Hughes, B. G., Cotton, F. A. and Zimmerman, R. (1965). *Inorg. Chem.* **4,** 1491.
1587. McCarley, R. E. and Roddy, J. W. (1964). *Inorg. Chem.* **3,** 54.
1588. McCarley, R. E. and Roddy, J. W. (1964). *Inorg. Chem.* **3,** 60.
1589. McCarley, R. E., Roddy, J. W. and Berry, K. O. (1964). *Inorg. Chem.* **3,** 50.
1590. McCarley, R. E. and Torp, B. A. (1963). *Inorg. Chem.* **2,** 540.
1591. McCarroll, W. H., Katz, L. and Ward, R. (1957). *J. Amer. Chem. Soc.* **79,** 5410.
1592. McCarroll, W. H., Ward, R. and Katz, L. (1956). *J. Amer. Chem. Soc.* **78,** 2909.
1593. McCleverty, J. A., Locke, J., Wharton, E. J. and Gerloch, M. (1968). *J. Chem. Soc.* (*A*), 816.
1594. McCleverty, J. A. and Wilkinson, G. (1961). *Chem. and Ind.*, 288.
1595. McClung, D. A., Dalton, L. R. and Brubaker, C. H. (1966). *Inorg. Chem.* **5,** 1985.
1596. McCullough, J. D. and Trueblood, K. N. (1959). *Acta Cryst.* **12,** 507.
1597. McDonald, G. D., Thompson, M. and Larsen, E. M. (1968). *Inorg. Chem.* **7,** 648.
1598. McDonald, T. R. R. and Spink, J. M. (1967). *Acta Cryst.* **23,** 944.
1599. McGarvey, B. R. (1966). *Inorg. Chem.* **5,** 476.
1600. Machin, D. J., Morris, D. F. C. and Short, E. L. (1964). *J. Chem. Soc.*, 4658.
1601. Machin, D. J. and Murray, K. S. (1967). *J. Chem. Soc.* (*A*), 1498.
1602. Machin, D. J., Murray, K. S. and Walton, R. A. (1968). *J. Chem. Soc.* (*A*), 195.
1603. Mackay, R. A. and Schneider, R. F. (1967). *Inorg. Chem.* **6,** 549.
1604. Mackay, R. A. and Schneider, R. F. (1968). *Inorg. Chem.* **7,** 455.
1605. McKechnie, J. S. and Paul, I. C. (1966). *J. Amer. Chem. Soc.* **88,** 5927.
1606. McKechnie, J. S. and Paul, I. C. (1967). *Chem. Comm.*, 747.
1607. McKeever, L. D., Waack, R., Doran, M. A. and Baker, E. B. (1969). *J. Amer. Chem. Soc.* **91,** 1057.
1608. Mackintosh, A. R. (1963). *J. Chem. Phys.* **38,** 1991.
1609. McLaren, E. H. and Helmholz, L. (1959). *J. Phys. Chem.* **63,** 1279.
1610. McNeill, W. and Conroy, L. E. (1962). *J. Chem. Phys.* **36,** 87.
1611. McPartlin, M., Mason, R. and Malatesta, L. (1969). *Chem. Comm.*, 334.
1612. McRae, V. M., Peacock, R. D. and Russell, D. R. (1969). *Chem. Comm.*, 62.
1613. McTaggart, F. K. and Turnbull, A. G. (1964). *Austral. J. Chem.* **17,** 727.
1614. McWhan, D. M. and Lundgren, G. (1963). *Acta Cryst.* **A16,** 36.
1615. McWhan, D. M. and Lundgren, G. (1966). *Inorg. Chem.* **5,** 284.
1616. Madan, S. K. and Donohue, A. M. (1966). *J. Inorg. Nuclear Chem.* **28,** 1303.

1617. Magee, J. and Richardson, E. (1960). *J. Inorg. Nuclear Chem.* **15,** 272.
1618. Magee, T. A., Matthews, C. N., Wang, T. S. and Wotiz, J. H. (1961). *J. Amer. Chem. Soc.* **83,** 3200.
1619. Magneli, A. (1948). *Acta Chem. Scand.* **2,** 501.
1620. Magneli, A. (1948). *Acta Chem. Scand.* **2,** 861.
1621. Magneli, A. (1949). *Arkiv Kemi* **1,** 269.
1622. Magneli, A. (1950). *Arkiv Kemi* **1,** 213.
1623. Magneli, A. (1950). *Arkiv Kemi* **1,** 223.
1624. Magneli, A. (1950). *Arkiv Kemi* **1,** 513.
1625. Magneli, A. (1951). *Acta Chem. Scand.* **5,** 670.
1626. Magneli, A. (1952). *Nature* **169,** 791; *Acta Chem. Scand.* **7,** 315 (1953).
1627. Magneli, A. (1953). *Acta Cryst.* **6,** 495.
1628. Magneli, A. (1956). *J. Inorg. Nuclear Chem.* **2,** 330.
1629. Magneli, A. and Andersson, G. (1955). *Acta Chem. Scand.* **9,** 1378.
1630. Magneli, A. and Andersson, G. (1955). *Acta Chem. Scand.* **9,** 1378.
1631. Magneli, A., Andersson, G. and Sundqvist, G. (1955). *Acta Chem. Scand.* **9,** 1402.
1632. Magneli, A. and Blomberg, B. (1951). *Acta Chem. Scand.* **5,** 372.
1633. Magneli, A., Blomberg, B., Kihlborg, L. and Sundqvist, G. (1955). *Acta Chem. Scand.* **9,** 1382.
1634. Magneli, A. and Nilsson, R. (1950). *Acta Chem. Scand.* **4,** 398.
1635. Magneli, A. and Nord, S. (1965). *Acta Chem. Scand.* **19,** 1510.
1636. Magnusson, W. L., Griswold, E. and Kleinberg, J. (1964). *Inorg. Chem.* **3,** 88.
1637. Mais, R. H. B., Owston, P. G. and Thompson, D. T. (1967). *J. Chem. Soc.* (*A*), 1735.
1638. Majumdar, A. K. and Bhattacharyya, R. G. (1967). *J. Inorg. Nuclear Chem.* **29,** 2359.
1639. Majumdar, A. K., Mukherjee, A. K. and Bhattacharya, R. G. (1964). *J. Inorg. Nuclear Chem.* **26,** 386.
1640. Mak, T. C. W. (1968). *Canad. J. Chem.* **46,** 3491.
1641. Malatesta, L. (1933). *Gazzetta* **69,** 408.
1642. Malik, S. A. and Weakley, T. J. R. (1967). *Chem. Comm.*, 1094; *J. Chem. Soc.* (*A*), 2647 (1968).
1643. Mallock, A. K. (1967). *Inorg. Nuclear Chem. Letters* **3,** 441.
1644. Mannerskantz, H. C. E. and Wilkinson, G. (1962). *J. Chem. Soc.*, 4454.
1645. Manning, A. R. (1967). *J. Chem. Soc.* (*A*), 1984.
1646. Marinder, B.-O. (1961). *Acta Chem. Scand.* **15,** 707; *Arkiv Kemi* **19,** 435.
1647. Marinder, B.-O, Dorm, E. and Seleborg, M. (1962). *Acta Chem. Scand.* **16,** 293.
1648. Marinder, B.-O. and Magneli, A. (1957). *Acta Chem. Scand.* **11,** 1635.
1649. Marinder, B.-O. and Magneli, A. (1958). *Acta Chem. Scand.* **12,** 1345.
1650. Maroni, V. A. and Spiro, T. G. (1968). *Inorg. Chem.* **7,** 183, 188, 193.
1651. Martin, H. A. and Jellinek F. (1966). *J. Organometallic Chem.* **6,** 293; **8,** 115 (1967).

1652. Martin, H. A., Lemaire, P. J. and Jellinek, F. (1968). *J. Organometallic Chem.* **14,** 149.
1653. Martin, R. L. and Winter, G. (1960). *Nature, Lond.* **188,** 313.
1654. Martin, R. L. and Winter, G. (1961). *J. Chem. Soc.*, 2947.
1655. Martin, R. L. and Winter, G. (1963). *Nature, Lond.* **197,** 687.
1656. Martin, R. L. and Winter, G. (1965). *J. Chem. Soc.*, 4709.
1657. Marzilli, P. A. and Buckingham, D. A. (1966). *Austral. J. Chem.* **19,** 2259.
1658. Marzluff, W. F. (1964). *Inorg. Chem.* **3,** 395.
1659. Maskill, R. and Pratt, J. M. (1968). *J. Chem. Soc. (A)*, 1914.
1660. Mason, S. F. and Wood, J. W. (1968). *Chem. Comm.*, 1512.
1661. Massart, R. and Souchay, P. (1963). *Compt. rend.* **256,** 4671; **257,** 1297.
1662. Mathern, R., Weiss, R. and Rohmer, R. (1969). *Chem. Comm.*, 70.
1663. Mathew, M. and Palenik, G. J. (1969). *Canad. J. Chem.* **47,** 705.
1664. Matijevic, E. and Kerker, M. (1959). *J. Amer. Chem. Soc.* **81,** 1307.
1665. Matkovic, B. and Grdenic, D. (1963). *Acta Cryst.* **16,** 456.
1666. Mattes, R. (1968). *Z. anorg. Chem.* **357,** 30.
1667. Mattes, R. (1969). *Z. anorg. Chem.* **364,** 279, 290.
1668. Matthews, J. D. and Swallow, A. G. (1969). *Chem. Comm.*, 882.
1669. Mays, M. J. and Pearson, S. M. (1968). *J. Chem. Soc. (A)*, 2291.
1670. Mays, M. J. and Pearson, S. M. (1969). *J. Chem. Soc. (A)*, 136.
1671. Mays, M. J. and Robb, J. D. (1968). *J. Chem. Soc. (A)*, 329.
1672. Mays, M. J. and Robb, J. D. (1969). *J. Chem. Soc. (A)*, 561.
1673. Medeiros, L. O. (1966). *J. Inorg. Nuclear Chem.* **28,** 599.
1674. Mehrotra, R. C. (1954). *J. Amer. Chem. Soc.* **76,** 2266.
1675. Mehrotra, R. C., Agrawal, M. M. and Kapoor, P. N. (1968). *J. Chem. Soc. (A)*, 2673.
1676. Mehrotra, R. C. and Kapoor, P. N. (1964). *J. Less-Common Metals* **7,** 98.
1677. Meier, J. and Schwarzenbach, G. (1958). *J. Inorg. Nuclear Chem.* **8,** 302.
1678. Melby, L. R. (1969). *Inorg. Chem.* **8,** 349.
1679. Mercer, M. (1967). *Chem. Comm.*, 119.
1680. Mercer, M. (1969). *J. Chem. Soc. (A)*, 2019.
1681. Mercer, M., Ouellette, T. J., Ratcliffe, C. T. and Sharp, D. W. A. (1969). *J. Chem. Soc. (A)*, 2232.
1682. Mercier, R. C. and Paris, M. R. (1964). *Compt. rend.* **259,** 2445.
1683. Meyer, G., Oosterom, J. F. and van Oeveren, W. J. (1961). *Rec. Trav. chem.* **80,** 502.
1684. Michael, A. and Murphy, A. (1910). *Ann. Chem.* **44,** 365.
1685. Miller, F. A. and Baer, W. K. (1961). *Spectrochim. Acta* **17,** 112.
1686. Miller, F. A., Carlson, G. L. and White, W. B. (1959). *Spectrochim. Acta* **15,** 709.
1687. Miller, F. A. and Cousins, L. R. (1957). *J. Chem. Phys.* **26,** 329.
1688. Minami, S., Takano, H. and Ishino, T. (1957). *Kogyo Kagaku Zasshi* **60,** 1406.
1689. Mills, O. S. and Redhouse, A. D. (1965). *Angew. Chem. Internat. Edn.* **4,** 1082; *J. Chem. Soc. (A)*, 642 (1968); 1274 (1969).
1690. Mitchell, P. C. H. (1961). *J. Inorg. Nuclear Chem.* **21,** 382.

1691. Mitchell, P. C. H. (1963). *J. Inorg. Nuclear Chem.* **25,** 963.
1692. Mitchell, P. C. H. (1964). *J. Inorg. Nuclear Chem.* **26,** 1967.
1693. Mitchell, P. C. H. (1966). *Quart. Rev.* **20,** 103.
1694. Mitchell, P. C. H. (1969). *J. Chem. Soc.* (*A*), 146.
1695. Mitchell, P. C. H. and Williams, R. J. P. (1962). *J. Chem. Soc.*, 4570.
1696. Mitschler, A., LeCarpentier, J. M. and Weiss, R. (1968). *Chem. Comm.*, 1260.
1697. Mittal, R. K. and Mehrotra, R. C. (1964). *Z. anorg. Chem.* **332,** 189.
1698. Miyake, A., Kondo, H. and Aoyama, M. (1969). *Angew. Chem. Internat. Edn.* **8,** 520.
1699. Moeller, C. W. (1957). *J. Chem. Phys.* **27,** 983.
1700. Moissan, H. (1902). *Compt. rend.* **134,** 212.
1701. Montgomery, H., Chastain, R. V., Natt, J. J., Witkowska, A. M. and Lingafelter, E. C. (1967). *Acta Cryst.* **22,** 775.
1702. Moody, G. J. and Selig, H. (1966). *J. Inorg. Nuclear Chem.* **28,** 2429.
1703. Mooney, R. C. L. (1949). *Acta Cryst.* **2,** 189.
1704. Moore, F. W. and Larson, M. L. (1967). *Inorg. Chem.* **6,** 998.
1705. Moore, F. W. and Rice, R. E. (1968). *Inorg. Chem.* **7,** 2510.
1706. Moore, P., Kettle, S. F. A. and Wilkins, R. G. (1966). *Inorg. Chem.* **5,** 220.
1707. Moore, R. V. and Tyree, S. Y. (1954). *J. Amer. Chem. Soc.* **76,** 5253.
1708. Morawietz, H. (1966). *J. Inorg. Nuclear Chem.* **28,** 941.
1709. Morette, A. (1938). *Compt. rend.* **207,** 1218.
1710. Morgan, G. T. and Castell, A. S. (1928). *J. Chem. Soc.*, 3252.
1711. Morgan, G. T. and Moss, H. W. (1914). *J. Chem. Soc.*, 78.
1712. Morgan, L. O. and Justus, N. L. (1956). *J. Amer. Chem. Soc.* **78,** 38.
1713. Mori, B., Gohring, J., Cassimatis, D. and Susz, B. P. (1962). *Helv. Chim. Acta* **45,** 77.
1714. Morin, F. J. (1959). *Phys. Rev. Letters* **3,** 34.
1715. Morino, Y. and Uehara, H. (1966). *J. Chem. Phys.* **45,** 4543.
1716. Morosin, B. (1966). *Acta Cryst.* **21,** 280.
1717. Morosin, B. and Narath, A. (1964). *J. Chem. Phys.* **40,** 1958.
1718. Morozov, I. S., Korshunov, B. G. and co-workers (1956). *Zhur. neorg. Khim.* **1,** 145, 1646; **2,** 1907 (1957); **3,** 1637 (1958); **7,** 1979 (1962).
1719. Morozov, I. S. and Li Ch'ih-fa (1963). *Russ. J. Inorg. Chem.* **8,** 1432.
1720. Morozov, I. S. and Lipatova, N. P. (1966). *Russ. J. Inorg. Chem.* **11,** 550.
1721. Morozov, I. S. and Lipatova, N. P. (1968). *Russ. J. Inorg. Chem.* **13,** 1101.
1722. Morozov, I. S. and Morozov, A. I. (1966). *Russ. J. Inorg. Chem.* **11,** 182.
1723. Mortimer, P. I. and Strong, M. A. (1965). *Austral. J. Chem.* **18,** 1579.
1724. Moss, D. S. and Sinha, S. P. (1969). *Z. Phys. Chem.* **63,** 190.
1725. Moss, J. R. and Shaw, B. L. (1968). *Chem. Comm.*, 632.
1726. Motov, D. L. and Ritter, M. P. (1968). *Russ. J. Inorg. Chem.* **13,** 879, 1339.
1727. Moureau, H. and Hamblet, C. H. (1935). *Compt. rend.* **200,** 2184; **59,** 33 (1937).
1728. Moureau, H., Sue, P. and Magat, M. (1947). "Contribution a étude de la structure moleculaire. Vol. Commemoratif Victor Henri". 125.
1729. Muetterties, E. L. (1960). *J. Amer. Chem. Soc.* **82,** 1082.

1730. Muetterties, E. L. (1965). *Inorg. Chem.* **4,** 769.
1731. Muetterties, E. L. (1966). *J. Amer. Chem. Soc.* **88,** 305.
1732. Muetterties, E. L. and Castle, J. E. (1961). *J. Inorg. Nuclear Chem.* **18,** 148.
1733. Muetterties, E. L. and Packer, K. J. (1964). *J. Amer. Chem. Soc.* **86,** 293.
1734. Muetterties, E. L. and Wright, C. M. (1965). *J. Amer. Chem. Soc.* **87,** 4706.
1735. Muha, G. M. and Vaughan, P. A. (1960). *J. Chem. Phys.* **33,** 194.
1736. Muir, K. W. and Ferguson, G. (1968). *J. Chem. Soc.* (*B*), 476.
1737. Muir, K. W., Ferguson, G. and Sim, G. A. (1968). *J. Chem. Soc.* (*B*), 467.
1738. Mukhtar, A. and Winand, R. (1965). *Compt. rend.* **260,** 3674.
1739. Muller, H. (1968). *J. Chem. Phys.* **49,** 475.
1740. Muller, J. and Thiele, K.-H. (1968). *Z. anorg. Chem.* **362,** 120.
1741. Muller, M. and Rohmer, R. (1967). *Bull. Soc. chim. France*, 925.
1742. Muller, M. and Rohmer, R. (1967). *Bull. Soc. chim. France*, 928.
1743. Murakami, M. and Hirahara, E. (1958). *J. Phys. Soc. Japan* **13,** 1407.
1744. Murdoch, H. D. (1965). *J. Organometallic Chem.* **4,** 119.
1745. Murdoch, H. D. and Henzi, R. (1966). *J. Organometallic Chem.* **5,** 552.
1746. Murray, J. G. (1959). *J. Amer. Chem. Soc.* **81,** 752.
1747. Murray, R. W. and Haendler, H. M. (1960). *J. Inorg. Nuclear Chem.* **14,** 135.
1748. Nabivanets, B. I. (1961). *Russ. J. Inorg. Chem.* **6,** 586.
1749. Nabivanets, B. I. (1962). *Russ. J. Inorg. Chem.* **7,** 1428.
1750. Nagasawa, H., Takeshita, S. K., Tomono, Y., Minomura, S. and Okai, B. (1964). *J. Phys. Soc. Japan* **19,** 2232.
1751. Namoradze, Z. G. and Zvyagintsev, O. E. (1939). *J. Appl. Chem.* (*U.S.S.R.*) **12,** 603.
1752. Nanda, R. K. and Wallbridge, M. G. H. (1964). *Inorg. Chem.* **3,** 1798.
1753. Nannelli, P. and Block, B. P. (1968). *Inorg. Chem.* **7,** 2423.
1754. Nannelli, P. and Block, B. P. (1969). *Inorg. Chem.* **8,** 1767.
1755. Narath, A. and Wallace, D. C. (1962). *Phys. Rev.* **127,** 724.
1756. Nast, R. and Kohl, H. (1963). *Z. anorg. Chem.* **320,** 135.
1757. Natta, G., Corradini, P. and Allegra, G. (1959). *Atti Accad. naz. Lincei, Rend. Classe Sci. fis. mat. nat.* **26,** 155.
1758. Natta, G., Corradini, P. and Allegra, G. (1961). *J. Polymer Sci.* **51,** 399.
1759. Natta, G., Corradini, P. and Bassi, I. W. (1958). *J. Amer. Chem. Soc.* **80,** 755.
1760. Natta, G., Corradini, P., Bassi, I. W. and Porri, L. (1958). *Atti Accad. naz. Lincei, Rend. Classe Sci. fis. mat. nat.* **24,** 121.
1761. Natta, G., Dall'asta, G., Mazzanti, G., Giannini, U. and Cesca, S. (1959). *Angew. Chem.* **71,** 205.
1762. Natta, G. and Mazzanti, G. (1960). *Tetrahedron* **8,** 86.
1763. Naumann, D. (1961). *Z. anorg. Chem.* **309,** 37.
1764. Naumann, A. W. and Hallada, C. J. (1964). *Inorg. Chem.* **3,** 70.
1765. Nayar, V. S. V. and Peacock, R. D. (1964). *J. Chem. Soc.*, 2827.
1766. Nazarova, L. A., Chernyaev, I. I. and Morozova, A. S. (1965). *Russ. J. Inorg. Chem.* **10,** 291.

1767. Nelson, W. H. and Tobins, R. S. (1963). *Inorg. Chem.* **2,** 985.
1768. Nelson, W. H. and Tobins, R. S. (1964). *Inorg. Chem.* **3,** 653.
1769. Nelson, W. H. and Tobins, R. S. (1964). *Canad. J. Chem.* **42,** 731.
1770. Nesmeyanov, A. N., Anisimov, K. N., Kolobova, N. E. and Beschastnov, A. S. (1964). *Doklady Akad. Nauk S.S.S.R.* **159,** 377.
1771. Nesmayanov, A. N., Gusev, A. I., Pasynskii, A. A., Anisimov, K. N., Kolobova, N. E. and Struckhov, Yu. T. (1968). *Chem. Comm.*, 1365.
1772. Nesmayanov, A. N., Gusev, A. I., Pasynskii, A. A., Anisimov, K. N., Kolobova, N. E. and Struchkov, Yu. T. (1969). *Chem. Comm.*, 277.
1773. Neumann, G. (1964). *Acta Chem. Scand.* **18,** 278.
1774. Neumann, H. M. and Cook, N. C. (1956). *J. Amer. Chem. Soc.* **79,** 3026.
1775. Newman, L., Lafleur, W. J., Brousaides, F. J. and Ross, A. M. (1958). *J. Amer. Chem. Soc.* **80,** 4491.
1776. Newnham, I. E. (1957). *J. Amer. Chem. Soc.* **79,** 5415.
1777. Newnham, I. E. and Watts, J. A. (1960). *J. Amer. Chem. Soc.* **82,** 2113.
1778. Newnham, R. E. and de Haan, Y. M. (1960). "Quart. Progress Report No. XXVI, Laboratory for Insulation Research, M.I.T., Cambridge, Mass. Quoted in reference 1027, p. 261.
1779. Newnham, R. E. and de Haan, Y. M. (1962). *Z. Krist.* **117,** 235.
1780. Newton, T. W. and Baker, F. B. (1964). *Inorg. Chem.* **3,** 569.
1781. Ngai, L. H., Stafford, F. E. and Schafer, L. (1969). *J. Amer. Chem. Soc.* **91,** 48.
1782. Nicholls, D. (1962). *J. Inorg. Nuclear Chem.* **24,** 1001.
1783. Nicholls, D. and Swindells, R. (1964). *J. Chem. Soc.*, 4204.
1784. Nicholls, D. and Wilkinson, D. N. (1969). *J. Chem. Soc.* (*A*), 1232.
1785. van Niekerk, J. N. and Shoening, F. R. L. (1953). *Acta Cryst.* **6,** 227.
1786. van Niekerk, J. N., Shoening, F. R. L. and de Wet, J. F. (1953). *Acta Cryst.* **6,** 501.
1787. Nigam, H. L., Nyholm, R. S. and Stiddard, M. H. B. (1960). *J. Chem. Soc.*, 1803.
1788. Nigam, H. L., Nyholm, R. S. and Stiddard, M. H. B. (1960). *J. Chem. Soc.*, 1806.
1789. Nikolaev, N. S. and Opalovskii, A. Z. (1959). *Zhur. neorg. Khim.* **4,** 1174.
1790. Nisel'son, L. A. (1957). *Zhur. neorg. Khim.* **2,** 816.
1791. Nisel'son, L. A. (1960). *Russ. J. Inorg. Chem.* **5,** 792.
1792. Nisel'son, L. A. and Petrusevich, I. V. (1960). *Russ. J. Inorg. Chem.* **5,** 120.
1793. Nisel'son, L. A., Pustil'nik, A. I. and Sokolova, T. D. (1964). *Russ. J. Inorg. Chem.* **9,** 574.
1794. Nisel'son, L. A. and Sokolova, T. D. (1965). *Russ. J. Inorg. Chem.* **10,** 9.
1795. Noble, A. M. and Winfield, J. M. (1969). *Chem. Comm.*, 151.
1796. Nordenjskold, I. (1901). *Berichte* **34,** 1572.
1797. Norin, R. (1963). *Acta Chem. Scand.* **17,** 1391; **20,** 871 (1966).
1798. Norin, R., Carlsson, M. and Elgquist, B. (1966). *Acta Chem. Scand.* **20,** 2892.
1799. Noth, H. and Hartwimmer, R. (1960). *Berichte* **93,** 2238.

1800. Novikov, G. I., Andrews, N. V. and Polyachenok, O. G. (1961). *Russ. J. Inorg. Chem.* **6,** 1019.
1801. Novikov, G. I. and Galitskii, N. V. (1965). *Russ. J. Inorg. Chem.* **10,** 313.
1802. Nunez, F. G. and Figueroa, E. (1938). *Compt. rend.* **206,** 437.
1803. Nyholm, R. S. and Sharpe, A. G. (1952). *J. Chem. Soc.*, 3579.
1804. Nyholm, R. S., Snow, M. R. and Stiddard, M. H. B. (1965). *J. Chem. Soc.*, 6564.
1805. Nyholm, R. S., Snow, M. R. and Stiddard, M. H. B. (1965). *J. Chem. Soc.*, 6570.
1806. O'Donnell, T. A. (1956). *J. Chem. Soc.*, 4681.
1807. O'Donnell, T. A. and Stewart, D. F. (1962). *J. Inorg. Nuclear Chem.* **24,** 309.
1808. O'Donnell, T. A. and Stewart, D. F. (1966). *Nature* **210,** 836.
1809. O'Donnell, T. A. and Stewart, D. F. (1966). *Inorg. Chem.* **5,** 1434.
1810. O'Donnell, T. A. and Wilson, P. W. (1968). *Austral. J. Chem.* **21,** 1415.
1811. Ogawa, S. (1960). *J. Phys. Soc. Japan* **15,** 1901.
1812. Opalovskii, A. A. and Batsanov, S. S. (1968). *Russ. J. Inorg. Chem.* **13,** 278.
1813. Opalovskii, A. A. and Khaldoyanidi, K. A. (1968). *Russ. J. Inorg. Chem.* **13,** 310.
1814. Opalovskii, A. A. and Samoilov, P. P. (1968). *Russ. J. Inorg. Chem.* **13,** 196.
1815. Opalovskii, A. A. and Shaburova, V. P. (1968). *Russ. J. Inorg. Chem.* **13,** 875.
1816. Oppegard, A. L., Smith, W. C., Muetterties, E. L. and Engelhardt, V. A. (1960). *J. Amer. Chem. Soc.* **82,** 3835.
1817. Oppermann, H. (1967). *Z. anorg. Chem.* **351,** 113.
1818. Oppermann, H. (1967). *Z. anorg. Chem.* **351,** 127.
1819. Oppermann, H. (1968). *Z. anorg. Chem.* **359,** 51.
1820. Osmond, W. P. (1966). *Proc. Phys. Soc.* **87,** 767.
1821. Ostertag, W. (1969). *Inorg. Chem.* **8,** 1372.
1822. Ostrowetsky, O. (1964). *Bull. Soc. chim. France*, 1003.
1823. Ostrowetsky, O. (1964). *Bull. Soc. chim. France*, 1012, 1018.
1824. Ott, F. (1912). *Z. Elektrochem.* **18,** 349.
1825. Ott, J. B., Coates, J. R., Jensen, R. J. and Mangelson, N. F. (1965). *J. Inorg. Nuclear Chem.* **27,** 2005.
1826. Ozerov, R. P., Gol'der, G. A. and Zhdanov, G. S. (1957). *Soviet Phys. Cryst.* **2,** 211.
1827. Ozin, G. A., Fowles, G. W. A., Tidmarsh, D. J. and Walton, R. A. (1969). *J. Chem. Soc.* (*A*), 642.
1828. Ozin, G. A. and Reynolds, D. J. (1969). *Chem. Comm.*, 884.
1829. Packer, K. J. and Muetterties, E. L. (1963). *J. Amer. Chem. Soc.* **85,** 3035.
1830. Pailleret, P., Borensztajn, J., Freundlich, W. and Rimsky, A. (1966). *Compt. rend.* **263C,** 1131.
1831. Pajdowski, L. (1966). *J. Inorg. Nuclear Chem.* **28,** 433.

1832. Pajdowski, L. and Jezowska-Trzebiatowska, B. (1966). *J. Inorg. Nuclear Chem.* **28,** 443.
1833. Palkin, A. P. and Chikanov, N. D. (1959). *Russ. J. Inorg. Chem.* **4,** 407; **7,** 1370, 2394, 2388 (1962).
1834. Palmer, K. J. (1938). *J. Amer. Chem. Soc.* **60,** 2360.
1835. Pande, K. C. and Mehrotra, R. C. (1958). *Chem. and Ind.*, 1198.
1836. Pankratova, L. N., Vlasov, L. G. and Lapitskii, A. V. (1964). *Russ. J. Inorg. Chem.* **9,** 954.
1837. Pantonin, J. A., Fischer, A. K. and Heintz, E. A. (1960). *J. Inorg. Nuclear Chem.* **14,** 145.
1838. Parish, R. V. (1969). Private Communication.
1839. Parish, R. V. (1966). *Spectrom. Acta* **22,** 1191.
1840. Parish, R. V. and Perkins, P. G. (1967). *J. Chem. Soc. (A)*, 345.
1841. Parish, R. V., Simms, P. G., Wells, M. A. and Woodward, L. A. (1968). *J. Chem. Soc. (A)*, 2882.
1842. Park, J. J., Glick, M. D. and Hoard, J. L. (1969). *J. Amer. Chem. Soc.* **91,** 301.
1843. Patil, H. R. H. and Graham, W. A. G. (1966). *Inorg. Chem.* **5,** 1401.
1844. Paul, R. C. and Chadha, S. L. (1969). *J. Inorg. Nuclear Chem.* **31,** 1679.
1845. Pauling, P. (1965). Private communication.
1846. Pauling, P. and Robertson, G. B. (1965). Private communication.
1847. Payne, D. S. (1956). *Rec. Trav. chim.* **75,** 620.
1848. Peacock, R. D. (1957). *Proc. Chem. Soc.*, 59.
1849. Peacock, R. D. (1957). *J. Chem. Soc.*, 4684.
1850. Peacock, R. D. (1960). *Progr. Inorg. Chem.* **2,** 193.
1851. Pearce, M. L. and McCabe, N. R. (1965). *J. Inorg. Nuclear Chem.* **27,** 1876.
1852. Pearson, A. D. (1958). *J. Phys. and Chem. Solids* **5,** 316.
1853. Pearson, I. M. and Garner, C. S. (1961). *J. Phys. Chem.* **65,** 690.
1854. Pechkovskii, V. V. and Vorob'ev, N. I. (1965). *Russ. J. Inorg. Chem.* **10,** 779.
1855. Pedersen, B. F. and Pedersen, B. (1963). *Acta Chem. Scand.* **17,** 557.
1856. Pence, H. E. and Selbin, J. (1969). *Inorg. Chem.* **8,** 353.
1857. Perakis, N. (1927). *J. Phys. Radium* **8,** 473.
1858. Perakis, N. and Wucher, J. (1952). *Compt. rend.* **235,** 354.
1859. Perlmutter-Hayman, B. (1965). *J. Phys. Chem.* **69,** 1736.
1860. Perloff, A. (1966). Doctoral Diss., Georgetown University, Washington, D.C.; *Diss. Abs.* **B27,** 2676.
1861. Perumareddi, R., Liehr, A. D. and Adamson, A. W. (1963). *J. Amer. Chem. Soc.* **85,** 249.
1862. Petersen, E. (1904). *Z. anorg. Chem.* **38,** 342.
1863. Petillon, F., Youinou, M.-T. and Guerchais, J. E. (1968). *Bull. Soc. chim. France*, 2375.
1864. Petit, M. and Massart, R. (1969). *Compt. rend.* **268C,** 1860.
1865. Pfeiffer, P., Hesse, T., Pfitzner, H., Scholl, W. and Thielert, H. (1937). *J. prakt. Chem.* **149,** 217.

1866. Pflugmacher, A., Cardruck, H. J. and Zucketto, M. (1958). *Naturwiss.* **45,** 490.
1867. Phillips, T., Sands, D. E. and Wagner, W. F. (1968). *Inorg. Chem.* **7,** 2295.
1868. Pinnavaia, T. J. and Fay, R. C. (1966). *Inorg. Chem.* **5,** 233.
1869. Pinsker, G. Z. (1966). *Soviet Phys. Cryst.* **11,** 634.
1870. Pinsker, G. Z. and Kuznetsov, V. G. (1968). *Soviet Phys. Cryst.* **13,** 56.
1871. Piovesana, O. and Selbin, J. (1969). *J. Inorg. Nuclear Chem.* **31,** 433.
1872. Piper, T. S. and Wilkinson, G. (1956). *J. Inorg. Nuclear Chem.* **2,** 38; **3,** 104.
1873. Piutti, A. (1879). *Gazzetta* **9,** 538.
1874. Plane, R. A. and Taube, H. (1952). *J. Phys. Chem.* **56,** 33.
1875. Plastas, H. J., Stewart, J. M. and Grim, S. O. (1969). *J. Amer. Chem. Soc.* **91,** 4326.
1876. van de Poel, J. and Neumann, H. M. (1968). *Inorg. Chem.* **7,** 2086.
1877. Papaconstantinou, E. and Pope, M. T. (1967). *Inorg. Chem.* **6,** 1152.
1878. Pope, M. T. and Papaconstantinou, E. (1967). *Inorg. Chem.* **6,** 1147.
1879. Pope, M. T. and Varga, G. M. (1966). *Chem. Comm.*, 653.
1880. Pope, M. T. and Varga, G. M. (1966). *Inorg. Chem.* **5,** 1249.
1881. Porai-Koshits, M. A. and Antsyshkina, A. R. (1962). *Doklady Akad. Nauk. S.S.S.R.* **146,** 1102.
1882. Pouraud, M. and Chaigneau, M. (1959). *Compt rend.* **249,** 2568.
1883. Powell, H. M. and Wells, A. F. (1935). *J. Chem. Soc.*, 1008.
1884. Prandtl, W. and Beyer, B. (1910). *Z. anorg. Chem.* **67,** 257.
1885. Prandtl, W. and Hess, L. (1913). *Z. anorg. Chem.* **82,** 103.
1886. Prasad, S. and Krishnaiah, K. S. R. (1960). *J. Indian Chem. Soc.* **37,** 588.
1887. Prasad, S. and Krishnaiah, K. S. R. (1960). *J. Indian Chem. Soc.* **37,** 681.
1888. Prasad, S. and Krishnaiah, K. S. R. (1961). *J. Indian Chem. Soc.* **38,** 153.
1889. Prasad, S. and Krishnaiah, K. S. R. (1961). *J. Indian Chem. Soc.* **38,** 177.
1890. Prasad, S. and Krishnaiah, K. S. R. (1961). *J. Indian Chem. Soc.* **38,** 182.
1891. Prasad, S. and Krishnaiah, K. S. R. (1961). *J. Indian Chem. Soc.* **38,** 352.
1892. Prasad, S. and Srivastava, R. C. (1965). *Z. anorg. Chem.* **337,** 221.
1893. Prasad, S. and Srivastava, R. C. (1965). *Z. anorg. Chem.* **340,** 325.
1894. Pratt, D. W. and Myers, R. J. (1967). *J. Amer. Chem. Soc.* **89,** 6470.
1895. Pregaglia, G., Andreetta, A., Ferrari, G. and Ugo, R. (1969). *Chem. Comm.*, 590.
1896. Preisler, E. and Glemser, O. (1960). *Z. anorg. Chem.* **303,** 316.
1897. Preiss, H. (1966). *Z. anorg. Chem.* **346,** 272.
1898. Preiss, H. (1968). *Z. anorg. Chem.* **362,** 13.
1899. Preston, F. J. and Reed, R. I. (1966). *Chem. Comm.*, 51.
1900. Priest, H. F. and Schumb, W. C. (1948). *J. Amer. Chem. Soc.* **70,** 3378.
1901. Prinz, R. and Werner, H. (1967). *Angew. Chem. Internat. Edn.* **6,** 91.
1902. Puri, D. M. and Mehrotra, R. C. (1962). *J. Indian Chem. Soc.* **39,** 499.
1903. Rabenau, A. (1967). *Angew. Chem. Internat. Edn.* **6,** 68.
1904. Racah, G. (1943). *J. Chem. Phys.* **11,** 214.
1905. Ragsdale, R. O. and Stewart, B. B. (1964). *Proc. Chem. Soc.*, 194.

1906. Ramaiah, K., Anderson, F. E. and Martin, D. F. (1964). *Inorg. Chem.* **3,** 296. Ramaiah, K. and Martin, D. F. (1965). *J. Inorg. Nuclear Chem.* **27,** 1663. Martin, D. F. and Ramaiah, K. (1965). *J. Inorg. Nuclear Chem.* **27,** 2027.
1907. Randic, M. (1960). *Croat. Chem. Acta* **32,** 189.
1908. Randic, M. (1962). *J. Chem. Phys.* **36,** 2094.
1909. Randic, M. and Maksic, Z. (1966). *Theor. Chim. Acta* **4,** 145.
1910. Rasmussen, P. G., Kuska, H. A. and Brubaker, C. H. (1965). *Inorg. Chem.* **4,** 343.
1911. Ray, T. C. and Westland, A. D. (1965). *Inorg. Chem.* **4,** 1501.
1912. Reichert, K. H. and Mallman, M. (1969). *Angew. Chem. Internat. Edn.* **8,** 217.
1913. Reid, A. F., Shannon, J. S., Swan, J. M. and Wailes, P. C. (1965). *Austral. J. Chem.* **18,** 173.
1914. Reid, A. F. and Sienko, M. J. (1967). *Inorg. Chem.* **6,** 321.
1915. Reid, A. F. and Wailes, P. C. (1964). *J. Organometallic Chem.* **2,** 329.
1916. Reid, A. F. and Wailes, P. C. (1965). *Austral. J. Chem.* **18,** 9.
1917. Remy, H. and May, I. (1961). *Naturwiss.* **48,** 524.
1918. Renz, C. (1903). *Z. anorg. Chem.* **36,** 103.
1919. Reubin, J. and Fiat, D. (1967). *Inorg. Chem.* **6,** 579.
1920. Reynolds, L. T. and Wilkinson, G. (1958). *J. Inorg. Nuclear Chem.* **9,** 86.
1921. Richards, S., Pedersen, B., Silverton, J. V. and Hoard, J. L. (1964). *Inorg. Chem.* **3,** 27.
1922. Ridgley, D. and Ward, R. (1955). *J. Amer. Chem. Soc.* **77,** 6132.
1923. Riley, R. F. and Ho, L. (1962). *J. Inorg. Nuclear Chem.* **24,** 1121.
1924. Rillema, D. P. and Brubaker, C. H. (1969). *Inorg. Chem.* **8,** 1645.
1925. Rillema, D. P., Reagan, W. J. and Brubaker, C. H. (1969). *Inorg. Chem.* **8,** 587.
1926. Rinke, K. and Schafer, H. (1967). *Angew. Chem. Internat. Edn.* **6,** 637.
1927. Ripan, R. and Puscasiu, M. (1968). *Z. anorg. Chem.* **358,** 82.
1928. Rivest, R. (1962). *Canad. J. Chem.* **40,** 2234.
1929. Rivet, E., Aubin, R. and Rivest, R. (1961). *Canad. J. Chem.* **39,** 2343.
1930. Robbins, G. D., Thoma, R. E. and Insley, H. (1965). *J. Inorg. Nuclear Chem.* **27,** 559.
1931. Robin, M. B. and Day, P. (1967). *Adv. Inorg. Chem. Radiochem.* **10,** 247.
1932. Robin, M. B. and Kuebler, N. A. (1965). *Inorg. Chem.* **4,** 978.
1933. Rogers, A. and Staples, P. J. (1965). *J. Chem. Soc.*, 6834.
1934. Rogers, D. B., Shannon, R. D., Sleight, A. W. and Gillson, J. L. (1969). *Inorg. Chem.* **8,** 841.
1935. Rollins, O. W. and Baker, L. C. W. (1969). *Inorg. Chem.* **8,** 397.
1936. Rollins, O. W. and Earley, J. E. (1959). *J. Amer. Chem. Soc.* **81,** 5571.
1937. Rollinson, C. L. and White, R. C. (1962). *Inorg. Chem.* **1,** 281.
1938. Rolsten, R. F. (1958). *J. Amer. Chem. Soc.* **80,** 2952.
1939. Rolsten, R. F. (1958). *J. Phys. Chem.* **62,** 126.
1940. Rolsten, R. F. and Sisler, H. H. (1957). *J. Amer. Chem. Soc.* **79,** 1068.
1941. Rolsten, R. F. and Sisler, H. H. (1957). *J. Amer. Chem. Soc.* **79,** 1819.

1942. Rolsten, R. F. and Sisler, H. H. (1957). *J. Amer. Chem. Soc.* **79,** 5891.
1943. Romanov, G. V. and Spiridonov, V. P. (1966). *J. Struct. Chem.* **7,** 816.
1944. Roscoe, H. E. (1869). *Ann. chim. Supplement* **7,** 70.
1945. Roscoe, H. E. (1878). *Chem. News* **37,** 25.
1946. Rosenheim, A., Abel, G. and Lewy, R. (1931). *Z. anorg. Chem.* **197,** 189.
1947. Rosenheim, A. and Koss, M. (1906). *Z. anorg. Chem.* **49,** 148.
1948. Rosenheim, A., Loewenstein, W. and Singer, L. (1903). *Berichte* **36,** 1833.
1949. Rosenheim, A. and Mong, H. (1925). *Z. anorg. Chem.* **148,** 25.
1950. Rossotti, F. J. C. and Rossotti, H. S. (1956). *Acta Chem. Scand.* **10,** 957.
1951. Rost, E. (1959). *J. Amer. Chem. Soc.* **81,** 3843.
1952. Roth, R. S. and Wadsley, A. D. (1965). *Acta Cryst.* **18,** 724.
1953. Roth, R. S. and Wadsley, A. D. (1965). *Acta Cryst.* **19,** 26, 32, 38, 42.
1954. Rowe, R. A., Jones, M. M., Bryant, B. E. and Fernelius, W. C. (1957). *Inorg. Synth.* **5,** 113.
1955. Rudorff, W. and Krug, D. (1964). *Z. anorg. Chem.* **329,** 211.
1956. Rudorff, W. and Luginsland, H.-H. (1964). *Z. anorg. Chem.* **334,** 125.
1957. Rudorff, W. and Marklin, J. (1964). *Z. anorg. Chem.* **334,** 142.
1958. Rudorff, W., Walter, G. and Stadler, J. (1958). *Z. anorg. Chem.* **297,** 1.
1959. Ruff, J. K. (1967). *Inorg. Chem.* **6,** 1502.
1960. Ruff, J. K. (1967). *Inorg. Chem.* **6,** 2080.
1961. Ruff, O. and Ascher, E. (1931). *Z. anorg. Chem.* **196,** 413.
1962. Ruff, O. and Eisner, F. (1907). *Berichte* **40,** 2926.
1963. Ruff, O. and Eisner, F. (1907). *Z. anorg. Chem.* **52,** 256.
1964. Ruff, O., Eisner, F. and Heller, W. (1907). *Z. anorg. Chem.* **52,** 256.
1965. Ruff, O. and Lickfett, H. (1911). *Berichte* **44,** 2534.
1966. Ruff, O. and Lickfett, H. (1911). *Berichte* **44,** 2539.
1967. Ruff, O. and Neumann, F. (1923). *Z. anorg. Chem.* **128,** 81.
1968. Ruff, O. and Schiller, E. (1911). *Z. anorg. Chem.* **72,** 329.
1969. Ruff, O. and Thomas, F. (1925). *Z. anorg. Chem.* **148,** 19.
1970. Ruff, O. and Zedner, J. (1909). *Berichte* **42,** 492.
1971. Russ, B. T. and Fowles, G. W. A. (1966). *Chem. Comm.*, 19.
1972. Russ, B. J. and Wood, J. S. (1966). *Chem. Comm.*, 745.
1973. Ryan, J. L. (1969). *Inorg. Chem.* **8,** 2058.
1974. Sabatini, A. and Bertini, I. (1966). *Inorg. Chem.* **5,** 204.
1975. Sacconi, L. and Cini, R. (1954). *J. Amer. Chem. Soc.* **76,** 4239.
1976. Safanov, V. V., Korshunov, B. G., Shevtsova, Z. N. and Bakum, S. I. (1964). *Russ. J. Inorg. Chem.* **9,** 914.
1977. Safonov, V. V., Korshunov, B. G., Shevtsova, Z. N. and Shadrova, L. G. (1964). *Russ. J. Inorg. Chem.* **9,** 763.
1978. Safonov, V. V., Korshunov, B. G., Shevtsova, Z. N. and Shadrova, L. G. (1965). *Russ. J. Inorg. Chem.* **10,** 359.
1979. Safanov, V. V., Korshunov, B. G. and Steblovskaya, S. N. (1966). *Russ. J. Inorg. Chem.* **11,** 1148.
1980. Safanov, V. V., Korshunov, B. G. and Zimina, T. N. (1966). *Russ. J. Inorg. Chem.* **11,** 488.

1981. Safanov, V. V., Korshunov, B. G., Zimina, T. N. and Shevtsova, Z. N. (1966). *Russ. J. Inorg. Chem.* **11,** 1146.
1982. Saillant, R., Hayden, J. L. and Wentworth, R. A. D. (1967). *Inorg. Chem.* **6,** 1497.
1983. Saillant, R. and Wentworth, R. A. D. (1968). *Inorg. Chem.* **7,** 1606.
1984. Saillant, R. and Wentworth, R. A. D. (1969). *Inorg. Chem.* **8,** 1226.
1985. Saillant, R. and Wentworth, R. A. D. (1969). *J. Amer. Chem. Soc.* **91,** 2174.
1986. Salzmann, J.-J. and Mosimann, P. (1967). *Helv. Chim. Acta* **50,** 1831.
1987. Sands, D. E. and Zalkin, A. (1959). *Acta Cryst.* **12,** 723.
1988. Sands, D. E., Zalkin, A. and Elson, R. E. (1959). *Acta Cryst.* **12,** 21.
1989. Sarry, B. (1967). *Angew. Chem. Internat. Edn.* **6,** 571.
1990. Sarry, B. and Dettke, M. (1963). *Angew. Chem. Internat. Edn.* **2,** 690.
1991. Sasaki, Y. (1961). *Acta Chem. Scand.* **15,** 175.
1992. Sasaki, Y., Lindqvist, I. and Sillen, L. G. (1959). *J. Inorg. Nuclear Chem.* **9,** 93.
1993. Sasaki, Y. and Sillen, L. G. (1964). *Acta Chem. Scand.* **18,** 1014.
1994. Sasaki, Y. and Sillen, L. G. (1968). *Arkiv Kemi* **29,** 253.
1995. Sathyanarayana, D. N. and Patel, C. C. (1965). *J. Inorg. Nuclear Chem.* **27,** 297.
1996. Sathyanarayana, D. N. and Patel, C. C. (1965). *J. Inorg. Nuclear Chem.* **27,** 2549.
1997. Savchenko, G. S. and Tananaev, I. V. (1946). *J. Appl. Chem.* (*U.S.S.R.*) **19,** 1093.
1998. Scaife, D. E. (1967). *Austral. J. Chem.* **20,** 845.
1999. Scane, J. G. (1967). *Acta Cryst.* **23,** 85.
2000. Scane, J. G. and Stephens, R. M. (1967). *Proc. Phys. Soc.* **92,** 833.
2001. Scavnicar, S. and Grdenic, D. (1955). *Acta Cryst.* **8,** 275.
2002. Scavnicar, S. and Herceg, M. (1966). *Acta Cryst.* **21,** A151.
2003. Scavnicar, S. and Prodic, B. (1965). *Acta Cryst.* **18,** 698.
2004. Schaefer, W. P. (1965). *Inorg. Chem.* **4,** 642.
2005. Schafer, H. (1955). *Angew. Chem.* **67,** 748.
2006. Schafer, H. and Bauer, D. (1965). *Z. anorg. Chem.* **340,** 62.
2007. Schafer, H., Bauer, D., Beckmann, W., Gerkin, R., Nieder-Vahrenholz, H.-G., Niehues, K.-J. and Scholz, H. (1964). *Naturwiss.* **51,** 241.
2008. Schafer, H. and Bayer, L. (1954). *Z. anorg. Chem.* **277,** 140.
2009. Schafer, H., Bayer, L. and Lehmann, H. (1952). *Z. anorg. Chem.* **268,** 268.
2010. Schafer, H. and Beckmann, W. (1966). *Z. anorg. Chem.* **347,** 225.
2011. Schafer, H. and Dohmann, K. D. (1959). *Z. anorg. Chem.* **300,** 1.
2012. Schafer, H. and Dohmann, K. D. (1961). *Z. anorg. Chem.* **311,** 134.
2013. Schafer, H. and Gerkin, R. (1962). *Z. anorg. Chem.* **317,** 105.
2014. Schafer, H., Gerkin, R. and Scholz, H. (1965). *Z. anorg. Chem.* **335,** 96.
2015. Schafer, H., Giegling, D. and Rinke, K. (1968). *Z. anorg. Chem.* **357,** 25.
2016. Schafer, H., Goser, C. and Bayer, L. (1951). *Z. anorg. Chem.* **265,** 258.
2017. Schafer, H. and Grau, L. (1954). *Z. anorg. Chem.* **275,** 198.
2018. Schafer, H. and Jori, M. (1954). *Z. anorg. Chem.* **277,** 341.
2019. Schafer, H. and Kahlenberg, F. (1960). *Z. anorg. Chem.* **305,** 291.

2020. Schafer, H. and Kahlenberg, F. (1960). *Z. anorg. Chem.* **305,** 327.
2021. Schafer, H. and Kahlenberg, F. (1958). *Z. anorg. Chem.* **294,** 242.
2022. Schafer, H. and Kahlenberg, F. (1960). *Z. anorg. Chem.* **305,** 178.
2023. Schafer, H. and Liedmeier, F. (1964). *J. Less-Common Metals* **6,** 307.
2024. Schafer, H., Schibilla, E., Gerken, R. and Scholz, H. (1964). *J. Less-Common Metals* **6,** 239.
2025. Schafer, H., Schneidereit, G. and Gerhardt, W. (1963). *Z. anorg. Chem.* **319,** 327.
2026. Schafer, H. and Schnering, H. G. (1964). *Angew Chem.* **76,** 833.
2027. Schafer, H., Schnering, H. G., Neihues, K.-J. and Nieder-Vahrenholz, H. G. (1965). *J. Less-Common Metals* **9,** 95.
2028. Schafer, H., v. Schnering, H. G., Simon, A., Giegling, D., Bauer, D., Siepmann, R. and Spreckelmeyer, B. (1965). *J. Less-Common Metals* **10,** 154.
2029. Schafer, H., v. Schnering, H. G., Tillack, J., Kuhnen, F., Wohrle, H. and Baumann, H. (1967). *Z. anorg. Chem.* **353,** 281.
2030. Schafer, H., Scholz, H. and Gerken, R. (1964). *Z. anorg. Chem.* **331,** 154.
2031. Schafer, H. and Sibbing, E. (1960). *Z. anorg. Chem.* **305,** 340.
2032. Schafer, H., Sibbing, E. and Gerkin, R. (1961). *Z. anorg. Chem.* **307,** 163.
2033. Schafer, H. and Siepmann, R. (1966). *J. Less-Common Metals* **11,** 76.
2034. Schafer, H. and Siepmann, R. (1968). *Z. anorg. Chem.* **357,** 273.
2035. Schafer, H. and Spreckelmeyer, B. (1966). *J. Less-Common Metals* **11,** 73.
2036. Schafer, H. and Tillack, J. (1964). *J. Less-Common Metals* **6,** 152.
2037. Schafer, H. and Wartenpfuhl, F. (1961). *J. Less-Common Metals* **3,** 29.
2038. Schafer, H. and Wartenpfuhl, F. (1961). *Z. anorg. Chem.* **308,** 282.
2039. Schafer, H., Wartenpfuhl, F. and Weise, E. (1958). *Z. anorg. Chem.* **295,** 268.
2040. Schafer, H., Wiese, U., Rinke, K. and Brendel, K. (1967). *Angew Chem. Internat. Edn.* **6,** 253.
2041. Schafer, H. and Zylka, L. (1965). *Z. anorg. Chem.* **338,** 309.
2042. Schaffer, C. E. (1958). *J. Inorg. Nuclear Chem.* **8,** 149.
2043. Schiller, K. and Thilo, E. (1961). *Z. anorg. Chem.* **310,** 261.
2044. Schindler, F. and Schmidbaur, H. (1967). *Angew. Chem. Internat. Edn.* **6,** 683.
2045. Schlaefer, H. (1957). *Z. phys. Chem.* **11,** 277.
2046. Schlafer, H. L. and Christ, K. (1967). *Z. anorg. Chem.* **349,** 289.
2047. Schlafer, H. L. and Fritz, H. P. (1967). *Spectrochim. Acta* **23A,** 1409.
2048. Schlafer, H. L. and Gotz, R. (1961). *Z. anorg. Chem.* **309,** 104.
2049. Schlafer, H. L. and Gotz, R. (1964). *Z. anorg. Chem.* **328,** 1.
2050. Schlafer, H. L. and Kling, O. (1959). *Z. anorg. Chem.* **302,** 1; **309,** 245 (1961).
2051. Schlafer, H. L. and Schroeder, W. (1966). *Z. anorg. Chem.* **347,** 45, 59.
2052. Schlafer, H. L. and Skoludek, H. (1962). *Z. anorg. Chem.* **316,** 15.
2053. Schlafer, H. L. and Wille, H.-W. (1964). *Z. anorg. Chem.* **327,** 253.
2054. Schlafer, H. L. and Wille, H.-W. (1967). *Z. anorg. Chem.* **351,** 279.
2055. Schmeisser, M. (1955). *Angew. Chem.* **67,** 493.

2056. Schmeisser, M. and Brandle, K. (1961). *Angew. Chem.* **73,** 388.
2057. Schmeisser, M. and Lutzow, D. (1954). *Angew. Chem.* **66,** 230.
2058. Schmitz-Dumont, O., Bruns, I. and Heckmann, I. (1953). *Z. anorg. Chem.* **271,** 347.
2059. Schmitz-Dumont, O. and Heckmann, I. (1952). *Z. anorg. Chem.* **267,** 277.
2060. Schmitz-Dumont, O. and Opgenhoff, P. (1954). *Z. anorg. Chem.* **275,** 21; **276,** 235.
2061. Schmitz-Dumont, O. and Ross, B. (1964). *Angew. Chem. Internat. Edn.* **3,** 315; *Z. anorg. Chem.* **342,** 82 (1966).
2062. Schneider, R. F. and Mackay, R. A. (1968). *J. Chem. Phys.* **48,** 843.
2063. v. Schnering, H. G. (1966). *Naturwiss.* **53,** 359.
2064. v. Schnering, H. G. and Beckmann, W. (1966). *Z. anorg. Chem.*, **64,** 231.
2065. v. Schnering, H. G., Brodersen, K., Moers, F., Breitbach, H. K. and Thiele, G. (1966). *J. Less-Common Metals* **11,** 288.
2066. v. Schnering, H. G. and Mertin, W. (1964). *Naturwiss.* **51,** 552.
2067. v. Schnering, H. G. and Wohrle, H. (1963). *Naturwiss.* **50,** 91.
2068. v. Schnering, H. G. and Wohrle, H. (1963). *Angew. Chem. Internat. Edn.* **2,** 558.
2069. v. Schnering, H. G., Wohrle, H. and Schafer, H. (1961). *Naturwiss.* **48,** 159.
2070. Scholder, R., Schwochow, F. and Schwarz, H. (1968). *Z. anorg. Chem.* **363,** 10.
2071. Schonberg, N. (1954). *Acta Chem. Scand.* **8,** 240.
2072. Schrauzer, G. N. and Mayweg, V. P. (1966). *J. Amer. Chem. Soc.* **88,** 3235.
2073. Schroder, F. (1965). *Naturwiss.* **52,** 389.
2074. Schroder, J. and Grewe, F. J. (1968). *Angew. Chem. Internat. Edn.* **7,** 132.
2075. Schulze, G. E. R. (1934). *Z. Krist.* **89,** 477.
2076. Schulz, C. O. and Stafford, F. E. (1968). *J. Phys. Chem.* **72,** 4686.
2077. Schumaker, E. and Taubenest, R. (1966). *Helv. Chim. Acta* **49,** 1447.
2078. Schumb, W. C. and Morehouse, C. K. (1947). *J. Amer. Chem. Soc.* **69,** 2696.
2079. Schumb, W. C. and Sundstrom, R. F. (1933). *J. Amer. Chem. Soc.* **55,** 596.
2080. Schunn, R. A., Fritchie, C. J. and Hewitt, C. T. (1966). *Inorg. Chem.* **5,** 892.
2081. Schwartz, D. and Bernd, P. (1964). *J. Less-Common Metals* **7,** 108.
2082. Schwartz, D. and Larson, B. (1963). *J. Less-Common Metals* **5,** 365.
2083. Schwarzenbach, G. and Geier, G. (1963). *Helv. Chim. Acta* **46,** 906.
2084. Schwarzenbach, G. and Parissakis, G. (1958). *Helv. Chim. Acta.* **41,** 2425.
2085. Schwarzenbach, G. and Sandera, J. (1953). *Helv. Chim. Acta* **36,** 1089.
2086. Seabaugh, P. W. and Corbett, J. D. (1965). *Inorg. Chem.* **4,** 176.
2087. Sears, D. R. and Burns, J. H. (1964). *J. Chem. Phys.* **41,** 3478.
2088. Seifert, H. J. (1962). *Z. anorg. Chem.* **317,** 123.
2089. Seifert, H. J. (1963). *Angew. Chem. Internat. Edn.* **2,** 558.
2090. Seifert, H. J. and Auel, T. (1968). *J. Inorg. Nuclear Chem.* **30,** 2081.
2091. Seifert, H. J. and Auel, T. (1968). *Z. anorg. Chem.* **360,** 50.
2092. Seifert, H. J. and Ehrlich, P. (1959). *Z. anorg. Chem.* **302,** 284.

2093. Seifert, H. J., Fink, H. and Just, E. (1968). *Naturwiss.* **55,** 297.
2094. Siefert, H. J. and Gerstenberg, B. (1962). *Z. anorg. Chem.* **315,** 56.
2095. Seifert, H. J. and Klatyk, K. (1964). *Z. anorg. Chem.* **334,** 113.
2096. Seifert, H. J. and Loh, H. W. (1966). *Inorg. Chem.* **5,** 1822.
2097. Seifert, H. J., Loh, H. W. and Jungnickel, K. (1968). *Z. anorg. Chem.* **360,** 62.
2098. Seip, H. M. and Seip, R. (1966). *Acta Chem. Scand.* **20,** 2698.
2099. Selbin, J. (1965). *Chem. Rev.* **65,** 153.
2100. Selbin, J. (1966). *Angew. Chem. Internat. Edn.* **5,** 712.
2101. Selbin, J. and Holmes, L. H. (1962). *J. Inorg. Nuclear Chem.* **24,** 1111.
2102. Selbin, J., Holmes, L. H. and McGlynn, S. P. (1963). *J. Inorg. Nuclear Chem.* **25,** 1359.
2103. Selbin, J., Maus, G. and Johnson, D. L. (1967). *J. Inorg. Nuclear Chem.* **29,** 1735.
2104. Selig, H. and Claassen, H. H. (1966). *J. Chem. Phys.* **44,** 1404.
2105. Senderoff, S. and Brenner, A. (1954). *J. Electrochem. Soc.* **101,** 28.
2106. Senderoff, S. and Labrie, R. J. (1955). *J. Electrochem. Soc.* **102,** 77.
2107. Senff, H. and Klemm, W. (1940). *J. prakt. Chem.* **54,** 73.
2108. Sense, K. A., Snyder, M. J. and Filbert, R. B. (1954). *J. Phys. Chem.* **58,** 995.
2109. Serpone, N. and Fay, R. C. (1967). *Inorg. Chem.* **6,** 1835.
2110. Shannon, R. D. (1968). *Solid State Comm.* **6,** 139.
2111. Sharpe, A. G. (1960). *Adv. Fluorine Chem.* **1,** 29.
2112. Sharpe, A. G. and Woolf, A. A. (1951). *J. Chem. Soc.*, 798.
2113. Shchukarev, S. A. and Perfilova, I. L. (1963). *Russ. J. Inorg. Chem.* **8,** 1100.
2114. Shchukarev, S. A., Smirnova, E. K. and Shemyakina, T. S. (1962). *Russ. J. Inorg. Chem.* **7,** 1147.
2115. Shchukarev, S. A., Tolmacheva, T. A. and Tsintsius, V. M. (1962). *Russ. J. Inorg. Chem.* **7,** 777.
2116. Shchukarev, S. A., Vasil'kova, I. V. and Korol'kov, D. V. (1964). *Russ. J. Inorg. Chem.* **9,** 980.
2117. Sheka, I. A. (1959). *Chem. Zvesti* **13,** 656.
2118. Sheka, I. A., Lastochkina, A. A. and Malinko, L. A. (1968). *Russ. J. Inorg. Chem.* **13,** 1531.
2119. Sheka, I. A., Voitovich, B. A. and Nisel'son, L. A. (1959). *Russ. J. Inorg. Chem.* **4,** 813.
2120. Sheldon, J. C. (1959). *Nature, Lond.* **184,** 1210.
2121. Sheldon, J. C. (1960). *J. Chem. Soc.*, 1007.
2122. Sheldon, J. C. (1960). *J. Chem. Soc.*, 3106.
2123. Sheldon, J. C. (1961). *J. Chem. Soc.*, 750.
2124. Sheldon, J. C. (1961). *Chem. and Ind.*, 323.
2125. Sheldon, J. C. (1962). *J. Chem. Soc.*, 410.
2126. Sheldon, J. C. (1963). *J. Chem. Soc.*, 4183.
2127. Sheldon, J. C. (1964). *J. Chem. Soc.*, 1287.
2128. Sheldon, J. C. (1964). *Austral. J. Chem.* **17,** 1191.

2129. Sheldon, J. C. and Tyree, S. Y. (1958). *J. Amer. Chem. Soc.* **80,** 4775.
2130. Shiloff, J. C. (1960). *J. Phys. Chem.* **64,** 1566.
2131. Shimao, E. (1967). *Nature, Lond.* **214,** 170; *Bull. Chem. Soc. Japan* **40,** 1609.
2132. Shriver, D. F. (1962). *J. Amer. Chem. Soc.* **84,** 4611.
2133. Siddiqui, M. T., Ahmad, N. and Rahman, S. M. F. (1965). *Z. anorg. Chem.* **336,** 110.
2134. Siegel, S. (1952). *Acta Cryst.* **5,** 683.
2135. Siegel, S. (1956). *Acta Cryst.* **9,** 684.
2136. Siegert, F. W. and de Liefde Meijer, H. J. (1968). *Rec. Trav. chim.* **87,** 1445.
2137. Siegert, F. W. and de Liefde Meijer, H. J. (1968). *J. Organometallic Chem.* **15,** 131.
2138. Sienko, M. J. (1959). *J. Amer. Chem. Soc.* **81,** 5556.
2139. Sienko, M. J. and Morehouse, S. M. (1963). *Inorg. Chem.* **2,** 485.
2140. Sienko, M. J. and Truong, T. B. N. (1961). *J. Amer. Chem. Soc.* **83,** 3939.
2141. Siepmann, R. and Schafer, H. (1965). *Naturwiss.* **52,** 344.
2142. Siepmann, R. and v. Schnering, H.-G. (1968). *Z. anorg. Chem.* **357,** 289.
2143. Siepmann, R., v. Schnering, H.-G. and Schafer, H. (1967). *Angew. Chem. Internat. Edn.* **6,** 657.
2144. Signer, R. and Gross, H. (1934). *Helv. Chim. Acta* **17,** 1076.
2145. Silverton, J. V. and Hoard, J. L. (1963). *Inorg. Chem.* **2,** 243.
2146. Sime, R. J. (1967). *Z. Krist.* **124,** 238.
2147. Simon, A. (1967). *Z. anorg. Chem.* **355,** 311.
2148. Simon, A. and v. Schnering, H. G. (1966) *J. Less-Common Metals* **11,** 31.
2149. Simon, A., v. Schnering, H. G. and Schafer, H. (1967). *Z. anorg. Chem.* **355,** 295.
2150. Simon, A., v. Schnering, H.-G. and Schafer, H. (1968). *Z. anorg. Chem.* **361,** 235.
2151. Simon, A., v. Schnering, H.-G., Wohrle, H. and Schafer, H. (1965). *Z. anorg. Chem.* **339,** 155.
2152. Simon, J. and Jahr, K. F. (1964). *Z. Naturforsch..* **19b,** 165.
2153. Simon, J. P. and Souchay, P. (1956). *Bull. Soc. chim. France,* 1402.
2154. Simons, E. L. (1964). *Inorg. Chem.* **3,** 1079.
2155. Simons, J. H. and Powell, M. G. (1945). *J. Amer. Chem. Soc.* **67,** 75.
2156. Simons, P. Y. and Dachille, F. (1967). *Acta Cryst.* **23,** 334.
2157. Singer, J. and Cromer, D. T. (1959). *Acta Cryst.* **12,** 719.
2158. Singer, N., Studd, B. F. and Swallow, A. G. (1970). *Chem. Comm.,* 342.
2159. Siratori, K. and Iida, S. (1962). *J. Phys. Soc. Japan* **17,** Supp. B–1, 208.
2160. Skapski, A. C., Troughton, P. G. H. and Sutherland, H. H. (1968). *Chem. Comm.,* 1418.
2161. Skinner, H. A. and Sutton, L. E. (1940). *Trans. Faraday Soc.* **36,** 668.
2162. Slawisch, A. (1969). *Naturwiss.* **56,** 369.
2163. Sleight, A. W. (1966). *Acta Chem. Scand.* **20,** 1102.
2164. Sleight, A. W. (1969). *Inorg. Chem.* **8,** 1764.
2165. Sloan, C. L. and Barber, W. A. (1959). *J. Amer. Chem. Soc.* **81,** 1364.

2166. Sloan, T. E. and Wojcicki, A. (1968). *Inorg. Chem.* **7,** 1268.
2167. Smale, A. (1967). *Monatsh.* **98,** 163.
2168. Smirnov, M. V. and Kudyakov, V. Ya. (1965). *Russ. J. Inorg. Chem.* **10,** 655.
2169. Smirnova, E. K., Vasil'kova, I. V. and Prokhorova, L. I. (1968). *Russ. J. Inorg. Chem.* **13,** 933.
2170. Smith, A. E., Schrauzer, G. N., Mayweg, V. P. and Heinrich, W. (1965). *J. Amer. Chem. Soc.* **87,** 5798.
2171. Smith, G. S., Johnson, Q. and Elson, R. E. (1967). *Acta Cryst.* **22,** 300.
2172. Smith, P. W. and Wedd, A. G. (1966). *J. Chem. Soc.* (*A*), 231; 1377 (1968).
2173. Sneeden, R. P. A., Glockling, F. and Zeiss, H. (with Bonfiglioli, R.) (1966). *J. Organometallic Chem.* **6,** 149.
2174. Sneeden, R. P. A. and Throndsen, H. P. (1966). *J. Organometallic Chem.* **6,** 542.
2175. Snow, M. R., Pauling, P. and Stiddard, M. H. B. (1969). *Austral. J. Chem.* **22,** 709.
2176. Snow, M. R. and Stiddard, M. H. B. (1965). *Chem. Comm.*, 580.
2177. Souchay, P. (1943). *Ann. Chim.* **18** (11), 73.
2178. Souchay, P. (1943). *Ann. Chim.* **18** (11), 169.
2179. Souchay, P. (1945). *J. Chim. phys.* **42,** 61.
2180. Souchay, P. (1947). *Ann. Chim.* **2,** 203.
2181. Souchay, P. (1963). "Polyanions et Polycations". Gauthier-Villars, Paris.
2182. Souchay, P. (1969). *Compt. rend.* **268C,** 804.
2183. Souchay, P. and Carpeni, G. (1946). *Bull. Soc. chim. France* **13,** 160.
2184. Souchay, P., Chauveau, F. and Courtin, P. (1968). *Bull. Soc. chim. France*, 2384.
2185. Souchay, P. and Cotnant, R. (1967). *Compt. rend.* **265C,** 723.
2186. Souchay, P. and Dubois, S. (1948). *Ann. Chim.* **3,** 88.
2187. Souchay, P. and Faucherre, J. (1951). *Bull. Soc. chim. France* **18,** 355.
2188. Souchay, P. and Herve, G. (1965). *Compt. rend.* **261C,** 2486.
2189. Souchay, P. and Massert, R. (1961). *Compt. rend.* **253,** 1699.
2190. Souchay, P. and Schaal, R. (1950). *Bull. Soc. chim. France* **17,** 824.
2191. Spacu, P. (1937). *Z. anorg. Chem.* **232,** 225.
2192. Sparks, J. T., Mead, W. and Komoto, T. (1962). *J. Phys. Soc. Japan* **17,** Suppl. B–1, 249.
2193. Spreckelmeyer, B. (1968). *Z. anorg. Chem.* **358,** 147.
2194. Spreckelmeyer, B. (1969). *Z. anorg. Chem.* **365,** 225.
2195. Spreckelmeyer, B. (1969). *Z. anorg. Chem.* **368,** 18.
2196. Spreckelmeyer, B. and Schafer, H. (1967). *J. Less-Common Metals* **13,** 122.
2197. Spreckelmeyer, B. and Schafer, H. (1967). *J. Less-Common Metals* **13,** 127.
2198. Spiro, T. G., Maroni, V. A. and Quicksall, C. O. (1969). *Inorg. Chem.* **8,** 2524.
2199. Spiro, T. G., Templeton, D. H. and Zalkin, A. (1968). *Inorg. Chem.* **7,** 2165.
2200. Spiro, T. G., Templeton, D. H. and Zalkin, A. (1969). *Inorg. Chem.* **8,** 856.

2201. Spitsyn, V. I., Kolli, I. D. and T'am Wen-hsia (1964). *Russ. J. Inorg. Chem.* **9,** 54.
2202. Spittle, H. M. and Wardlaw, W. (1929). *J. Chem. Soc.*, 792.
2203. Stahler, A. (1904). *Berichte* **37,** 4411.
2204. Stammreich, H., Kawai, K. and Tavares, Y. (1959). *Spectrochim. Acta* **15,** 438.
2205. Stanescu, D. (1969). *Z. anorg. Chem.* **366,** 104.
2206. Star, C., Bitter, F. and Kaufmann, A. R. (1940). *Phys. Rev.* **58,** 977.
2207. Stasicka, Z., Samotus, A. and Jakob, W. (1966). *Roczniki Chem.* **40,** 967.
2208. Steele, M. C. (1957). *Austral. J. Chem.* **10,** 367.
2209. Steele, M. C. (1957). *Austral. J. Chem.* **10,** 489.
2210. Stiefel, E. I., Dori, Z. and Gray, H. B. (1967). *J. Amer. Chem. Soc.* **89,** 3353.
2211. Stiefel, E. I., Eisenberg, R., Rosenberg, R. C. and Gray, H. B. (1966). *J. Amer. Chem. Soc.* **88,** 2956.
2212. Stiefel, E. I. and Gray, H. B. (1965). *J. Amer. Chem. Soc.* **87,** 4012.
2213. Steinfink, H. and Burns, J. H. (1964). *Acta Cryst.* **7,** 823.
2214. Stephenson, N. C. (1966). *Acta Cryst.* **20,** 59.
2215. Stephenson, N. C. and Wadsley, A. D. (1965). *Acta Cryst.* **19,** 241.
2216. Stephenson, T. A., Bannister, E. and Wilkinson, G. (1964). *J. Chem. Soc.*, 2538.
2217. Stephenson, T. A. and Wilkinson, G. (1966). *J. Inorg. Nuclear Chem.* **28,** 2285.
2218. Stevenson, D. L. and Dahl, L. F. (1967). *J. Amer. Chem. Soc.* **89,** 3721.
2219. Stewart, R. P. and Treichel, P. M. (1968). *Inorg. Chem.* **7,** 1942.
2220. Stezowski, J. J. and Eick, H. A. (1969). *J. Amer. Chem. Soc.* **91,** 2890.
2221. Stiddard, M. H. B. (1962). *J. Chem. Soc.*, 4712.
2222. Stiddard, M. H. B. (1963). *J. Chem. Soc.*, 756.
2223. Stiddard, M. H. B. and Townsend, R. E. (1969). *J. Chem. Soc.* (*A*), 2355.
2224. Stock, H. P. and Jahr, K. F. (1963). *Z. Naturforsch.* **18b,** 1134.
2225. Stomberg, R. (1963). *Acta Chem. Scand.* **17,** 1563.
2226. Stomberg, R. (1964). *Arkiv Kemi* **22,** 29.
2227. Stomberg, R. (1964). *Arkiv Kemi* **22,** 49.
2228. Stomberg, R. (1965). *Arkiv Kemi* **23,** 401.
2229. Stomberg, R. (1965). *Arkiv Kemi* **24,** 47.
2230. Stomberg, R. (1965). *Arkiv Kemi* **24,** 111.
2231. Stomberg, R. (1965). *Arkiv Kemi* **24,** 283.
2232. Stomberg, R. (1968). *Acta Chem. Scand.* **22,** 1076.
2233. Stomberg, R. and Ainelem, I.-B. (1968). *Acta Chem. Scand.* **22,** 1439.
2234. Stomberg, R. and Brossett, C. (1960). *Acta Chem. Scand.* **14,** 441.
2235. Strahle, J. and Barnighausen, H. (1966). *Angew. Chem. Internat. Edn.* **5,** 417.
2236. Strahle, J. and Dehnicke, K. (1965). *Z. anorg. Chem.* **338,** 287.
2237. Straumanis, M. E. and Ejima, T. (1962). *Acta Cryst.* **15,** 404.
2238. Straumanis, M. E. and Hsu, S. S. (1950). *J. Amer. Chem. Soc.* **72,** 4027.

2239. Strohmeier, W. and von Hobe, D. (1961). *Berichte* **94,** 761. Strohmeier, W., Gerlach, K. and von Hobe, D. (1961). *Berichte* **94,** 164. Strohmeier, W., Guttenberger, J. F., Blumenthal, H. and Albert, G. (1966). *Berichte* **99,** 3419.
2240. Strunz, H. (1943). *Naturwiss.* **31,** 89.
2241. Struss, A. W. and Corbett, J. D. (1969). *Inorg. Chem.* **8,** 227.
2242. Stubbin, P. M. and Mellor, D. P. (1948). *Proc. Roy. Soc. N.S.W.* **82,** 225.
2243. Stucky, G. and Rundle, R. E. (1964). *J. Amer. Chem. Soc.* **86,** 4821.
2244. Studd, B. F. and Swallow, A. G. (1968). *J. Chem. Soc.* (*A*), 1961.
2245. Sturm, B. J. (1962). *Inorg. Chem.* **1,** 665.
2246. Sue, P. S. (1939). *Bull. Soc. chim. France* **6,** 830.
2247. Summers, L., Uloth, R. H. and Holmes, A. (1955). *J. Amer. Chem. Soc.* **77,** 3604.
2248. Sundholm, A., Andersson, S., Magneli, A. and Marinder, B.-O. (1958). *Acta Chem. Scand.* **12,** 1343.
2249. Surtees, J. R. (1965). *Chem. Comm.*, 567.
2250. Sutton, G. J. (1959). *Austral. J. Chem.* **12,** 122.
2251. Sutton, G. J. (1962). *Austral. J. Chem.* **15,** 232.
2252. Swallow, A. G., Ahmed, F. R. and Barnes, W. H. (1966). *Acta Cryst.* **21,** 397.
2253. Swaroop, B. and Flengas, S. N. (1964). *Canad. J. Phys.* **42,** 1886.
2254. Swaroop, B. and Flengas, S. N. (1964). *Canad. J. Chem.* **42,** 1495.
2255. Swaroop, B. and Flengas, S. N. (1965). *Canad. J. Chem.* **43,** 2115.
2256. Swinehart, J. H. and Castellan, G. W. (1964). *Inorg. Chem.* **3,** 278.
2257. Sytnik, A. A., Furman, A. A. and Kulyasova, A. S. (1966). *Russ. J. Inorg. Chem.* **11,** 543.
2258. Taft, J. C. and Jones, M. M. (1960). *J. Amer. Chem. Soc.* **82,** 4196.
2259. Takada, T. and Kawai, N. (1962). *J. Phys. Soc. Japan* **17,** Supp. B–1, 691.
2260. van Tamelen, E. E., Fechter, R. B., Schneller, S. W., Boche, G., Greeley, R. H. and Akermark, B. (1969). *J. Amer. Chem. Soc.* **91,** 1551.
2261. Tananaev, I. V. and Guzeeva, L. S. (1966). *Russ. J. Inorg. Chem.* **11,** 587, 590.
2262. Tanner, K. N. and Duncan, A. B. T. (1951). *J. Amer. Chem. Soc.* **73,** 1164.
2263. Tapscott, R. E., Belford, R. L. and Paul, I. C. (1968). *Inorg. Chem.* **7,** 356.
2264. Tate, D. P., Augl, J. M., Ritchey, W. M., Ross, B. L. and Grasselli, J. G. (1964). *J. Amer. Chem. Soc.* **86,** 3261.
2265. Taylor, J. C. and Mueller, M. H. (1965). *Acta Cryst.* **19,** 536.
2266. Taylor, J. C., Mueller, M. H. and Hitterman, R. L. (1966). *Acta Cryst.* **20,** 842.
2267. Tebbe, F. N. and Muetterties, E. L. (1968). *Inorg. Chem.* **7,** 172.
2268. Templeton, D. H. and Zalkin, A. (1963). *Acta Cryst.* **16,** 762.
2269. Terao, N. (1963). *Japan J. Appl. Phys.* **2,** 565.
2270. Terezakis, E. G. and Carlin, R. L. (1967). *Inorg. Chem.* **6,** 2125.
2271. Teubin, J. H. and de Liefde Meijer, H. J. (1969). *J. Organometallic Chem.* **17,** 87.
2272. Teufer, G. (1962). *Acta Cryst.* **15,** 1187.

2273. Thiele, K.-H. and Jacob, K. (1968). *Z. anorg. Chem.* **356,** 195.
2274. Thiele, K.-H. and Muller, J. (1968). *Z. anorg. Chem.* **362,** 113.
2275. Thomas, A. W. and Owens, H. S. (1935). *J. Amer. Chem. Soc.* **57,** 1825.
2276. Thomas, I. M. (1961). *Canad. J. Chem.* **39,** 1386.
2277. Thomas, J. C. (1956). *Chem. and Ind.*, 1388.
2278. Thompson, H. B. and Bartell, L. S. (1968). *Inorg. Chem.* **7,** 488.
2279. Tillack, J. (1968). *Z. anorg. Chem.* **357,** 11.
2280. Tillack, J., Eckerlin, P. and Dettingmeijer, J. H. (1966). *Angew. Chem. Internat. Edn.* **5,** 421.
2281. Tillack, J. and Kaiser, R. (1968). *Angew. Chem. Internat. Edn.* **7,** 294.
2282. Tillack, J. and Kaiser, R. (1969). *Angew. Chem. Internat. Edn.* **8,** 142.
2283. Titze, H. (1969). *Acta Chem. Scand.* **23,** 399.
2284. Tobias, R. J. (1965). *Canad. J. Chem.* **43,** 1222.
2285. Tolmacheva, T. A., Tsintsius, V. M. and Yudovich, E. E. (1966). *Russ. J. Inorg. Chem.* **11,** 249.
2286. Tong, J. Y. and Johnson, R. L. (1966). *Inorg. Chem.* **5,** 1902.
2287. Tong, J. Y. and King, E. L. (1953). *J. Amer. Chem. Soc.* **75,** 6180.
2288. Toptygina, G. M. and Barskaya, I. B. (1965). *Russ. J. Inorg. Chem.* **10,** 1226.
2289. Tourne, C. (1968). *Compt. rend.* **266C,** 702.
2290. Tourne, C. M. and Souchay, P. (1966). *Compt. rend.* **263C,** 1142.
2291. Tourne, G. and Tourne, C. (1964). *Compt. rend.* **258,** 4556; **262C,** 648 (1966).
2292. Tourne, C. and Tourne, G. (1969). *Bull. Soc. chim. France*, 1124.
2293. Tracy, J. W., Gregory, N. W. and Lingafelter, E. C. (1962). *Acta Cryst.* **15,** 672.
2294. Tracy, J. W., Gregory, N. W., Lingafelter, E. C., Dunitz, J. D., Mez, H.-C., Rundle, R. E., Scheringer, C., Yakel, H. L. and Wilkinson, M. K. (1961). *Acta Cryst.* **14,** 927.
2295. Tracy, J. W., Gregory, N. W., Stewart, J. M. and Lingafelter, E. C. (1962). *Acta Cryst.* **15,** 460.
2296. Traube, W., Burmeister, E. and Stahn, R. (1925). *Z. anorg. Chem.* **147,** 50.
2297. Treichel, P. M., Morris, J. H. and Stone, F. G. A. (1963). *J. Chem. Soc.*, 720.
2298. Treichel, P. M. and Shubkin, R. L. (1967). *Inorg. Chem.* **6,** 1328.
2299. Treichel, P. M. and Werber, G. P. (1968). *J. Organometallic Chem.* **12,** 479.
2300. Treichel, P. M. and Wilkes, G. R. (1966). *Inorg. Chem.* **5,** 1182.
2301. Trevorrow, L. E. (1958). *J. Phys. Chem.* **62,** 362.
2302. Trevorrow, L. E., Fisher, J. and Steunenberg, R. K. (1957). *J. Amer. Chem. Soc.* **79,** 5167.
2303. Trost, W. R. (1952). *Canad. J. Chem.* **30,** 835.
2304. Tsang, W. S., Meek, D. W. and Wojcicki, A. (1968). *Inorg. Chem.* **7,** 1263.
2305. Tsigdinos, G. A. and Hallada, C. J. (1968). *Inorg. Chem.* **7,** 437.
2306. Tsumura, R. and Hagihara, N. (1965). *Bull. Chem. Soc. Japan* **38,** 861.
2307. Tsumura, R. and Hagihara, N. (1965). *Bull. Chem. Soc. Japan* **38,** 1901.

2308. Tuck, D. G. and Faithful, B. D. (1965). *J. Chem. Soc.*, 5753.
2309. Tulinsky, A., Worthington, C. R. and Pignataro, A. (1959). *Acta Cryst.* **12,** 623, 626, 634.
2310. Turnbull, A. G. and Watts, J. A. (1963). *Austral. J. Chem.* **16,** 947.
2311. Ueki, T., Zalkin, A. and Templeton, D. H. (1966). *Acta Cryst.* **20,** 836.
2312. Uemeda, J., Kusumoto, H., Narita, K. and Yamada, E. (1965). *J. Chem. Phys.* **42,** 1458.
2313. Vandeven, D., Pouchard, M. and Hagenmuller, P. (1966). *Compt. rend.* **263C,** 228.
2314. van Haagen, W. K. (1910). *J. Amer. Chem. Soc.* **32,** 729.
2315. Varga, L. P. and Freund, H. (1962). *J. Phys. Chem.* **66,** 21.
2316. Vasil'kova, I. V. and Perfilova, I. L. (1965). *Russ. J. Inorg. Chem.* **10,** 1248.
2317. Vaughan, P. A., Sturdivant, J. H. and Pauling, L. (1950). *J. Amer. Chem. Soc.* **72,** 5477.
2318. Velikodnyi, Yu. A., Kovba, L. M. and Trunov, V. K. (1967). *Russ. J. Inorg. Chem.* **12,** 1699.
2319. Villadsen, J. (1959). *Acta Chem. Scand.* **13,** 2146.
2320. Voitovich, B. A. and Barabanova, A. S. (1961). *Zhur. neorg. Khim.* **6,** 2098.
2321. Von Beck, H. P. (1931). *Z. anorg. Chem.* **196,** 85.
2322. Vorob'ev, S. P. and Davydov, I. P. (1966). *Russ. J. Inorg. Chem.* **11,** 1087.
2323. Vorres, K. S. and Donohue, J. (1955). *Acta Cryst.* **8,** 25.
2324. Vorres, K. S. and Dutton, F. B. (1955). *J. Amer. Chem. Soc.* **77,** 2019.
2325. Vrieze, K. (1964). Thesis, University of London.
2326. Waber, J. T. and Cromer, D. T. (1965). *J. Chem. Phys.* **42,** 4116.
2327. Wadsley, A. D. (1955). *Acta Cryst.* **8,** 695.
2328. Wadsley, A. D. (1957). *Acta Cryst.* **10,** 261.
2329. Wadsley, A. D. (1961). *Acta Cryst.* **14,** 660, 664.
2330. Wadsley, A. D. (1964). *Z. Krist.* **120,** 396.
2331. Walton, R. A. and Brisdon, B. J. (1967). *Spectrochim. Acta* **23A,** 2489.
2332. Walton, R. A. and Edwards, D. A. (1968). *Spectrochim. Acta* **24A,** 833.
2333. Waltz, W. L., Adamson, A. W. and Fleischauer, P. D. (1967). *J. Amer. Chem. Soc.* **89,** 3923.
2334. Ward, B. G. and Stafford, F. E. (1968). *Inorg. Chem.* **7,** 2569.
2335. Wardlaw, W. and Harding, A. J. I. (1926). *J. Chem. Soc.*, 1592.
2336. Wardlaw, W. and Wormell, R. L. (1924). *J. Chem. Soc.*, 2370; 130 (1927).
2337. Wardlaw, W. and Webb, H. W. (1930). *J. Chem. Soc.*, 2100.
2338. Wardlaw, W. and Wormell, R. L. (1927). *J. Chem. Soc.*, 1087.
2339. Warekois, E. P. (1960). *J. Appl. Phys.* **31,** 346S.
2340. von Wartenburg, H. (1941). *Z. anorg. Chem.* **247,** 135.
2341. Watenpaugh, K. and Caughlan, C. N. (1966). *Inorg. Chem.* **5,** 1782.
2342. Watenpaugh, K. and Caughlan, C. N. (1967). *Chem. Comm.*, 76.
2343. Watenpaugh, K. and Caughlan, C. N. (1967). *Inorg. Chem.* **6,** 963.
2344. Waters, J. H., Williams, R., Gray, H. B., Schrauzer, G. N. and Finck, H. W. (1964). *J. Amer. Chem. Soc.* **86,** 4198.
2345. Waters, T. N. (1964). *Chem. and Ind.*, 713.

2346. Watkins, E. D., Cunningham, J. A., Phillips, T., Sands, D. E. and Wagner, W. F. (1969). *Inorg. Chem.* **8,** 29.
2347. Watson, W. H. and Waser, J. (1958). *Acta Cryst.* **11,** 689.
2348. Watt, G. W. and Baker, W. A. (1961). *J. Inorg. Nuclear Chem.* **22,** 49.
2349. Watt, G. W., Baye, L. J. and Drummond, F. O. (1966). *J. Amer. Chem. Soc.* **88,** 1138.
2350. Watt, G. W. and Drummond, F. O. (1966). *J. Amer. Chem. Soc.* **88,** 5926.
2351. Watts, J. A. (1966). *Inorg. Chem.* **5,** 281.
2352. Waugh, J. L. T., Shoemaker, D. P. and Pauling, L. (1954). *Acta Cryst.* **7,** 438.
2353. Webster, M. (1966). *Chem. Rev.* **66,** 87.
2354. Wechter, M. A., Shanks, H. R. and Voigt, A. F. (1968). *Inorg. Chem.* **7,** 845.
2355. Weidleim, J. and Dehnicke, K. (1966). *Z. anorg. Chem.* **348,** 278.
2356. Weinland, R. F. and Fiederer, M. (1906). *Berichte* **39,** 4042; **40,** 2090 (1907).
2357. Weinland, R. F. and Knoll, W. (1905). *Z. anorg. Chem.* **44,** 81.
2358. Weinland, R. F. and Storz, L. (1906). *Berichte* **39,** 3056; *Z. anorg. Chem.* **54,** 223 (1907).
2359. Weiss, A., Nagorsen, G. and Weiss, A. (1953). *Z. anorg. Chem.* **274,** 151.
2360. Weiss, A. and Weiss, A. (1960). *Angew. Chem.* **72,** 413.
2361. Weiss, E., Alsdorf, H. and Kuhr, H. (1967). *Angew. Chem. Internat. Edn.* **6,** 801.
2362. Weiss, E. and Hubel, W. (1959). *J. Inorg. Nuclear Chem.* **11,** 42.
2363. Weiss, E. and Lucken, E. A. C. (1964). *J. Organometallic Chem.* **2,** 197.
2364. Weiss, G. (1969). *Z. anorg. Chem.* **368,** 279.
2365. Weiss, R., Fischer, J. and Chevrier, B. (1965). *Compt. rend.* **260,** 3401; *Acta Cryst.* **20,** 534.
2366. Wells, A. F. (1938). *Z. Krist.* **100,** 189.
2367. Wendlandt, W. W. and Sveum, L. K. (1966). *J. Inorg. Nuclear Chem.* **28,** 393.
2368. Wendling, E. (1965). *Bull. Soc. chim. France*, 437.
2369. Wendling, E. (1967). *Bull. Soc. chim. France*, 16.
2370. Wendling, E. and Rohmer, R. (1964). *Bull. Soc. chim. France*, 360.
2371. Wentworth, R. A. D. and Brubaker, C. H. (1962). *Inorg. Chem.* **1,** 971; **2,** 551 (1963).
2372. Wentworth, R. A. D. and Brubaker, C. H. (1964). *Inorg. Chem.* **3,** 47.
2373. Wentworth, R. A. D. and Saillant, R. (1967). *Inorg. Chem.* **6,** 1436.
2374. Werner, A. (1907). *Berichte* **40,** 2103, 4834.
2375. Werner, H., Prinz, R., Bundschuh, E. and Deckelmann, K. (1966). *Angew. Chem. Internat. Edn.* **5,** 606.
2376. Werner, R. P. M., Filbey, A. H. and Manastyrskyj, S. A. (1964). *Inorg. Chem.* **3,** 298.
2377. Werner, R. P. M. and Podall, H. E. (1961). *Chem. and Ind.*, 144.
2378. Wessel, G. J. and IJdo, D. J. W. (1957). *Acta Cryst.* **10,** 466.
2379. Westland, A. D. and Westland, L. (1965). *Canad. J. Chem.* **43,** 426.

2380. Westman, S. (1961). *Acta Chem. Scand.* **15,** 217.
2381. Whitaker, A. and Jeffery, J. W. (1967). *Acta Cryst.* **23,** 977.
2382. Whittaker, A. G. and Yost, D. M. (1949). *J. Chem. Phys.* **17,** 188.
2383. Whittaker, M. P., Asay, J. and Eyring, E. M. (1966). *J. Phys. Chem.* **70,** 1005.
2384. Wilford, J. B. and Powell, H. M. (1969). *J. Chem. Soc.* (*A*), 8.
2385. Wilford, J. B., Whitla, A. and Powell, H. M. (1967). *J. Organometallic Chem.* **8,** 495.
2386. Wilhelmi, K.-A. (1967). *Arkiv Kemi* **26,** 149.
2387. Wilhelmi, K.-A. and Jonsson, O. (1958). *Acta Chem. Scand.* **12,** 1532.
2388. Wilkinson, G. (1954). *J. Amer. Chem. Soc.* **76,** 209.
2389. Wilkinson, G. and Birmingham, J. M. (1954). *J. Amer. Chem. Soc.* **76,** 4281.
2390. Wilkinson, G., Cotton, F. A. and Birmingham, J. M. (1956). *J. Inorg. Nuclear Chem.* **2,** 95.
2391. Wilkinson, G., Pauson, P. L., Birmingham, J. M. and Cotton, F. A. (1953). *J. Amer. Chem. Soc.* **75,** 1011.
2392. Wilkinson, G. and Piper, T. S. (1956). *J. Inorg. Nuclear Chem.* **3,** 104.
2393. Wilkinson, M. K., Wollan, E. O., Child, H. R. and Cable, J. W. (1961). *Phys. Rev.* **121,** 74.
2394. Willey, G. R. (1968). *J. Amer. Chem. Soc.* **90,** 3362.
2395. Williams, M. B. and Hoard, J. L. (1942). *J. Amer. Chem. Soc.* **64,** 1139.
2396. Williams-Wynn, D. A. (1959). *J. Phys. Chem.* **63,** 2065.
2397. Willis, B. T. M. and Rooksby, H. P. (1952). *Proc. Phys. Soc.* **65B,** 950.
2398. Wilmarth, W. K., Graff, H. and Gustin, S. T. (1956). *J. Amer. Chem. Soc.* **78,** 2683.
2399. Wilson, F. C. and Shoemaker, D. P. (1957). *J. Chem. Phys.* **27,** 809.
2400. Wilson, L. J. and Rose, N. J. (1968). *J. Amer. Chem. Soc.* **90,** 6041.
2401. Wilson, R. L. and Duffey, G. H. (1961). *J. Chem. Phys.* **35,** 568.
2402. Wing, R. M. and Callahan, K. P. (1969). *Inorg. Chem.* **8,** 871.
2403. Winstein, S., Kaesz, H. D., Kreiter, C. G. and Friedrich, E. C. (1965). *J. Amer. Chem. Soc.* **87,** 3267.
2404. Winter, G. (1966). *Inorg. Nuclear Chem. Letters* **2,** 161.
2405. Wiseman, E. L. and Gregory, N. W. (1949). *J. Amer. Chem. Soc.* **71,** 2344.
2406. Witters, R. D. and Caughlan, C. N. (1965). *Nature, Lond.* **205,** 1312.
2407. Wojcicki, A. and Farona, M. F. (1964). *J. Inorg. Nuclear Chem.* **26,** 2289.
2408. Wold, A., Kunnmann, W., Arnott, R. J. and Ferretti, A. (1964). *Inorg. Chem.* **3,** 545.
2409. Wohler, F. (1824). *Pogg. Ann.* **2,** 350.
2410. Wolfe, C. W., Block, M. L. and Baker, L. C. W. (1955). *J. Amer. Chem. Soc.* **77,** 2200.
2411. Wood, J. S. (1968). *Inorg. Chem.* **7,** 852.
2412. Wood, J. S. and Greene, P. T. (1969). *Inorg. Chem.* **8,** 491.
2413. Wuthrich, K. and Connick, R. E. (1967). *Inorg. Chem.* **6,** 583.
2414. Yahia, J. and Frederikse, H. P. R. (1961). *Phys. Rev.* **123,** 1257.
2415. Yamamoto, A. and Kambara, S. (1959). *J. Amer. Chem. Soc.* **81,** 2663.

2416. Yamamoto, A., Ookawa, M. and Ikeda, S. (1969). *Chem. Comm.*, 841.
2417. Yamatera, H. and Nakatsu, K. (1954). *Bull. Chem. Soc. Japan* **27**, 244.
2418. Yannoni, N. F. (1961). Thesis, Boston University, quoted in Rasmussen, P. G. (1967). *J. Chem. Educ.* **44**, 277.
2419. Yatsimirskii, K. B. and Kalinina, V. E. (1964). *Russ. J. Inorg. Chem.* **9**, 611.
2420. Yatsimirskii, K. B. and Romanov, V. F. (1964). *Russ. J. Inorg. Chem.* **9**, 856.
2421. Young, R. C. (1931). *J. Amer. Chem. Soc.* **53**, 2148.
2422. Young, R. C. (1932). *J. Amer. Chem. Soc.* **54**, 1402.
2423. Young, R. C. (1932). *J. Amer. Chem. Soc.* **54**, 4515.
2424. Young, R. C. and Brubaker, C. H. (1952). *J. Amer. Chem. Soc.* **74**, 4967.
2425. Young, R. C. and Hastings, T. J. (1942). *J. Amer. Chem. Soc.* **64**, 1740.
2426. Young, R. C. and Laudise, R. A. (1956). *J. Amer. Chem. Soc.* **78**, 4861.
2427. Young, R. C. and Schumb, W. C. (1930). *J. Amer. Chem. Soc.* **52**, 4233.
2428. Zachariasen, W. H. (1951). *Acta Cryst.* **4**, 408.
2429. Zachariasen, W. H. (1954). *Acta Cryst.* **7**, 783.
2430. Zachariasen, W. H. (1954). *Acta Cryst.* **7**, 792.
2431. Zachariasen, W. H. and Plettinger, H. A. (1959). *Acta Cryst.* **12**, 526.
2432. Zado, F. (1963). *J. Inorg. Nuclear Chem.* **25**, 1115.
2433. Zaitsev, L. M., Bochkarev, G. S. and Kozhenkova, V. N. (1965). *Russ. J. Inorg. Chem.* **10**, 590.
2434. Zalkin, A., Forrester, J. D. and Templeton, D. H. (1963). *J. Chem. Phys.* **39**, 2881.
2435. Zalkin, A. and Sands, D. E. (1958). *Acta Cryst.* **11**, 615.
2436. Zalkin, A. and Templeton, D. H. (1964). *J. Chem. Phys.* **40**, 501.
2437. Zalkin, A., Templeton, D. H. and Karraker, D. G. (1969). *Inorg. Chem.* **8**, 2680.
2438. Zannetti, P. (1967). *Ricerca sci.* **37**, 291.
2439. Zeiss, H. and Herwig, W. (1957). *J. Amer. Chem. Soc.* **79**, 6561; **81**, 4798 (1959).
2440. Zeiss, H. and Sneedon, R. P. A. (1967). *Angew. Chem. Internat. Edn.* **6**, 435.
2441. Zeiss, H. and Tsutsui, M. (1959). *J. Amer. Chem. Soc.* **81**, 1367.
2442. Zeitler, V. A. and Brown, C. A. (1957). *J. Amer. Chem. Soc.* **79**, 4616, 4618.
2443. Zharouskii, F. G. (1957). *Zhur. neorg. Khim.* **2**, 623.
2444. Zucchini, U., Giannini, U., Albizzati, E. and D'Angelo, R. (1969). *Chem. Comm.*, 1174.
2445. Zvara, I. and Tarasov, L. K. (1962). *Zhur. neorg. Khim.* **7**, 2665.

Author Index

Numbers in italics refer to pages on which references are listed at the end of the book.

A

B

C

D

F

G

H

K

L

M

O

P

Q

R

S

U

V

W

Y

Z

Subject Index

D

E

H

I

M